Undergraduate Texts in Mathematics

Springer
New York
Berlin
Heidelberg
Barcelona
Budapest
Hong Kong
London
Milan
Paris
Santa Clara
Singapore
Tokyo

Undergraduate Texts in Mathematics

Anglin: Mathematics: A Concise History and Philosophy.
Readings in Mathematics.
Anglin/Lambek: The Heritage of Thales.
Readings in Mathematics.
Apostol: Introduction to Analytic Number Theory. Second edition.
Armstrong: Basic Topology.
Armstrong: Groups and Symmetry.
Axler: Linear Algebra Done Right.
Beardon: Limits: A New Approach to Real Analysis.
Bak/Newman: Complex Analysis. Second edition.
Banchoff/Wermer: Linear Algebra Through Geometry. Second edition.
Berberian: A First Course in Real Analysis.
Brémaud: An Introduction to Probabilistic Modeling.
Bressoud: Factorization and Primality Testing.
Bressoud: Second Year Calculus.
Readings in Mathematics.
Brickman: Mathematical Introduction to Linear Programming and Game Theory.
Browder: Mathematical Analysis: An Introduction.
Buskes/van Rooij: Topological Spaces: From Distance to Neighborhood.
Cederberg: A Course in Modern Geometries.
Childs: A Concrete Introduction to Higher Algebra. Second edition.
Chung: Elementary Probability Theory with Stochastic Processes. Third edition.
Cox/Little/O'Shea: Ideals, Varieties, and Algorithms. Second edition.
Croom: Basic Concepts of Algebraic Topology.
Curtis: Linear Algebra: An Introductory Approach. Fourth edition.
Devlin: The Joy of Sets: Fundamentals of Contemporary Set Theory. Second edition.
Dixmier: General Topology.
Driver: Why Math?
Ebbinghaus/Flum/Thomas: Mathematical Logic. Second edition.
Edgar: Measure, Topology, and Fractal Geometry.
Elaydi: Introduction to Difference Equations.
Exner: An Accompaniment to Higher Mathematics.
Fine/Rosenberger: The Fundamental Theory of Algebra.
Fischer: Intermediate Real Analysis.
Flanigan/Kazdan: Calculus Two: Linear and Nonlinear Functions. Second edition.
Fleming: Functions of Several Variables. Second edition.
Foulds: Combinatorial Optimization for Undergraduates.
Foulds: Optimization Techniques: An Introduction.
Franklin: Methods of Mathematical Economics.
Gordon: Discrete Probability.
Hairer/Wanner: Analysis by Its History.
Readings in Mathematics.
Halmos: Finite-Dimensional Vector Spaces. Second edition.
Halmos: Naive Set Theory.
Hämmerlin/Hoffmann: Numerical Mathematics.
Readings in Mathematics.
Hijab: Introduction to Calculus and Classical Analysis.
Hilton/Holton/Pedersen: Mathematical Reflections: In a Room with Many Mirrors.
Iooss/Joseph: Elementary Stability and Bifurcation Theory. Second edition.
Isaac: The Pleasures of Probability.
Readings in Mathematics.

(continued after index)

Rudolf Lidl   Günter Pilz

Applied Abstract Algebra

Second Edition

With 112 illustrations

Rudolf Lidl
DVC Office
University of Tasmania
Launceston, Tasmania 7250
Australia

Günter Pilz
Institut für Mathematik
Universität Linz
A-4040 Linz
Austria

Mathematics Subject Classification (1991): 05-01, 06-01, 08-01, 12-01, 13-01, 16-01, 20-01, 68-01, 93-01

Library of Congress Cataloging-in-Publication Data
Lidl, Rudolf.
Applied abstract algebra / Rudolf Lidl, Günter Pilz. — 2nd ed.
p. cm. — (Undergraduate texts in mathematics)
Includes bibliographical references and index.

1. Algebra, Abstract. I. Pilz, Günter, 1945– . II. Title.
III. Series.
QA162.L53 1997 97-22883
512′.02—dc21

Printed on acid-free paper.

Production managed by Steven Pisano; manufacturing supervised by Joe Quatela.
Photocomposed by Integre Technical Publishing Co., Inc., Albuquerque, NM.

Printed in the United States of America.

9 8 7 6 5 4 3 2 1

ISBN 978-1-4419-3117-7

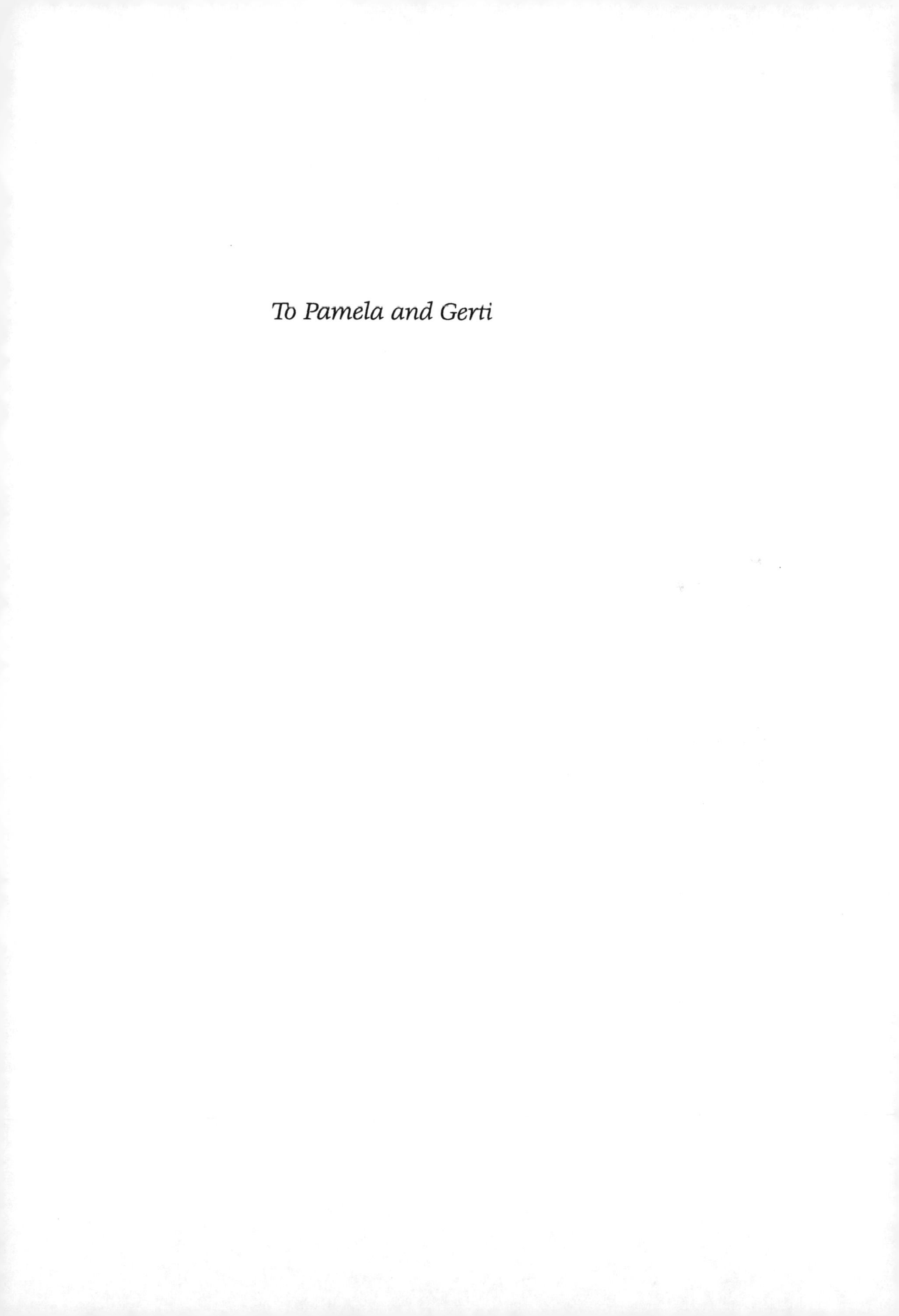

To Pamela and Gerti

Preface

Algebra is beautiful. It is so beautiful that many people forget that algebra can be very useful as well. It is still the case that students who have studied mathematics quite often enter graduate studies or enter employment without much knowledge of the applicability of the algebraic structures they have studied.

The aim of this book is to convey to senior undergraduate students, graduate students, and lecturers/instructors the fact that concepts of abstract algebra encountered previously in a first algebra course can be used in many areas of applications. Of course, in order to apply algebra, we first need some theory which then can be applied. Hence we tried to blend the theory and its applications so that the reader can experience both parts.

This book assumes knowledge of the material covered in a course on linear algebra and, preferably, a first course in (abstract) algebra covering the basics of groups, rings, and fields, although this book will provide the necessary definitions and brief summaries of the main results that will be required from such a course in algebra.

This second edition includes major changes to the first edition, published in 1984: it contains corrections and, as we believe, substantial improvements to the first four chapters of the first edition. It includes a largely new chapter on Cryptology (Chapter 5) and an enlarged chapter on Applications of Groups (Chapter 6). An extensive Chapter 7 has been added to survey other (mostly quite recent) applications, many of which were not included in the first edition. An interdependence chart of the material in the sections is presented below.

For a one-semester course (2–3 hours per week) on Applied Algebra or Discrete Mathematics, we recommend the following path: §§1, 2, 3, 4, 6–17, 21, 22, 23, and selected topics in Chapter 7 chosen by the instructor.

As in the first edition, we again emphasize the inclusion of worked-out examples and computational aspects in presenting the material. More than 500 exercises accompany the 40 sections. A separate solution manual for all these exercises is available from the publisher. The book also includes some historical notes and extensive references for further reading.

The text should be useful to mature mathematics students, to students in computer or information science with an interest and background knowledge in algebra, and to physical science or engineering students with a good knowledge in linear and some abstract algebra. Many of the topics covered are relevant to and have connections with computer science, computer algebra, physical sciences, and technology.

It is a great pleasure to acknowledge the assistance of colleagues and friends at various stages of preparing this second edition. Most of all, we would like to express our sincere appreciation to Franz Binder, who prepared many drafts and the final version of the entire book with LaTeX. Through his expertise in algebra, he was able to suggest many improvements and provided valuable information on many topics. Many useful suggestions and comments were provided by: E. Aichinger (Linz, Austria), G. Birkenmeier (Lafayette, Louisiana), J. Ecker (Linz, Austria), H. E. Heatherly (Lafayette, Louisiana), H. Kautschitsch (Klagenfurt, Austria), C. J. Maxson (College Station, Texas), W. B. Müller (Klagenfurt, Austria), G. L. Mullen (University Park, Pennsylvania), C. Nöbauer, P. Paule (Linz, Austria), A. P. J. Van der Walt (Stellenbosch, South Africa), and F. Winkler (Linz, Austria). Special thanks are due to L. Shevrin and I. O. Koryakov (Ekaterinenburg, Russia) for preparing a Russian translation of the first edition of our text. Their comments improved the text substantially. We also wish to thank Springer-Verlag, especially Mr. Thomas von Foerster, Mr. Steven Pisano, and Mr. Brian Howe, for their kind and patient cooperation.

October 1997 R.L. and G.P.

Interdependence Chart

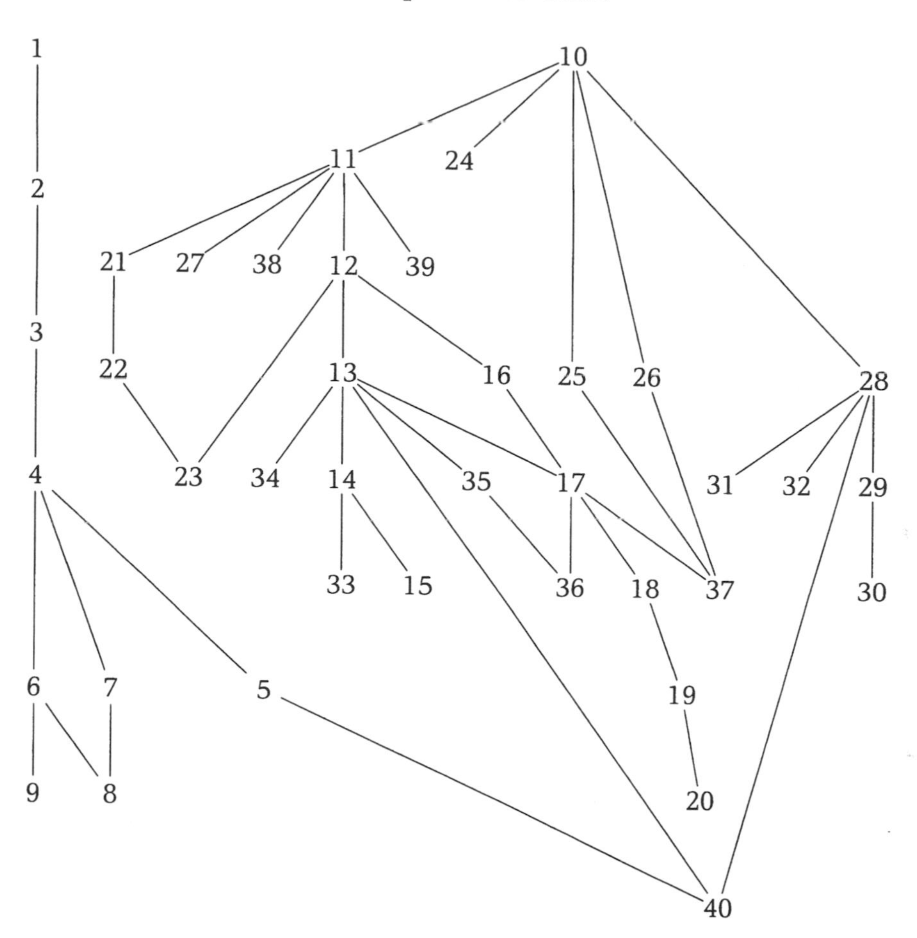

Among the numerous general texts on algebra, we mention Birkhoff & MacLane (1977), Childs (1995), Herstein (1975), Jacobson (1985), and Lang (1984). Application-oriented books include Biggs (1985), Birkhoff & Bartee (1970), Bobrow & Arbib (1974), Cohen, Giusti & Mora (1996), Dorninger & Müller (1984), Fisher (1977), Gilbert (1976), Prather (1976), Preparata & Yeh (1973), Spindler (1994), and Stone (1973). A survey of the present "state of the art" in algebra is Hazewinkel (1996) (with several more volumes to follow). Historic notes on algebra can be found in Birkhoff (1976). Applications of linear algebra (which are not covered in this book) can be found in Goult (1978), Noble & Daniel (1977), Rorres & Anton (1984), and Usmani (1987). Lipschutz (1976) contains a large collection of Exercises. Good books on computational aspects ("Computer Algebra") include Geddes, Czapor & Labahn (1993), Knuth (1981), Lipson (1981), Sims (1984), Sims (1994), and Winkler (1996).

Contents

Preface vii

List of Symbols xv

1 Lattices **1**
1 Properties and Examples of Lattices 1
2 Distributive Lattices 16
3 Boolean Algebras 19
4 Boolean Polynomials 26
5 Ideals, Filters, and Equations 34
6 Minimal Forms of Boolean Polynomials 40
Notes 51

2 Applications of Lattices **55**
7 Switching Circuits 55
8 Applications of Switching Circuits 62
9 More Applications of Boolean Algebras 78
Notes 93

3 Finite Fields and Polynomials **95**
10 Some Group Theory 95
11 Rings and Polynomials 109

12 Fields 124
13 Finite Fields 138
14 Irreducible Polynomials over Finite Fields 153
15 Factorization of Polynomials over Finite Fields 166
Notes 176

4 Coding Theory **183**
16 Introduction to Coding 183
17 Linear Codes 192
18 Cyclic Codes 205
19 Special Cyclic Codes 222
20 Decoding BCH Codes 229
Notes 236

5 Cryptology **239**
21 Classical Cryptosystems 240
22 Public Key Cryptosystems 255
23 Discrete Logarithms and Other Ciphers 266
Notes 279

6 Applications of Groups **283**
24 Fast Adding 284
25 Pólya's Theory of Enumeration 287
26 Image Understanding 302
27 Symmetry Groups 314
Notes 329

7 Further Applications of Algebra **331**
28 Semigroups 333
29 Semigroups and Automata 342
30 Semigroups and Formal Languages 350
31 Semigroups and Biology 356
32 Semigroups and Sociology 360
33 Linear Recurring Sequences 365
34 Fast Fourier Transforms 379
35 Latin Squares 388
36 Block Designs 399
37 Hadamard Matrices, Transforms, and Networks 413

38 Gröbner Bases for Algebraic and Differential Equations . 426
39 Systems Theory 434
40 Abstract Data Types 447
Notes 458

Bibliography **459**

Index **475**

List of Symbols

$a\,R\,b$, $a \not R\, b$	3	$(\mathrm{GL}(n, \mathbb{R}), \cdot)$	97	R^X	110
$[a]$	3	D_n	97	$R_1 \oplus R_2$	114
$\mathcal{P}(S)$	3	$S \leq G$	99	$\bigoplus_{i=1}^{n} R_i$	114
$\mathbb{N}$	8	$\langle X \rangle$	99, 335	$R[x]$	114
$\mathbb{B}$	20	$G = G_1 \oplus G_2$	100	$R[[x]]$	114
$\cong_{\mathrm{b}}$	21	$\bigoplus_{i \in I} G_i$	100	$\deg p$	115
P_n	26	$\prod_{i \in I} G_i$	100	$\gcd(f, g)$	116
$P_n(B)$	27	$G \hookrightarrow G'$	100	$\bar{p}$	119
$\bar{p}_B$	26	$\operatorname{Ker} f$	101	$P(R)$	119
N_{d}	29	$\operatorname{Im} f$	101	$R[x, y]$	122
$F_n(B)$	24	$\mathbb{Z}_n$	102	$R(x)$	125
$I \trianglelefteq B$	34	$\equiv_n$	102	$R\langle x \rangle$	126
$\ker h$	101	$N \trianglelefteq G$	103	$F(A)$, $F(a)$	129
b	99	$I \trianglelefteq R$	112	$F(a_1, \ldots, a_n)$	129
$p \to q$	81	$[G : N]$	104	$[K : F]$	130
$C \mathbin{\triangle} B$	83	$C(G)$	107	$\mathbb{F}_{p^n}$	140
$(G, \circ)$	96	G_n	108	$\operatorname{ind}_\beta(b)$	142
$\mathbb{Z}$, $\mathbb{Q}$, $\mathbb{R}$, $\mathbb{C}$, $n\mathbb{Z}$	96	$(R, +, \cdot)$	109	$\mathbb{F}_{p^\infty}$	143
S_X, S_n	97	$\mathbb{M}_n(R)$	110	$\varphi(n)$	144

Symbol	Page
Q_n	145
$\mu(n)$	147
$\log_\beta(b)$	142
$\mathrm{Tr}_{F/K}(\alpha)$	150
m_α	154
C_s	156
$d_{\min}$	188
$\lfloor x \rfloor$	189
$A(n, d)$	190
$\mathrm{mld}(\mathbf{H})$	196
$S(\mathbf{y})$	199
V_n	206
$C^\perp$, $h^\perp$	212
D_n	217
$\lambda(n)$	262
$J(a, n)$	266
$\lceil x \rceil$	285
$P(n)$	289
$\mathrm{sign}(\pi)$	290
A_n	290
$x \sim_G y$	292
$\mathrm{Orb}(x)$	292
$\mathrm{Stab}(x)$	292
$\mathrm{Fix}(g)$	294
$Z(G)$	296
$\mathbf{R}_\theta$	306
$\mathbf{S}_\theta$	307
$S(M)$	314
$O_2(\mathbb{R})$	315
$S((M_i)_{i \in I})$	316
$F[G]$	320
χ_i	321
G_S	335
$\mathbb{G}_n$	336
$R \diamond S$	337
R^{t}	337
A_*	338
$[X, R]$	339
X^*	340
M_S	345
$\mathcal{A}_1 \times \mathcal{A}_2$	346
$\mathcal{A}_1 \vdash\!\dashv \mathcal{A}_2$	346
$S_1 \mid S_2$	347
$\mathcal{A}_1 \mid \mathcal{A}_2$	347
$S \wr T$	347
$-\!\triangleright$	351
$=\!\triangleright$	351
$L(\mathcal{G})$	352
$W(\mathcal{A})$	353
$S(\mathcal{G})$	361
$\hat{p}$	382
$\mathbf{D}_n$, $\mathbf{D}_{n,\omega}$	383
$\widehat{\mathbf{a}}$	383
$F\mathbb{Z}_n$	387
$a *_t b$	403
$\mathbf{H}_0, \mathbf{H}_1, \ldots$	415
$\mathbf{S}_n$	418
$\mathbf{J}_n$	418
$RM(m)$	419
$P(\mathbf{H})$	423
$[\mathbf{F}, \mathbf{G}, \mathbf{H}, \mathbf{K}]$	435
Φ_{t_1,t_2}	437
$\mathfrak{R}(f)$	439
$\mathcal{K}(\tau)$	448
$\mathcal{K}_E$	452
$\mathrm{Th}(\mathcal{K})$	452

1 CHAPTER Lattices

In 1854, George Boole (1815–1864) introduced an important class of algebraic structures in connection with his research in mathematical logic. His goal was to find a mathematical model for human reasoning. In his honor these structures have been called Boolean algebras. They are special types of lattices. It was E. Schröder, who about 1890 considered the lattice concept in today's sense. At approximately the same time, R. Dedekind developed a similar concept in his work on groups and ideals. Dedekind defined, in modern terminology, modular and distributive lattices, which are types of lattices of importance in applications. The rapid development of lattice theory proper started around 1930, when G. Birkhoff made major contributions to the theory.

Boolean lattices or Boolean algebras may be described as the richest and, at the same time, the most important lattices for applications. Since they are defined as distributive and complemented lattices, it is natural to consider some properties of distributive and complemented lattices first.

§1 Properties and Examples of Lattices

We know from elementary arithmetic that for any two natural numbers a, b there is a largest number d which divides both a and b, namely the greatest common divisor $\gcd(a, b)$ of a and b. Also, there is a smallest

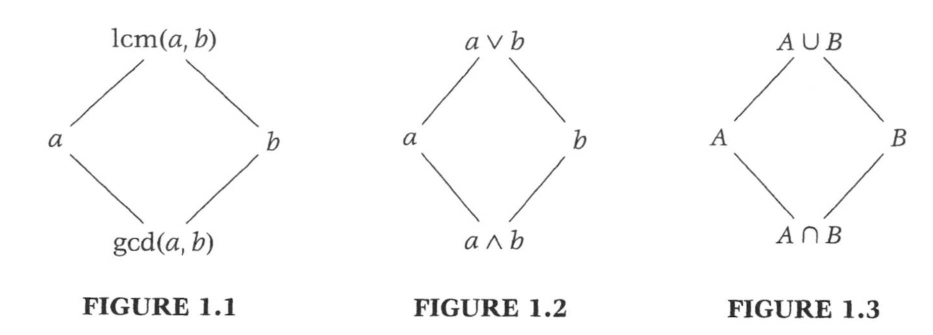

FIGURE 1.1 **FIGURE 1.2** **FIGURE 1.3**

number m which is a multiple of both a and b, namely the least common multiple $m = \operatorname{lcm}(a, b)$. This is pictured in Figure 1.1.

Turning to another situation, given two statements a, b, there is a "weakest" statement implying both a and b, namely the statement "a and b," which we write as $a \wedge b$. Similarly, there is a "strongest" statement which is implied by a and b, namely "a or b," written as $a \vee b$. This is pictured in Figure 1.2.

A third situation arises when we study sets A, B. Again, there is a largest set contained in A and B, the intersection $A \cap B$, and a smallest one containing both A and B, the union $A \cup B$. We get the similar diagram in Figure 1.3.

It is typical for modern mathematics that seemingly different areas lead to very similar situations. The idea then is to extract the common features in these examples, to study these features, and to apply the resulting theory to many different areas. This is very economical: proving one single general theorem automatically yields theorems in all areas to which the theory applies. And usually we discover many more new areas of application as the theory builds up.

We are going to do exactly that for the three examples we have met above. In the first two chapters, we shall study collections of items which have something like a "greatest lower bound" and a "least upper bound"; we shall call them *lattices*. Before doing so, we need a short "warm-up" to get fit for the theory to come.

One of the important concepts in all of mathematics is that of a relation. Of particular interest are equivalence relations, functions, and order relations. Here we concentrate on the latter concept and recall from an introductory mathematics course:

Let A and B be sets. A ***relation*** R ***from*** A ***to*** B is a subset of $A \times B$, the cartesian product of A and B. Relations from A to A are called relations ***on*** A, for short. If $(a, b) \in R$, we write $a \, R \, b$ and say that "a is in relation R to b." Otherwise, we write $a \not R \, b$. If we consider a set A together with a relation R, we write (A, R).

A relation R on a set A may have some of the following properties:

R is ***reflexive*** if $a \, R \, a$ for all $a \in A$;

R is ***symmetric*** if $a \, R \, b$ implies $b \, R \, a$ for all $a, b \in A$;

R is ***antisymmetric*** if $a \, R \, b$ and $b \, R \, a$ imply $a = b$ for all $a, b \in A$;

R is ***transitive*** if $a \, R \, b$ and $b \, R \, c$ imply $a \, R \, c$ for all $a, b, c \in A$.

A reflexive, symmetric, and transitive relation R is called an ***equivalence relation***. In this case, for any $a \in A$, $[a] := \{b \in A \mid a \, R \, b\}$ is called the ***equivalence class of*** a.

1.1. Definition. A reflexive, antisymmetric, and transitive relation R on a set A is called a ***partial order*** (***relation***). In this case, (A, R) is called a ***partially ordered set*** or ***poset***.

Partial order relations describe "hierarchical" situations; usually we write $\leq$ or $\subseteq$ instead of R. Partially ordered finite sets $(A, \leq)$ can be graphically represented by ***Hasse diagrams***. Here the elements of A are represented as points in the plane and if $a \leq b$, $a \neq b$ (in which case we write $a < b$), we draw b higher up than a and connect a and b with a line segment. For example, the Hasse diagram of the poset $(\mathcal{P}(\{1, 2, 3\}), \subseteq)$ is shown in Figure 1.4, where $\mathcal{P}(S)$ denotes the ***power set*** of S, i.e., the set of all subsets of S.

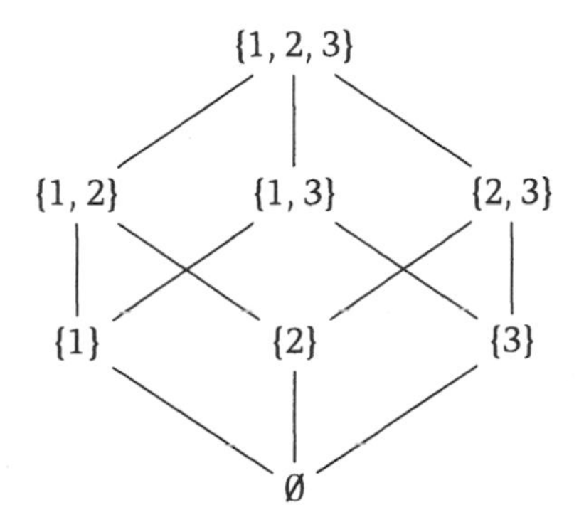

FIGURE 1.4 $(\mathcal{P}(\{1, 2, 3\}), \subseteq)$.

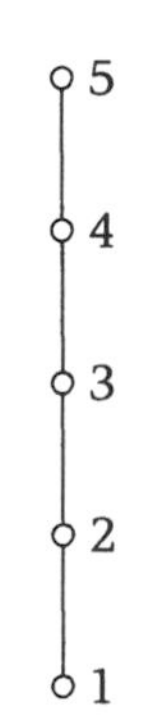

FIGURE 1.5 $(\{1, 2, 3, 4, 5\}, \leq)$.

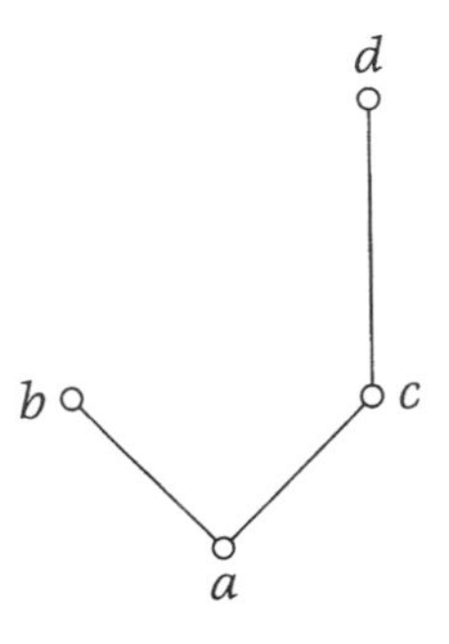

FIGURE 1.6

Here we do not draw a line from $\emptyset$ to {1,2}, because this line already exists via {1} or {2}, etc. The Hasse diagram of $(\{1, 2, 3, 4, 5\}, \leq)$, where $\leq$ means "less than or equal to," is shown in Figure 1.5. The difference between these two examples can be expressed by the following definition.

1.2. Definition. A partial order relation $\leq$ on A is called a ***linear order*** if for each $a, b \in A$ either $a \leq b$ or $b \leq a$. In this case, $(A, \leq)$ is called a ***linearly ordered set***, or a ***chain***.

For example, $(\{1, 2, 3, 4, 5\}, \leq)$ is a chain, while $(\mathcal{P}(\{1, 2, 3\}), \subseteq)$ is not.

If R is a relation from A to B, then R^{-1}, defined by $(a, b) \in R^{-1}$ iff $(b, a) \in R$, is a relation from B to A, called the ***inverse relation*** of R. If $(A, \leq)$ is a partially ordered set, then $(A, \geq)$ is a partially ordered set, and $\geq$ is the inverse relation to $\leq$.

Let $(A, \leq)$ be a poset. We say, a is a greatest element if "all other elements are smaller." More precisely, $a \in A$ is called a ***greatest element*** of A if for all $x \in A$ we have $x \leq a$. The element b in A is called a ***smallest element*** of A if $b \leq x$ for all $x \in A$. The element $c \in A$ is called a ***maximal element*** of A if "no element is bigger," i.e., $c \leq x$ implies $c = x$ for all $x \in A$; similarly, $d \in A$ is called a ***minimal element*** of A if $x \leq d$ implies $x = d$ for all $x \in A$. It can be shown that $(A, \leq)$ has at most one greatest and one smallest element. However, there may be none, one, or several maximal or minimal elements. Every greatest element is maximal and every smallest element is minimal. For instance, in the poset of Figure 1.6, a is a minimal and smallest element; b and d are maximal, but there is no greatest element.

1.3. Definitions. Let $(A, \leq)$ be a poset and $B \subseteq A$.

(i) $a \in A$ is called an ***upper bound*** of B if $b \leq a$ for all $b \in B$.

(ii) $a \in A$ is called a ***lower bound*** of B if $a \leq b$ for all $b \in B$.

(iii) The greatest amongst the lower bounds of B, whenever it exists, is called the ***infimum*** of B, and is denoted by $\inf B$.

(iv) The least upper bound of B, whenever it exists, is called the ***supremum*** of B, and is denoted by $\sup B$.

For instance, if $(A, \leq) = (\mathbb{R}, \leq)$ and B is the interval $[0, 3)$, then $\inf B = 0$ and $\sup B = 3$. Thus the infimum (supremum) of B may or may not be an element of B. If $B' = \mathbb{N}$, $\inf B' = 1$, but $\sup B'$ does not exist.

If $B = \{a_1, \ldots, a_n\}$, then we write $\inf(a_1, \ldots, a_n)$ and $\sup(a_1, \ldots, a_n)$ instead of $\inf\{a_1, \ldots, a_n\}$ and $\sup\{a_1, \ldots, a_n\}$, respectively.

The following statement can neither be proved nor can it be refuted (it is undecidable). It is an additional axiom, that may be used in mathematical arguments (we usually do so without any comment), and it is equivalent to the Axiom of Choice.

1.4. Axiom (*Zorn's Lemma*). *If $(A, \leq)$ is a poset such that every chain of elements in A has an upper bound in A, then A has at least one maximal element.*

In general, not every subset of a poset $(L, \leq)$ has a supremum or an infimum. We study those posets more closely which are axiomatically required to have a supremum and infimum for certain families of subsets.

1.5. Definition. A poset $(L, \leq)$ is called ***lattice ordered*** if for every pair x, y of elements of L their supremum and infimum exist.

1.6. Remarks.

(i) Every chain is lattice ordered.

(ii) In a lattice ordered set $(L, \leq)$ the following statements can be easily seen to be equivalent for all x and y in L:

(a) $x \leq y$;

(b) $\sup(x, y) = y$;

(c) $\inf(x, y) = x$. □

There is another (yet equivalent) approach, which does not use order relations, but algebraic operations instead.

1.7. Definition. An (***algebraic***) ***lattice*** $(L, \wedge, \vee)$ is a set L with two binary operations $\wedge$ (***meet***) and $\vee$ (***join***) (also called intersection or product and union or sum, respectively) which satisfy the following laws for all $x, y, z \in L$:

(L1) $x \wedge y = y \wedge x$, $\quad x \vee y = y \vee x$,
(L2) $x \wedge (y \wedge z) = (x \wedge y) \wedge z$, $\quad x \vee (y \vee z) = (x \vee y) \vee z$,
(L3) $x \wedge (x \vee y) = x$, $\quad x \vee (x \wedge y) = x$.

Two applications of (L3), namely $x \wedge x = x \wedge (x \vee (x \wedge x)) = x$, lead to the additional laws

(L4) $x \wedge x = x$, $\quad x \vee x = x$.

(L1) is the ***commutative law***, (L2) is the ***associative law***, (L3) is the ***absorption law***, and (L4) is the ***idempotent law***.

Sometimes we read $x \vee y$ and $x \wedge y$ as "x vee y" and "x wedge y." The connection between lattice ordered sets and algebraic lattices is as follows.

1.8. Theorem.

(i) *Let $(L, \leq)$ be a lattice ordered set. If we define*

$$x \wedge y := \inf(x, y), \quad x \vee y := \sup(x, y),$$

then $(L, \wedge, \vee)$ is an algebraic lattice.

(ii) *Let $(L, \wedge, \vee)$ be an algebraic lattice. If we define*

$$x \leq y :\Longleftrightarrow x \wedge y = x,$$

then $(L, \leq)$ is a lattice ordered set.

Proof.

(i) Let $(L, \leq)$ be a lattice ordered set. For all $x, y, z \in L$ we have:

(L1) $x \wedge y = \inf(x, y) = \inf(y, x) = y \wedge x$,
$x \vee y = \sup(x, y) = \sup(y, x) = y \vee x$.

(L2) $x \wedge (y \wedge z) = x \wedge \inf(y, z) = \inf(x, \inf(y, z)) = \inf(x, y, z)$
$= \inf(\inf(x, y), z) = \inf(x, y) \wedge z = (x \wedge y) \wedge z$,
and similarly $x \vee (y \vee z) = (x \vee y) \vee z$.

(L3) $x \wedge (x \vee y) = x \wedge \sup(x, y) = \inf(x, \sup(x, y)) = x$,
$x \vee (x \wedge y) = x \vee \inf(x, y) = \sup(x, \inf(x, y)) = x$.

(ii) Let $(L, \wedge, \vee)$ be an algebraic lattice. Clearly, for all x, y, z in L:

- $x \wedge x = x$ and $x \vee x = x$ by (L4); so $x \leq x$, i.e., $\leq$ is reflexive.

- If $x \leq y$ and $y \leq x$, then $x \wedge y = x$ and $y \wedge x = y$, and by (L1) $x \wedge y = y \wedge x$, so $x = y$, i.e., $\leq$ is antisymmetric.
- If $x \leq y$ and $y \leq z$, then $x \wedge y = x$ and $y \wedge z = y$. Therefore

$$x = x \wedge y = x \wedge (y \wedge z) = (x \wedge y) \wedge z = x \wedge z,$$

  so $x \leq z$ by (L2), i.e., $\leq$ is transitive.

Let $x, y \in L$. Then $x \wedge (x \vee y) = x$ implies $x \leq x \vee y$ and similarly $y \leq x \vee y$. If $z \in L$ with $x \leq z$ and $y \leq z$, then $(x \vee y) \vee z = x \vee (y \vee z) = x \vee z = z$ and so $x \vee y \leq z$. Thus $\sup(x, y) = x \vee y$. Similarly $\inf(x, y) = x \wedge y$. Hence $(L, \leq)$ is a lattice ordered set. □

1.9. Remark. It follows from Remark 1.6 that Theorem 1.8 yields a one-to-one relationship between lattice ordered sets and algebraic lattices. Therefore we shall use the term ***lattice*** for both concepts. The number $|L|$ of elements of L denotes the ***cardinality*** (or the ***order***) of the lattice L.

If N is a subset of a poset, then $\bigvee_{x \in N} x$ and $\bigwedge_{x \in N} x$ denote the supremum and infimum of N, respectively, whenever they exist. We say that the supremum of N is the join of all elements of N and the infimum is the meet of all elements of N.

In Definition 1.7, we have seen that for each of the laws (L1)–(L3), two equations are given. This leads to

1.10 (*Duality Principle*). *Any "formula" involving the operations $\wedge$ and $\vee$ which is valid in any lattice $(L, \wedge, \vee)$ remains valid if we replace $\wedge$ by $\vee$ and $\vee$ by $\wedge$ everywhere in the formula. This process of replacing is called* ***dualizing***.

The validity of this assertion follows from the fact that any formula in a lattice that can be derived using (L1)–(L3) remains correct if we interchange $\wedge$ and $\vee$, $\leq$ and $\geq$, respectively, everywhere in the formula, because every dual of a condition in (L1)–(L3) holds, too. This is very convenient, since we only have to prove "one-half" of the results (see, e.g., 1.13 and 1.14).

1.11. Definition. If a lattice L contains a smallest (greatest) element with respect to $\leq$, then this uniquely determined element is called the ***zero element*** (***unit element***, respectively), denoted by 0 (by 1). The elements 0 and 1 are called ***universal bounds***. If they exist, L is called ***bounded***.

Every finite lattice L is bounded (see Exercise 6). If a lattice is bounded (by 0 and 1), then every x in L satisfies $0 \leq x \leq 1$, $0 \wedge x = 0$, $0 \vee x = x$, $1 \wedge x = x$, $1 \vee x = 1$. We consider some examples of lattices.

Set	$\leq$	$x \wedge y$	$x \vee y$	0	1
M	linear order	$\min(x, y)$	$\max(x, y)$	smallest element	greatest element
$\mathbb{N}$	"divides"	$\gcd(x, y)$	$\operatorname{lcm}(x, y)$	1	does not exist
$\mathcal{P}(M)$	$\subseteq$	$X \cap Y$	$X \cup Y$	$\emptyset$	M
$\times_{i \in I} M_i$	componentwise	componentwise	componentwise	$(\ldots, 0, \ldots)$	$(\ldots, 1, \ldots)$
all subspaces of V	$\subseteq$	$X \cap Y$	$X + Y$	$\{\mathbf{0}\}$	V

FIGURE 1.7

1.12. Examples. Let M and M_i, $i \in I$, be linearly ordered sets with smallest element 0 and greatest element 1, let $\mathbb{N}$ be the set $\{1, 2, \ldots\}$ of natural numbers, and let V be a vector space. Then Figure 1.7 lists some important lattices.

Theorem 1.8 and Remark 1.9 enable us to represent any lattice as a special poset or as an algebraic structure using operation tables. In Figures 1.8 and 1.9, we present the Hasse diagrams of all lattices with at most six elements. V_i^n denotes the ith lattice with n elements. Figure 1.6 shows an example of a poset which is not a lattice (since $\sup(b, c)$ did not exist).

In Figure 1.10, we give the ***operation tables*** for the lattice V_4^5, which table all $x \wedge y$ and $x \vee y$, for x, y in the lattice. Observe that all entries in these tables must again belong to the lattice.

1.13. Lemma. *In every lattice L the operations $\wedge$ and $\vee$ are isotone, i.e., $y \leq z \Longrightarrow x \wedge y \leq x \wedge z$ and $x \vee y \leq x \vee z$.*

Proof. $y \leq z \Longrightarrow x \wedge y = (x \wedge x) \wedge (y \wedge z) = (x \wedge y) \wedge (x \wedge z) \Longrightarrow x \wedge y \leq x \wedge z$. The second formula is verified by duality. □

1.14. Theorem. *The elements of an arbitrary lattice satisfy the following* ***distributive inequalities****:*

$$\begin{aligned} x \wedge (y \vee z) &\geq (x \wedge y) \vee (x \wedge z), \\ x \vee (y \wedge z) &\leq (x \vee y) \wedge (x \vee z). \end{aligned} \tag{1.1}$$

Proof. From $x \wedge y \leq x$ and $x \wedge y \leq y \leq y \vee z$ we get $x \wedge y \leq x \wedge (y \vee z)$, and similarly $x \wedge z \leq x \wedge (y \vee z)$. Thus $x \wedge (y \vee z)$ is an upper bound for both $x \wedge y$ and $x \wedge z$; therefore $x \wedge (y \vee z) \geq (x \wedge y) \vee (x \wedge z)$. The second inequality follows by duality. □

We can construct new lattices from given ones by forming substructures, homomorphic images, and products.

1.15. Definition. A subset S of a lattice L is called a ***sublattice*** of L if S is a lattice with respect to the restrictions of $\wedge$ and $\vee$ from L to S.

Obviously, a subset S of L is a sublattice of the lattice L if and only if S is "closed" with respect to $\wedge$ and $\vee$ (i.e., $s_1, s_2 \in S \Longrightarrow s_1 \wedge s_2 \in S$ and $s_1 \vee s_2 \in S$). We note that a subset S of a lattice L can be a lattice with respect to the partial order of L without being a sublattice of L (see Example 1.16(iii) below).

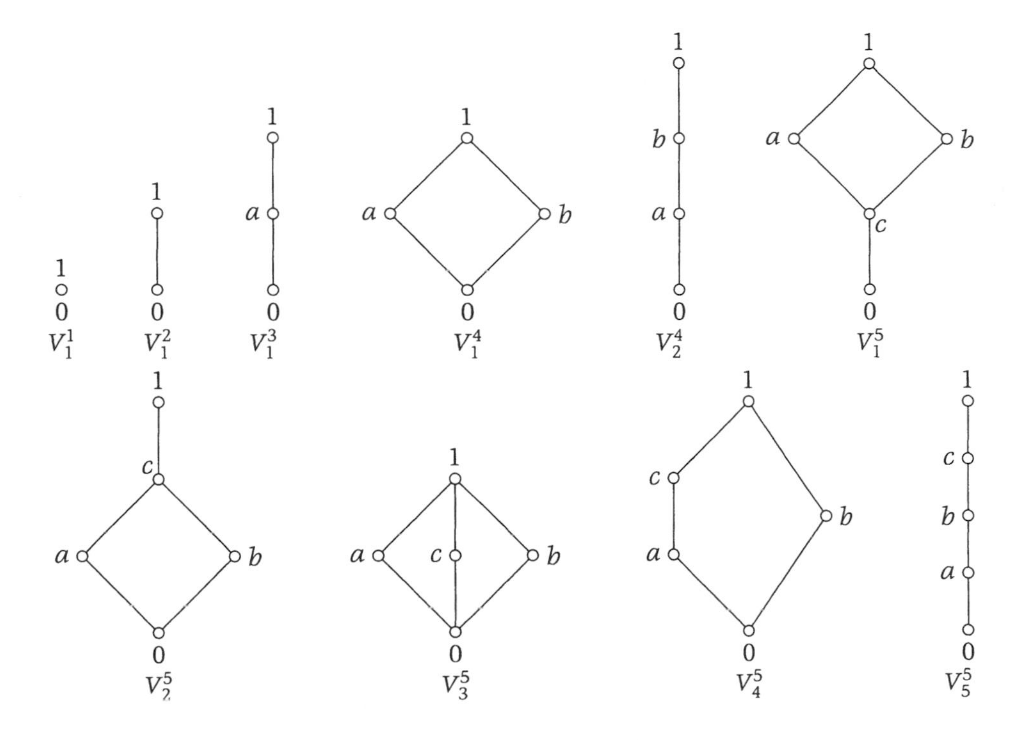

FIGURE 1.8 Hasse diagrams of all lattices with at most five elements.

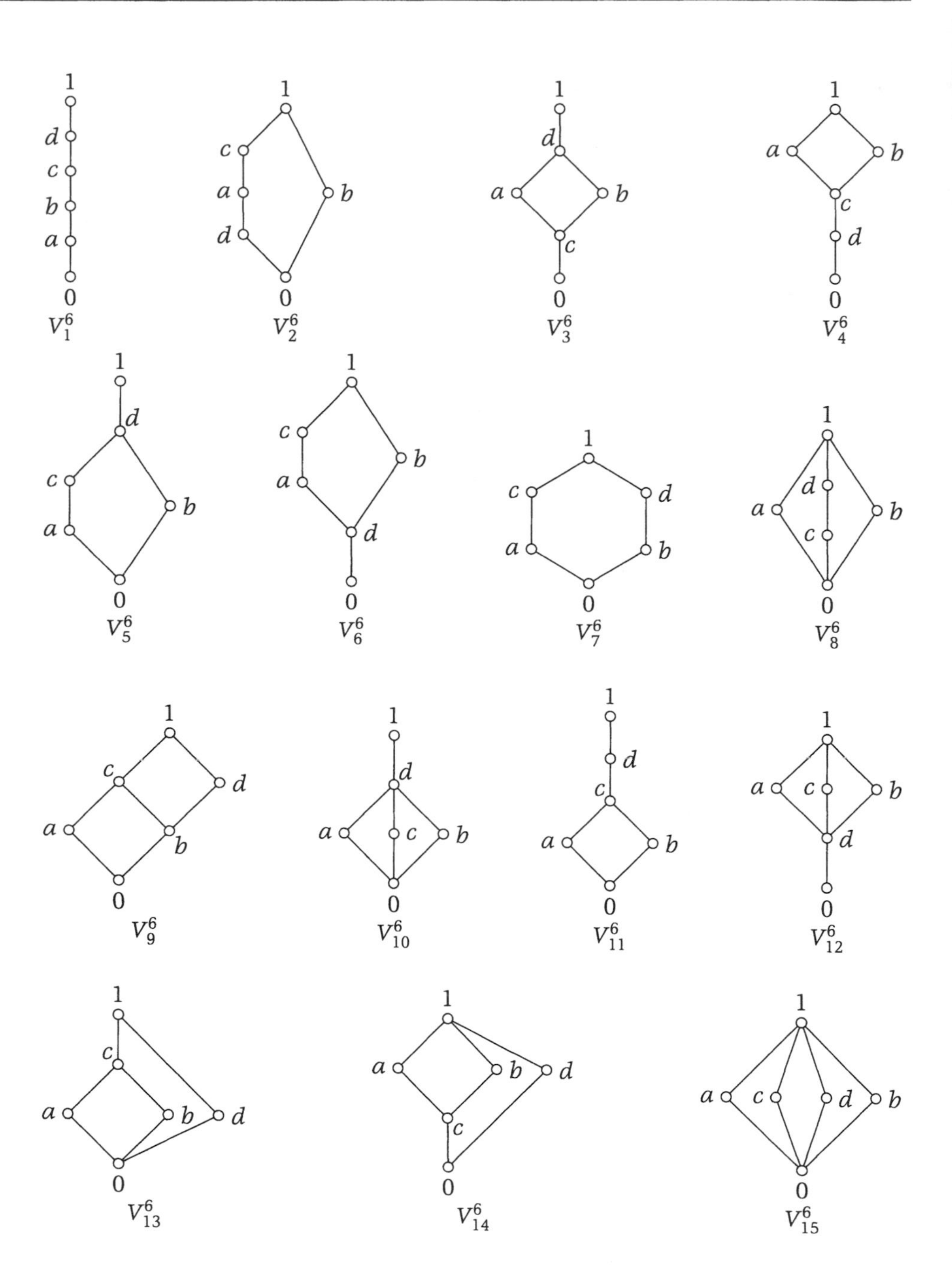

FIGURE 1.9 Hasse diagrams of all lattices with six elements.

$\wedge$	0	a	b	c	1
0	0	0	0	0	0
a	0	a	0	c	a
b	0	0	b	0	b
c	0	c	0	c	c
1	0	a	b	c	1

$\vee$	0	a	b	c	1
0	0	a	b	c	1
a	a	a	1	a	1
b	b	1	b	1	1
c	c	a	1	c	1
1	1	1	1	1	1

FIGURE 1.10

1.16. Examples.

(i) Every singleton of a lattice L is a sublattice of L.

(ii) For any two elements x, y in a lattice L, the ***interval***

$$[x, y] := \{a \in L \mid x \leq a \leq y\}$$

is a sublattice of L.

(iii) Let L be the lattice of all subsets of a vector space V and let S be the set of all subspaces of V. Then S is a lattice with respect to inclusion but not a sublattice of L.

1.17. Definitions. Let L and M be lattices. A mapping $f\colon L \to M$ is called a:

(i) ***join-homomorphism*** if $f(x \vee y) = f(x) \vee f(y)$;

(ii) ***meet-homomorphism*** if $f(x \wedge y) = f(x) \wedge f(y)$;

(iii) ***order-homomorphism*** if $x \leq y \Longrightarrow f(x) \leq f(y)$;

hold for all $x, y \in L$. We call f a ***homomorphism*** (or ***lattice homomorphism***) if it is both a join- and a meet-homomorphism. Injective, surjective, or bijective (lattice) homomorphisms are called (lattice) ***monomorphisms, epimorphisms***, or ***isomorphisms***, respectively. If f is a homomorphism from L to M, then $f(L)$ is called a ***homomorphic image*** of L; it is a sublattice of M (see Exercise 11). If there is an isomorphism from L to M, then we say that L and M are ***isomorphic*** and denote this by $L \cong M$.

It can be easily shown that every join-(or meet-)homomorphism is an order-homomorphism. The converse, however, is not true (why?). The relationship between the different homomorphisms is symbolized in Figure 1.11.

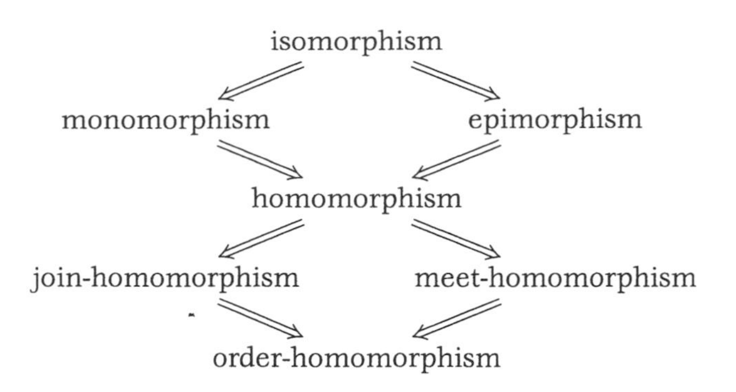

FIGURE 1.11

For example, 1, a, b, 0 and u, s, t, r are isomorphic under the isomorphism $0 \mapsto r$, $a \mapsto s$, $b \mapsto t$, $1 \mapsto u$. The map $0 \mapsto r$, $a \mapsto t$, $b \mapsto s$, $1 \mapsto u$ is another isomorphism. Observe that in Figures 1.8 and 1.9 we have in fact listed only all nonisomorphic lattices of orders up to 6. Observe that there are already infinitely many different lattices with one element. As another example, V_1^3 is isomorphic (but not equal) to 1, b, 0. We see, in most cases it makes sense to identify isomorphic lattices.

1.18. Example. Let L_1, L_2, and L_3 be the lattices with Hasse diagrams of Figure 1.12, respectively. We define:

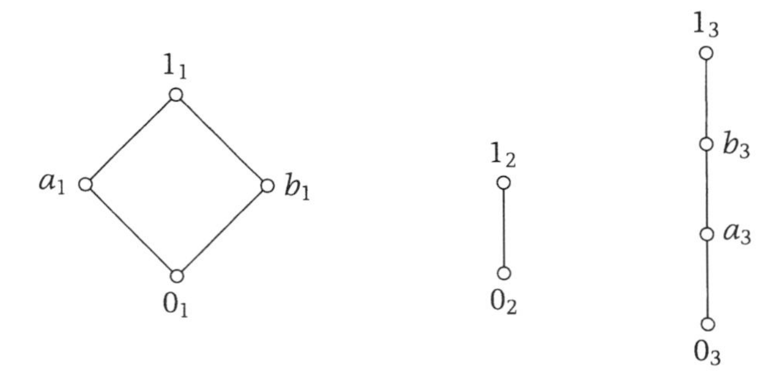

FIGURE 1.12

$$f\colon L_1 \to L_2;\quad f(0_1) = f(a_1) = f(b_1) = 0_2,\ f(1_1) = 1_2;$$
$$g\colon L_1 \to L_2;\quad g(0_1) = 0_2,\ g(1_1) = g(a_1) = g(b_1) = 1_2;$$
$$h\colon L_1 \to L_3;\quad h(0_1) = 0_3,\ h(a_1) = a_3,\ h(b_1) = b_3,\ h(1_1) = 1_3.$$

These three mappings are order-homomorphisms. f is even a meet-homomorphism, since

$$f(a_1 \wedge b_1) = f(0_1) = 0_2 = f(a_1) \wedge f(b_1), \text{ etc.}$$

However, f is not a homomorphism, since

$$f(a_1 \vee b_1) = f(1_1) = 1_2 \quad \text{and} \quad f(a_1) \vee f(b_1) = 0_2.$$

Dually, g is a join-homomorphism, but not a homomorphism. h is neither a meet- nor a join-homomorphism, since

$$h(a_1 \wedge b_1) = h(0_1) = 0_3 \quad \text{and} \quad h(a_1) \wedge h(b_1) = a_3 \wedge b_3 = a_3,$$
$$h(a_1 \vee b_1) = h(1_1) = 0_3 \quad \text{and} \quad h(a_1) \vee h(b_1) = a_3 \vee b_3 = b_3.$$

1.19. Definition. Let L and M be lattices. The set of ordered pairs

$$\{(x, y) \mid x \in L,\ y \in M\}$$

with operations $\vee$ and $\wedge$ defined by

$$(x_1, y_1) \vee (x_2, y_2) := (x_1 \vee x_2, y_1 \vee y_2),$$
$$(x_1, y_1) \wedge (x_2, y_2) := (x_1 \wedge x_2, y_1 \wedge y_2),$$

is the ***direct product*** of L and M, in symbols $L \times M$, also called the ***product lattice***. It should be clear how to define the direct product of more than two lattices.

It is verified easily that $L \times M$ is a lattice in the sense of Definition 1.7. The partial order of $L \times M$ which results from the correspondence in 1.8(1.8) satisfies

$$(x_1, y_1) \le (x_2, y_2) \iff x_1 \le x_2 \text{ and } y_1 \le y_2. \tag{1.2}$$

1.20. Example. The direct product of the lattices L and M can graphically be described in terms of the Hasse diagrams at the top of the following page.

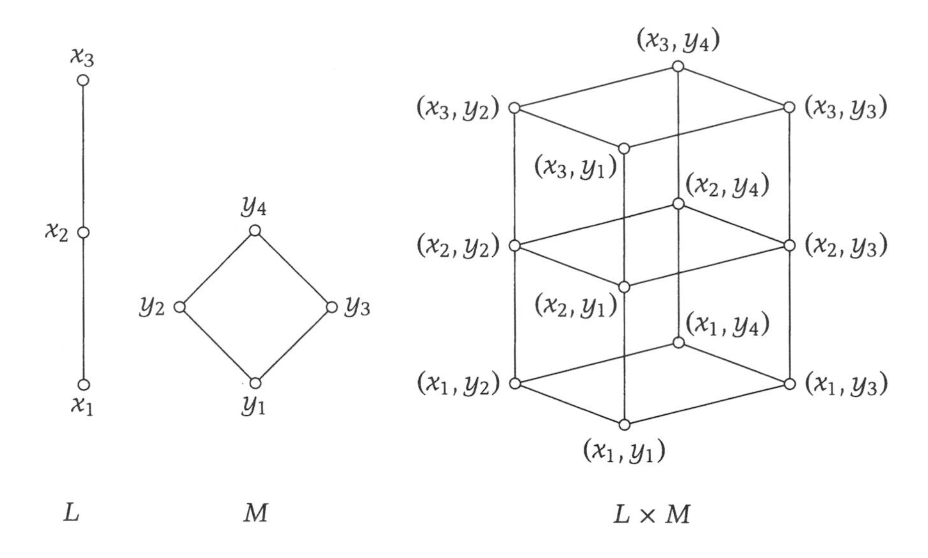

Exercises

1. Determine all the partial orders and their Hasse diagrams on the set $L = \{a, b, c\}$. Which of them are chains?
2. Give an example of a poset which has exactly one maximal element but does not have a greatest element.
3. Let $(\mathbb{R}, \leq)$ be the poset of all real numbers and let $A = \{x \in \mathbb{R} \mid x^3 < 3\}$. Is there an upper bound (or lower bound) or a supremum (or infimum) of A?
4. Show that the examples in 1.12 are really lattices and prove the indicated properties.
5. Is $F := \{A \subseteq \mathbb{N} \mid A \text{ finite}\}$ a sublattice of $\mathcal{P}(\mathbb{N})$? Is F bounded? Is $\mathcal{P}(\mathbb{N})$ bounded?
6. Prove that any finite lattice is bounded. Find a lattice without a zero and a unit element.
7. More generally, prove that in a lattice $(L, \leq)$ every finite nonempty subset S has a least upper bound and a greatest lower bound.

8. Let $\mathbb{C}$ be the set of complex numbers $z = x + iy$ where x and y are in $\mathbb{R}$. Define a partial order $\subseteq$ on $\mathbb{C}$ as in Equation 1.2 by: $x_1 + iy_1 \subseteq x_2 + iy_2$ if and only if $x_1 \leq x_2$ and $y_1 \leq y_2$. Is this a linear order? Is there a minimal or a maximal element in $(\mathbb{C}, \subseteq)$? How does $\subseteq$ compare with the ***lexicographic order*** $\preceq$ defined by $x_1 + iy_1 \preceq x_2 + iy_2$ if and only if $x_1 < x_2$, or $x_1 = x_2$ and $y_1 \leq y_2$?

9. An ***isomorphism*** of posets is a bijective order-homomorphism, whose inverse is also an order-homomorphism. Prove: If f is an isomorphism of a poset L onto a poset M, and if L is a lattice, then M is also a lattice, and f is an isomorphism of the lattices.

10. Let $(C[a, b], \max, \min)$ be the lattice of continuous real-valued functions on a closed interval $[a, b]$ and let $D[a, b]$ be the set of all differentiable functions on $[a, b]$. Show by example that $D[a, b]$ is not a sublattice of $C[a, b]$.

11. Let f be a monomorphism from the lattice L into the lattice M. Show that L is isomorphic to a sublattice of M.

12. Which of the lattices in Figure 1.8 are isomorphic to a sublattice of V_{14}^{6} (see Figure 1.9)?

13. Let $D(k)$ be the set of all positive divisors of $k \in \mathbb{N}$. Show that $(D(k), \gcd, \operatorname{lcm})$ is a lattice. Construct the Hasse diagrams of the lattices $D(20)$ and $D(21)$, find isomorphic copies in Figures 1.8 and 1.9, and show that $D(20) \times D(21)$ is isomorphic to $D(420)$.

14. Determine the operation tables for $\wedge$ and $\vee$ for the lattice with Hasse diagram

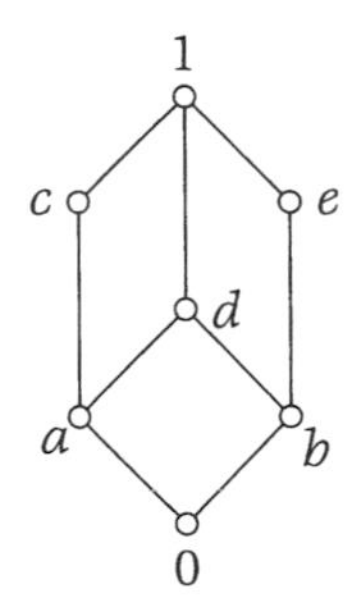

15. Prove: In any lattice L, we have

$$\big((x \wedge y) \vee (x \wedge z)\big) \wedge \big((x \wedge y) \vee (y \wedge z)\big) = x \wedge y \quad \text{for all } x, y, z \in L.$$

16. Let C_1 and C_2 be the finite chains $\{0, 1, 2\}$ and $\{0, 1\}$, respectively. Draw the Hasse diagram of the product lattice $C_1 \times C_2 \times C_2$.

17. Let L be a sublattice of M and let $f\colon M \to N$ be a homomorphism. If M is bounded, does this also apply to L and N (cf. Exercise 5)?

§2 Distributive Lattices

We now turn to special types of lattices, with the aim of defining very "rich" types of algebraic structures, Boolean algebras.

2.1. Definition. A lattice L is called ***distributive*** if the laws

$$x \vee (y \wedge z) = (x \vee y) \wedge (x \vee z),$$
$$x \wedge (y \vee z) = (x \wedge y) \vee (x \wedge z),$$

hold for all $x, y, z \in L$. These equalities are called ***distributive laws***.

Due to Exercise 9, the two distributive laws are equivalent, so it would be enough to require just one of them.

2.2. Examples.

(i) $(\mathcal{P}(M), \cap, \cup)$ is a distributive lattice.

(ii) $(\mathbb{N}, \gcd, \operatorname{lcm})$ is a distributive lattice (see Exercise 4).

(iii) The "diamond lattice" V_3^5 and the "pentagon lattice" V_4^5 are not distributive: In V_3^5, $a \vee (b \wedge c) = a \neq 1 = (a \vee b) \wedge (a \vee c)$, while in V_4^5, $a \vee (b \wedge c) = a \neq c = (a \vee b) \wedge (a \vee c)$.

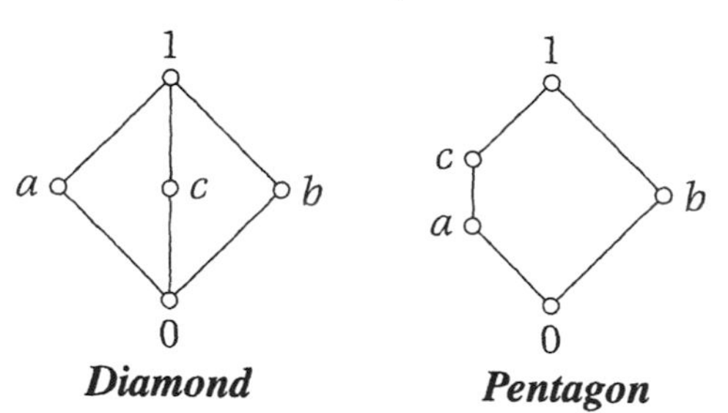

Diamond ***Pentagon***

These are the smallest nondistributive lattices.

2.3. Theorem. *A lattice is distributive if and only if it does not contain a sublattice isomorphic to the diamond or the pentagon.*

A lattice which "contains" the diamond or the pentagon must clearly be nondistributive. The converse needs much more work (see, e.g., Szász (1963)). As an application of 2.3 we get:

2.4. Corollary. *Every chain is distributive lattice.*

2.5. Example. The lattice with Hasse diagram

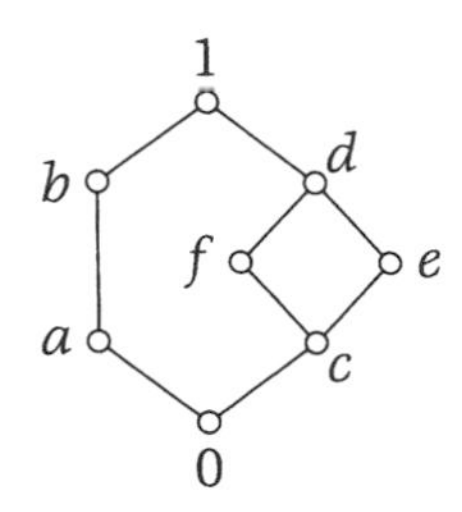

cannot be distributive since it contains the pentagon $\{0, a, e, b, 1\}$ as a sublattice.

2.6. Theorem. *A lattice L is distributive if and only if the* ***cancellation rule*** $x \wedge y = x \wedge z$, $x \vee y = x \vee z \Longrightarrow y = z$ *holds for all* $x, y, z \in L$.

Proof. Exercise 6.

2.7. Definition. A lattice L with 0 and 1 is called ***complemented*** if for each $x \in L$ there is at least one element y such that $x \wedge y = 0$ and $x \vee y = 1$. Each such y is called a ***complement*** of x.

2.8. Examples.

(i) Let $L = \mathcal{P}(M)$. Then $B = M \setminus A$ is the uniquely determined complement of A.

(ii) In a bounded lattice, 1 is a complement of 0, and 0 is a complement of 1.

(iii) Not every lattice with 0 and 1 is complemented. For instance, a in
1
a does not have a complement. In fact, every chain with more
0
than two elements is not complemented.

(iv) The complement need not be unique: a in the diamond has the two complements b and c.

(v) Let L be the lattice of subspaces of the vector space $\mathbb{R}^2$. If T is a complement of a subspace S, then $S \cap T = \{0\}$ and $S + T = \mathbb{R}^2$. Hence a complement is a complementary subspace. If $\dim S = 1$, then S has infinitely many complements, namely all subspaces T such that $S \oplus T = \mathbb{R}^2$. Therefore L cannot be distributive, as the following theorem shows.

2.9. Theorem and Definition. *If L is a distributive lattice, then each $x \in L$ has at most one complement. We denote it by x'.*

Proof. Suppose $x \in L$ has two complements y_1 and y_2. Then $x \vee y_1 = 1 = x \vee y_2$ and $x \wedge y_1 = 0 = x \wedge y_2$; thus $y_1 = y_2$ because of 2.6. □

Complemented distributive lattices will be studied extensively in the following sections.

2.10. Definition. Let L be a lattice with zero. $a \in L$ is called an ***atom*** if $a \neq 0$ and if for all $b \in L: 0 < b \leq a \Longrightarrow b = a$.

2.11. Definition. $a \in L$ is called ***join-irreducible*** if for all $b, c \in L$

$$a = b \vee c \Longrightarrow a = b \text{ or } a = c.$$

Otherwise a is called ***join-reducible***.

2.12. Lemma. *Every atom of a lattice with zero is join-irreducible.*

Proof. Let a be an atom and let $a = b \vee c$, $a \neq b$. Then $a = \sup(b, c)$; so $b \leq a$. Therefore $b = 0$ and $a = c$. □

2.13. Lemma. *Let L be a distributive lattice and let $p \in L$ be join-irreducible with $p \leq a \vee b$. Then $p \leq a$ or $p \leq b$.*

Proof. $p \leq a \vee b$ means $p = p \wedge (a \vee b) = (p \wedge a) \vee (p \wedge b)$. Since p is join-irreducible, $p = p \wedge a$ or $p = p \wedge b$, i.e., $p \leq a$ or $p \leq b$. □

2.14. Definitions. If $x \in [a, b] = \{v \in L \mid a \leq v \leq b\}$ and $y \in L$ with $x \wedge y = a$ and $x \vee y = b$, then y is called a ***relative complement*** of x with respect to $[a, b]$. If all intervals $[a, b]$ in a lattice L are complemented, then L is called ***relatively complemented***. If L has a zero element and all $[0, b]$ are complemented, then L is called ***sectionally complemented***.

Exercises

1. Prove the generalized distributive inequality for lattices:

$$y \wedge \left(\bigvee_{i=1}^{n} x_i \right) \geq \bigvee_{i=1}^{n} (y \wedge x_i).$$

2. Let L be a distributive lattice with 0 and 1. Prove: If a has a complement a', then

$$a \vee (a' \wedge b) = a \vee b.$$

3. Which of the lattices in Figures 1.8 and 1.9 are distributive? Complemented?
4. Show that the set $\mathbb{N}$, ordered by divisibility, is a distributive lattice. Is it complemented? Consider the same questions for $\mathbb{N}_0$.
5. Let S be an arbitrary set and D a distributive lattice. Show that the set of all functions from S to D is a distributive lattice, where $f \leq g$ means $f(x) \leq g(x)$ for all x.
6. Prove Theorem 2.6.
7. Show that sublattices, homomorphic images, and direct products of distributive lattices are again distributive.
8. Does the analogous question to Exercise 7 hold for complemented lattices? (Cf. Exercises 1.5 and 1.17.)
9. Show that the two distributive laws in Definition 2.1 are equivalent.
10. Let L be the lattice $(\mathbb{N}_0, \gcd, \mathrm{lcm})$. Determine the atoms in L. Which elements are join-irreducible?
11. Same question as in Exercise 10, now for $L = (\mathcal{P}(M), \cap, \cup)$.
12. Show by example that relative complements are not always unique.
13. If L and M are isomorphic lattices and L is distributive (complemented, sectionally complemented), show that this applies to M as well.

§3 Boolean Algebras

Boolean algebras are special lattices which are useful in the study of logic, both digital computer logic and that of human thinking, and of switching circuits. This latter application was initiated by C. E. Shannon, who showed that fundamental properties of electrical circuits of bistable elements can be represented by using Boolean algebras. We shall consider such applications in Chapter 2.

3.1. Definition. A complemented distributive lattice is called a ***Boolean algebra*** (or a ***Boolean lattice***).

Distributivity in a Boolean algebra guarantees the uniqueness of complements (see 2.9).

3.2. Notation. From now on, in Chapters 1 and 2, B will denote a set with the two binary operations $\wedge$ and $\vee$, with zero element 0 and a unit element 1, and the unary operation of complementation $'$, in short $B = (B, \wedge, \vee, 0, 1, ')$ or $B = (B, \wedge, \vee)$, or simply B.

3.3. Examples.

(i) $(\mathcal{P}(M), \cap, \cup, \emptyset, M, ')$ is the Boolean algebra of the power set of a set M. Here $\cap$ and $\cup$ are the set-theoretic operations intersection and union, and the complement is the set-theoretic complement, namely $M \setminus A = A'$; $\emptyset$ and M are the "universal bounds." If M has $n (\in \mathbb{N}_0)$ elements, then $\mathcal{P}(M)$ consists of 2^n elements.

(ii) Let $\mathbb{B}$ be the lattice V_1^2, where the operations are defined by

$\wedge$	0	1
0	0	0
1	0	1

$\vee$	0	1
0	0	1
1	1	1

	$'$
0	1
1	0

Then $(\mathbb{B}, \wedge, \vee, 0, 1, ')$ is a Boolean algebra. If $n \in \mathbb{N}$, we can turn $\mathbb{B}^n$ into a Boolean algebra via 1.19:

$$(i_1, \ldots, i_n) \wedge (j_1, \ldots, j_n) := (i_1 \wedge j_1, \ldots, i_n \wedge j_n),$$
$$(i_1, \ldots, i_n) \vee (j_1, \ldots, j_n) := (i_1 \vee j_1, \ldots, i_n \vee j_n),$$
$$(i_1, \ldots, i_n)' := (i_1', \ldots, i_n'),$$

and $0 = (0, \ldots, 0)$, $1 = (1, \ldots, 1)$.

(iii) More generally, any direct product of Boolean algebras is a Boolean algebra again (see Exercise 3).

3.4. Theorem (*De Morgan's Laws*). *For all x, y in a Boolean algebra, we have*

$$(x \wedge y)' = x' \vee y' \quad \textit{and} \quad (x \vee y)' = x' \wedge y'.$$

Proof. We have

$$(x \wedge y) \vee (x' \vee y') = (x \vee x' \vee y') \wedge (y \vee x' \vee y')$$
$$= (1 \vee y') \wedge (1 \vee x') = 1 \wedge 1 = 1,$$

and similarly, $(x \wedge y) \wedge (x' \vee y') = (x \wedge y \wedge x') \vee (x \wedge y \wedge y') = 0$. This implies that $x' \vee y'$ is the complement of $x \wedge y$. The second formula follows dually. □

3.5. Corollary. *In a Boolean algebra B we have for all $x, y \in B$,*

$$x \leq y \iff x' \geq y'.$$

Proof. $x \leq y \iff x \vee y = y \iff x' \wedge y' = (x \vee y)' = y' \iff x' \geq y'$. □

3.6. Theorem. *In a Boolean algebra B we have for all $x, y \in B$,*

$$x \leq y \iff x \wedge y' = 0 \iff x' \vee y = 1 \iff x \wedge y = x$$
$$\iff x \vee y = y.$$

Proof. See Exercise 4.

3.7. Definition. Let B_1 and B_2 be Boolean algebras. Then the mapping $f\colon B_1 \to B_2$ is called a ***(Boolean) homomorphism*** from B_1 into B_2 if f is a (lattice) homomorphism and for all $x \in B$ we have $f(x') = (f(x))'$.

Analogously, we can define Boolean monomorphisms and isomorphisms as in 1.17. If there is a Boolean isomorphism between B_1 and B_2, we write $B_1 \cong_b B_2$. The simple proofs of the following properties are left to the reader.

3.8. Theorem. *Let $f\colon B_1 \to B_2$ be a Boolean homomorphism. Then:*

(i) $f(0) = 0$, $f(1) = 1$;

(ii) *for all $x, y \in B_1$, $x \leq y \Longrightarrow f(x) \leq f(y)$;*

(iii) *$f(B_1)$ is a Boolean algebra and a "Boolean subalgebra" (which is defined as expected) of B_2.*

3.9. Examples.

(i) If $M \subset N$, then the map $f\colon \mathcal{P}(M) \to \mathcal{P}(N)$; $A \mapsto A$ is a lattice monomorphism but not a Boolean homomorphism, since for $A \in \mathcal{P}(M)$ the complements in M and N are different. Also, $f(1) = f(M) = M \neq N =$ the unit element in $\mathcal{P}(N)$.

(ii) If $M = \{1, \ldots, n\}$, then $\{0,1\}^n$ and $\mathcal{P}(M)$ are Boolean algebras, and the map $f : \{0,1\}^n \to \mathcal{P}(M)$, $(i_1, \ldots, i_n) \mapsto \{k \mid i_k = 1\}$ is a Boolean isomorphism. It is instructive to do the proof as an exercise.

(iii) More generally, let X be any set, A a subset of X, and let

$$\chi_A\colon\ X \to \{0,1\}; \qquad x \mapsto \begin{cases} 1 & \text{if } x \in A, \\ 0 & \text{if } x \notin A, \end{cases}$$

be the ***characteristic function*** of A. Then $h\colon \mathcal{P}(X) \to \{0,1\}^X$; $A \mapsto \chi_A$ is a Boolean isomorphism, so

$$\mathcal{P}(X) \cong_b \{0,1\}^X.$$

3.10. Theorem. *Let L be a lattice. Then the following implications hold:*

(i) *L is a Boolean algebra $\Longrightarrow$ L is relatively complemented;*

(ii) *L is relatively complemented $\Longrightarrow$ L is sectionally complemented;*

(iii) *L is finite and sectionally complemented $\Longrightarrow$ every nonzero element a of L is a join of finitely many atoms.*

Proof.

(i) Let L be a Boolean algebra and let $a \leq x \leq b$. Define $y := b \wedge (a \vee x')$. Then y is a complement of x in $[a, b]$, since

$$x \wedge y = x \wedge (b \wedge (a \vee x')) = x \wedge (a \vee x') = (x \wedge a) \vee (x \wedge x') = x \wedge a = a$$

and

$$\begin{aligned} x \vee y &= x \vee (b \wedge (a \vee x')) = x \vee ((b \wedge a) \vee (b \wedge x')) = x \vee (b \wedge x') \\ &= (x \vee b) \wedge (x \vee x') = b \wedge 1 = b. \end{aligned}$$

Thus L is relatively complemented.

(ii) If L is relatively complemented, then every $[a, b]$ is complemented; thus every interval $[0, b]$ is complemented, i.e., L is sectionally complemented.

(iii) Let $\{p_1, \dots, p_n\}$ be the set of atoms $\leq a \in L$ and let $b = p_1 \vee \dots \vee p_n$. Now $b \leq a$, and if we suppose that $b \neq a$, then b has a nonzero complement, say c, in $[0, a]$. Let p be an atom $\leq c$, then $p \in \{p_1, \dots, p_n\}$ and thus $p = p \wedge b \leq c \wedge b = 0$, which is a contradiction. Hence $a = b = p_1 \vee \dots \vee p_n$. □

Finite Boolean algebras can be characterized as follows:

3.11. Theorem (*Representation Theorem*). *Let B be a finite Boolean algebra, and let A denote the set of all atoms in B. Then B is isomorphic to $\mathcal{P}(A)$, i.e.,*

$$(B, \wedge, \vee) \cong_b (\mathcal{P}(A), \cap, \cup).$$

Proof. Let $v \in B$ be an arbitrary element and let $A(v) := \{a \in A \mid a \leq v\}$. Then $A(v) \subseteq A$. Define

$$h\colon B \to \mathcal{P}(A); \qquad v \mapsto A(v).$$

We show that h is a Boolean isomorphism. First we show that h is a Boolean homomorphism: For an atom a and for $v, w \in V$ we have

$$a \in A(v \wedge w) \iff a \leq v \wedge w \iff a \leq v \text{ and } a \leq w$$
$$\iff a \in A(v) \cap A(w),$$

which proves $h(v \wedge w) = h(v) \cap h(w)$. Similarly,

$$a \in A(v \vee w) \iff a \leq v \vee w \iff a \leq v \text{ or } a \leq w$$
$$\iff a \in A(v) \cup A(w);$$

here, the second equivalence follows from 2.12 and 2.13. Finally,

$$a \in A(v') \iff a \leq v' \iff a \wedge v' = 0 \iff a \not\leq v$$
$$\iff a \in A \setminus A(v);$$

here, the second equivalence follows from 3.6. Note that $h(0) = \emptyset$ and 0 is the unique element which is mapped to $\emptyset$. Since B is finite we are able to use Theorem 3.10 to verify that h is bijective. We know that every $v \in B$ can be expressed as a join of finitely many atoms: $v = a_1 \vee \cdots \vee a_n$ with all atoms $a_i \leq v$. Let $h(v) = h(w)$, i.e., $A(v) = A(w)$. Then $a_i \in A(v)$ and $a_i \in A(w)$. Therefore $a_i \leq w$, and thus $v \leq w$. Reversing the roles of v and w yields $v = w$, and this shows that h is injective.

To show that h is surjective we verify that for each $C \in \mathcal{P}(A)$ there is some $v \in B$ such that $h(v) = C$. Let $C = \{c_1, \ldots, c_n\}$ and $v = c_1 \vee \cdots \vee c_n$. Then $A(v) \supseteq C$, hence $h(v) \supseteq C$. Conversely, if $a \in h(v)$, then a is an atom with $a \leq v = c_1 \vee \cdots \vee c_n$. Therefore $a \leq c_i$, for some $i \in \{1, \ldots, n\}$, by 2.12 and 2.13. So $a = c_i \in C$. Altogether this implies $h(v) = A(v) = C$. □

3.12. Theorem. *The cardinality of a finite Boolean algebra B is always of the form 2^n, and B then has precisely n atoms. Any two Boolean algebras with the same finite cardinality are isomorphic.*

Proof. The first assertion follows immediately from 3.11. If B_1 and B_2 have the same cardinality $m \in \mathbb{N}$, then m is of the form 2^n, and B_1, B_2 have both n atoms. So $B_1 \cong_b \mathcal{P}(\{1, \ldots, n\}) \cong_b B_2$ by 3.9(ii), hence $B_1 \cong_b B_2$. □

In this way we have also seen:

3.13. Theorem. *For every finite Boolean algebra $B \neq \{0\}$ there is some $n \in \mathbb{N}$ with*

$$B \cong_b \{0, 1\}^n = \mathbb{B}^n.$$

3.14. Examples.

(i) The lattice of the divisors of 30, i.e., the Boolean algebra $B = (\{1,2,3,5,6,10,15,30\}, \gcd, \operatorname{lcm}, 1, 30,$ complement with respect to 30), has $8 = 2^3$ elements and is therefore isomorphic to the lattice of the power set $\mathcal{P}(\{a,b,c\})$.

(ii) We sketch the Hasse diagrams of all nonisomorphic Boolean algebras of orders < 16:

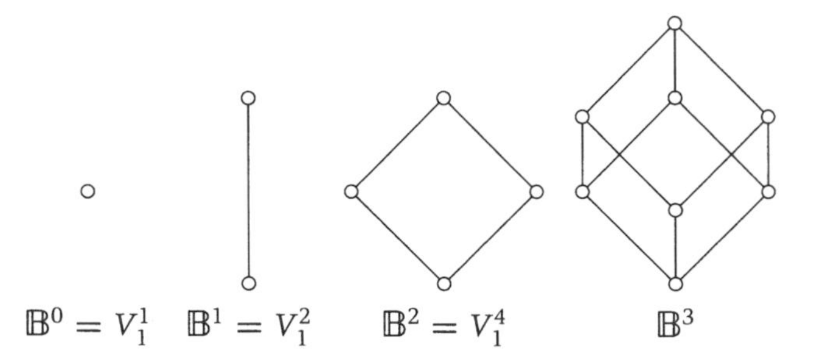

3.15. Remark. The identification of an arbitrary Boolean algebra with a power set as in 3.11 is not always possible in the infinite case (see Exercise 9). Similar to the proof of 3.11 it can be shown that for every (not necessarily finite) Boolean algebra B there is a set M and a Boolean monomorphism ("Boolean embedding") from B to $\mathcal{P}(M)$. This is called ***Stone's Representation Theorem***.

3.16. Definition and Theorem. Let B be a Boolean algebra and let X be any set. For mappings f and g from X into B we define

$$\begin{aligned} f \wedge g\colon X \to B; &\quad x \mapsto f(x) \wedge g(x); \\ f \vee g\colon X \to B; &\quad x \mapsto f(x) \vee g(x); \\ f'\colon X \to B; &\quad x \mapsto (f(x))'; \\ f_0\colon X \to B; &\quad x \mapsto 0; \\ f_1\colon X \to B; &\quad x \mapsto 1; \end{aligned}$$

for all $x \in X$. Then the set B^X of all mappings from X into B is a Boolean algebra (cf. Exercise 2.5). In particular, if $X = B^n$ (the n-fold cartesian product of B), then we obtain the Boolean algebra $B^{B^n} =: F_n(B)$ of all functions from B^n to B, which will be crucial in the following sections.

Exercises

1. Give twenty examples of lattices which are not Boolean algebras.
2. How many Boolean algebras are there with four elements 0, 1, a, and b?
3. Show that the direct product of Boolean algebras is again a Boolean algebra.
4. Prove Theorem 3.6.
5. Demonstrate in detail how an interval $[a, b]$ in a Boolean algebra B can be made into a Boolean algebra.
6. Show that $(B, \gcd, \operatorname{lcm})$ is a Boolean algebra if B is the set of all positive divisors of 110.
7. Show that $(\{1, 2, 3, 6, 9, 18\}, \gcd, \operatorname{lcm})$ does not form a Boolean algebra for the set of positive divisors of 18.
8. Prove that the lattice of all positive divisors of $n \in \mathbb{N}$ is a Boolean algebra with respect to lcm and gcd if and only if the prime factor decomposition of n does not contain any squares.
9. Let $B = \{X \subseteq \mathbb{N} \mid X \text{ is finite or its complement in } \mathbb{N} \text{ is finite}\}$. Show that B is a Boolean subalgebra of $\mathcal{P}(\mathbb{N})$ which cannot be Boolean isomorphic to some $\mathcal{P}(M)$.
10. Consider the set $\mathcal{M}$ of $n \times n$ matrices $\mathbf{X} = (x_{ij})$ whose entries x_{ij} belong to a Boolean algebra $B = (B, \wedge, \vee, 0, 1, ')$. Define two operations on $\mathcal{M}$:

$$\mathbf{X} \vee \mathbf{Y} := (x_{ij} \vee y_{ij}), \qquad \mathbf{X} \wedge \mathbf{Y} := (x_{ij} \wedge y_{ij}).$$

Show that $\mathcal{M}$ is a Boolean algebra. Furthermore, let

$$\mathbf{I} = \begin{pmatrix} 1 & 0 & \cdots & 0 \\ 0 & 1 & \cdots & 0 \\ \vdots & \vdots & \ddots & \vdots \\ 0 & 0 & \cdots & 1 \end{pmatrix}$$

and consider the subset $\mathcal{N}$ of $\mathcal{M}$ consisting of all $\mathbf{X} \in \mathcal{M}$ with the property $\mathbf{X} \geq \mathbf{I}$. (Here $\mathbf{X} \geq \mathbf{Y} \iff x_{ij} \geq y_{ij}$ for all i, j.) Show that $\mathcal{N}$ is a sublattice of $\mathcal{M}$.

11. Let $\mathcal{N}$ be as in Exercise 10, define matrix multiplication as usual, and show that $\mathcal{N}$ is closed with respect to multiplication, i.e., $\mathbf{X}, \mathbf{Y} \in \mathcal{N} \Longrightarrow \mathbf{XY} \in \mathcal{N}$. Prove: If $\mathbf{X} \in \mathcal{N}$, then $\mathbf{X} \leq \mathbf{X}^2 \leq \cdots \leq \mathbf{X}^{n-1} = \mathbf{X}^n$. We shall return to these ideas in connection with sociology (see 32.7).

§4 Boolean Polynomials

We introduce Boolean polynomials and polynomial functions in a form which is well suited for applications described in Chapter 2. The notion of a Boolean polynomial is defined recursively.

4.1. Definition. Let $X = \{x_1, \dots, x_n\}$ be a set of n symbols (called "indeterminates" or "variables," although they are determined and do not vary) which does not contain the symbols 0 and 1. The ***Boolean polynomials*** in X are the objects which can be obtained by finitely many successive applications of:

(i) $x_1, x_2, \dots, x_n$, and $0, 1$ are Boolean polynomials;

(ii) if p and q are Boolean polynomials, then so are the formal strings $p \wedge q, p \vee q, p'$.

Two polynomials are ***equal*** if their sequences of symbols are identical. We denote the set of all Boolean polynomials in $\{x_1, \dots, x_n\}$ by P_n.

4.2. Remark. Note that $0'$ is not the same as 1, for example. Also, $x_1 \wedge x_2 \neq x_2 \wedge x_1$, and so on.

4.3. Example. Some examples of Boolean polynomials over $\{x_1, x_2\}$ are $0, 1, x_1, x_1 \vee 1, x_1 \wedge x_2, x_1 \vee x_2, x_1', x_1' \wedge x_2, x_1' \wedge (x_2 \vee x_1)$.

Since every Boolean polynomial over $x_1, \dots, x_n$ can be regarded as a Boolean polynomial over $x_1, \dots, x_n, x_{n+1}$, we have

$$P_1 \subset P_2 \subset \cdots \subset P_n \subset P_{n+1} \subset \cdots.$$

Note that P_n is not a Boolean algebra. Of course, we want for instance $x_1 \wedge x_2$ and $x_2 \wedge x_1$ to be related. Therefore we introduce the concept of polynomial functions as follows.

4.4. Definition. Let B be a Boolean algebra, let B^n be the direct product of n copies of B, and let p be a Boolean polynomial in P_n. Then

$$\bar{p}_B \colon B^n \to B; \quad (a_1, \dots, a_n) \mapsto \bar{p}_B(a_1, \dots, a_n),$$

is called the ***Boolean polynomial function*** induced by p on B. Here $\bar{p}_B(a_1, \dots, a_n)$ is the element in B which is obtained from p by replacing each x_i by $a_i \in B$, $1 \leq i \leq n$.

The following example shows that two different Boolean polynomials can have the same Boolean polynomial function. Again, $\mathbb{B}$ denotes the Boolean algebra $\{0, 1\}$ with the usual operations.

4.5. Example. Let $n = 2$, $p = x_1 \wedge x_2$, $q = x_2 \wedge x_1$. Then

$$\bar{p}_{\mathbb{B}}\colon \mathbb{B}^2 \to \mathbb{B}; \qquad (0,0) \mapsto 0, (0,1) \mapsto 0, (1,0) \mapsto 0, (1,1) \mapsto 1,$$
$$\bar{q}_{\mathbb{B}}\colon \mathbb{B}^2 \to \mathbb{B}; \qquad (0,0) \mapsto 0, (0,1) \mapsto 0, (1,0) \mapsto 0, (1,1) \mapsto 1.$$

Therefore $\bar{p}_{\mathbb{B}} = \bar{q}_{\mathbb{B}}$.

Let B be a Boolean algebra. Using the notation introduced in 4.4, we define

4.6. Definition. $P_n(B) := \{\bar{p}_B \mid p \in P_n\}$.

4.7. Theorem. *Let B be a Boolean algebra; then the set $P_n(B)$ is a Boolean algebra and a subalgebra of the Boolean algebra $F_n(B)$ of all functions from B_n into B.*

Proof. We have to verify that $P_n(B)$ is closed with respect to join, meet, and complement of functions (as defined in 3.16) and that $P_n(B)$ contains f_0 and f_1. For $\wedge$, we get for all $a_1, \dots, a_n \in B$:

$$\begin{aligned}(\bar{p}_B \wedge \bar{q}_B)(a_1, \dots, a_n) &= \bar{p}_B(a_1, \dots, a_n) \wedge \bar{q}_B(a_1, \dots, a_n) \\ &= \overline{(p \wedge q)}_B(a_1, \dots, a_n).\end{aligned}$$

This implies that for all $\bar{p}_B, \bar{q}_B \in P_n(B)$, $\bar{p}_B \wedge \bar{q}_B = \overline{(p \wedge q)}_B \in P_n(B)$. For $\vee$ and $'$ we proceed similarly. Also, $\bar{0} = f_0$, $\bar{1} = f_1$. □

4.8. Definition. Two Boolean polynomials $p, q \in P_n$ are ***equivalent*** (in symbols $p \sim q$) if their Boolean polynomial functions on $\mathbb{B}$ are equal, i.e.,

$$p \sim q :\Longleftrightarrow \bar{p}_{\mathbb{B}} = \bar{q}_{\mathbb{B}}.$$

4.9. Theorem.

(i) *The relation $\sim$ in 4.8 is an equivalence relation on P_n.*

(ii) *$P_n/\!\sim$ is a Boolean algebra with respect to the usual operations on equivalence classes, namely $[p] \wedge [q] := [p \wedge q]$ and $[p] \vee [q] := [p \vee q]$. Also,*

$$P_n/\!\sim\; \cong_b P_n(\mathbb{B}).$$

Proof.

(i) We have $p \sim p$ for all $p \in P_n$, since $\bar{p}_{\mathbb{B}} = \bar{p}_{\mathbb{B}}$. For all p, q, r in P_n we have

$$p \sim q \Longrightarrow \bar{p}_{\mathbb{B}} = \bar{q}_{\mathbb{B}} \Longrightarrow \bar{q}_{\mathbb{B}} = \bar{p}_{\mathbb{B}} \Longrightarrow q \sim p$$

and, similarly, $p \sim q$ and $q \sim r \implies p \sim r$.

(ii) The operations $\wedge, \vee$ in $P_n/\sim$ are well defined: If $[p_1] = [p_2]$ and $[q_1] = [q_2]$, then $p_1 \sim p_2$ and $q_1 \sim q_2$. Since $\overline{(p_1 \wedge q_1)}_{\mathbb{B}} = \overline{(p_2 \wedge q_2)}_{\mathbb{B}}$, we get $p_1 \wedge q_1 \sim p_2 \wedge q_2$ and hence $[p_1 \wedge q_1] = [p_2 \wedge q_2]$. Similarly, we get $[p_1 \vee q_1] = [p_2 \vee q_2]$. It is easy to verify that $P_n/\sim$ is a lattice. For instance, since $p \wedge q \sim q \wedge p$, we have $[p] \wedge [q] = [p \wedge q] = [q \wedge p] = [q] \wedge [p]$. We define the mapping $h\colon P_n(\mathbb{B}) \to P_n/\sim$ by $h(\bar{p}_{\mathbb{B}}) := [p]$. Since

$$\bar{p}_{\mathbb{B}} = \bar{q}_{\mathbb{B}} \iff p \sim q \iff [p] = [q],$$

h is well defined and injective. By the definition of $\wedge, \vee$ in $P_n/\sim$, h is a lattice homomorphism, and clearly $h(\bar{0}_{\mathbb{B}}) = 0$, $h(\bar{1}_{\mathbb{B}}) = 1$. Since $h(\bar{p}_{\mathbb{B}}) \wedge h(\bar{p'}_{\mathbb{B}}) = h(\bar{p}_{\mathbb{B}} \wedge \bar{p'}_{\mathbb{B}}) = h(\bar{0}_{\mathbb{B}}) = 0$ and $h(\bar{p}_{\mathbb{B}}) \vee h(\bar{p'}_{\mathbb{B}}) = 1$, h is a Boolean homomorphism. By definition, h is surjective. So h is a Boolean isomorphism and by Exercise 2.13, $P_n/\sim$ is a Boolean algebra. □

For equivalent polynomials, the corresponding polynomial functions coincide on any Boolean algebra, not only on $\mathbb{B}$:

4.10. Theorem. *Let $p, q \in P_n$; $p \sim q$ and let B be an arbitrary Boolean algebra. Then $\bar{p}_B = \bar{q}_B$.*

Proof. From 3.13 and 3.15, we may assume that B is a Boolean subalgebra of $\mathcal{P}(X) \cong_b \mathbb{B}^X$ for some set X. So it suffices to prove the theorem for $\mathbb{B}^X$. We know (from the definition) that

$$\begin{aligned} p \sim q &\iff \bar{p}_{\mathbb{B}} = \bar{q}_{\mathbb{B}} \\ &\iff \bar{p}_{\mathbb{B}}(i_1, \dots, i_n) = \bar{q}_{\mathbb{B}}(i_1, \dots, i_n) \text{ for all } i_1, \dots, i_n \in \mathbb{B}. \end{aligned}$$

Let $f_1, \dots, f_n \in \mathbb{B}^X$ and let $x \in X$. Then we get $(\bar{p}_{\mathbb{B}^M}(f_1, \dots, f_n))(x) = \bar{p}_{\mathbb{B}}(f_1(x), \dots, f_n(x)) = \bar{q}_{\mathbb{B}}(f_1(x), \dots, f_n(x)) = (\bar{q}_{\mathbb{B}^X}(f_1, \dots, f_n))(x)$. Hence $\bar{p}_{\mathbb{B}^X} = \bar{q}_{\mathbb{B}^X}$. □

From now on, we simply write $\bar{p}_{\mathbb{B}^X}$ as $\bar{p}$ if the domain of $\bar{p}$ is clear. We frequently want to replace a given polynomial p by an equivalent polynomial which is of simpler or more systematic form. This is achieved by considering so-called normal forms. The collection of normal forms is a system of representatives for the equivalence classes of P_n.

4.11. Definition. $N \subseteq P_n$ is called a ***system of normal forms*** if:

(i) every $p \in P_n$ is equivalent to some $q \in N$;

(ii) for all $q_1, q_2 \in N$, $q_1 \neq q_2$ implies $q_1 \not\sim q_2$.

4.12. Notation. To simplify notation, we shall from now on write $p+q$ for $p \vee q$ and pq for $p \wedge q$.

In order to get such a system of normal forms we look for instance at the function induced by $p = x_1x_2'x_3'$. We see that $\bar{p}$ has value 1 only at $(1, 0, 0)$, and is zero everywhere else. Similarly, $q = x_1x_2'x_3' + x_1x_2x_3$ has value 1 exactly at $(1, 0, 0)$ and $(1, 1, 1)$. So for every sum of products of x_1 (or x_1'), x_2 (or x_2'), ..., x_n (or x_n'), we know immediately the values of the induced polynomial function. If f is any function from $\mathbb{B}^n$ into $\mathbb{B}$, we look at each $(b_1, \ldots, b_n)$ with $f(b_1, \ldots, b_n) = 1$ and write down the term $x_1^{b_1}x_2^{b_2} \cdots x_n^{b_n}$, where $x_i^1 := x_i$ and $x_i^0 := x_i'$. The sum

$$p = \sum_{f(b_1,\ldots,b_n)=1} x_1^{b_1}x_2^{b_2} \cdots x_n^{b_n}$$

obviously induces $\bar{p} = f$ and is the *only* sum of terms of the type $x_1^{c_1} \cdots x_n^{c_n}$ with this property. We almost have a system of normal forms now. To make things more uniform, we replace each summand $x_1^{b_1} \cdots x_n^{b_n}$ in p by $1x_1^{b_1} \cdots x_n^{b_n}$, and add all the other terms $x_1^{c_1} \cdots x_n^{c_n}$ that do not appear in p as $0x_1^{c_1} \cdots x_n^{c_n}$. Varying these "coefficients" for each term we induce each function $\mathbb{B}^n \to \mathbb{B}$ precisely once. We have shown:

4.13. Theorem and Definition. *The collection N_d of all polynomials of the form*

$$\sum_{(i_1,\ldots,i_n)\in\mathbb{B}^n} d_{i_1 i_2 \cdots i_n} x_1^{i_1} x_2^{i_2} \cdots x_n^{i_n},$$

where each $d_{i_1 \cdots i_n}$ is 0 or 1, is a system of normal forms in P_n, called the ***system of disjunctive normal forms****. Each summand is called a* ***minterm****.*

Each of the 2^n summands $x_1^{i_1} \cdots x_n^{i_n}$ can have a coefficient $d_{i_1 \cdots i_n} = 0$ or $= 1$, hence we get

4.14. Corollary and Definition.

(i) N_d has 2^{2^n} elements; hence P_n splits into 2^{2^n} different equivalence classes.

(ii) If $p \in P_n$, we write down the function table of $\bar{p}$ and get from this the "representative" $p_d \in N_d$ (i.e., the unique polynomial in N_d that is equivalent to p) in the way described before in 4.13. We call p_d the ***disjunctive normal form of*** p.

4.15. Example. We take a close look at P_2:

$0x_1'x_2' + 0x_1'x_2 + 0x_1x_2' + 0x_1x_2, \dots, 0, \dots$	$\leftarrow$ class #1,
$0x_1'x_2' + 0x_1'x_2 + 0x_1x_2' + 1x_1x_2, \dots, x_1x_2, \dots$	$\leftarrow$ class #2,
$0x_1'x_2' + 0x_1'x_2 + 1x_1x_2' + 0x_1x_2, \dots, x_1x_2', \dots$	$\leftarrow$ class #3,
....	...
....	...
....	...
$1x_1'x_2' + 1x_1'x_2 + 1x_1x_2' + 1x_1x_2, \dots, 1, \dots$	$\leftarrow$ class $\#2^{2^2}=16$.

Note that each class is infinite, and the p_d's are "quite long." So, in lazy moments, we omit the terms with coefficient 0 and do not write down the coefficient 1 for the other summands. Note: $1x_1x_2$ and x_1x_2 are *not* the same (but they are equivalent). So the "lazy picture" would list other representatives (except in the last row):

$\dots, 0, \dots$	$\leftarrow$ class # 1
$\dots, x_1x_2, \dots$	$\leftarrow$ class # 2
$\dots, x_1x_2', \dots$	$\leftarrow$ class # 3
$\dots, x_1x_2' + x_1x_2, \dots$	$\leftarrow$ class # 4
$\dots, x_1'x_2, \dots$	$\leftarrow$ class # 5
$\dots, x_1'x_2 + x_1x_2, \dots$	$\leftarrow$ class # 6
$\dots, x_1'x_2 + x_1x_2', \dots$	$\leftarrow$ class # 7
$\dots, x_1'x_2 + x_1x_2' + x_1x_2, \dots$	$\leftarrow$ class # 8
$\dots, x_1'x_2', \dots$	$\leftarrow$ class # 9
$\dots, x_1'x_2' + x_1x_2, \dots$	$\leftarrow$ class # 10
$\dots, x_1'x_2' + x_1x_2', \dots$	$\leftarrow$ class # 11
$\dots, x_1'x_2' + x_1x_2' + x_1x_2, \dots$	$\leftarrow$ class # 12
$\dots, x_1'x_2' + x_1'x_2, \dots$	$\leftarrow$ class # 13
$\dots, x_1'x_2' + x_1'x_2 + x_1x_2, \dots$	$\leftarrow$ class # 14

$\dots, x_1'x_2' + x_1'x_2 + x_1x_2', \dots$	$\leftarrow$ class # 15
$\dots, x_1'x_2' + x_1'x_2 + x_1x_2' + x_1x_2, \dots$	$\leftarrow$ class # 16.

Still, some of these new representatives might look unnecessarily long. For instance, we could represent class 4 also simply by x_1, since $x_1x_2' + x_1x_2 \sim x_1(x_2' + x_2) \sim x_1 1 \sim x_1$. In this way we arrive at "very simple" representatives:

$\dots, 0, \dots$	$\leftarrow$ class # 1
$\dots, x_1x_2, \dots$	$\leftarrow$ class # 2
$\dots, x_1x_2', \dots$	$\leftarrow$ class # 3
$\dots, x_1, \dots$	$\leftarrow$ class # 4
$\dots, x_1'x_2, \dots$	$\leftarrow$ class # 5
$\dots, x_2, \dots$	$\leftarrow$ class # 6
$\dots, x_1'x_2 + x_1x_2', \dots$	$\leftarrow$ class # 7
$\dots, x_1 + x_2, \dots$	$\leftarrow$ class # 8
$\dots, x_1'x_2', \dots$	$\leftarrow$ class # 9
$\dots, x_1'x_2' + x_1x_2, \dots$	$\leftarrow$ class # 10
$\dots, x_2', \dots$	$\leftarrow$ class # 11
$\dots, x_1 + x_2', \dots$	$\leftarrow$ class # 12
$\dots, x_1', \dots$	$\leftarrow$ class # 13
$\dots, x_1' + x_2, \dots$	$\leftarrow$ class # 14
$\dots, x_1' + x_2', \dots$	$\leftarrow$ class # 15
$\dots, 1, \dots$	$\leftarrow$ class # 16.

We will return to the question of minimizing such sums of products later on and will give an algorithm for this in §6. This algorithm produces precisely the short representatives listed above.

4.16. Example. We want to find the disjunctive normal form of $p = ((x_1 + x_2)'x_1 + x_2''')' + x_1x_2 + x_1x_2'$. We list the values of $\bar{p}$; first, $\bar{p}(0,0) =$

$((0+0)'0+0''')'+00+00'=0$, and so on:

b_1	b_2	$\bar{p}(b_1,b_2)$
0	0	0
0	1	1
1	0	1
1	1	1

$$\begin{aligned} \text{So} \quad p_d &= 0x_1'x_2' + 1x_1'x_2 + 1x_1x_2' + 1x_1x_2 && \text{(in full form)} \\ \text{or} \quad &= x_1'x_2 + x_1x_2' + x_1x_2 && \text{(in lazy form),} \end{aligned}$$

which could still be further reduced to $x_1 + x_2$.

So we see how we can simplify a "nasty" polynomial p:

1. Bring p into disjunctive normal form p_d.
2. Reduce p_d with our forthcoming algorithm.

Of course, "simplify" and "reduce" mean that we replace p by an equivalent polynomial in "shorter form."

4.17. Example. Find a Boolean polynomial p that induces the function f:

b_1	b_2	b_3	$f(b_1,b_2,b_3)$
0	0	0	1 ←
0	0	1	0
0	1	0	0
0	1	1	1 ←
1	0	0	1 ←
1	0	1	0
1	1	0	0
1	1	1	0

Solution. We only look at the marked lines and get a polynomial p immediately:

$$p = x_1'x_2'x_3' + x_1'x_2x_3 + x_1x_2'x_3'.$$

The first and third summand can be combined:

$$\begin{aligned} p &\sim x_1'x_2'x_3' + x_1x_2'x_3' + x_1'x_2x_3 \sim (x_1' + x_1)x_2'x_3' + x_1'x_2x_3 \\ &\sim x_2'x_3' + x_1'x_2x_3 =: q, \end{aligned}$$

so q is also a solution to our problem, i.e., $\bar{q} = \bar{p} = f$.

From 3.16 and 4.14 we get

4.18. Corollaries.

(i) $|P_n/{\sim}| = 2^{2^n}$ and $P_n(\mathbb{B}) = F_n(\mathbb{B})$.

(ii) If $|B| = m > 2$, then $|P_n(B)| = |P_n/{\sim}| = 2^{2^n} < m^{m^n} = |F_n(B)|$; so $P_n(B) \subset F_n(B)$.

(iii) If $p = \sum_{(i_1\dots,i_n)\in\mathbb{B}^n} d_{i_1\cdots i_n} x_1^{i_1}\cdots x_n^{i_n}$, then $d_{i_1\cdots i_n} = \bar{p}(i_1,\dots,i_n)$.

(iv) If $p \in P_n$, then $p \sim \sum_{\bar{p}(i_1,\dots,i_n)=1} x_1^{i_1}\cdots x_n^{i_n}$.

The result (i) means that every function from $\mathbb{B}^n$ into $\mathbb{B}$ is a Boolean polynomial function. We therefore say that $\mathbb{B}$ is ***polynomially complete***. Part (ii) tells us that $\mathbb{B}$ is the only polynomially complete Boolean algebra; and (iii) tells us again how to find the disjunctive normal form of a given polynomial.

By interchanging the roles of $0, 1$ and $+, \cdot$, we get another normal form N_c, which consists of polynomials that are products of sums. The polynomials in N_c are said to be in ***conjunctive normal form*** and are of the shape

$$p = \prod_{(i_1,\dots,i_n)\in\mathbb{B}^n} (d_{i_1,\dots,i_n} + x_1^{i_1} + \cdots + x_n^{i_n}).$$

A way to come from the disjunctive to the conjunctive normal form is to write p as $(p')'$, expand p' by using de Morgan's laws (3.4), and negate this again. For example

$$\begin{aligned} p = x_1'x_2 + x_1x_2' &\sim (x_1'x_2 + x_1x_2')'' \sim ((x_1'x_2 + x_1x_2')')' \\ &\sim ((x_1 + x_2')(x_1' + x_2))' \sim (x_1x_1' + x_1x_2 + x_2'x_1' + x_2'x_2)' \\ &\sim (x_1x_2 + x_2'x_1')' \sim (x_1' + x_2')(x_1 + x_2). \end{aligned}$$

Exercises

1. Find all $p \in P_2$ which can be obtained by one application of 4.1(ii) from $0, 1, x_1, x_2$.
2. Let $B = \mathcal{P}(\{a, b\})$ and $p = x_1 \wedge x_2'$. Determine $\bar{p}_B$ by tabulating its values.
3. With B as in Exercise 2, find the number of elements in $P_5(B)$.
4. With B as above, find the size of $F_5(B)$.

5. Are $x_1(x_2+x_3)'+x_1'+x_3'$ and $(x_1x_3)'$ equivalent?
6. Simplify the following Boolean polynomials:

   (i) $xy+xy'+x'y$;

   (ii) $xy'+x(yz)'+z$.

7. Let $f\colon \mathbb{B}^3 \to \mathbb{B}$ have the value 1 precisely at the arguments $(0,0,0)$, $(0,1,0)$, $(0,1,1)$, $(1,0,0)$. Find a Boolean polynomial p with $\bar{p}=f$ and try to simplify p.
8. Find the disjunctive normal form of:

   (i) $x_1(x_2+x_3)'+(x_1x_2+x_3')x_1$;

   (ii) $\big((x_2+x_1x_3)(x_1+x_3)x_2\big)'$.

9. Find the conjunctive normal form of $(x_1+x_2+x_3)(x_1x_2+x_1'x_3)'$.

§5 Ideals, Filters, and Equations

For applications in §9 we shall need a few further concepts from the theory of Boolean algebras. Some of the following terms may also be defined and studied in more general lattices. In the case of Boolean algebras, however, we obtain some simplifications; therefore we restrict our considerations to this case. Those who know some basic ring theory will find the following concept familiar. We still write $b+c$ and bc instead of $b \vee c$ and $b \wedge c$.

5.1. Definition. Let B be a Boolean algebra. $I \subseteq B$ is called an ***ideal*** in B, in symbols $I \trianglelefteq B$, if I is nonempty and if for all $i,j \in I$ and $b \in B$

$$ib \in I \text{ and } i+j \in I.$$

If we set $b=i'$, we see that 0 must be in I. Next we consider some useful characterizations of ideals. As expected, the ***kernel*** of a Boolean homomorphism $h\colon B_1 \to B_2$ is defined as $\ker h := \{b \in B_1 \mid h(b)=0\}$.

5.2. Theorem. *Let B be a Boolean algebra and let I be a nonempty subset of B. Then the following conditions are equivalent:*

(i) $I \trianglelefteq B$;

(ii) *For all $i,j \in I$ and $b \in B$: $i+j \in I$ and $b \le i \Longrightarrow b \in I$;*

(iii) *I is the kernel of a Boolean homomorphism from B into another Boolean algebra.*

Proof.

(i) $\Longrightarrow$ (ii) is almost trivial since $b \leq i$ implies that $b = bi \in I$.

(ii) $\Longrightarrow$ (iii): Let I be as in (ii), and define a relation $\sim_I$ on B via $b_1 \sim_I b_2 :\Longleftrightarrow b_1 + b_2 \in I$. On the equivalence classes $[b]$ of $B/\sim_I$ we define $[b_1]+[b_2] := [b_1+b_2]$ and $[b_1]\cdot[b_2] := [b_1b_2]$. These new operations are well defined: suppose $[b_1] = [c_1]$ and $[b_2] = [c_2]$. Then $b_1 \sim_I c_1$ and $b_2 \sim_I c_2$, so $b_1 + c_1 \in I$ and $b_2 + c_2 \in I$. But then $b_1 + b_2 + c_1 + c_2 \in I$, so $[b_1+b_2] = [c_1+c_2]$. Also, $b_1b_2 + c_1c_2 \leq b_1 + b_2 + c_1 + c_2$, so $b_1b_2 + c_1c_2 \in I$, whence $[b_1b_2] = [c_1c_2]$. It is easy to check that $(B/\sim_I, +, \cdot)$ is a Boolean algebra with zero $[0]$ and unit $[1]$, called the ***Boolean factor algebra*** and abbreviated by B/I. Now the map $h\colon B \to B/I;\ b \mapsto [b]$ is easily seen to be a Boolean homomorphism. Its kernel consists of all b with $[b] = [0]$, i.e., $\ker h = \{b \mid b + 0 \in I\} = I$.

(iii) $\Longrightarrow$ (i): Let $I = \ker h$, where $h\colon B \to B'$ is a Boolean homomorphism. If $i \in I$ and $b \in B$, then $h(ib) = h(i)h(b) = 0h(b) = 0$, so $ib \in I$. Similarly, $i_1, i_2 \in I$ implies $i_1 + i_2 \in I$. □

5.3. Examples. Let B be a Boolean algebra.

(i) For $B = \mathcal{P}(M)$, the set $\{A \subseteq M \mid A \text{ finite}\}$ forms an ideal of B.

(ii) For any $N \subseteq M$, we have $\mathcal{P}(N) \trianglelefteq \mathcal{P}(M)$.

(iii) $\{0\}$ and B are ideals of B, all other ideals are called ***proper ideals*** of B.

(iv) For any $[p] \in P_n/\sim$, $([p]) := \{[p][q] \mid q \in P_n\}$ is an ideal in $P_n/\sim$. □

The examples in 5.3(ii)–(iv) are generated by a single element. In general, if b is an element in a Boolean algebra B, the set (b) of all "multiples" bc of b for any $c \in B$, is called a ***principal ideal***. An ideal $M \neq B$ in B is ***maximal*** when the only ideals of B containing M are M and B itself.

5.4. Theorem. *Let B be a Boolean algebra and let $b \in B$. Then the principal ideal generated by b is*

$$(b) = \{a \in B \mid a \leq b\}.$$

The proof follows immediately from 5.2(ii). The example given in 5.3(i) is not a principal ideal, not even a "finitely generated" ideal if M is an infinite set. The ideals in 5.3(ii)–(iv) are principal ideals.

We now "dualize" the concept of an ideal.

5.5. Definition. Let B be a Boolean algebra and let $F \subset B$. F is called a ***filter*** (or ***dual ideal***) if $F \neq \emptyset$ and if for all $f, g \in F$ and $b \in B$

$$fg \in F \text{ and } f + b \in F.$$

In $\mathcal{P}(\mathbb{N})$, $F = \{A \subseteq \mathbb{N} \mid A' \text{ is finite}\}$ is the filter of ***cofinite*** subsets of $\mathbb{N}$; this filter is widely used in convergence studies in analysis. The connection with 5.3(i) is motivation for the following theorem, the proof of which is left as Exercise 2.

5.6. Theorem. *Let B be a Boolean algebra, and let $I, F \subseteq B$.*

(i) *If $I \trianglelefteq B$, then $I' := \{i' \mid i \in I\}$ is a filter in B.*

(ii) *If F is a filter in B, then $F' := \{f' \mid f \in F\}$ is an ideal in B.*

This theorem enables us to dualize 5.2.

5.7. Theorem. *Let B be a Boolean algebra and let $F \subseteq B$. Then the following conditions are equivalent:*

(i) *F is a filter in B.*

(ii) *For all $f, g \in F$ and for all $b \in B$: $fg \in F$ and $b \leq f \Longrightarrow f \in F$.*

(iii) *There is a Boolean homomorphism h from B into another Boolean algebra such that $F = \{b \in B \mid h(b) = 1\}$.*

Maximal ideals (filters) can be characterized in a very simple way.

5.8. Theorem. *Let B be a Boolean algebra. An ideal (filter) M in B is maximal if and only if for any $b \in B$ either $b \in M$ or $b' \in M$, but not both, hold.*

Proof. It is sufficient to prove the theorem for ideals. Suppose for every $b \in B$ we have $b \in M$ or $b' \in M$, but not both. If J were an ideal in B that properly contained M, then for $j \in J \setminus M$, both j and j' are in J, which would imply $1 \in J$, and then $J = B$. Conversely, let M be a maximal ideal in B and $b_0 \in B$ such that neither b_0 nor b_0' are in M. Then $J := \{b + m \mid m \in M, b \in B, b \leq b_0\}$ is an ideal, generated by $M \cup \{b_0\}$, which contains M properly. Since M is maximal, we have $J = B$. Then there exist $m \in M$ and $b \in B$ with $b \leq b_0$ and $b + m = b_0'$. This implies $b_0'(b + m) = b_0'b_0'$, which means $b_0'm = b_0'$; therefore $b_0' \leq m$ and by 5.2(ii) $b_0' \in M$, a contradiction. □

We can show (using Zorn's Lemma) that any proper ideal (filter) is contained in a maximal ideal (filter); see Exercise 7. Maximal filters are called ***ultrafilters***. Ultrafilters in $\mathcal{P}(M)$ are, e.g., $F_m := \{A \subseteq M \mid m \in A\}$, for a fixed $m \in M$, called ***fixed ultrafilters***. If M is finite, all ultrafilters are fixed (Exercise 6). If M is infinite, $F_c := \{A \subseteq M \mid A' \text{ is finite}\}$ is a proper filter of $\mathcal{P}(M)$. From the remark above, there is an ultrafilter $\mathcal{M}$ containing F_c, and this ultrafilter is clearly not of the form F_m, for any $m \in M$. Nobody has ever determined $\mathcal{M}$ explicitly.

Finally, we study equations in Boolean algebras. In order to do this we have to clarify the term "equation." We wish to know under which conditions two Boolean polynomial functions $\bar{p}, \bar{q}$ have the same value. We cannot, in general, speak of the equation $p = q$, since $p = q$ holds, by definition, only if p and q are identical. Therefore we have to be a little more careful (extensive treatment of this topic can be found in Lausch & Nöbauer (1973)).

5.9. Definition. Let p and q be Boolean polynomials in P_n. Then the pair (p, q) is called an ***equation***; $(b_1, \dots, b_n) \in B^n$ is called a ***solution*** of this equation in a Boolean algebra B if $\bar{p}_B(b_1, \dots, b_n) = \bar{q}_B(b_1, \dots, b_n)$. A ***system of equations*** is a set of equations $\{(p_i, q_i) \mid i \in I\}$; a ***solution of a system*** is a common solution of all equations (p_i, q_i).

Often we shall write $p = q$ instead of (p, q) in case no confusion can arise. For instance, $x_1'x_2 + x_3 = x_1(x_2 + x_3)$ is an equation and $(1, 0, 1)$ is a solution of the equation since the polynomial functions corresponding to both polynomials have value 1 at $(1, 0, 1)$. However, $x_1 + x_1' = 0$ is an equation which does not have a solution. For the following, it is convenient to transform an equation $p = q$ into an equation of the form $r = 0$.

5.10. Theorem. *The equations $p = q$ and $pq' + p'q = 0$ have the same solutions.*

Proof. Let B be a Boolean algebra and let $(b_1, \dots, b_n) \in B^n$. Then for $a := \bar{p}_B(b_1, \dots, b_n)$ and $b := \bar{q}_B(b_1, \dots, b_n)$ we have

$$a = b \iff 0 = (a + b)(a' + b') = aa' + ab' + a'b + bb' = ab' + a'b,$$

which proves the theorem. □

Using this theorem we are able to transform the system of equations $\{(p_i, q_i) \mid 1 \leq i \leq m\}$ into a single equation

$$p_1q_1' + p_1'q_1 + p_2q_2' + p_2'q_2 + \cdots + p_mq_m' + p_m'q_m = 0. \tag{5.1}$$

If we express the left-hand side in conjunctive normal form, we see that (5.1) has a solution if at least one factor has value 0, since for this m-tuple the whole expression on the left-hand side of (5.1) has value 0. In this way we obtain all solutions of the original system of equations.

5.11. Example. We wish to solve the system $\{(x_1x_2, x_1x_3+x_2), (x_1+x_2', x_3)\}$. Or, written more loosely, the system

$$\begin{aligned} x_1x_2 &= x_1x_3 + x_2, \\ x_1 + x_2' &= x_3. \end{aligned}$$

Using 5.10 we can transform this system into a single equation of the form

$$(x_1x_2)(x_1x_3 + x_2)' + (x_1x_2)'(x_1x_3 + x_2) + (x_1 + x_2')x_3' + (x_1 + x_2')'x_3 = 0.$$

If we express the left-hand side in conjunctive normal form, we obtain the equation

$$(x_1 + x_2 + x_3')(x_1' + x_2' + x_3') = 0,$$

which has the same set of solutions as the original system. The solutions over $B = \mathbb{B}$ are the zeros of the first or second factor, namely all $(a_1, a_2, a_3) \in \mathbb{B}^3$, such that

$$a_1 + a_2 + a_3' = 0, \quad \text{i.e.,} \quad a_1 = a_2 = 0, \quad a_3 = 1,$$

or with

$$a_1' + a_2' + a_3' = 0, \quad \text{i.e.,} \quad a_1 = a_2 = a_3 = 1.$$

Our system of equations therefore has exactly two solutions in $\mathbb{B}$, namely $(0, 0, 1)$ and $(1, 1, 1)$.

A connection between ideals and equations is given by

5.12. Theorem. *Let $p_j = 0$ ($j \in J$) be a system of equations over a Boolean algebra B with all $p_j \in P_n$, and let I be the set of all finite sums $b_1\bar{p}_1 + \cdots + b_n\bar{p}_n$ with $b_j \in B$. Then I is an ideal in $P_n(B)$ and $(a_1, \ldots, a_n)$ is a solution of $p_j = 0$ ($j \in J$) iff $\bar{\imath}(a_1, \ldots, a_n) = 0$ for all $i \in I$.*

The proof should be given in Exercise 12. We shall return to a similar idea in 38.2.

Exercises

1. Prove that in a finite Boolean algebra every ideal is a principal ideal.
2. Let B be a Boolean algebra. Prove: F is a filter in B if and only if $F' := \{x' \mid x \in F\}$ is an ideal of B.
3. Find all ideals and filters of the Boolean algebra $\mathcal{P}(\{1,2,3\})$. Which of the filters are ultrafilters? Are they fixed?
4. Prove: A nonempty subset I of a Boolean algebra B is an ideal if and only if

$$i \in I, j \in I \iff i + j \in I.$$

5. Let S be a subset of a Boolean algebra B. Prove:

   (i) The intersection J of all ideals I containing S is an ideal containing S.

   (ii) J in (i) consists of all elements of the form

   $$b_1 s_1 + \cdots + b_n s_n, \qquad n \geq 1,$$

   where $s_1, \ldots, s_n \in S$, and $b_1, \ldots, b_n$ are arbitrary elements of B.

   (iii) J in (i) consists of the set A of all a such that

   $$a \leq s_1 + \cdots + s_n, \qquad n \geq 1,$$

   where $s_1, \ldots, s_n$ are arbitrary elements of S.

   (iv) J in (i) is a proper ideal of B if and only if

   $$s_1 + \cdots + s_n \neq 1$$

   for all $s_1, \ldots, s_n \in S$.

6. Show that all ultrafilters in a finite Boolean algebra are fixed.
7. By use of Zorn's Lemma, prove that each proper filter in a Boolean algebra B is contained in an ultrafilter. (This is called the ***Ultrafilter Theorem***.)
8. Solve the system of equations

$$x_1 x_3 = x_2 + x_1 x_2 x_3',$$
$$x_1' + x_3 = x_1 x_2.$$

9. Show that the equation $a + x = 1$ in a Boolean algebra B has the general solution $x = a' + u$, where u is an arbitrary element in B.
10. Show that the equation $a + x = b$ has a solution iff $a \leq b$, and if this condition holds, show that the general solution is of the form
$$x = (u + a')b.$$
11. Prove: The general solution of $x + y = c$ is of the form
$$x = c(u + v'), \qquad y = c(u' + v).$$
12. Prove Theorem 5.12.
13. Use Exercises 12 and 1 to show again that equations over a finite Boolean algebra can be reduced to a single one.
14. Is every ideal (filter) of B a sublattice of B? Is it a Boolean subalgebra?

§6 Minimal Forms of Boolean Polynomials

We have seen in the previous section that it is possible and desirable to simplify a given Boolean polynomial by using the axioms of a Boolean algebra. For this process of simplification it is often difficult to decide which axioms should be used and in which order they should be used. There are several systematic methods to simplify Boolean polynomials. Many of these methods have the disadvantage that the practical implementation is impossible when the number of indeterminates of the polynomial is too large. This problem area in the theory of Boolean algebras is called the ***optimization*** or ***minimization problem*** for Boolean polynomials; it is of importance in applications such as the simplification of switching circuits (see Chapter 2).

We shall discuss the simplification of Boolean polynomials, especially the reduction of polynomials to a "minimal form" with respect to a suitably chosen minimality condition. Our considerations will be restricted to a special minimal condition for sum-of-product expressions. We define a ***literal*** to be any variable x_i, either complemented or not, and 0 and 1. Let d_f denote the total number of literals in a sum-of-product expression of f. Let e_f be the number of summands in f. We say such an f is ***simpler*** than a sum-of-product expression g if $e_f < e_g$, or $e_f = e_g$ and $d_f < d_g$, and that f is ***minimal*** if there is no simpler sum-of-product expression which is equivalent to f. In other words, we are looking for the "shortest" sum-of-product expression with the smallest possible number of literals, which is equivalent to f. Such a minimal form is not always unique (see, e.g., 6.6). We shall describe one amongst several methods for the simplification process. It is based on work by Quine and has been improved by McCluskey, so it is called the ***Quine-McCluskey method***. A Boolean polynomial in this section is also called an "expression."

6.1. Definition. An expression p ***implies*** an expression q if for all $b_1, \ldots, b_n \in \mathbb{B}$, $\bar{p}_{\mathbb{B}}(b_1, \ldots, b_n) = 1 \Longrightarrow \bar{q}_{\mathbb{B}}(b_1, \ldots, b_n) = 1$. In this case, p is called an ***implicant*** of q.

6.2. Definition. A ***product expression*** (briefly a product) α is an expression in which $+$ does not occur. A ***prime implicant*** for an expression p is a product expression α which implies p, but which does not imply p if one factor in α is deleted. A product whose factors form a subset of the factors of another product is called ***subproduct*** of the latter.

6.3. Example. x_1x_3 is a subproduct of $x_1x_2x_3$ and also of $x_1x_2'x_3$ and implies the expression

$$p = x_1x_2x_3 + x_1x_2'x_3 + x_1'x_2'x_3'$$

because $\overline{x_1x_3}(1, i_2, 1) = 1$, and $\bar{p}(1, i_2, 1) = 1$ as well; $\overline{x_1x_3}$ gives 0 for the other arguments. Neither x_1 nor x_3 imply p, since for example we have $\bar{x}_1(1,1,0) = 1$, but $\bar{p}(1,1,0) = 0$. Therefore x_1x_3 is a prime implicant.

6.4. Theorem. *A polynomial $p \in P_n$ is equivalent to the sum of all prime implicants of p.*

Proof. Let q be the sum of all prime implicants p_a of p. If $\bar{q}(b_1, \ldots, b_n) = 1$ for $(b_1, \ldots, b_n) \in \mathbb{B}^n$, then there is a prime implicant p_α with $\bar{p}_\alpha(b_1, \ldots, b_n) = 1$. Since p_α implies p, $\bar{p}(b_1, \ldots, b_n) = 1$ is a consequence.

Conversely, suppose that $\bar{p}(b_1, \ldots, b_n) = 1$ and $s := x_1^{b_1} \cdots x_n^{b_n}$. Then s is an implicant of p. In s we remove all those $x_i^{b_i}$ for which $\bar{p}(b_1, \ldots, b_{i-1}, b_i', b_{i+1}, \ldots, b_n) = 1$. The remaining product r still implies p, but does not imply p anymore if a further factor is removed. Therefore r is a prime implicant for p with $\bar{r}(b_1, \ldots, b_n) = 1$. Hence $\bar{q}(b_1, \ldots, b_n) = 1$. □

A sum of prime implicants of p is called ***irredundant*** if it is equivalent to p, but does not remain equivalent if any one of its summands is omitted. A minimal sum-of-product expression must be irredundant. In order to determine a minimal expression we therefore determine the set of irredundant expressions and amongst them we look for the one with the least number of literals.

Prime implicants are obtained by starting with the disjunctive normal form d for the Boolean polynomial p and applying the rule

$$yz + yz' \sim y,$$

(from left to right) wherever possible. More generally, we use

$$\alpha\beta\gamma + \alpha\beta'\gamma \sim \alpha\gamma, \tag{6.1}$$

where α, β, and γ are product expressions. The set of all expressions which cannot be simplified any more by this procedure is the set of prime implicants. The sum of these prime implicants is equivalent to p.

6.5. Example. Let p be the Boolean polynomial (now in w, x, y, z instead of x_1, x_2, x_3, x_4) whose disjunctive normal form d is given by

$$d = wxyz' + wxy'z' + wx'yz + wx'yz' + w'x'yz + w'x'yz' + w'x'y'z.$$

We use the idempotent law and (6.1) for the $\binom{7}{2} = 21$ pairs of products in d (as far as this is possible) and by doing this we "shorten" these products. For instance, the first and second product expressions of d yield wxz' by using (6.1). If a product expression is used once or more often for the simplification, it is ticked. Since $+$ is idempotent, an expression can be used any number of times and one tick suffices. In this way all the expressions are ticked which contain other product expressions and therefore cannot be prime implicants. Altogether, this process leads

$$\begin{array}{llllll}
\text{from} & wxyz' & \text{and} & wxy'z' & \text{to} & wxz', \\
\text{from} & wx'yz & \text{and} & wx'yz' & \text{to} & wx'y, \\
\text{from} & wxyz' & \text{and} & wx'yz' & \text{to} & wyz', \\
\text{from} & w'x'yz & \text{and} & w'x'yz' & \text{to} & w'x'y, \\
\text{from} & w'x'yz & \text{and} & w'x'y'z & \text{to} & w'x'z, \\
\text{from} & wx'yz & \text{and} & w'x'yz & \text{to} & x'yz, \\
\text{from} & wx'yz' & \text{and} & w'x'yz' & \text{to} & x'yz'.
\end{array}$$

In general, this procedure is repeated over and over again using only the ticked expressions (which become shorter and shorter). The other ones are prime implicants and remain unchanged.

In our example, the second round of simplifications yields:

$$\begin{array}{llllll}
\text{from} & wx'y & \text{and} & w'x'y & \text{to} & x'y, \\
\text{from} & x'yz & \text{and} & x'yz' & \text{to} & x'y.
\end{array}$$

These four expressions $wx'y$, $w'x'y$, $x'yz$, and $x'yz'$ are ticked. The remaining ones, namely wxz', wyz', and $w'x'z$ cannot be simplified. Hence p is equivalent to a sum of prime implicants:

$$p \sim wxz' + wyz' + w'x'z + x'y.$$

McCluskey improved this method, which leads us to the general ***Quine-McCluskey algorithm***. We use the polynomial of Example 6.5 to describe the procedure.

Step 1. Represent all product expressions in terms of zero-one-sequences, such that x_i' and x_i are denoted by 0 and 1, respectively. Missing variables are indicated by a dash, e.g., $w'x'y'z$ is 0001, $w'x'z$ is 00−1.

Step 2. The product expressions, regarded as binary n-tuples, are partitioned into classes according to their numbers of ones. We sort the classes according to increasing numbers of ones. In our example,

$w'x'y'z$	0	0	0	1
$w'x'yz'$	0	0	1	0
$w'x'yz$	0	0	1	1
$wx'yz'$	1	0	1	0
$wxy'z'$	1	1	0	0
$wx'yz$	1	0	1	1
$wxyz'$	1	1	1	0

Step 3. Each expression with r ones is added to each expression containing $r+1$ ones. If we use (6.1), we can simplify expressions in neighboring equivalence classes. We only have to compare expressions in neighboring classes with dashes in the same position. If two of these expressions differ in exactly one position, then they are of the forms $p = i_1 i_2 \cdots i_r \cdots i_n$ and $q = i_1 i_2 \cdots i_r' \cdots i_n$, where all i_k are in $\{0, 1, -\}$, and i_r is in $\{0, 1\}$, respectively. Then (6.1) reduces p and q to $i_1 i_2 \cdots i_{r-1} - i_{r+1} \cdots i_n$, and p and q are ticked. In our example this yields

0	0	−	1	
0	0	1	−	√
−	0	1	0	√
−	0	1	1	√
1	0	1	−	√
1	−	1	0	
1	1	−	0.	

The expressions with ticks are not prime implicants and will be subject to further reduction. They yield the single expression

$$- \quad 0 \quad 1 \quad -$$

Thus we have found all prime implicants, namely

$$\begin{array}{ccccl} 0 & 0 & - & 1 & w'x'z \\ 1 & - & 1 & 0 & wyz' \\ 1 & 1 & - & 0 & wxz' \\ - & 0 & 1 & - & x'y. \end{array}$$

As the sum of all prime implicants is not necessarily in minimal form (because some summands might be superfluous), we perform the last step in the procedure.

Step 4. Since the sum of all the prime implicants of p is equivalent to p by 6.4, for each product expression in the disjunctive normal form d of p there must be a prime implicant which is a subproduct of this product expression. This is determined by establishing a table of prime implicants. The heading elements for the columns are the product expressions in d, and at the beginning of the rows we have the prime implicants calculated in Step 3. A cross $\times$ is marked off at the intersection of the ith row and jth column if the prime implicant in the ith row is a subproduct of the product expression in the jth column. A product expression is said to ***cover*** another product expression if it is a subproduct of the latter one. In order to find a minimal sum of prime implicants, which is equivalent to d, we choose a subset of the set of prime implicants in such a way that each product expression in d is covered by at least one prime implicant of the subset. A prime implicant is called a ***main term*** if it covers a product expression which is not covered by any other prime implicant; the sum of the main terms is called the ***core***.

Either the summands of the core together cover all product expressions in d; then the core is already the (unique) minimal form of d. Otherwise, we denote by $q_1, \ldots, q_k$ those product expressions which are not covered by prime implicants in the core. The prime implicants not in the core are denoted by $p_1, \ldots, p_m$. We form a second table with index elements q_j for the columns and index elements p_i for the rows. The mark $\times$ is placed in the entry (i, j) indicating that p_i covers q_j. We then try to find a minimal subcollection of $p_1, \ldots, p_m$ which covers all of $q_1, \ldots, q_k$ and add them to the core. If the second table is small, such a minimal subcollection is easily found. Otherwise, there exists another algorithm to achieve this (see Hohn (1970)). This finally yields a minimal form of d, but it is not necessarily unique.

In our example, we get the following table of prime implicants:

	0001	0010	0011	1010	1100	1011	1110
00 − 1	×		×				
1 − 10				×			×
11 − 0					×		×
−01−		×	×	×		×	

The core is given by the sum of those prime implicants which are the only ones to cover a summand in d, namely by the sum of 00 − 1, −01−, and 11 − 0. This sum already covers all summands in d, so the minimal form of d is given by the core $w'x'z+y'z+wxz'$. The prime implicant wyz' was superfluous.

6.6. Example. Determine the minimal form of p, which is given in disjunctive normal form

$$\begin{aligned} p = {} & v'w'x'y'z' + v'w'x'yz' + v'w'xy'z' + v'w'xyz' + v'wx'y'z + v'wx'yz' \\ & + v'wxy'z + v'wxyz' + v'wxyz + vw'x'y'z' + vw'x'y'z + vw'xy'z \\ & + vwx'yz' + vwxy'z' + vwxyz' + vwxyz. \end{aligned}$$

Steps 1 and 2.

							row numbers
0 ones	0	0	0	0	0	√	(1)
1 one	0	0	0	1	0	√	(2)
	0	0	1	0	0	√	(3)
	1	0	0	0	0	√	(4)
2 ones	0	0	1	1	0	√	(5)
	0	1	0	0	1	√	(6)
	0	1	0	1	0	√	(7)
	1	0	0	0	1	√	(8)
3 ones	0	1	1	0	1	√	(9)
	0	1	1	1	0	√	(10)
	1	0	1	0	1	√	(11)
	1	1	0	1	0	√	(12)
	1	1	1	0	0	√	(13)
4 ones	0	1	1	1	1	√	(14)
	1	1	1	1	0	√	(15)
5 ones	1	1	1	1	1	√	(16)

Step 3. Combination of rows (i) and (j) yields the following simplifications:

(1)(2)	0	0	0	–	0	$\checkmark$
(1)(3)	0	0	–	0	0	$\checkmark$
(1)(4)	–	0	0	0	0	J
(2)(5)	0	0	–	1	0	$\checkmark$
(2)(7)	0	–	0	1	0	$\checkmark$
(3)(5)	0	0	1	–	0	$\checkmark$
(4)(8)	1	0	0	0	–	I
(5)(10)	0	–	1	1	0	$\checkmark$
(6)(9)	0	1	–	0	1	H
(7)(10)	0	1	–	1	0	$\checkmark$
(7)(12)	–	1	0	1	0	$\checkmark$
(8)(11)	1	0	–	0	1	G
(9)(14)	0	1	1	–	1	F
(10)(14)	0	1	1	1	–	$\checkmark$
(10)(15)	–	1	1	1	0	$\checkmark$
(12)(15)	1	1	–	1	0	$\checkmark$
(13)(15)	1	1	1	–	0	E
(14)(16)	–	1	1	1	1	$\checkmark$
(15)(16)	1	1	1	1	–	$\checkmark$

Repeating this step by combining the rows as indicated gives

(1)(2), (3)(5)	0	0	–	–	0	D
(2)(5), (7)(10)	0	–	–	1	0	C
(7)(10), (12)(15)	–	1	–	1	0	B
(10)(15), (14)(16)	–	1	1	1	–	A

The marking of the expression by $\checkmark$ or letters $A, B, \dots$ is, of course, done after the simplification processes. Having found the prime implicants we denote them by $A, B, \dots, J$.

Step 4. We give the table of prime implicants, where the first "row" represents the product expressions of d as binary 5-tuples in column form.

		(1)	(2)	(3)	(4)	(5)	(6)	(7)	(8)	(9)	(10)	(11)	(12)	(13)	(14)	(15)	(16)
		0	0	0	1	0	0	0	1	0	0	1	1	1	0	1	1
		0	0	0	0	0	1	1	0	1	1	0	1	1	1	1	1
		0	0	1	0	1	0	0	0	1	1	1	0	1	1	1	1
		0	1	0	0	1	0	1	0	0	1	0	1	0	1	1	1
		0	0	0	0	0	1	0	1	1	0	1	0	0	1	0	1
−111−	A										×				×	×	×
−1−10	B							×			×		×			×	
0−−10	C		×			×		×			×						
00−−0	D	×	×	×		×											
111−0	E													×		×	
011−1	F									×					×		
10−01	G								×			×					
01−01	H						×			×							
1000−	I				×				×								
−0000	J	×			×												

The core, i.e., the sum of the main terms, is $D+H+G+B+E+A$ (in our short notation). Column (4) is the only product expression that is not covered by the core; it is denoted by q_1. The prime implicants p_i which are not in the core are C, F, I, and J. The new table is of the form

		(4)
0−−10	C	
011−1	F	
1000−	I	×
−0000	J	×

This means that the minimal form is

$$D+H+G+B+E+A+I$$

if we use I; it is

$$D+H+G+B+E+A+J$$

if we choose J. In our usual notation the first minimal form of p is

$$v'w'z' + v'wy'z + vw'y'z + wyz' + vwxz' + wxy + vw'x'y'. \qquad \square$$

We refer to Hohn (1970) for a detailed description of this method, more detailed proofs included.

We now briefly describe a different way of representing Boolean polynomials, namely via ***Karnaugh diagrams***. We explain them by using the polynomial $p = x_1 x_2$.

row	b_1	b_2	minterm	$\bar{p}(b_1, b_2) = b_1 b_2$
(1)	1	1	$x_1 x_2$	1
(2)	1	0	$x_1 x_2'$	0
(3)	0	1	$x_1' x_2$	0
(4)	0	0	$x_1' x_2'$	0

The fourth column is the unique minterm which has the value 1 for the given assignment of input variables and value 0 everywhere else. The Karnaugh diagram consists of a b_1 and b_1' column and a b_2 and b_2' row for two input variables b_1, b_2.

	b_1	b_1'
b_2	(1)	(3)
b_2'	(2)	(4)

Each section in the intersection of a row and a column corresponds to a minterm. In the case of the function $\bar{p}$ above, the shaded section has value 1; the others have values 0.

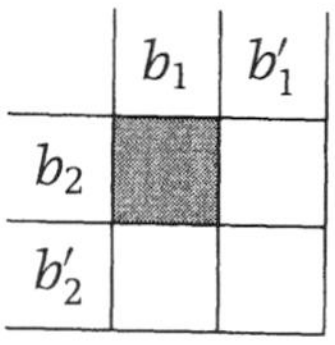

	b_1	b_1'
b_2	(shaded)	
b_2'		

Karnaugh diagrams with three input variables b_1, b_2, b_3 can be presented as follows:

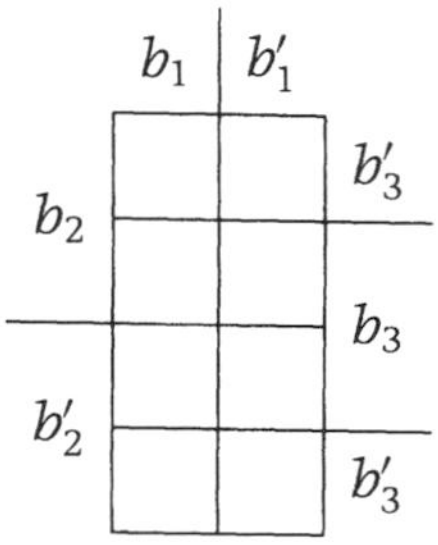

	b_1	b_1'	
b_2			b_3'
b_2			b_3
b_2'			b_3
b_2'			b_3'

Karnaugh diagrams for four input variables are of the following form (called the ***standard square*** SQ):

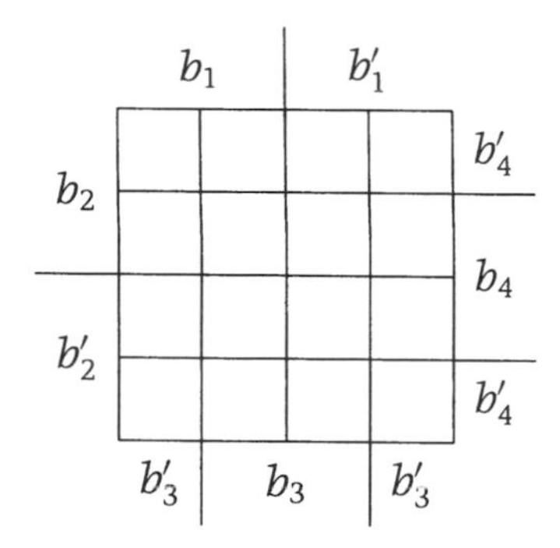

The standard square enables us to construct Karnaugh diagrams with more than four input variables as shown in Figure 6.1.

We give examples of the Karnaugh diagrams of some polynomials in x_1 and x_2:

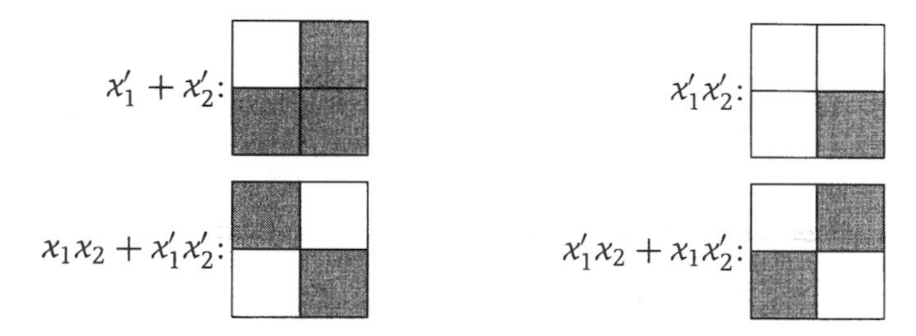

Karnaugh diagrams can be used to simplify Boolean polynomials. The following idea is fundamental: we try to collect as many portions of the diagram as possible to form a block; these represent "simple" polynomials or polynomial functions. Here we may use parts of the diagram more than once, since the polynomials corresponding to blocks are connected by $+$.

5 Variables:

b_5	b_5'
SQ	SQ

6 Variables:

	b_5	b_5'
b_6	SQ	SQ
b_6'	SQ	SQ

FIGURE 6.1

As an example we consider simplifying the polynomial

$$p = (x_1 + x_2)(x_1 + x_3) + x_1x_2x_3.$$

Its Karnaugh diagram is:

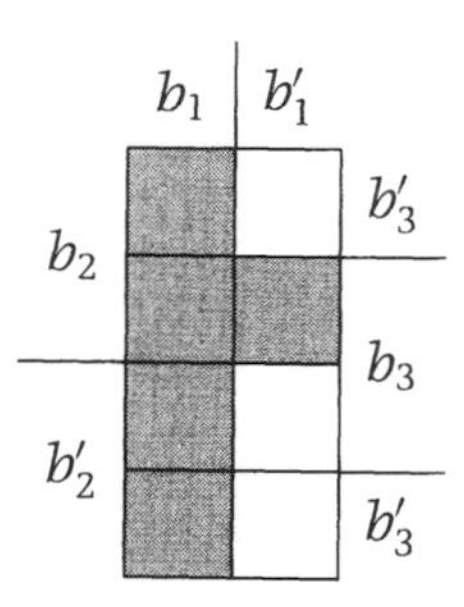

The diagram consists of the block formed by squares (1), (3), (5), (7) and the block formed by (3), (4). The first block represents x_1, the second x_2x_3. Thus $p \sim x_1 + x_2x_3$.

Exercises

1. Is $x_1x_2x_3' + x_1x_2'x_3 + x_1'x_2x_3 + x_1'x_2'x_3' + x_1'x_3$ irredundant?
2. Find all prime implicants of $xy'z + x'yz' + xyz' + xyz$ and form the corresponding prime implicant table.
3. Find three prime implicants of $xy + xy'z + x'y'z$.
4. Use the Quine-McCluskey method to find the minimal form of $wx'y'z + w'xy'z' + wx'y'z' + w'xyz + w'x'y'z' + wxyz + wx'yz + w'xyz' + w'x'yz'$.
5. Determine the prime implicants of $f = w'x'y'z' + w'x'yz' + w'xy'z + w'xyz' + w'xyz + wx'y'z' + wx'yz + wxy'z + wxyz + wxyz'$ by using Quine's procedure. Complete the minimizing process of f by using the Quine-McCluskey method.
6. Find the disjunctive normal form of f and simplify it:

$$f = x'y + x'y'z + xy'z' + xy'z.$$

7. Use a Karnaugh diagram to simplify:

   (a) $x_1x_2x_3' + x_1'x_2x_3' + (x_1 + x_2'x_3')'(x_1 + x_2 + x_3)' + x_3(x_1' + x_2)$;

   (b) $x_1x_2x_3 + x_2x_3x_4 + x_1'x_2x_4' + x_1'x_2x_3x_4' + x_1'x_2'x_4'$.
8. Find the minimal forms for $x_3(x_2 + x_4) + x_2x_4' + x_2'x_3'x_4$ using the Karnaugh diagrams.

9. Find simple functions for the following Karnaugh diagrams:

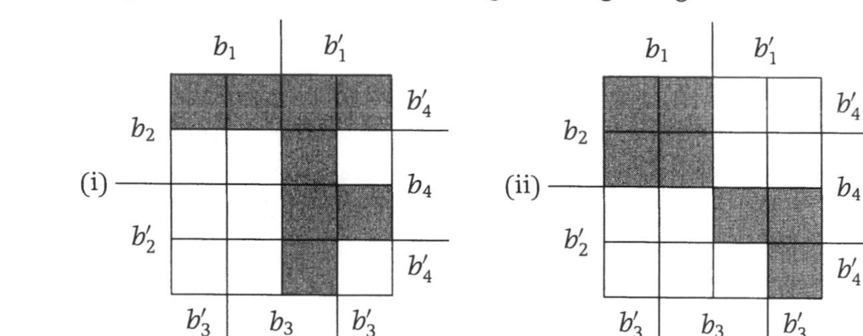

Notes

Standard reference books on lattice theory are Birkhoff (1967), Szász (1963), Grätzer (1978), Halmos (1967), and Grätzer (1971). The latter book is a more advanced book on the subject. Several of the books on applied algebra also contain sections on lattices; most have a chapter on Boolean algebras, we mention a few of them: Birkhoff & Bartee (1970), Dornhoff & Hohn (1978), Gilbert (1976), Gill (1976), Fisher (1977), and Prather (1976). A comprehensive work on Boolean algebras is Monk & Bonnet (1989). We shall refer again in the Notes to Chapter 2 to some of these books.

The history of the concept of a lattice and the development of lattice theory from the early beginnings in the nineteenth century up to the concept of universal algebra (see §40) is beautifully traced in Mehrtens (1979). Here we indicate some of the highlights in the history of lattices.

In 1847, G. Boole wrote an epoch-making little book *The Mathematical Analysis of Logic*, in which logic is treated as a purely formal system and the interpretation in ordinary language comes afterward. His next book *Investigation of the Law of Thought* (1854) contains the concept of Boolean algebra. We shall describe these connections in §9.

George Boole's calculus of logic centered on the formal treatment of logic by means of mathematical (especially algebraic) methods and on the description of logical equations. Following Boole, a school of English mathematicians, Schröder, and also Whitehead, developed the axiomatization of operations (conjunction, disjunction, negation); on the other hand, Pierce and Schröder created the axiomatics of order, with inclusion as the fundamental term. In 1904, E. V. Huntington studied the two systems of axioms and thus started the treatment of Boolean algebras as mathematical structures apart from logic.

Another approach to lattices was taken by R. Dedekind, who transfered the divisibility relation on $\mathbb{N}$ to ideals in rings (see §11), and reformulated gcd and lcm as set-theoretic operations. Thus the lattice structure appeared in several concrete applications. In 1897, Dedekind arrived at the abstract concept of a lattice, which he called "Dualgruppe"; but this abstract axiomatic foundation of lattice theory remained unnoticed.

Some 30 years later, several mathematicians formulated the lattice concept anew. The axiomatic method was accepted by then and the time was ripe for lattices. Garrett Birkhoff published his very important paper on lattice theory in 1933, in which he introduced "lattices." In following papers he deepened some aspects and widened the area of applications. Other authors or contributors to lattice theory at that time were Karl Menger, Fritz Klein, and Oystein Ore. Schröder's lattice concepts derived from his work on the algebra of logic and abstract arithmetic, Dedekind's lattice concepts originated in the structure of his algebraic number theory.

Boole used distributivity of meet with respect to join, which had been noted by J. Lambert before him. He worked with sets and denoted the meet of x and y by xy, the join by $x + y$ if x and y are disjoint. Similar to Leibniz he interpreted the inclusion relation as $xy = x$, which easily gave him the classical rules of syllogism. Jevons then extended the operation join to arbitrary x and y; de Morgan and later Pierce proved the duality relations called De Morgan's laws.

Most of the nineteenth-century logicians did not show much interest in applying their findings to mathematics. One reason for this was the lack of the use of variables and quantifiers, which were introduced by Frege and C. S. Pierce. Peano, among others, introduced the symbols $\cup, \cap, -$ for join, meet, and difference of sets, respectively. Ore published a paper in 1935, which was one of the fundamental papers on lattice theory. The following years saw many contributions to the subject and work on varied applications of lattices, e.g., in group theory, projective geometry, quantum mechanics, functional analysis, measure and integration theory (see Mehrtens (1979, pp. 202–203)).

In the years 1933–1937, M. H. Stone developed important results on Boolean algebras, which he interpreted as special rings, namely Boolean rings, and developed their ideal theory. Other fundamental questions tackled by Stone were the representation of Boolean algebras and applications of Boolean algebras in topology. From then on, lattice theory expanded steadily into a healthy and vigorous discipline of its own. It is, however, not completely accepted that lattices are part of algebra. For instance, several of the most influential or most popular algebra texts do not include lattices at all, e.g., van der Waerden (1970), Herstein (1975), Lang (1984), Rédei (1967); on the other hand, some important texts do include lattices, e.g., Birkhoff & MacLane (1977), MacLane & Birkhoff (1967), Jacobson (1985), Kurosh (1965).

In the theory of lattices a class of lattices that include the distributive lattices are the modular lattices. Here a lattice L is called ***modular*** if for all $x, y, z \in L$ the modular equation

$$x \leq z \Longrightarrow x \vee (y \wedge z) = (x \vee y) \wedge z$$

holds. For example, the lattice of all subspaces of a vector space is modular, but not distributive. It can be shown that a lattice is modular if and only if none of its sublattices is isomorphic to the pentagon lattice V_4^5.

There are several other algebraic structures which can be mentioned in connection with lattices. For instance, a ***semilattice*** is an ordered set in which any two elements have an infimum. An example of a semilattice is the set of a man and all his male descendants, where the order relation is defined as "is ancestor of." Semilattices are used in the developmental psychology of Piaget. A semilattice could also be defined as an idempotent and commutative semigroup.

A ***quasilattice*** is a set with a reflexive and transitive relation, where to any two elements in the set there exist an infimum and a supremum. It can be shown that quasilattices can also be described as sets with two binary operations $\vee$ and $\wedge$, which satisfy the associative and absorption laws, but not the commutative law. These structures were studied in the context of quantum physics (and called ***skew lattices***, ***Schrägverbände***).

There are other methods available for finding minimal forms. Dornhoff & Hohn (1978) describe a method for finding all prime implicants and for finding minimal forms which is based on work by Reusch (1975). A tutorial on this topic is Reusch & Detering (1979).

CHAPTER 2 Applications of Lattices

One of the most important applications of lattice theory and also one of the oldest applications of modern algebra is the use of Boolean algebras in modeling and simplifying switching or relay circuits. This application will be described in §7. It should be noted that the algebra of switching circuits is presented here not only because of its importance today but also for historical reasons and because of its elegant mathematical formulation. The same theory will also describe other systems, e.g., plumbing systems, road systems with traffic lights, etc. Then we consider propositional logic and indicate connections to probability theory.

§7 Switching Circuits

The main aspect of the algebra of switching circuits is to describe electrical or electronic switching circuits in a mathematical way or to design a diagram of a circuit with given properties. Here we combine electrical or electronic switches into series or parallel circuits. Such switches or contacts are switching elements with two states (open/closed), e.g., mechanical contacts, relays, semiconductors, photocells, or transistors. The type of the two states depends on these switching elements; we can

FIGURE 7.1

$- S_1 — S_2 -$

FIGURE 7.2 Series connection.

S_1

S_2

FIGURE 7.3 Parallel connection.

consider conductor–nonconductor elements, charged–uncharged, positively magnetized–negatively magnetized, etc. We shall use the notation introduced in Chapter 1.

Electrical switches or contacts can be symbolized in a ***switching*** or ***circuit diagram*** or ***contact sketch***, as shown in Figure 7.1.

Such a switch can be bi-stable, either "open" or "closed". Sometimes open and closed switches are symbolized as

The basic assumption is that for current to flow through a switch or for a contact to be established it is necessary that the switch be closed.

The symbol —S_1'— (complementation) indicates a switch which is open iff —S_1— (a switch appearing elsewhere in the circuit) is closed. In other words, S_1 and S_1' constitute two switches which are linked, in the sense that their states are related in this way. Similarly, if S_1 appears in two separate places in a circuit, it means that there are two separate switches linked, so as to ensure that they are always either both open or both closed. In Figure 7.2 we have "current" if and only if both S_1 and S_2 are closed. In 7.3 we have "current" if and only if at least one of S_1 and S_2 is closed. These properties of electrical switches motivate the following definitions, which give a connection between electrical switches and the elements of a Boolean algebra. Let $X_n := \{x_1, \ldots, x_n\}$.

7.1. Definitions.

(i) Each $x_1, \ldots, x_n \in X_n$ is called a ***switch***.

(ii) Every $p \in P_n$ is called a ***switching circuit***.

(iii) x_i' is called the ***complementation switch*** of x_i.

(iv) $x_i x_j$ is called the ***series connection*** of x_i and x_j.

(v) $x_i + x_j$ is called the ***parallel connection*** of x_i and x_j.

(vi) For $p \in P_n$ the corresponding polynomial function $\bar{p} \in P_n(\mathbb{B})$ is called the ***switching function*** of p.

(vii) $\bar{p}(a_1, \dots, a_n)$ is called the ***value*** of the switching circuit p at $a_1, \dots, a_n \in \mathbb{B}$. The a_i are called ***input variables***.

Switches and switching circuits in the sense of 7.1, the mathematical models of circuits, can be graphically represented by using contact diagrams. Instead of S_i we use x_i according to 7.1. For instance, the polynomial (i.e., the circuit) $x_1x_2 + x_1x_3$ can be represented as

$$
\begin{array}{c}
x_1 \text{ --- } x_2 \\
x_1 \text{ --- } x_3
\end{array}
$$

The electrical realization would be

$$
\begin{array}{c}
S_1 \text{ --- } S_2 \\
S_1 \text{ --- } S_3
\end{array}
$$

Another method of representation is as a switching or circuit diagram. This shows the circuit in terms of a "box", which converts input variables into values:

$$
\begin{array}{ccccc}
 & a_1 \bullet\!\!- & & & \\
\text{input variables} & a_2 \bullet\!\!- & p \in P_n & \longrightarrow & \bar{p}(a_1, \dots, a_n) \\
a_i \in \mathbb{B} & \vdots & & & \\
 & a_n \bullet\!\!- & & &
\end{array}
$$

For the example given above we have the diagram

$$
\begin{array}{cccc}
a_1 \bullet\!\!- & & & \\
a_2 \bullet\!\!- & p & \longrightarrow & a_1a_2 + a_1a_3 \\
a_3 \bullet\!\!- & & &
\end{array}
$$

$\bar{p}(a_1, \dots, a_n) = 1$ (or $= 0$) means that the circuit p conducts current (or does not conduct current).

Thus it is possible to model electrical circuits by using Boolean polynomials. Here different electrical circuits operate "identically" if their values are the same for all possible combinations of the input variables. This means for the corresponding polynomials p, q that $\bar{p}_{\mathbb{B}} = \bar{q}_{\mathbb{B}}$; that is, $p \sim q$.

In order to find a possible simplification of an electrical circuit retaining its original switching properties we can look for a "simple" Boolean polynomial which is equivalent to the original polynomial. This can be done by transposing the given polynomial into disjunctive normal form and then applying the Quine-McCluskey algorithm (see §6).

At this point we mention that in this way we can construct and simplify a wide variety of flow diagrams with "barriers", not only electrical circuits, e.g., pipe systems (for water, oil, or gas pipes) and traffic diagrams (with traffic lights as switches). We only describe some aspects of the electrical interpretation of the situation as given above.

7.2. Examples.

(i) The diagram for the switching circuit $p = x_1(x_2(x_3 + x_4) + x_3(x_5 + x_6))$ is given in Figure 7.4.

(ii) Figure 7.5 determines the switching circuit $p = x_1(x_2'(x_6 + x_3(x_4 + x_5')) + x_7(x_3 + x_6)x_8')$.

Nowadays, electrical switches are of less importance than semiconductor elements. These elements are types of electronic blocks which are predominant in the logical design of digital building components of electronic computers. In this context the switches are represented as

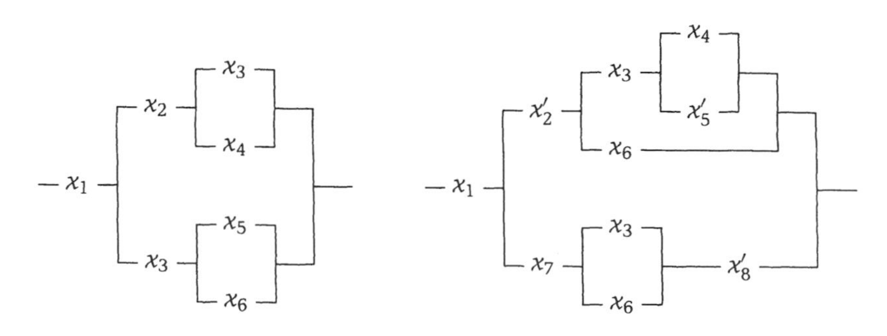

FIGURE 7.4 FIGURE 7.5

so-called ***gates***, or combinations of gates. We call this the ***symbolic representation***. Thus a gate (or a combination of gates) is a polynomial p which has, as values in $\mathbb{B}$, the elements obtained by replacing x_i by $a_i \in \mathbb{B}$ for each i. We also say that the gate is a realization of a switching function. If $\bar{p}(a_1, \dots, a_n) = 1$ (or 0), we have current (or no current) in the switching circuit p. We define some special gates.

7.3. Definition.

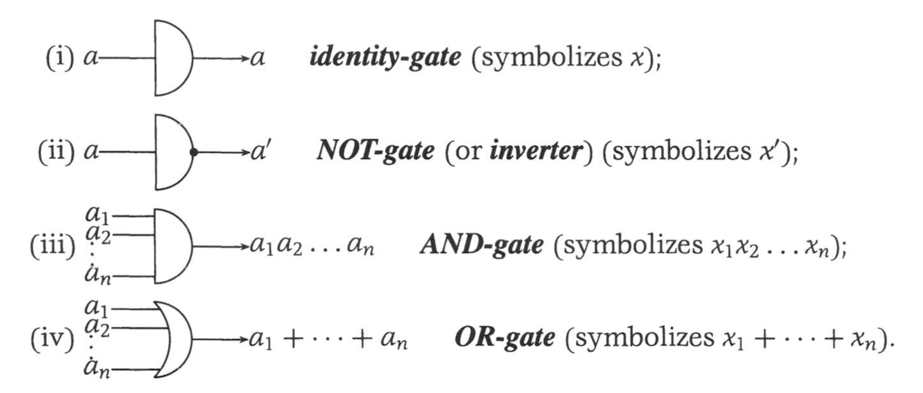

A briefer notation for the NOT-gate is to draw a black disc immediately before or after one of the other gates to indicate an inverter, e.g.,

a_1, a_2 → $(a_1 a_2)'$ and a_1, a_2 → $a_1 a_2'$

Also, switches like [gate diagram] should be clear. In propositional logic (see §9) the three polynomials $x_1' + x_2$ (***subjunction***), $(x_1 + x_2)'$ (***Pierce-operation***), and $(x_1 x_2)'$ (***Sheffer-operation***) are also of importance.

7.4. Definitions.

(i) a_1, a_2 → $a_1' + a_2$ ***subjunction-gate***;

(ii) a_1, a_2 → $(a_1 + a_2)'$ ***NOR-gate***;

(iii) a_1, a_2 → $(a_1 a_2)'$ ***NAND-gate***.

7.5. Examples.

(i) The symbolic representation of $p = (x_1'x_2)' + x_3$ (using 7.3) is

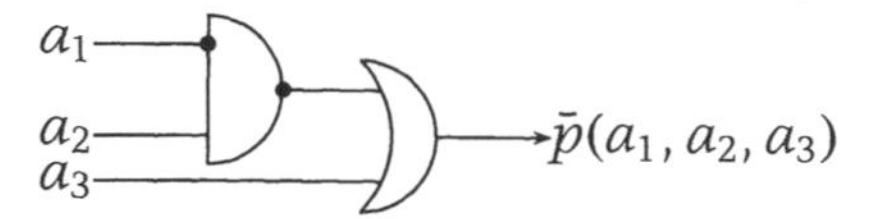

(ii) The polynomial p which corresponds to the diagram

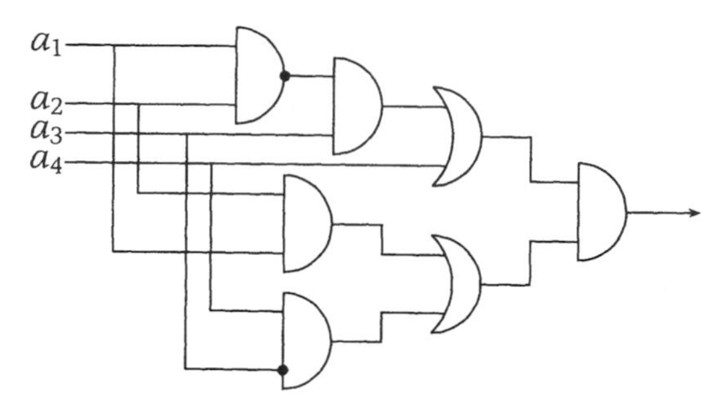

is $p = ((x_1x_2)'x_3 + x_4)(x_1x_2 + x_3'x_4)$. Now, using §6, we simplify p to $q = x_1x_2x_4 + x_3'x_4$:

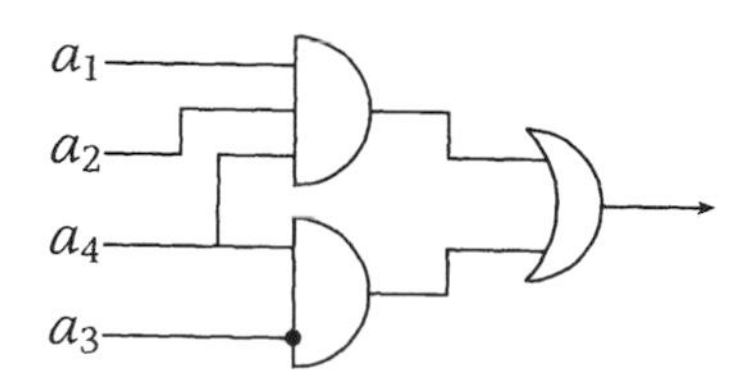

We see that $p \neq q$, since q is much simpler and cheaper than p. But since $\bar{p} = \bar{q}$, the circuits operate in the same way. This clearly shows the difference between equality and equivalence of polynomials!

7.6. Example. In 4.18(i) we noted that there are $2^{2^2} = 16$ Boolean polynomial functions on $\mathbb{B}$ in the case $n = 2$. The function value table for all these polynomial functions, written as columns, is as follows:

b_1	b_2	$\bar{p}_1$	$\bar{p}_2$	$\bar{p}_3$	$\bar{p}_4$	$\bar{p}_5$	$\bar{p}_6$	$\bar{p}_7$	$\bar{p}_8$	$\bar{p}_9$	$\bar{p}_{10}$	$\bar{p}_{11}$	$\bar{p}_{12}$	$\bar{p}_{13}$	$\bar{p}_{14}$	$\bar{p}_{15}$	$\bar{p}_{16}$
0	0	0	0	0	0	0	0	0	0	1	1	1	1	1	1	1	1
0	1	0	0	0	0	1	1	1	1	0	0	0	0	1	1	1	1
1	0	0	0	1	1	0	0	1	1	0	0	1	1	0	0	1	1
1	1	0	1	0	1	0	1	0	1	0	1	0	1	0	1	0	1

The minimal forms of the polynomials $p_1, \ldots, p_{16}$ inducing $\bar{p}_1, \ldots, \bar{p}_{16}$ are precisely the ones listed in the third table of 4.15.

From these sixteen polynomial functions, eight are very important in the algebra of switching circuits and get special names.

7.7. Definitions.

$\bar{p}_2 \ldots$ ***AND-function*** $\bar{p}_9 \ldots$ ***NOR-function***
$\bar{p}_3 \ldots$ ***inhibit-function*** $\bar{p}_{10} \ldots$ ***equivalence-function***
$\bar{p}_7 \ldots$ ***antivalence-function*** $\bar{p}_{14} \ldots$ ***implication-function***
$\bar{p}_8 \ldots$ ***OR-function*** $\bar{p}_{15} \ldots$ ***NAND-function***

Exercises

1. Determine the symbolic representation of the circuit given by

$$p = (x_1 + x_2 + x_3)(x_1' + x_2)(x_1 x_3 + x_1' x_2)(x_2' + x_3).$$

2. Determine the Boolean polynomial p of the circuit

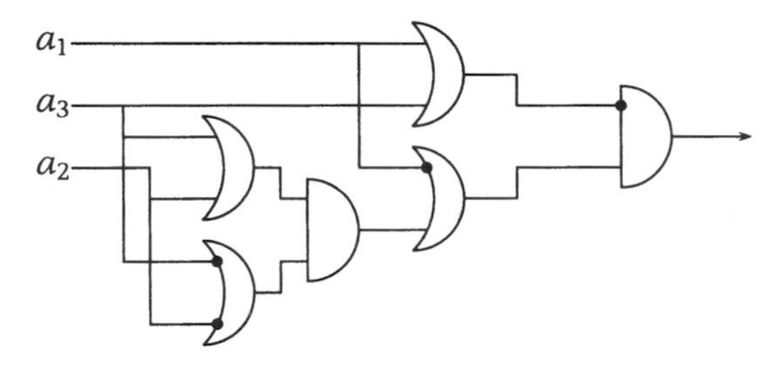

3. Find the symbolic representation of a simple circuit for which the binary polynomial function f in four variables is defined as follows: f is 0 at $(0,0,1,0)$, $(0,0,1,1)$, $(0,1,1,0)$, $(0,1,1,1)$, $(1,0,0,0)$, $(1,0,0,1)$, $(1,1,0,0)$, $(1,1,0,1)$, $(1,1,1,1)$, and has value 1 otherwise.
4. Find the symbolic gate representation of the contact diagram in Figure 7.6.
5. Simplify $p = (x_1 + x_2)(x_1 + x_3) + x_1 x_2 x_3$.
6. Determine which of the following contact diagrams give equivalent circuits:

x_1' / x_2' — x_2' (i)

x_1' — x_2 / x_1 — x_2' (ii)

x_1' / x_2' — x_1 / x_2 (iii)

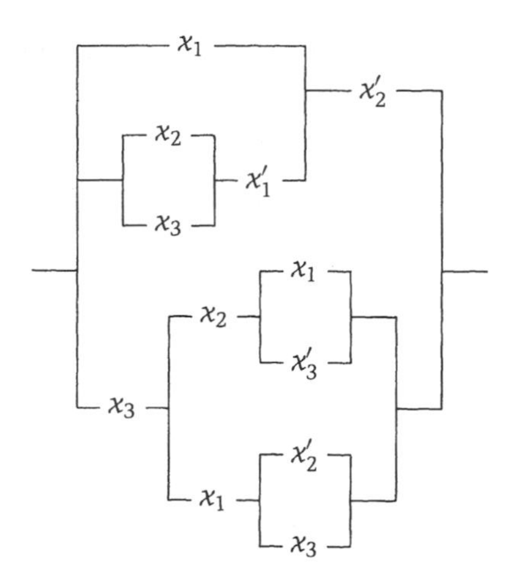

FIGURE 7.6

7. Clearly, $p = x_1x_2+x_1x_3+x_1x_4$ is in minimal form. Also $p \sim q := x_1(x_2+x_3+x_4)$. How many gates do we need for p, how many for q? Why then do we say that p is minimal?

§8 Applications of Switching Circuits

We describe some applications by examples.

8.1. Example. In a large room there are electrical switches next to the three doors to operate the central lighting. The three switches operate alternatively, i.e., each switch can switch on or switch off the lights. We wish to determine the switching circuit p, its symbolic representation, and contact diagram. Each switch has two positions: either on or off. We denote the switches by x_1, x_2, x_3 and the two possible states of the switches x_i by $a_i \in \{0, 1\}$.

The light situation in the room is given by the value $\bar{p}(a_1, a_2, a_3) = 0$ $(= 1)$ if the lights are off (are on, respectively). We arbitrarily choose $\bar{p}(1, 1, 1) = 1$.

(i) If we operate one or all three switches, then the lights go off, i.e., we have $\bar{p}(a_1, a_2, a_3) = 0$ for all (a_1, a_2, a_3) which differ in one or in three places from $(1, 1, 1)$.

(ii) If we operate two switches, the lights stay on, i.e., we have $\bar{p}(a_1, a_2, a_3) = 1$ for all those (a_1, a_2, a_3) which differ in two places from $(1, 1, 1)$.

This yields the following table of function values:

a_1	a_2	a_3	minterms	$\bar{p}(a_1, a_2, a_3)$
1	1	1	$x_1x_2x_3$	1
1	1	0	$x_1x_2x_3'$	0
1	0	1	$x_1x_2'x_3$	0
1	0	0	$x_1x_2'x_3'$	1
0	1	1	$x_1'x_2x_3$	0
0	1	0	$x_1'x_2x_3'$	1
0	0	1	$x_1'x_2'x_3$	1
0	0	0	$x_1'x_2'x_3'$	0

From this table we can derive the disjunctive normal form for the switching circuit p as in 4.18(iii):

$$p = x_1x_2x_3 + x_1x_2'x_3' + x_1'x_2x_3' + x_1'x_2'x_3,$$

p is already in minimal form. Thus we obtain the symbolic representation:

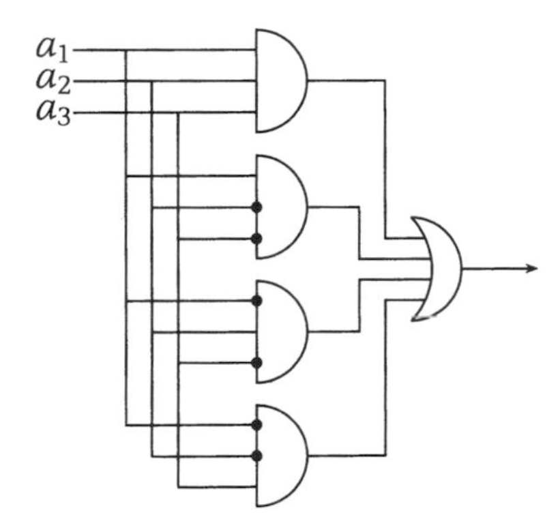

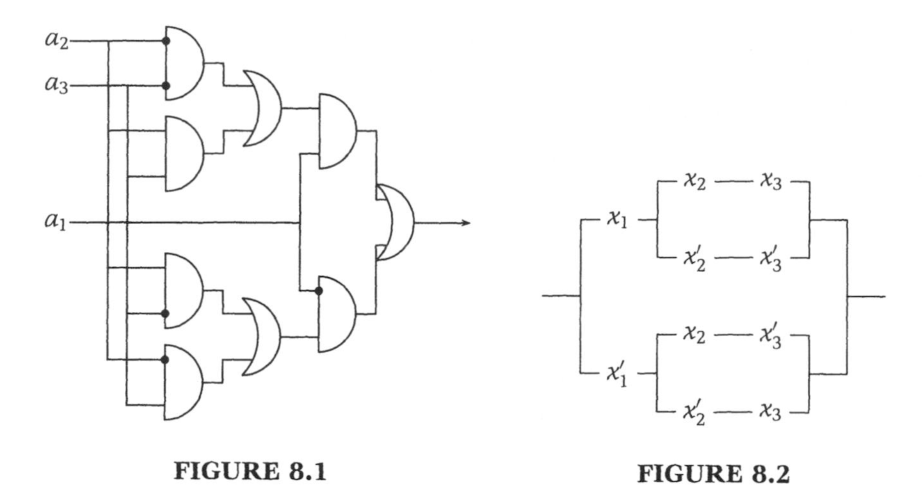

FIGURE 8.1 **FIGURE 8.2**

This switching circuit can also be represented in terms of antivalence and equivalence switches (see 7.6 and 7.7):

$$\begin{aligned} p &= x_1x_2x_3 + x_1x_2'x_3' + x_1'x_2x_3' + x_1'x_2'x_3 \\ &\sim (x_1\underbrace{(x_2x_3 + x_2'x_3')}_{\substack{\text{equivalence}\\ \text{of } x_2 \text{ and } x_3}}) + (x_1'\underbrace{(x_2x_3' + x_2'x_3)}_{\substack{\text{antivalence}\\ \text{of } x_2 \text{ and } x_3}}). \end{aligned}$$

This solution is represented symbolically in Figure 8.1. A circuit diagram is in Figure 8.2.

8.2. Example. In fast printers for computers, and in machines for paper production or in machines with high speed of paper transport, careful control of the paper movements is essential. We draw a schematic model of the method of paper transportation and the control mechanism (see Figure 8.3). The motor operates the pair of cylinders (1), which transports the paper strip (2). This paper strip forms a light barrier for the lamp (3). If the paper strip breaks, the photo cell (4) receives light and passes on an impulse which switches off the motor. The light in lamp (3) can vary in its brightness or it can fail, therefore a second photo cell (5) supervises the brightness of lamp (3). The work of the lamp is satisfactory as long as its brightness is above a given value a. If the brightness falls below a, but remains above a minimum value b, then the diminished brightness is indicated by a warning lamp (6). In this case the transportation mech-

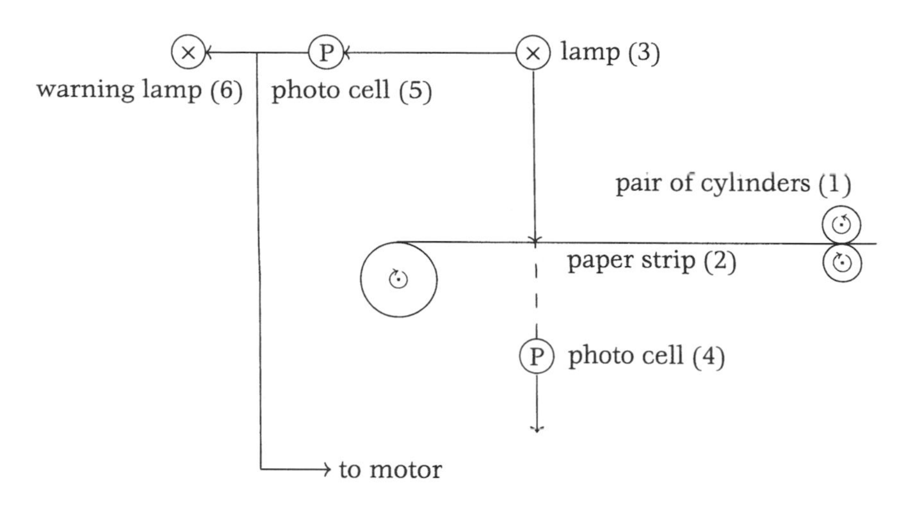

FIGURE 8.3

anism still operates. If the brightness of the lamp goes below b, then the photo cell (4) cannot work satisfactorily and the motor is switched off. We obtain the switching circuit and its symbolic representation, using the following notation:

$a_1 = 1$ if brightness of (3) $> a$;
$a_1 = 0$ if brightness of (3) $\leq a$;
$a_2 = 1$ if brightness of (3) $> b$;
$a_2 = 0$ if brightness of (3) $\leq b$;
$a_3 = 1$ if paper strip is broken;
$a_3 = 0$ if paper strip is unbroken. Note that $b < a$.

Thus we need a function value table for the state of the motor (say $\bar{p}_1(a_1, a_2, a_3)$) and one for the warning lamp (say $\bar{p}_2(a_1, a_2, a_3)$). We define

$\bar{p}_1(a_1, a_2, a_3) = 1 :$ motor operates;
$\bar{p}_1(a_1, a_2, a_3) = 0 :$ motor is switched off;
$\bar{p}_2(a_1, a_2, a_3) = 1 :$ warning lamp (6) operates;
$\bar{p}_2(a_1, a_2, a_3) = 0 :$ warning lamp (6) does not operate.

Therefore the values of the functions can be summarized

a_1	a_2	a_3	$\bar{p}_1(a_1, a_2, a_3)$	$\bar{p}_2(a_1, a_2, a_3)$
1	1	1	0	0
1	1	0	1	0
~~1~~	~~0~~	~~1~~		
~~1~~	~~0~~	~~0~~		
0	1	1	0	1
0	1	0	1	1
0	0	1	0	0
0	0	0	0	0

According to our definitions, the case $a_1 = 1$, $a_2 = 0$ cannot occur, and we might assign arbitrary values for $\bar{p}_1$ and $\bar{p}_2$ in the table above (***Don't-care combinations***). But this assignment might be important for getting the minimal form. So a really careful treatment would consist of minimizing the circuit for all possible assignments for the don't-care combinations, and to choose that assignment which produced the shortest minimal form. But here we just do it in a way that assigns 0 to these impossible cases (see Exercise 7).

This yields as the disjunctive normal form of the polynomial expressing the switching circuit $p_1 = x_1x_2x_3' + x_1'x_2x_3' \sim x_2x_3'$. For p_2 we obtain the disjunctive normal form $p_2 = x_1'x_2x_3 + x_1'x_2x_3' \sim x_1'x_2$. We see that the state of the motor is independent of a_1 and the state of the warning lamp is independent of a_3. The symbolic representation is

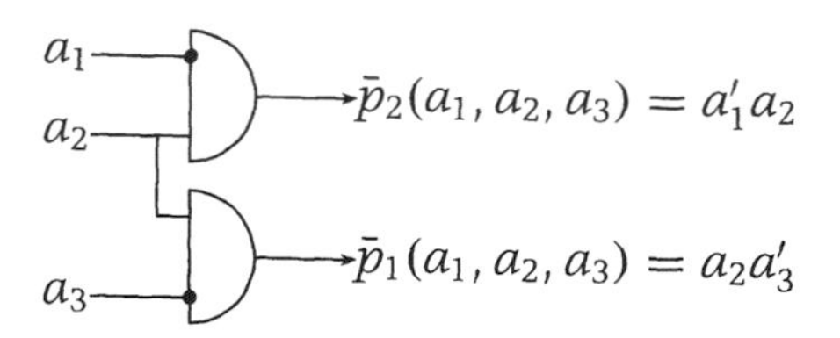

8.3. Example. A motor is supplied by three generators. The operation of each generator is monitored by a corresponding switching element which closes a circuit as soon as a generator fails. We demand the following conditions from the electrical monitoring system:

(i) A warning lamp lights up if one or two generators fail.
(ii) An acoustic alarm is initiated if two or all three generators fail.

We determine a symbolic representation as a mathematical model of this problem. Let $a_i = 0$ denote that generator i is operating, $i \in \{1, 2, 3\}$; $a_i = 1$ denotes that generator i does not operate. The table of function values has two parts $\bar{p}_1(a_1, a_2, a_3)$ and $\bar{p}_2(a_1, a_2, a_3)$, defined by:

$$\begin{aligned}
\bar{p}_1(a_1, a_2, a_3) = 1: &\quad \text{acoustic alarm sounds;}\\
\bar{p}_1(a_1, a_2, a_3) = 0: &\quad \text{acoustic alarm does not sound;}\\
\bar{p}_2(a_1, a_2, a_3) = 1: &\quad \text{warning lamp lights up;}\\
\bar{p}_2(a_1, a_2, a_3) = 0: &\quad \text{warning lamp is not lit up}
\end{aligned}$$

Then we obtain the following table for the function values:

a_1	a_2	a_3	$\bar{p}_1(a_1, a_2, a_3)$	$\bar{p}_2(a_1, a_2, a_3)$
1	1	1	1	0
1	1	0	1	1
1	0	1	1	1
1	0	0	0	1
0	1	1	1	1
0	1	0	0	1
0	0	1	0	1
0	0	0	0	0

For p_1 we choose the disjunctive normal form, namely

$$p_1 = x_1x_2x_3 + x_1x_2x_3' + x_1x_2'x_3 + x_1'x_2x_3.$$

This can be simplified by using rules of a Boolean algebra:

$$p_1 \sim x_1x_2 + x_2x_3 + x_1x_3.$$

For p_2 we choose the conjunctive normal form, which is preferable if there are many 1's as function values:

$$p_2 = (x_1 + x_2 + x_3)(x_1' + x_2' + x_3').$$

The symbolic representation is

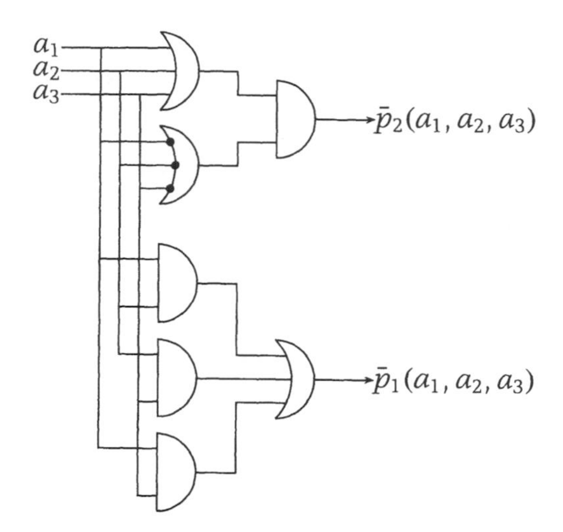

One of the applications of Boolean algebras is the simplification of electromechanical or electronic switching circuits. In order to economize, it is often useful to construct switching circuits in such a way that the costs for their technical realization are as small as possible, e.g., that a minimal number of gates is used. Unfortunately, it is often difficult to decide from the diagram of a switching circuit whether its technical implementation is simple. Also, the simplest and most economical switching circuit may not necessarily be a series-parallel connection (see Exercise 7.7), in which case switching algebra is not of much help. Some methods of simplification other than the Quine-McCluskey algorithm are discussed in Dornhoff & Hohn (1978) and also in Hohn (1970).

8.4. Remark. A switching circuit p can be simplified by our methods, as follows:

(i) It can be simplified according to the laws of a Boolean algebra (e.g., by applying the distributive, idempotent, absorption, and de Morgan laws).

(ii) Sometimes calculating the dual $d(p)$ of p and simplifying the dual yields a simple expression.

(iii) We can also determine the minimal form of p by using the method of Quine and McCluskey. Recall that this algorithm can only be started if p is in disjunctive normal form.

(iv) Use Karnaugh diagrams.

8.5. Examples. We give an example for the first two methods mentioned in 8.4.

(i) $p = (x_1' + x_2 + x_3 + x_4)(x_1' + x_2 + x_3 + x_4')(x_1' + x_2' + x_3 + x_4')$.
$\sim (x_1' + x_2 + x_3)(x_1' + x_3 + x_4')$
$\sim x_1' + x_3 + x_2 x_4'$.

In here, we have used the fact $(\alpha + \beta)(\alpha + \beta') = \alpha$ twice.

(ii) $p = ((x_1 + x_2)(x_1 + x_3)) + (x_1 x_2 x_3)$
$\sim \underbrace{((x_1 + x_2) + (x_1 x_2 x_3))}_{=:p_1}\underbrace{((x_1 + x_3) + (x_1 x_2 x_3))}_{=:p_2}$.

Let d denote "dual of". We have $d(p_1) = (x_1 x_2)(x_1 + x_2 + x_3) \sim x_1 x_2$. Therefore $d(d(p_1)) \sim x_1 + x_2$. Also, $d(p_2) = (x_1 x_3)(x_1 + x_2 + x_3) \sim x_1 x_3$. Thus $d(d(p_2)) \sim x_1 + x_3$. Altogether we have $p \sim p_1 p_2 \sim (x_1 + x_2)(x_1 + x_3) \sim x_1 + x_2 x_3$.

We consider two more examples of applications (due to Dokter & Steinhauer (1972)).

8.6. Example. An elevator services three floors. On each floor there is a call-button C to call the elevator. It is assumed that at the moment of call the cabin is stationary at one of the three floors. Using these six input variables we want to determine a control which moves the motor M in the right direction for the current situation. One, two, or three call-buttons may be pressed simultaneously; so there are eight possible combinations of calls, the cabin being at one of the three floors. Thus we have to consider $8 \cdot 3 = 24$ combinations of the total of $2^6 = 64$ input variables. We use the following notation: $a_i := c_i$ (for $i = 1, 2, 3$) for the call-signals. $c_i = 0$ (or 1) indicates that no call (or a call) comes from floor i. $a_4 := f_1$, $a_5 := f_2$, $a_6 := f_3$ are position signals; $f_i = 1$ means the elevator cabin is on floor i. $\bar{p}_1(a_1, \ldots, a_6) =: M\uparrow$, $\bar{p}_2(a_1, \ldots, a_6) =: M\downarrow$ indicate the direction of movement to be given to the motor; then the signal $M\uparrow = 1$ means movement of the motor upward, etc. The output signals (function values) of the motor are determined as follows. If there is no call for the cabin the motor does not operate. If a call comes from the floor where the cabin is at present, again the motor does not operate. Otherwise, the motor follows the direction of the call. The only exception is the case when the cabin is at the second floor and there are two simultaneous calls from the

Call			Floor			Direction of motor	
c_1	c_2	c_3	f_1	f_2	f_3	$M\uparrow$	$M\downarrow$
1	1	1	1	0	0	0	0
1	1	0	1	0	0	0	0
1	0	1	1	0	0	0	0
1	0	0	1	0	0	0	0
0	1	1	1	0	0	1	0
0	1	0	1	0	0	1	0
0	0	1	1	0	0	1	0
0	0	0	1	0	0	0	0
1	1	1	0	1	0	0	0
1	1	0	0	1	0	0	0
1	0	1	0	1	0	0	1
1	0	0	0	1	0	0	1
0	1	1	0	1	0	0	0
0	1	0	0	1	0	0	0
0	0	1	0	1	0	1	0
0	0	0	0	1	0	0	0
1	1	1	0	0	1	0	0
1	1	0	0	0	1	0	1
1	0	1	0	0	1	0	0
1	0	0	0	0	1	0	1
0	1	1	0	0	1	0	0
0	1	0	0	0	1	0	1
0	0	1	0	0	1	0	0
0	0	0	0	0	1	0	0

FIGURE 8.4

third and first floor. We agree that the cabin goes down first. Figure 8.4 shows the table of function values. From this table we derive the switching circuits p_1 for $M\uparrow$ and p_2 for $M\downarrow$ in disjunctive normal form. Here x_i are replaced by C_i for $i = 1, 2, 3$ and by F_{i-3} for $i = 4, 5, 6$.

$$\begin{aligned} p_1 \sim\ & C_1'C_2C_3F_1F_2'F_3' + C_1'C_2C_3'F_1F_2'F_3' \\ & + C_1'C_2'C_3F_1F_2'F_3' + C_1'C_2'C_3F_1'F_2F_3'. \end{aligned}$$

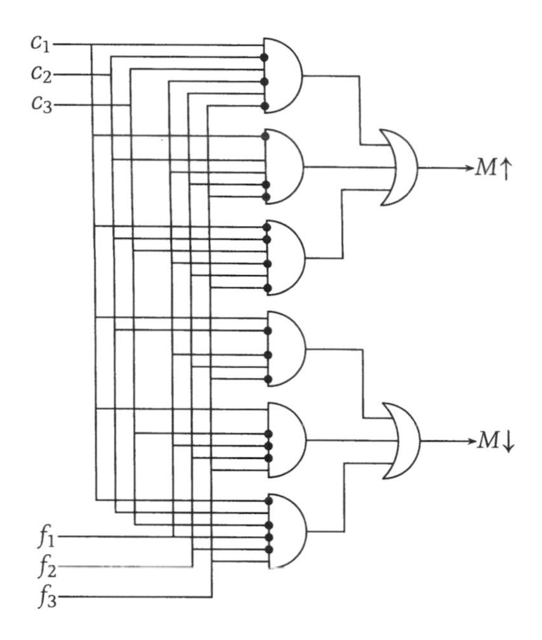

FIGURE 8.5

The first and third minterms are complementary with respect to C_2 and can be combined. This gives:

$$p_1 \sim C_1'C_2C_3'F_1F_2'F_3' + C_1'C_3F_1F_2'F_3' + C_1'C_2'C_3F_1'F_2F_3'.$$

For $M\downarrow$ we obtain

$$p_2 \sim C_1C_2'C_3F_1'F_2F_3' + C_1C_2'C_3'F_1'F_2F_3' + C_1C_2C_3'F_1'F_2'F_3 + C_1C_2'C_3'F_1'F_2'F_3 + C_1'C_2C_3'F_1'F_2'F_3.$$

The first two minterms are complementary with respect to C_3, the third and fourth minterm are complementary with respect to C_2. Simplification gives

$$p_2 \sim C_1C_2'F_1'F_2F_3' + C_1C_3'F_1'F_2'F_3 + C_1'C_2C_3'F_1'F_2'F_3.$$

The two switching circuits enable us to design the symbolic representation of Figure 8.5 (we have six NOT-gates, six AND-gates, and two OR-gates). □

Observe that in Examples 8.2, 8.3, and 8.6 we had not only a switching circuit, but a ***switching network***, which differs from a circuit by having multiple outputs:

$$\text{Inputs}\quad \begin{array}{l} a_1 \bullet\!\!- \\ a_2 \bullet\!\!- \\ \vdots \\ a_n \bullet\!\!- \end{array} \quad \begin{array}{c}\text{Switching}\\ \text{network}\end{array} \quad \begin{array}{l} \longrightarrow b_1 \\ \longrightarrow b_2 \\ \vdots \\ \longrightarrow b_m \end{array} \quad \text{Outputs}$$

Optimizing such a network reduces to the minimization of all circuits $\begin{array}{l} a_1 \bullet\!\!- \\ \vdots \\ a_n \bullet\!\!- \end{array} \longrightarrow b_i$; we have precisely done that in these examples.

As another example of applications of this type we consider the addition of binary numbers with half-adders and adders. Decimals can be represented in terms of quadruples of binary numbers; such a quadruple is called a ***tetrad***. Each digit of a decimal gets assigned a tetrad; thus we use ten different tetrads corresponding to $0, 1, 2, \ldots, 9$. Using four binary positions we can form $2^4 = 16$ tetrads. Since we need only ten tetrads to represent $0, 1, \ldots, 9$, there are six superfluous tetrads, which are called ***pseudotetrads***. A binary coded decimal then uses the following association between $0, 1, \ldots, 9$ and tetrads:

		a_3	a_2	a_1	a_0	$\bar{p}(a_0, a_1, a_2, a_3)$
pseudotetrades		1	1	1	1	1
		1	1	1	0	1
		1	1	0	1	1
		1	1	0	0	1
		1	0	1	1	1
		1	0	1	0	1
decimals	9	1	0	0	1	0
	8	1	0	0	0	0
	7	0	1	1	1	0
	6	0	1	1	0	0
	5	0	1	0	1	0
	4	0	1	0	0	0
	3	0	0	1	1	0
	2	0	0	1	0	0
	1	0	0	0	1	0
	0	0	0	0	0	0

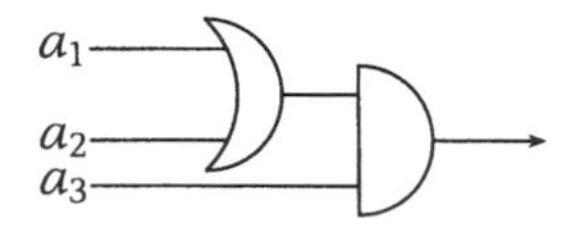

FIGURE 8.6

$\bar{p}(a_0, a_1, a_2, a_3) = 1$ denotes the pseudotetrads. We have to evaluate $\bar{p}(a_0, a_1, a_2, a_3)$ to find out if the result of a computing operation is a pseudotetrad.

We represent p in disjunctive normal form:

$$p = x_3x_2x_1x_0 + x_3x_2x_1x_0' + x_3x_2x_1'x_0 + x_3x_2x_1'x_0' + x_3x_2'x_1x_0 + x_3x_2'x_1x_0'.$$

The pairs of minterms 1 and 2, 3 and 4, 5 and 6 are complementary with respect to x_0 and can be simplified:

$$\begin{aligned} p &\sim x_3x_2x_1 + x_3x_2x_1' + x_3x_2'x_1 \\ &\sim x_3x_2x_1 + x_3x_2x_1 + x_3x_2x_1' + x_3x_2'x_1 \\ &\sim (x_3x_2x_1 + x_3x_2x_1') + (x_3x_2x_1 + x_3x_2'x_1) \\ &\sim x_3x_2 + x_3x_1 \sim x_3(x_2 + x_1). \end{aligned}$$

This result indicates that determining if a tetrad with four positions a_0, a_1, a_2, a_3 is a pseudotetrad is independent of a_0. If we use the a_i as inputs, then Figure 8.6 indicates the occurrence of a pseudotetrad.

8.7. Example (*Half-Adders*). We describe the addition of two binary numbers. In order to add two single digit binary numbers a_1 and a_2 we have to consider a carry $\bar{c}(a_1, a_2)$ in case $a_1 = a_2 = 1$. The table of the function values for the sum $\bar{s}(a_1, a_2)$ is as follows:

a_1	a_2	$\bar{s}(a_1, a_2)$	$\bar{c}(a_1, a_2)$
1	1	0	1
1	0	1	0
0	1	1	0
0	0	0	0

s and c have the disjunctive normal forms $x_1x_2'+x_1'x_2$ and x_1x_2, respectively. Thus the corresponding network is

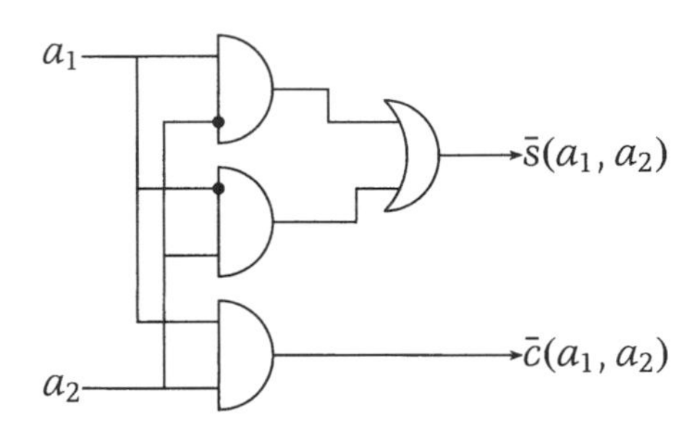

To obtain a simpler network we modify s according to the axioms of a Boolean algebra:

$$\begin{aligned} s = x_1x_2' + (x_1'x_2) &\sim ((x_1' + x_2)' + (x_1 + x_2')') \\ &\sim ((x_1' + x_2)(x_1 + x_2'))' \sim (x_1'x_1 + x_2x_1 + x_1'x_2' + x_2x_2')' \\ &\sim (x_1x_2 + x_1'x_2')' \sim (x_1x_2)'(x_1'x_2')' \sim c'(x_1 + x_2). \end{aligned}$$

This leads to a circuit called the half-adder:

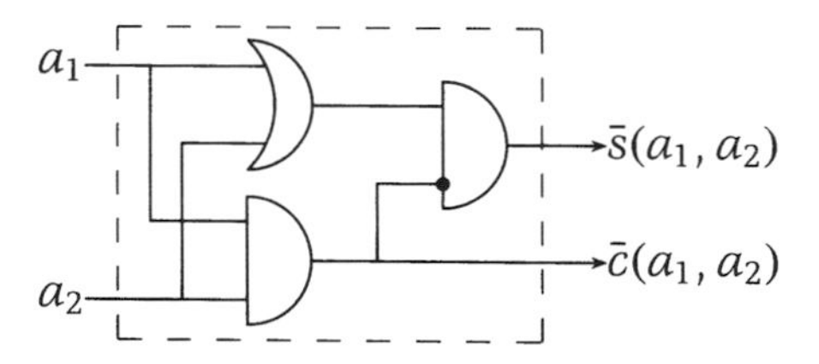

Symbolically we write

a_1 HA $\bar{s}(a_1, a_2)$
a_2 $\bar{c}(a_1, a_2)$

8.8. Example. ***Full-adders*** can add three one-digit binary numbers. Let a_1, a_2, a_3 denote three numbers. Then we can summarize all possible cases in the following table:

a_1	a_2	a_3	$\bar{s}(a_1, a_2, a_3)$	$\bar{c}(a_1, a_2, a_3)$
1	1	1	1	1
1	1	0	0	1
1	0	1	0	1
1	0	0	1	0
0	1	1	0	1
0	1	0	1	0
0	0	1	1	0
0	0	0	0	0

Next we consider partial sums of two summands:

a_2	a_3	$\bar{s}_1(a_2, a_3)$	$\bar{c}_1(a_2, a_3)$
1	1	0	1
1	0	1	0
0	1	1	0
0	0	0	0
1	1	0	1
1	0	1	0
0	1	1	0
0	0	0	0

a_1	$\bar{s}_1(a_2, a_3)$	$\bar{s}(a_1, a_2, a_3)$	$\bar{c}_2(a_1, \bar{s}_1(a_2, a_3))$
1	0	1	0
1	1	0	1
1	1	0	1
1	0	1	0
0	0	0	0
0	1	1	0
0	1	1	0
0	0	0	0

$\bar{c}_1(a_2, a_3)$	$\bar{c}_2(a_1, \bar{s}_1(a_2, a_3))$	$\bar{c}(a_1, a_2, a_3)$
1	0	1
0	1	1
0	1	1
0	0	0
1	0	1
0	0	0
0	0	0
0	0	0

From these tables we derive: a_2 and a_3 are inputs of a half-adder with outputs $\bar{s}_1(a_2, a_3)$ and $\bar{c}_1(a_2, a_3)$. The output $\bar{s}_1(a_2, a_3)$ together with a_1 forms inputs of a second half-adder, whose outputs are $\bar{s}(a_1, a_2, a_3)$ and $\bar{c}_2(a_1, \bar{s}_1(a_2, a_3))$. Here $\bar{s}(a_1, a_2, a_3)$ is the final sum. Finally, disjoining $\bar{c}_1(a_2, a_3)$ and $\bar{c}_2(a_1, \bar{s}_1(a_2, a_3))$ yields $\bar{c}(a_1, a_2, a_3)$. Hence a full-adder is composed of half-adders in the form:

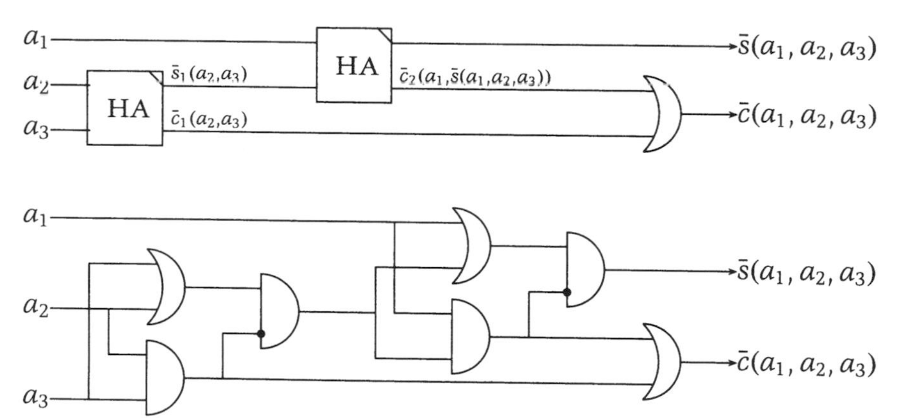

A symbol for the full-adder is:

8.9. Example. Addition of two binary numbers $a_2a_1a_0$ and $b_2b_1b_0$ is performed according to three steps:

(i) A half-adder calculates the provisional sums s_i and carry c_i', $i = 0, 1, 2$.

(ii) A second half-adder combines s_i with c_{i-1} $(i = 1, 2)$ to the final sum x_i in position i. There may be a carry c_i''.

(iii) Either c_i' or c_i'' is 1. An OR-gate generates the carry c_i for the following position $i+1$.

The corresponding diagram should be determined in Exercise 8.

We have seen that an arbitrary electrical circuit or network can be obtained from suitable compositions of series and parallel circuits. Any function from $\{0,1\}^n$ into $\{0,1\}$, thus any switching function of an electrical circuit, is a Boolean polynomial function; therefore any electrical circuit can be described by a Boolean polynomial in its mathematical model.

Sometimes, the inputs $a_1, \dots, a_n$ in a switching circuit are divided into two parts: $a_1, \dots, a_k$ and $a_{k+1}, \dots, a_n$. The first part is interpreted as "programming variables", while the others are variables as before. Depending on the values of $a_1, \dots, a_k$, the circuit receiving the inputs $a_{k+1}, \dots, a_n$ works in a specified way. We explain this by two examples:

8.10. Example. We want a circuit p to compute a_2a_3 if $a_1 = 0$ and to compute $a_2 + a_3$ if $a_1 = 1$:

a_1	a_2	a_3	$\bar{p}(a_1, a_2, a_3)$
0	0	0	0
0	0	1	0
0	1	0	0
0	1	1	1
1	0	0	0
1	0	1	1
1	1	0	1
1	1	1	1

Of course, $p = x_1'x_2x_3 + x_1x_2'x_3 + x_1x_2x_3' + x_1x_2x_3 \sim x_1'x_2 + x_1'x_3 + x_2x_3$ does the job.

8.11. Example (*Chakrabarti's Cell*). For $k = 2$ and $n = 5$, we want, NOR(a_3, a_4, a_5) for $(a_1, a_2) = (0, 0)$, OR(a_3, a_4, a_5) for $(a_1, a_2) = (0, 1)$, NAND(a_3, a_4, a_5) for $(a_1, a_2) = (1, 0)$, and AND(a_3, a_4, a_5) for $(a_1, a_2) = (1, 1)$ (see 7.7). Then $p = x_2'x_3'x_4'x_5' + x_1x_2'x_3' + x_1x_2'x_4' + x_1x_2'x_5' + x_1'x_2x_3 + x_1'x_2x_4 + x_1'x_2x_5 + x_2x_3x_4x_5$ describes this circuit (see Exercise 9).

The use of these "programmable" circuits is clear: with identical building blocks we can construct arbitrarily complex switching circuits and networks.

Exercises

1. In a production process there are three motors operating, but only two are allowed to operate at the same time. Design a switching circuit which prevents that more than two motors can be switched on simultaneously.
2. Design a switching circuit that enables you to operate one lamp in a room from four different switches in that room.
3. A voting-machine for three voters has three YES-NO switches. Current is in the circuit precisely when YES has a majority. Draw the contact diagram and the symbolic representation by gates and simplify it.
4. An oil pipeline has three pipelines b_1, b_2, b_3 which feed it. Design a plan for switching off the pipeline at three points S_1, S_2, S_3 such that oil flows in the following two situations: S_1 and S_3 are both open or both closed but S_2 is open; S_1 is open and S_2, S_3 are closed.
5. A hall light is controlled by two switches, one upstairs and one downstairs. Design a circuit so that the light can be switched on or off from the upstairs or the downstairs.
6. Let T_1 and T_2 be two telephones which are arranged in such a way that T_2 cannot be used unless T_1 is engaged but T_2 is not cut off when T_1 is not engaged. A light L is switched on whenever both T_1 and T_2 are engaged but does not switch off until both telephones are disengaged. Construct a suitable switching network and obtain Boolean polynomials for the circuits for T_2 and L.
7. Compute the minimal circuits for the other three assignments to the don't-care combinations in Example 8.2. Which one gives the "really" shortest form?
8. Draw the diagram in Example 8.9.
9. Show that p in Example 8.11 describes the Chakrabarti-cell in a minimal form.

§9 More Applications of Boolean Algebras

In this section we apply our knowledge of Boolean algebras to logic, set and probability theory, and finally to quantum physics.

The "algebra of logic" represents one of the early applications of Boolean algebras. Logic is concerned with studying and analyzing models of thoughts, arguments, and conclusions by making extensive use of symbols. This was the historical origin and the initial purpose for the foundation of the algebra named after G. Boole.

9.1. Informal Definition. A ***proposition*** is a "sentence" for which "it makes sense" to ask if it is true or false.

The careful reader will have noticed that 9.1 is not really a definition but rather a description of the term proposition. At the basis of the concept of propositions we have the two-value principle (also called "***principle of the excluded middle***" or "***tertium non datur***"), a principle which goes back to the classical propositional logic of Aristotle. It means that each proposition must be either true of false, there is no other possibility "in between".

9.2. Examples. The following are propositions:

(i) 7 is a prime number.

(ii) $2 + 3 = 4$.

(iii) There exists extra-terrestrial life.

The following are not propositions:

(iv) Be quiet!

(v) Two dogs on the way to yesterday.

The following are doubtful:

(vi) Here is xbx! (It depends if you are able to assign a "meaning" to "xbx".)

(vii) It rains. (When, where, how long,...? Try for a moment to define "rain"; it is harder than you might have thought in the beginning! See also the lines before 9.10.)

Propositions will be denoted by capital letters $A, B, C, \ldots$. They can be compounded in several ways, e.g., by "and", "or", "if... / then...". We can also obtain a new proposition from a given one by negating it. The truth value "true" of a proposition will be denoted by 1 and the truth value "false" by 0. Truth tables will be used to define the compound proposition by describing its truth value according to the truth values of the propositions involved in the combination.

9.3. Definitions. Let A and B be propositions. The ***negation*** (or ***complementation***) of A (in symbols: $\neg A$) is the proposition "not A", i.e., $\neg A$ is true if and only if A is false. The ***conjunction*** of A and B (in symbols: $A \wedge B$) is true if and only if A as well as B are true. The ***disjunction*** of A and B (in symbols: $A \vee B$) is the proposition "A or B", i.e., $A \vee B$ is true if and only if either A or B or both are true.

9.4. Examples. We give some examples of truth tables:

A	$\neg A$
1	0
0	1

A	B	$A \wedge B$	$A \vee B$	$\neg A \vee B$	$(\neg A \vee B) \wedge (\neg B \vee A)$
1	1	1	1	1	1
1	0	0	1	0	0
0	1	0	1	1	0
0	0	0	0	1	1

It is most natural to expect that the propositions form a lattice w.r.t. $\wedge$ and $\vee$. The difficulty comes from a rather unexpected side. We cannot even form *sets* of propositions as long as we have not defined when two propositions are equal. And that poses a real problem:

Let, for instance, A be the proposition "It rains now and the wind blows", and let B be "It's now windy and rainy". Should A and B be equal? Two frequent answers are:

- "Yes, because when A is true, so is B, and conversely". But then, *all* true propositions would be the same, and there would only be two different propositions on Earth.
- "No, because the wording is different". Well, but then there is no hope for laws like commutativity.

There seems to be no way out. Let's start again, and give new names to the concepts defined in §4:

9.5. Definitions. Take $n \in \mathbb{N}$, and $x_1, \ldots, x_n$ as distinct "symbols" as in 4.1. Let $\mathbb{B}$ denote the Boolean algebra $\{0, 1\}$.

(i) $x_1, \ldots, x_n$ are called ***propositional variables***.

(ii) Each $p \in P_n$ is called a ***propositional form***.

(iii) $\bar{p}_{\mathbb{B}} \in P_n(\mathbb{B})$ is called the ***truth function*** of p.

(iv) $P_n/\sim$ is called (in this context) the ***propositional algebra***; here $\sim$ denotes the equivalence of polynomials (see 4.8).

Thus a propositional form is nothing else than a Boolean polynomial, but in this section we are heading for another interpretation. Let, for example, p be the polynomial $x_1x_2+x_1x_3'$, and let A_1, A_2, A_3 be propositions (don't form a set out of A_1, A_2, A_3!). We then might denote $(A_1 \wedge A_2) \vee (A_1 \wedge \neg A_2)$ by $\bar{p}(A_1, A_2, A_3)$; so we might use Boolean polynomials to get "compound propositions". And we can use the Quine-McCluskey method to simplify complicated statements (see Example 9.9 below).

9.6. Definitions. $p \in P_n$ is called a ***tautology*** if the Boolean polynomial function $\bar{p}$ is always 1, i.e., $p \sim 1$. A propositional form p is called a ***contradiction*** if the Boolean polynomial function $\bar{p}$ is always 0, i.e., $p \sim 0$.

The normal forms of Boolean polynomials enable us to decide whether a given proposition is a tautology. Corollary 4.18 shows:

9.7. Theorem.

(i) *A propositional form is a tautology if and only if its disjunctive normal form has all coefficients equal to* 1.

(ii) *A proposition is a contradiction if and only if its disjunctive normal form has all coefficients equal to* 0.

A certain propositional form is of special importance in logic and gets a name of its own.

9.8. Definition. $p' + q =: p \to q$ is called the ***subjunction*** of p and q (or "if-then" operation).

Note that $(p \to q) \wedge (q \to p)$ holds if and only if $p \sim q$. Also, $pq \to p$ is always true. One of the problems of propositional logic is to study and simplify propositions and to determine their truth values. In practical applications it is sometimes useful to describe the logical framework in complicated treaties, rules, or laws, and to determine their truth values by applying methods of propositional algebra. A simple example will suffice to demonstrate a possible use.

9.9. Example. We determine if the following argument is correct: "If the workers in a company do not go on strike, then a necessary and sufficient condition for salary increases is that the hours of work increase. In case of a salary increase there will be no strikes. If working hours increase, there will be no salary increase. Therefore salaries will not be increased."

The compound statement can be formulated as follows:

$$\big((\neg S \to ((I \to W) \wedge (W \to I))\big) \wedge (I \to \neg S) \wedge (W \to \neg I)\big) \to \neg I,$$

where S denotes "strike", I "salary increase", and W "increase of working hours". We abbreviate the compound statement by $\big((1) \wedge (2) \wedge (3)\big) \to \neg I$.

Solution #1 Use truth tables:

S	I	W	(1)	(2)	(3)	$(1) \wedge (2) \wedge (3)$	$\big((1) \wedge (2) \wedge (3)\big) \to \neg I$
0	0	0	1	1	1	1	1
0	0	1	0	1	1	0	1
0	1	0	0	1	1	0	1
0	1	1	1	1	0	0	1
1	0	0	1	1	1	1	1
1	0	1	1	1	1	1	1
1	1	0	1	0	1	0	1
1	1	1	1	0	0	0	1

Hence the statement is correct.

Solution #2 We assume there is a truth assignment making $\big((1) \wedge (2) \wedge (3)\big)$ true and $\neg I$ false. For such an assignment, $\neg S \to \big((I \to W) \wedge (W \to I)\big)$, $I \to \neg S$, $W \to \neg I$ are true and $\neg I$ is false; therefore I is true. $I \to \neg S$ is true implies that $\neg S$ is true. Hence, by the truth of the first assumption above, we know that $(I \to W) \wedge (W \to I)$ is true. I is true; therefore W must be true. But since $W \to \neg I$ is true, $\neg I$ is true, which is a contradiction to the assumption. Therefore the original argument is correct.

Solution #3 The Boolean polynomial

$$p = (x_1' \to (x_2 \sim x_3))(x_2 \to x_1')(x_3 \to x_2')$$

"resembles" the statement $\big((1) \wedge (2) \wedge (3)\big)$. From the table of solution #1 we see that $p \sim x_1'x_2'x_3' + x_1x_2'x_3' + x_1x_2'x_3$. Quine-McCluskey readily gives us

000 ✓	−00
100 ✓	10−
101 ✓	

Hence $p \sim x_2'x_3' + x_1x_2' \sim x_2'(x_3' + x_1) \sim x_2'(x_3 \to x_1)$, which clearly implies x_2'. Note that this solution is not shorter than solution #1, but we see from this method that the complicated statement $((1)\wedge(2)\wedge(3))$ is the "same" as the much simpler "There is no salary increase and an increase in working hours implies a strike." □

Sentences such as "x is a natural number", where we can substitute "things" for x which make this proposition true or false (e.g., $x = 3$ or $x = \sqrt{3}$, respectively), are called ***predicates***. More precisely, let X be a set of "variables" which can "assume values" in a set U disjoint from X. Then a ***predicate in** X **over** U* is a "sentence which turns into a proposition whenever each $x \in X$ is replaced by some u_x in U." In the example above, $X = \{x\}$ and $U = \mathbb{R}$. An example of a predicate in $X = \{x, y\}$ over $U = \mathbb{R}$ is $x^2 + y^2 = y^2 + 1$, which is true precisely for the replacements $u_x = 1$ and $u_x = -1$ for x, no matter what we take for u_y. A remarkable other example is $x^2 + y^2 = y^2 + x^2$, which yields a true statement for all replacements of x and y. Such a predicate is called a ***law***.

We shall not pursue this topic, but the reader should be warned that many "statements" are predicates rather than propositions. For instance, "It rains" is true or false depending where you are and when you formulate this sentence. So this should better read as "At time t it rains at the place with coordinates (x, y, z)", which is a predicate in $X = \{t, x, y, z\}$ over $\mathbb{R}$. And we still have to define what it means to "rain" ....

The ideas developed for simplifying propositions can also be applied to simplify set constructions. The following example will show how this works.

9.10. Example. Let A, B, C, D be sets. Simplify

$$M = (A \cap (B \setminus C)) \cup (B \setminus C) \cup ((D \setminus A) \cap (D \setminus B) \cap C),$$

where $B \mathbin{\triangle} C$ denotes the ***symmetric difference*** $(B \setminus C) \cup (C \setminus B)$ of B and C. Solution: Let

$$\begin{array}{lll} b_1 = 1 & \text{if} & x \in A \\ b_2 = 1 & \text{if} & x \in B \\ b_3 = 1 & \text{if} & x \in C \\ b_4 = 1 & \text{if} & x \in D \end{array} \qquad (\text{and } b_i = 0 \text{ otherwise}).$$

The "corresponding" Boolean polynomial of M is

$$p = x_1x_2x_3' + x_2x_3' + x_2'x_3 + x_4x_1'x_4x_2'x_3$$

since, e.g., $x \in D \backslash A$ means $(x \in D) \wedge \neg(x \in A)$. Simplification gives $p \sim x_2x_3' + x_2'x_3$, from which we get

$$M = (B \backslash C) \cup (C \backslash B) = B \triangle C. \qquad \square$$

If we want to prove set-theoretical identities like $A_1 \setminus (A_2 \cap A_3) = (A_1 \backslash A_2) \cup (A_1 \backslash A_3)$, we often draw a "Venn-diagram"

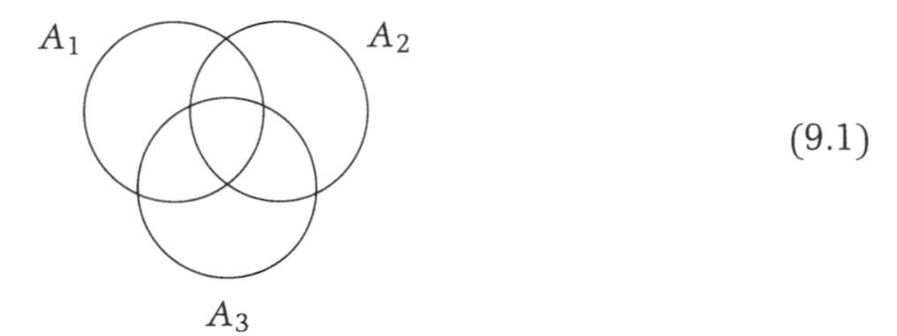

for the sets A_1, A_2, A_3, and shade the areas which belong to $A_1 \setminus (A_2 \cap A_3)$ and $(A_1 \backslash A_2) \cup (A_1 \backslash A_3)$, respectively. In both cases, we get

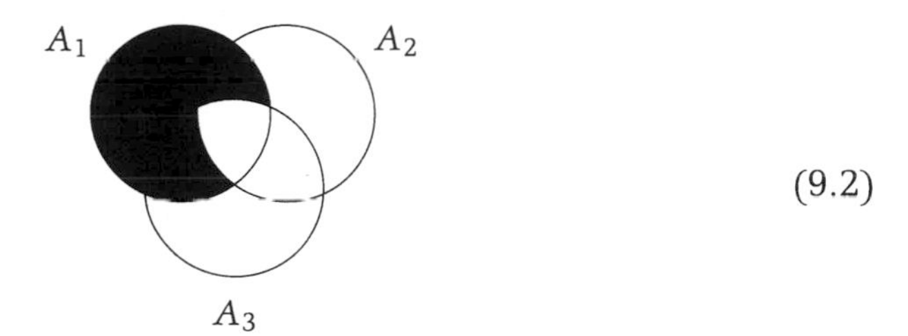

Does this really constitute a proof for this identity? Surprisingly, the answer is "yes". Note that (9.1) divides the plane into $2^3 = 8$ sections:

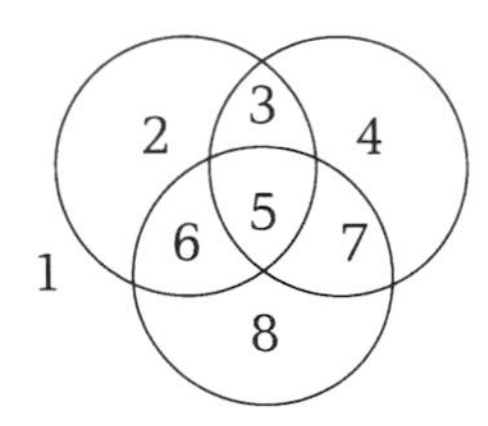

Each of these sections stands for a line in a truth table; for example, section 6 above stands for "$x \in A_1 \wedge x \notin A_2 \wedge x \in A_3$", i.e., for the row of 1 0 1 in a truth table like in Example 9.4. So if we draw A_1, A_2, A_3 (in general: $A_1, A_2, , \ldots, A_n$) so that it divides the plane into 8 (in general: 2^n) sections, we really can "prove set identities by shading". Note that we cannot conduct proofs with sketches like

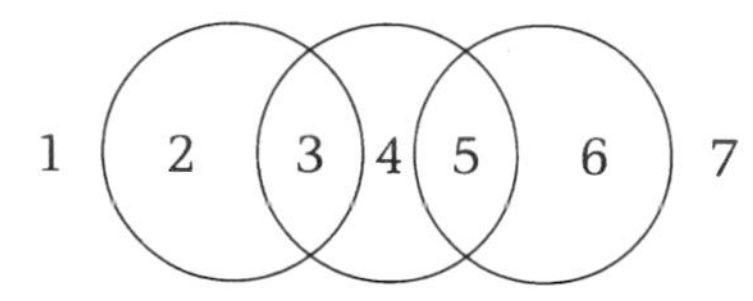

For more on this subject, see Gumm & Poguntge (1981).

Next, we consider the connections between lattices and probability theory. At the foundations of probability theory and statistics we have the analysis of "experiments with random outcome". Here we mean "experiments" or observations of experiments whose outcome is not predetermined. This may be the case because of lack of knowledge of the circumstances of the experiment, because the situation is too complicated (see 9.11(i)), or because of real randomness (see 9.11(ii)); we could also have a situation where a random outcome is quite likely (see 9.11(iii)).

9.11. Examples.

(i) Let an experiment consist of casting a die once. If the die is completely homogeneous and symmetrical, then we may assume that any of the numbers $1, 2, \ldots, 6$ have the same chance of occurring. If the die is not "ideal", then one (or more) numbers will occur more frequently.

(ii) The experiment consists of counting the number of α-particles which are emitted by a radioactive substance during one second. The outcome of this experiment is (presently thought to be) truly random.

(iii) Consider the values of the first, second, third... decimal place of a real number, given in decimal notation. If x is rational, then x has a periodic decimal expansion and in the sequence of the numbers in the various decimal places there is nothing random. However, if x is irrational, then the numbers in the decimal places are randomly (?) distributed.

These examples show that the term "randomness" is not an easy one. The question if and how much the outcome of an experiment is random can be answered by statistical methods in such a way that we try to confirm or contradict the assumption of randomness by using a series of tests, e.g., homogeneity of a die tested by a series of casts. In general an experiment has many possible outcomes not all of which are of interest in a special situation. Moreover, some or all combinations of outcomes may be of interest, like "the die shows more than four points", etc.

We shall construct a mathematical model to study random experiments similar to the model constructed for switching circuits. This mathematical model will depend on the main aspects of the situation we are interested in.

9.12. Model I for Random Experiments. In a random experiment let Ω be the set of all (interesting) outcomes of the experiment. The elements of Ω are called ***samples*** and Ω is called the ***sample space*** of the experiment. Combinations of outcomes of an experiment can be modeled as subsets of the sample space. For example, if $\Omega = \{1, 2, \ldots, 6\}$ is the sample space for tossing a die once, then the combination "the die shows more than four points" of possible outcomes can be described by $\{5, 6\}$. If K_1, K_2 denote two combinations of outcomes (described by the subsets A_1, A_2 of Ω), then "K_1 and K_2" (described by $A_1 \cap A_2$), "K_1 or K_2" (described by $A_1 \cup A_2$), and "not K_1" (described by A_1') are also combinations of outcomes. These three "operations" on outcome combinations (which interest us) should again be interesting outcome combinations. This leads to the concept of Boolean algebras and to the second stage of our model building process.

9.13. Model II for Random Experiments. A ***random experiment*** is modeled by the pair $(\Omega, \mathcal{A})$. Here Ω is the sample space and $\mathcal{A}$ is a Boolean subalgebra of $\mathcal{P}(\Omega)$. The elements of $\mathcal{A}$ (corresponding to the interesting combinations of outcomes) are called ***events***. $\mathcal{A}$ is called the ***algebra of events*** over the sample space Ω.

9.14. Examples.

(i) In tossing a die let all outcomes and all combinations be of interest. Then a mathematical model is given by $(\Omega, \mathcal{P}(\Omega))$, where $\Omega = \{1, 2, \ldots, 6\}$.

(ii) In tossing a die assume we are only interested whether the points are less than 3 or greater than or equal to 3. In this case a model is given by $(\Omega, \mathcal{P}(\Omega))$, where $\Omega = \{a, b\}$, a means "points value < 3", b means "points value ≥ 3".

(iii) In tossing two dice, a suitable mathematical model could be $(\Omega, \mathcal{P}(\Omega))$, where $\Omega = \{1, 2, \ldots, 6\} \times \{1, 2, \ldots, 6\}$. The event $\{(4, 6), (5, 5), (6, 4)\}$ can be interpreted as the outcome combination "the sum of points is 10".

(iv) If we consider the experiment of counting α-particle emissions during one second (see 9.11(ii)), then $(\mathbb{N}_0, \mathcal{A})$ would be a suitable model where $\mathcal{A} = \{A \subseteq \mathbb{N}_0 \mid A \text{ is finite or } \mathbb{N}_0 \backslash A \text{ is finite}\}$ (see Exercise 3.9).

9.15. Definition. Let $\mathcal{A}$ be the algebra of events over the sample space Ω. If $\mathcal{A}$ (as a Boolean algebra) is generated by a subset $\mathcal{E}$, which means that the smallest Boolean algebra in $\mathcal{A}$ which contains $\mathcal{E}$ is $\mathcal{A}$ itself, then the elements of $\mathcal{E}$ are called ***elementary events***.

In the Examples 9.14 we can take one-element subsets as elementary events. If we chose $(\mathbb{N}_0, \mathcal{P}(\mathbb{N}_0))$ as a model in 9.14(iv), we would have to choose a much more complicated system of elementary events.

So far we have not mentioned "probability". It is useful to restrict the definition of an algebra of events. There are good reasons why we would like to have unions and intersections in $\mathcal{A}$ for countably many sets $A_1, A_2, \ldots$. Thus we have to compromise between arbitrary Boolean algebras (in which any two-element set, therefore also any finite set, has a supremum and an infimum) and those Boolean algebras, which are ***complete lattices*** (where every subset has a supremum and an infimum):

9.16. Definition. A Boolean algebra B is called a ***σ-algebra*** if every countable subset of B has a supremum and an infimum in B.

For example, $\mathcal{P}(M)$ is a σ-algebra for any set M. The Boolean algebra $\mathcal{A}$ in 9.14(iv) is not a σ-algebra since the family $\{\{0\}, \{1\}, \{2\}, \ldots\}$ has no supremum in $\mathcal{A}$. We refer to Halmos (1967, pp. 97–103) for the representability of any σ-algebra as a σ-algebra in a suitable $\mathcal{P}(M)$, factored by a suitable "σ-ideal" ("Theorem of Loomis").

9.17. Definition. Let $\mathcal{A}$ be an algebra of events on the sample space Ω. If $\mathcal{A}$ is a σ-algebra, then $(\Omega, \mathcal{A})$ is called a ***measurable space***.

Now we are able to "measure".

9.18. Definitions. Let B be a σ-algebra. A ***measure*** on B is a mapping μ from B into $\{x \in \mathbb{R} \mid x \geq 0\} \cup \{\infty\}$ with the following properties:

(i) $\mu(b) < \infty$ for at least one $b \in B$.

(ii) If $b_1, b_2, \ldots$ are countably many elements in B with $b_i b_j = 0$ for $i \neq j$ and if b is their supremum, then

$$\mu(b) = \sum_{i=1}^{\infty} \mu(b_i) \qquad (\sigma\text{-additivity}).$$

If, moreover, $\mu(1) = 1$, then μ is called a ***probability measure*** and $\mu(b)$ is the ***probability*** of $b \in B$. The pair (B, μ) is called a ***measure space***. If μ is a probability measure, then (B, μ) is also called a ***probability space***.

9.19. Examples. Let $M \neq \emptyset$ be a finite set. Then μ, defined by $\mu(A) := |A|$, is a measure on $\mathcal{P}(M)$. If $|M| \geq 2$, then μ is not a probability measure. $P(A) := |A|/|M|$ defines a probability measure P on $\mathcal{P}(M)$. If $B = \mathcal{P}(\{1, 2, \dots, 6\})$, see 9.14(i), then we have, e.g.,

$$P(\{5, 6\}) = \frac{|\{5, 6\}|}{|\{1, 2, \dots, 6\}|} = \frac{2}{6} = \frac{1}{3}.$$

Now we are able to conclude our model.

9.20. Model III for Random Experiments. The triplet $(\Omega, \mathcal{A}, P)$ is called a ***probabilistic model of a random experiment*** if Ω is the set of all outcomes (which interest us), $\mathcal{A}$ is a suitable σ-algebra in $\mathcal{P}(\Omega)$, and P is a probability measure on $\mathcal{A}$.

In this case, $(\Omega, \mathcal{A})$ is a measurable space and $(\mathcal{A}, P)$ is a probability space. The question of whether a given probability measure P is the "correct" measure to model a random experiment is one of the central questions of mathematical statistics. We cannot go into these problems here and refer the reader to the literature.

In case of a finite sample space Ω, one is usually best served by the σ-algebra $\mathcal{A} = \mathcal{P}(\Omega)$. However, for infinite sample spaces, there are some problems: if we choose $\mathcal{A}$ "too small" (e.g., as in 9.14(iv)), then we do not obtain a σ-algebra; if we choose $\mathcal{A}$ "too large" as, for instance, in $\mathcal{A} = \mathcal{P}(\Omega)$, it often happens that we cannot have a probability measure on $\mathcal{A}$, as it happens for $\Omega = \mathbb{R}$. Therefore we have to compromise. For $\Omega = \mathbb{R}$ and, more generally, for $\Omega = \mathbb{R}^n$ and its subsets, we obtain a solution to this problem as follows.

9.21. Definition. Let $\Omega \subseteq \mathbb{R}^n$. We consider the σ-algebra $\mathcal{B}$ in $\mathcal{P}(\mathbb{R}^n)$ which is generated by the set of all cartesian products of open intervals. $\mathcal{B}$ is called the ***σ-algebra of Borel sets in $\mathbb{R}^n$***. The ***σ-algebra of reduced Borel sets on*** Ω is defined as $\mathcal{B}_\Omega := \{B \cap \Omega \mid B \in \mathcal{B}\}$.

We obtain a measure space $(\mathcal{B}, \mu)$ if we define μ on the product of open intervals as follows:

$$\mu((a_1, b_1) \times (a_2, b_2) \times \cdots \times (a_n, b_n)) := (b_1 - a_1)(b_2 - a_2) \cdots (b_n - a_n)$$

and extend μ to all subsets of $\mathcal{B}$ in an obvious way. If we define

$$\mu((a_1, b_1) \times (a_2, b_2) \times \cdots \times (a_n, b_n)) := g_1(b_1 - a_1)g_2(b_2 - a_2) \cdots g_n(b_n - a_n)$$

with suitable weights g_i, we can obtain a probability space. For more information we have to refer to the literature on probability theory.

We summarize some of the basic properties of probability measures.

9.22. Theorem. *Let $(\mathcal{A}, P)$ be a probability space. Then, for all $b_1, b_2, b \in \mathcal{A}$, we get:*

(i) $b_1 b_2 = 0 \Longrightarrow P(b_1 + b_2) = P(b_1) + P(b_2)$;
(ii) $P(b_1 + b_2) = P(b_1) + P(b_2) - P(b_1 b_2)$;
(iii) $b_1 \leq b_2 \Longrightarrow P(b_1) \leq P(b_2)$;
(iv) $P(b) \in [0, 1]$;
(v) $P(b') = 1 - P(b)$.

Proof.

(i) This is a special case of the σ-additivity.
(ii) We have $b_1 + b_2 = b_1 + b_1' b_2$ with $b_1(b_1' b_2) = 0$. Also, $b_1' b_2 + b_1 b_2 = b_2$ with $(b_1' b_2)(b_1 b_2) = 0$. Hence $P(b_1 + b_2) = P(b_1) + P(b_1' b_2) = P(b_1) + P(b_2) - P(b_1 b_2)$.
(iii) $b_1 \leq b_2 \Longrightarrow b_2 = b_1 + b_1' b_2$ (with $b_1(b_1' b_2) = 0$). Therefore $P(b_2) = P(b_1) + P(b_1' b_2) \geq P(b_1)$.
(iv) Follows from (iii), since $P(b) \leq P(1) = 1$ for all $b \in \mathcal{A}$.
(v) $1 = b + b'$ with $bb' = 0$. Therefore $1 = P(1) = P(b) + P(b')$. □

Finally we mention a completely different situation where lattices appear "in nature". We consider a "classical mechanical system", such as our system of nine planets. Each planet can be described by its position (three local coordinates). Thus the system can be described as a "point" in $\mathbb{R}^{27}$ and $\mathbb{R}^{27}$ is called the ***phase space*** of the system. Any property of the system (e.g., "the distance between Jupiter and Saturn is less than k kilometers") determines a subset of $\mathbb{R}^{27}$; in the example this is $\{(x_1, \ldots, x_{27}) \mid (x_{13} - x_{16})^2 + (x_{14} - x_{17})^2 + (x_{15} - x_{18})^2 \leq k^2\}$ if (x_{13}, x_{14}, x_{15}) gives the position of Jupiter and (x_{16}, x_{17}, x_{18}) the position of Saturn. Conversely, there is the question of whether we can assign a relevant physical property to any of the subsets of $\mathbb{R}^{27}$, or to more general phase spaces. It seems to make sense to assume that we can assign sensible physical properties to the Borel sets in $\mathbb{R}^{27}$. Thus the physical system of our planets can be studied by means of the σ-algebra of the Borel sets on $\mathbb{R}^{27}$.

In microcosms we have some problems, since not all observables like place, impulse, spin, energy, etc., of a quantum-theoretical system can be precisely measured at the same time. We know this from Heisenberg's uncertainty principle. In this case it is advisable to choose an infinite-dimensional separable Hilbert space as our phase space. This is a vector space H with an inner product $\langle .,. \rangle$ such that any Cauchy sequence h_n, which is characterized by $\lim_{n,m\to\infty}\langle h_n - h_m, h_n - h_m \rangle = 0$, converges to some $h \in H$. Moreover, H is supposed to have a countable subset B such that any $h \in H$ is a limit of sequences of elements in B. Then the observable properties of a quantum system correspond to the closed subspaces of H. Here lattices of the following type arise:

9.23. Definitions. Let L be a bounded lattice. L is called an ***orthocomplemented lattice*** if for any $v \in L$ there is precisely one $v^\perp \in L$ (called the ***orthocomplement*** of v) such that for all $v, w \in L$:

(i) $(v^\perp)^\perp = v$;

(ii) $(v + w)^\perp = v^\perp w^\perp$ and $(vw)^\perp = v^\perp + w^\perp$;

(iii) $vv^\perp = 0$ and $v + v^\perp = 1$.

An orthocomplemented lattice L is called ***orthomodular*** if the ***orthomodular identity***

$$v \le w \Longrightarrow w = v + wv^\perp$$

is satisfied. The "Chinese Lantern" V_{15}^6 is orthomodular.

9.24. Example. Let L be the lattice of all subspaces of a finite-dimensional inner product space. L is orthomodular and the orthocomplement of a subspace U of L is the orthogonal complement $U^\perp := \{x \mid \langle x, u\rangle = 0 \text{ for all } u \in U\}$, the set of all vectors x orthogonal to U. Note that L is in general not distributive, so orthomodular lattices might be viewed as "non-distributive Boolean algebras".

We verify immediately:

9.25. Theorem. *Let H be a separable Hilbert space. Then the closed subspaces of H form a complete orthocomplemented lattice.*

The fact that some of the observables of a quantum-theoretic system can be measured simultaneously and others cannot, can be expressed in the following definitions.

9.26. Definitions. Let L be an orthomodular lattice and $v, w \in L$. The elements v and w are called ***orthogonal*** if $v \leq w^{\perp}$. Moreover, v and w are called ***simultaneously verifiable*** if there are pairwise orthogonal elements $a, b, c \in L$ such that $v = ac$, $w = bc$. We denote this by $v \leftrightarrow w$.

The relation $\leftrightarrow$ is reflexive and symmetric, but in general not transitive.

9.27. Definition. Let L be an orthomodular lattice. The element $z \in L$ is called ***central*** if $z \leftrightarrow v$ for all $v \in L$.

In all classical mechanical systems all observables are simultaneously measurable and therefore we obtain Boolean algebras. The next and last theorem shows that quantum mechanics in this sense can also be regarded as an extension of classical mechanics and quantum logic as an extension of classical logic.

9.28. Theorem. *Let L be an orthomodular lattice. Then the central elements in L form a Boolean algebra.*

This theory goes deep into lattice theory and quantum mechanics.

Exercises

1. Determine the truth function of $((x_1 \vee x_2) \vee x_3) \vee (x_1 \wedge x_3)$.
2. Which of the following are tautologies:

   (i) $x_1 \to (x_2 \wedge (x_1 \to x_2))$;

   (ii) $x_1 \to (x_1 \vee x_2)$;

   (iii) $(x_1 \to x_2) \vee (x_2 \to x_3)$.
3. Determine whether the following statements are consistent by representing the sentences as compound propositions: "An economist says that either devaluation will occur or if exports increase, then price controls will have to be imposed. If devaluation does not occur, then exports will not increase. If price controls are imposed the exports will increase."

4. A politician says in four of his election speeches:
   "Either full employment will be maintained or taxes must not be increased";
   "Since politicians have to worry about people, taxes have to be increased";
   "Either politicians worry about people or full employment cannot be maintained";
   "It is not true that full employment has increased taxes as a consequence".

   (i) Are these four statements consistent?

   (ii) Are the first three statements consistent or are these three statements put together nonsensical?

5. Each of the objects A, B, C is either green or red or white. Of the following statements one is true and four are false.

   (i) B is not green and C is not white.

   (ii) C is red and (iv) is true.

   (iii) Either A is green or B is red.

   (iv) Either A is red or (i) is false.

   (v) Either A is white or B is green.

   Determine the color of each object.

6. A random experiment consists of going to the doctor's surgery to find out how long one has to wait to get attended to. Give a model for this experiment.

7. An experiment consists of tossing a die on a table and measuring the distance of the die from the edge of the table. Give a model.

8. In the models of Exercises 6 and 7, state a system of elementary events for each experiment.

9. Let $\mathcal{B}$ be the σ-algebra of Borel sets on $\mathbb{R}$. What are the atoms? Is every set in $\mathcal{B}$ the union of atoms? Are all finite subsets elements of $\mathcal{B}$? Is the set of positive real numbers in $\mathcal{B}$? Is $\mathbb{Q} \in \mathcal{B}$?

10. Give a probability measure P for 9.14(iii). Interpret $A = \{1\} \times \{1, 2, \ldots, 6\}$ and $B = \{(1, 1), (2, 2), \ldots, (6, 6)\}$.

11. Verify that the lattice of all subspaces of a finite-dimensional real Euclidean space is an orthomodular lattice.

12. Construct an example of an orthocomplemented lattice with six elements that is not orthomodular.

13. Let L be orthocomplemented and $a, b, c \in L$. Show that if a is orthogonal to b, then a is orthogonal to bc, too. Also, if a is orthogonal to both b and c, then a is orthogonal to $b + c$.
14. Let L be an orthocomplemented lattice. Show that if L is orthomodular, then $a \leq b$ implies the existence of some $c \in L$ such that a and c are orthogonal and $a + c = b$.
15. Draw a sketch of four sets A_1, A_2, A_3, A_4 such that we can prove set identities involving four sets.

Notes

Some standard introductory textbooks on Boolean algebras and applications are Hohn (1970), Mendelson (1970), and Whitesitt (1961).

The collection of survey articles by Abbott (1970) contains applications of lattices to various areas of mathematics. It includes a paper on orthomodular lattices, one on geometric lattices, a general survey on "what lattices can do for you" within mathematics, and a paper on universal algebra. More recent books with a similar goal are Kalmbach (1983) and Monk & Bonnet (1989). Most of the books on applied algebra consider Boolean algebras in an introductory way and have applications to switching circuits, simplification methods, and logic: Birkhoff & Bartee (1970), Dornhoff & Hohn (1978), Fisher (1977), Gilbert (1976), Prather (1976), and Preparata & Yeh (1973)

Further texts describing the applications given in this chapter and several additional examples are Dokter & Steinhauer (1972), Dworatschek (1970), Harrison (1970), Perrin, Denquette & Dalcin (1972), and Peschel (1971). A comprehensive book on discrete and switching functions (algebraic theory and applications) is Davio, Deschaps & Thayse (1978). Rudeanu (1974) deals with polynomial equations over Boolean algebras.

As to the development of logic, the great merit of the fundamental work of Aristotle is that Aristotle succeeded in describing and systematizing in a set of rules the process of reaching logical conclusions from assumptions. Aristotle focused his attention mainly on a certain type of logical relations, the syllogisms. Leibniz tried to give Aristotelian logic an algebraic form. He used the notation AB for the conjunction of two terms, noted the idempotent law, and knew that "every A is a B" can be written as $A = AB$. He noted that the calculus of logic applied to statements or expressions as well. However, he did not go far enough to develop Boolean algebra as an adequate tool. Leibniz's interest in this context was mainly in transcribing the rules of syllogism into his notation. Boolean algebras reduced an important class of logical questions to a simple algebraic form and gave an algorithm of solving them mechanically.

The treatment of the power set of a set as a Boolean algebra was initially considered independently by C. S. Pierce and E. Schröder. The success of Boolean algebras in simplifying and clarifying many logical questions encouraged mathematical logicians to try to formalize all mathematical reasoning. Regarding texts on mathematical logic we mention, for example, Barnes & Mack (1975).

In 1900, at the International Congress of Mathematicians in Paris, D. Hilbert gave an address entitled "Mathematical Problems", in which he listed 23 problems or problem areas in mathematics. Problem 6 refers to the question of axiomatization of physics and probability theory, which prompted S. N. Bernstein (in 1917) and R. von Mises (in 1919) to use algebra in the foundations of probability theory. Kolmogorov (in 1933) based probability theory on the concept of measure theory and σ-algebras. An introductory text on probability theory based on Boolean σ-algebras is Franser (1976).

A survey article by Holland in Abbott's (1970) collection gives an excellent description of the history and development of orthomodular lattices and also includes a Bibliography on the subject. Varadarajan (1968) is an introduction to quantum theory, von Neumann (1955) is a classic on quantum mechanics, and Kalmbach (1983) is a standard recent work on orthomodular lattices.

Further applications of lattices and Boolean algebras are due to Zelmer & Stancu (1973), who try to describe biosystems (e.g., organisms and their environment) axiomatically and interpret them in terms of lattice theory. Friedell (1967) gives lattice theoretical interpretations to social hierarchies interpreted as partial orders. Wille (1982) describes the use of lattice theory to study hierarchies of concepts, and Wille (1980) shows how lattices are related to music.

3 CHAPTER Finite Fields and Polynomials

Finite fields give rise to particularly useful and, in our view, beautiful examples of the applicability of algebra. Such applications arise both within mathematics and in other areas; for example, in communication theory, in computing, and in statistics. In this chapter we present the basic properties of finite fields, with special emphasis on polynomials over these fields. The simplest finite field is the field $\mathbb{F}_2$ consisting of 0 and 1, with binary addition and multiplication as operations. Many of the results for $\mathbb{F}_2$ can be extended to more general finite fields.

In this chapter, §10 and §11 contain summaries of the necessary requirements from group theory and from ring theory, in particular polynomial rings. For a detailed exposition see any book on modern algebra or on group/ring theory. The core of the present chapter consists of §13 on finite fields and §14 on irreducible polynomials. We apply some of the results herein to the problem of factorization of polynomials over finite fields (§15).

§10 Some Group Theory

In the first two chapters we have studied lattices as algebraic structures with two "operations," join and meet. We now turn to other structures with one or two operations, which will remind us more of numbers and how to do computations with them.

Adding two numbers means that one computes a new number out of two given ones. The same point of view applies to subtraction, multiplication, and so on. Hence a "way of computing," we call it a ***binary operation*** on a set S, is simply a map from $S \times S$ to S.

10.1. Definition. A ***group*** is a set G together with a binary operation $\circ\colon G \times G \to G$ on G with the following properties. We write $g \circ h$ instead of $\circ(g, h)$.

(i) $\circ$ is ***associative***, i.e., $f \circ (g \circ h) = (f \circ g) \circ h$ for all $f, g, h \in G$.

(ii) There exists a ***neutral element*** $n \in G$, i.e., $n \circ g = g \circ n = g$ for all $g \in G$.

(iii) For every $g \in G$ there is some $h \in G$ with $g \circ h = h \circ g = n$, where n is the neutral element.

We then frequently write $(G, \circ)$, or simply G, to denote a group.

The Bibliography contains a small sample of the great number of excellent texts on group theory. We develop this theory only far enough to be able to discuss its applications. More on groups will come in Chapter 6.

The neutral element in 10.1(ii) is easily seen to be unique; for let n' be also neutral then $n = n \circ n' = n'$. Also, for each $g \in G$ there exists precisely one $h \in G$ with $g \circ h = h \circ g = n$ (see Exercise 1). We then call h the ***inverse*** to g and denote it by $h = g^{-1}$ (unless indicated otherwise). If $(G, \circ)$ also fulfills the law

$$g \circ h = h \circ g \qquad \text{for all } g, h \in G,$$

then $(G, \circ)$ is called a ***commutative*** or ***abelian group***, in honor of N. H. Abel (1802–1829) who used groups in order to show that for equations of degree ≥ 5 there cannot be any "solution formula." Similar to lattices (see 1.9), the number $|G|$ of elements of G is called the ***order*** of the group $(G, \circ)$.

10.2. Examples.

(i) The integers $\mathbb{Z}$, the set $\mathbb{Q}$ of all rational numbers, the collection $\mathbb{R}$ of all real numbers, the complex numbers $\mathbb{C}$, and $n\mathbb{Z} := \{nz \mid z \in \mathbb{Z}\}$ for $n \in \mathbb{N}$ are all abelian groups of infinite order w.r.t. addition $+$. The neutral element is always 0, the inverse of g is $-g$. Note that $1\mathbb{Z} = \mathbb{Z}$ and that $2\mathbb{Z}$ is the set of even numbers.

(ii) The nonzero elements of $\mathbb{Q}$, $\mathbb{R}$, and $\mathbb{C}$ form abelian "multiplicative" groups $(\mathbb{Q}^*, \cdot)$, $(\mathbb{R}^*, \cdot)$, $(\mathbb{C}^*, \cdot)$ of infinite order. The neutral element is 1, the inverse of g is $\frac{1}{g}$.

(iii) Let X be a nonempty set and $S_X := \{f\colon X \to X \mid f \text{ is bijective}\}$. If $\circ$ denotes the composition of functions, $(S_X, \circ)$ is a group with neutral element id_X; the inverse to f is the inverse map f^{-1}. If $|X| = n \in \mathbb{N}$, then S_X is a group of order $n!$; it is nonabelian (except for $n \leq 2$), called the ***symmetric group*** on X, and also denoted by S_n.

(iv) Similarly, the set of all nonsingular $n \times n$-matrices ($n > 1$) over the reals forms a nonabelian group $(\mathrm{GL}(n, \mathbb{R}), \cdot)$ w.r.t. matrix multiplication, called the ***general linear group***. This group has infinite order, neutral element $\mathbf{I}$ (the identity matrix), and $\mathbf{A}^{-1}$ is the inverse of the matrix, $\mathbf{A} \in \mathrm{GL}(n, \mathbb{R})$.

(v) A rather "exotic" group which relates to Chapter 1 is the power set $\mathcal{P}(X)$ of a set X w.r.t. the symmetric difference Δ (see 9.10). The neutral element is the empty set, and every element (= subset of X) S is inverse to itself since $S \,\Delta\, S = \emptyset$ for all $S \in \mathcal{P}(X)$. The group $(\mathcal{P}(X), \Delta)$ is abelian and of order $2^{|X|}$ if $|X|$ is finite (otherwise it is of infinite order).

(vi) The set D_n of all rotations and reflections of a regular n-gon consists of the n rotations id, $a :=$ rotation by $\frac{360°}{n}$ about the center, $a \circ a =: a^2 =$ rotation by $2\frac{360°}{n}$, ..., a^{n-1}, and n reflections on the n "symmetry axes." If b is one of these reflections, then

$$D_n = \{\mathrm{id}, a, a^2, \ldots, a^{n-1}, b, a \circ b =: ab, a^2 b, \ldots, a^{n-1} b\}.$$

We can check that $(D_n, \circ)$ is a group of order $2n$, called the ***dihedral group*** of degree n. We have $ba = a^{n-1}b$, so D_n is nonabelian if $n \geq 3$. If $n = 4$, for instance, D_n consists of the rotations by 0°, 90°, 180°, and 270°, and of the reflections on the dotted axes:

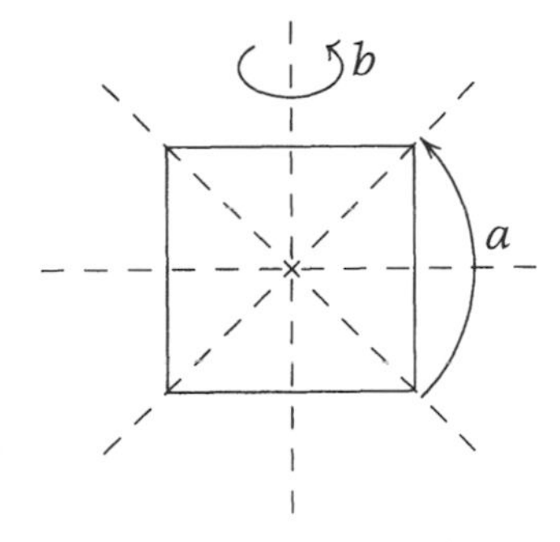

$$D_4 = \{\mathrm{id}, a, a^2, a^3, b, ab, a^2b, a^3b\} \quad \text{and} \quad ba = a^3b.$$

(vii) A "very small" example is $(\{0\}, +)$, a group of order 1.

(viii) $\{0, 1, 2, \ldots, 9\}$ does not form a group w.r.t. addition because, e.g., $7 + 3 = 10 \notin \{0, 1, 2, \ldots, 9\}$.

(ix) $\mathbb{Z}$ is not a group w.r.t. subtraction because subtraction is not associative. For instance, $3 - (7 - 9) \neq (3 - 7) - 9$.

If $G = \{g_1, g_2, \ldots, g_m\}$ is a finite group, then we can describe the group $(G, \circ)$ via its ***group table***, similar to Figure 1.10,

$\circ$	g_1	$\cdots$	g_j		g_m
g_1	$g_1 \circ g_1$	$\cdots$	$g_1 \circ g_j$	$\cdots$	$g_1 \circ g_m$
$\vdots$	$\cdots$	$\cdots$	$\cdots$	$\cdots$	$\cdots$
g_i	$g_i \circ g_1$	$\cdots$	$g_i \circ g_j$	$\cdots$	$g_i \circ g_m$
$\vdots$	$\cdots$	$\cdots$	$\cdots$	$\cdots$	$\cdots$
g_m	$g_m \circ g_1$	$\cdots$	$g_m \circ g_j$	$\cdots$	$g_m \circ g_m$

where all "products" $g_i \circ g_j$ are again in G. Associativity is not easily visible, but the row and the column of the identity must coincide with the margin row (column). Since each equation $x \circ g_i = g_j$ is uniquely solvable as $x = g_j \circ g_i^{-1}$, each row in the group table contains each element of G exactly once, and the same applies to the columns.

For example, the group table of $(\mathcal{P}(\{a, b\}), \Delta)$ is given by

Δ	$\emptyset$	$\{a\}$	$\{b\}$	$\{a, b\}$
$\emptyset$	$\emptyset$	$\{a\}$	$\{b\}$	$\{a, b\}$
$\{a\}$	$\{a\}$	$\emptyset$	$\{a, b\}$	$\{b\}$
$\{b\}$	$\{b\}$	$\{a, b\}$	$\emptyset$	$\{a\}$
$\{a, b\}$	$\{a, b\}$	$\{b\}$	$\{a\}$	$\emptyset$

From the group table we also see immediately if the group is abelian: this happens iff the group table is symmetric w.r.t. the "main diagonal."

Group tables for all groups of order ≤ 32, along with a lot of most useful information, can be found in Thomas & Wood (1980). For sets and for lattices, we have some most important concepts: subsets (sublattices), cartesian (direct) products, maps (homomorphisms), and factor sets (lattices). We shall now transfer these concepts to groups.

10.3. Definition. Let $(G, \circ)$ be a group and $S \subseteq G$ so that $S \circ S := \{s_1 \circ s_2 \mid s_1, s_2 \in S\} \subseteq S$. If $(S, \circ)$ is itself a group, then it is called a ***subgroup*** of $(G, \circ)$; we denote this by $S \leq G$. If $S \neq G$, then we write $S < G$.

Note the difference between $\leq$ and $\subseteq$. We always have $G \leq G$ and $\{n\} \leq G$ if n is the neutral element. More interesting situations are $(n\mathbb{Z}, +) < (\mathbb{Z}, +) < (\mathbb{Q}, +) < (\mathbb{R}, +) < (\mathbb{C}, +)$ and $(\mathbb{Q}^*, \cdot) < (\mathbb{R}^*, \cdot) < (\mathbb{C}^*, \cdot)$. Also, if $X \subseteq Y$, $(\mathcal{P}(X), \Delta) \leq (\mathcal{P}(Y), \Delta)$, which we also write as $\mathcal{P}(X) \leq \mathcal{P}(Y)$ if no confusion can arise. The following result gives a quick way of recognizing subgroups:

10.4. Theorem (*Subgroup Criterion*). *Let $(G, \circ)$ be a group and S a nonempty subset of G. Then S is a subgroup of G if $s \circ t^{-1} \in S$ for all $s, t \in S$.*

The proof should be given as Exercise 17. Note that the neutral element n of G is also in S via $n = s \circ s^{-1}$ for $s \in S$. Hence every subgroup of G has the same neutral element as G. Similarly, if $s \in S \leq G$, the inverse of s in S is the same as the inverse of s in G.

If S_i $(i \in I)$ are subgroups of $(G, \circ)$ and $S := \bigcap_{i \in I} S_i$, let $s, t \in S$. Then $s, t \in S_i$ for all i, hence $s \circ t^{-1} \in S_i$ for all i, hence $s \circ t^{-1} \in S$. Since $n \in S$, we have shown (using the subgroup criterion 10.4):

10.5. Theorem. *The intersection of subgroups of G is again a subgroup of G.*

This gives rise to another important concept. Let $(G, \circ)$ be a group and X a subset of G. By 10.5, the intersection $\langle X \rangle$ of all subgroups of G containing X is again a subgroup, and obviously it is the smallest subgroup containing X. We call $\langle X \rangle$ the ***subgroup generated by*** X. If $X = \{x_1, \dots, x_n\}$, we write $\langle X \rangle = \langle x_1, \dots, x_n \rangle$.

The subgroup $\langle 1 \rangle$ generated by 1 in $(\mathbb{Q}, +)$ is $\mathbb{Z}$, 2 generates $\{2^z \mid z \in \mathbb{Z}\}$ in $(\mathbb{Q}^*, \cdot)$, and so on. It is a little less trivial that in $(\mathbb{Z}, +)$ we have $\langle 14, 16, 20 \rangle = \langle 2 \rangle = 2\mathbb{Z}$, for instance.

There a general principle applies. Take any kind of "algebraic structure" A (lattices, Boolean algebras, groups, vector spaces,...) and a collection $\mathcal{C}$ of subsets which is closed w.r.t. intersections and with $A \in \mathcal{C}$. If $X \subseteq A$, then $\langle X \rangle_{\mathcal{C}} = \bigcap C$, where the intersection runs over all $C \in \mathcal{C}$ which contain X, is called the element $\in \mathcal{C}$ ***generated*** by X. Hence we can speak of "generated sublattices," "generated Boolean subalgebras," "generated subgroups," "generated subspaces" (i.e., linear hulls), and later about "generated subrings," "generated ideals," "generated subfields," and so on.

If $C \in \mathcal{C}$ can be generated by a finite set X, we call it ***finitely generated***. A group G which can be generated by a single element g is called ***cyclic*** and g is called a ***generator***. We shall see shortly what all cyclic groups look like (see 10.19).

Now we turn to products.

10.6. Theorem and Definition. *Let $(G_1, \circ_1)$ and $(G_2, \circ_2)$ be groups and let $G := G_1 \times G_2$ be their cartesian product. On G we define an operation $\circ$ via*

$$(g_1, g_2) \circ (h_1, h_2) := (g_1 \circ_1 h_1, g_2 \circ_2 h_2).$$

Then $(G, \circ)$ is again a group, called the ***direct product*** *of G_1 and G_2, and is denoted by $G = G_1 \times G_2$ or $G = G_1 \oplus G_2$.*

The proof that $(G, \circ)$ is a group is an easy exercise. The neutral element is (n_1, n_2), where n_i is neutral in G_i $(i = 1, 2)$. The inverse of (g_1, g_2) is (g_1^{-1}, g_2^{-1}). It is easy to see that $G_1 \times G_2$ is abelian iff G_1 and G_2 are abelian.

It is clear how to define the direct product of more than two groups. If infinitely many groups G_i $(i \in I)$ are involved, we write $\prod_{i \in I} G_i$ for their product. The subgroup $\bigoplus_{i \in I} G_i$ consisting of all elements $(\ldots, g_i, \ldots)$, where only finitely many of the g_i are not neutral, is called the ***direct sum*** of the groups G_i. Clearly, there is no difference between direct sums and direct products if I is finite.

Now we turn to "group-homomorphisms."

10.7. Definitions. Let $(G, \circ)$ and $(G', \circ')$ be groups. A map $f\colon G \to G'$ is called a ***homomorphism*** if $f(g \circ h) = f(g) \circ' f(h)$ holds for all $g, h \in G$. As in 1.17 (and the following lines), injective, surjective, and bijective homomorphisms are called ***monomorphism***, ***epimorphisms***, and ***isomorphisms***, respectively. Monomorphisms are also called ***embeddings***; if there exists an embedding from G to G', we say that G can be ***embedded*** in G', and denote this by $G \hookrightarrow G'$. If there exists an isomorphism from G to G', we call G and G' ***isomorphic*** and denote this by $G \cong G'$.

So isomorphic groups just differ in the "names" for the elements, the rule of calculation is the same in G and G', since the homomorphism property in 10.7 says that it makes no difference if we compose g and h first and then send the result to G', or if we first map g and h into G' and composes them there. Again, the concepts defined in 10.7 make sense for all algebraic structures which we shall meet on our way. As well as generated substructures, direct products, etc., they are part of the theory of "universal algebra." We shall describe applications of universal algebras in §40. The following properties are easy to check.

10.8. Theorem. *Let $f\colon G \to G'$ be a homomorphism and let n, n' be the neutral elements of G, G'.*

(i) $f(n) = n'$;

(ii) $f(g^{-1}) = f(g)^{-1}$ *for all* $g \in G$;

(iii) $\operatorname{Ker} f := \{g \in G \mid f(g) = n'\} \leq G$;

(iv) f *is a monomorphism* $\iff \operatorname{Ker} f = \{n\}$;

(v) $\operatorname{Im} f := \{f(g) \mid g \in G\} \leq G'$;

(vi) f *is an epimorphism* $\iff \operatorname{Im} f = G'$.

$\operatorname{Ker} f$ is called the ***kernel*** of f and $\operatorname{Im} f = f(G)$ is the ***image*** of f. If $G \hookrightarrow G'$ by some monomorphism f, then $G \cong f(G) \leq G'$, which means that by "changing the name of the elements," G is made into a subgroup $f(G)$ of G'; this explains the notation of "embedding":

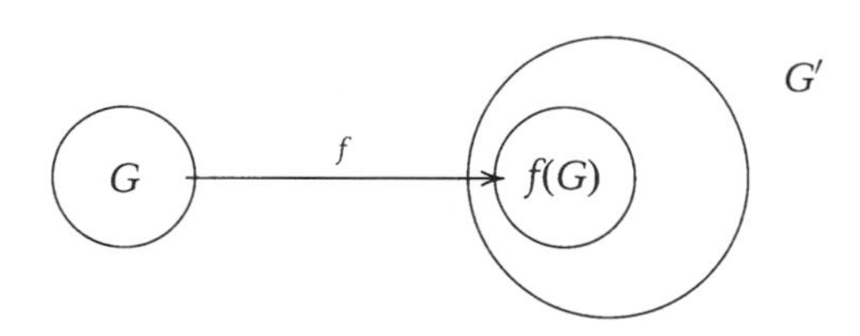

FIGURE 10.1

Next, we look at factors. Let $(G, \circ)$ be a group and let $\sim$ be an equivalence relation in G. Then we can form the equivalence classes $[g] := \{h \in G \mid h \sim g\}$ of the factor set $G/\sim$. It is tempting, but very dangerous, to define an operation $\circledcirc$ in $G/\sim$ via $[g] \circledcirc [h] := [g \circ h]$. Why is it dangerous? If $g \sim g'$ and $h \sim h'$, then $[g] = [g']$ and $[h] = [h']$. But can we be sure that $[g] \circledcirc [h] = [g \circ h]$ and $[g'] \circledcirc [h'] = [g' \circ h']$ will be the same, i.e., that $g \circ h \sim g' \circ h'$? We look at an example. Let $(G, \circ) = (\mathbb{Z}, +)$ and let $g \sim g' \iff |g| = |g'|$. Then, for instance, $3 \sim 3$ and $4 \sim -4$, but $7 = 3 + 4 \not\sim 3 + (-4) = -1$. So $[3] + [4] = [3] + [-4]$ would have to be equal to the two different values [7] and [−1] at the same time, which contradicts the definitions of + as a function.

10.9. Definition. Let $(G, \circ)$ be a group and $\sim$ an equivalence relation in G. Then $\sim$ is called ***compatible*** with $\circ$ or a ***congruence relation*** (or ***congruence***) if, for all $g, g', h, h' \in G$,

$$g \sim g' \text{ and } h \sim h' \Longrightarrow g \circ h \sim g' \circ h'.$$

10.10. Theorem and Definition. *Let $\sim$ be a congruence in $(G, \circ)$. Then we can define the operation $\circledcirc$ on $G/\sim$ via*

$$[g] \circledcirc [h] := [g \circ h]$$

and $(G/\sim, \circledcirc)$ is again a group, called the ***factor group*** *of $(G, \circ)$ w.r.t. $\sim$, it is briefly denoted by $G/\sim$. The neutral element is $[n]$, and $[g]^{-1} = [g^{-1}]$.*

If G is abelian, the same applies to $G/\sim$. For us, the most important examples are the following.

10.11. Example. $\mathbb{Z}_n$. Let n be a positive integer and let $\equiv_n$ denote the following relation on $\mathbb{Z}$: $a \equiv_n b :\Longleftrightarrow n$ divides $a-b$ (denoted by $n \mid a-b$). Then $\equiv_n$ is an equivalence relation on $\mathbb{Z}$, called the ***congruence modulo*** n. The equivalence classes are denoted by $[a]$, or by $[a]_n$. We also write $a \equiv b \pmod n$ for $a \equiv_n b$. Thus $[a]_n$ is the set of all integers z which on division by n give the same remainder as a gives. Therefore $[a]$ is also called the ***residue class of*** a ***modulo*** n. We have

$$\begin{aligned}
[0] &= \{0,\ n,\ -n,\ 2n,\ -2n,\ 3n,\ -3n,\ \ldots\},\\
[1] &= \{1,\ n{+}1,\ -n{+}1,\ 2n{+}1,\ -2n{+}1,\ 3n{+}1,\ -3n{+}1,\ \ldots\},\\
[2] &= \{2,\ n{+}2,\ -n{+}2,\ 2n{+}2,\ -2n{+}2,\ 3n{+}2,\ -3n{+}2,\ \ldots\},\\
&\ \vdots\\
[n{-}1] &= \{n{-}1,\ n{+}n{-}1,\ -n{+}n{-}1, 2n{+}n{-}1,\ -2n{+}n{-}1,\ 3n+n-1,\ \ldots\}.
\end{aligned}$$

There are no more: $[n] = [0]$, $[n+1] = [1], \ldots$. In general, $[k] = [a]$ for $a \in \{0, 1, \ldots, n-1\}$, if k divided by n gives the remainder a. The equivalence relation $\equiv_n$ on $\mathbb{Z}$ satisfies

$$a \equiv_n b \text{ and } c \equiv_n d \Longrightarrow a + c \equiv_n b + d \text{ and } ac \equiv_n bd. \tag{10.1}$$

In particular, $\equiv_n$ is a congruence in $(\mathbb{Z}, +)$. Hence we can define the operation $\oplus$ on $\mathbb{Z}_n := \{[0], \ldots, [n-1]\}$ by

$$[a] \oplus [b] = [a + b],$$

and from now on we will simply write $[a]+[b]$ or even $a+b$ if the context makes it clear. For instance, the group table of $\mathbb{Z}_5 = \{0, 1, 2, 3, 4\}$ is given by

+	0	1	2	3	4
0	0	1	2	3	4
1	1	2	3	4	0
2	2	3	4	0	1
3	3	4	0	1	2
4	4	0	1	2	3

In 10.11, $\equiv_n$ is related to a subgroup of $(\mathbb{Z}, +)$, namely to $[0]_n = n\mathbb{Z}$. In fact, all congruences come from certain subgroups.

10.12. Theorem. *Let $(G, \circ)$ be a group.*

(i) *If $\sim$ is a congruence in $(G, \circ)$, then the class $[n]$ of the neutral element is a subgroup N of G with*

$$g \circ m \circ g^{-1} \in N \quad \textit{for all} \quad m \in N \quad \textit{and} \quad g \in G. \tag{10.2}$$

(ii) *Conversely, if $N \leq G$ fulfills (10.2), then*

$$g \sim_N h :\iff g \circ h^{-1} \in N$$

gives a congruence $\sim_N$ on $(G, \circ)$.

It is instructive to prove this as Exercise 6. The implications (i)→(ii) and (ii)→(i) in 10.12 are "inverse to each other": if we start with $\sim$, and find N as in 10.12(i), then $\sim_N = \sim$, and similarly for the other "path."

10.13. Definition. A subgroup N of a group G with (10.2) is called a ***normal subgroup***, and we denote this by $N \trianglelefteq G$.

10.14. Examples.

(i) Every subgroup of an abelian group is normal.

(ii) For every homomorphism f, the kernel $\operatorname{Ker} f$ is normal (this does not hold for the image, though).

(iii) Let $N := \{\mathbf{A} \in \mathrm{GL}(n, \mathbb{R}) \mid \det \mathbf{A} = 1\}$. Since $\det(\mathbf{AB}^{-1}) = (\det \mathbf{A})(\det \mathbf{B})^{-1}$, $N < \mathrm{GL}(n, \mathbb{R})$ by the subgroup criterion 10.4. Also, 10.2 is satisfied since if $\mathbf{A} \in N$, $\mathbf{B} \in \mathrm{GL}(n, \mathbb{R})$ we have $\det(\mathbf{BAB}^{-1}) = (\det \mathbf{B}) \cdot 1 \cdot (\det \mathbf{B})^{-1} = 1$.

(iv) $\{n\}$ and G are always normal subgroups of G.

10.15. Notation. If the congruence $\sim$ in $(G, \circ)$ "comes" from the normal subgroup N as $\sim\, = \,\sim_N$, according to 10.12, we write $G/{\sim_N}$ as G/N.

10.16. Theorem (*Homomorphism Theorem for Groups*). *Let $(G, \circ)$ be a group.*

(i) *If $N \trianglelefteq G$, then $f\colon G \to G/N;\ g \mapsto [g]$ is an epimorphism.*

(ii) *Conversely, if $e\colon G \to G'$ is an epimorphism, then $\operatorname{Ker} e \trianglelefteq G$ and $G/\operatorname{Ker} e \cong G'$.*

While (i) is easy to check directly, $f\colon G/\operatorname{Ker} e \to G';\ [g] \mapsto e(g)$ will do the job in (ii). The reader is strongly encouraged to work out the details.

The last results tell us that congruences, normal subgroups, and homomorphic images are "all the same" (in the sense described in 10.12 and 10.16). If you know one of them, you know the two others.

Let us go back to 10.12. Even if N is not normal, $\sim_N$ in 10.12 turns out to be an equivalence relation. What do the equivalence classes $[g]$ look like? Now $[g] = \{h \in G | h \sim_N g\} = \{h \in G | h \circ g^{-1} \in N\} = \{h \in G | h \in N \circ g\} = N \circ g$, where $N \circ g := \{x \circ g \mid x \in N\}$. So the equivalence classes $[g]$ are of the form $N \circ g$ and are called ***cosets***. As usual for equivalence classes, two cosets $N \circ g_1$ and $N \circ g_2$ are either equal (iff $g_1 \sim_N g_2$) or else disjoint. This will be the reason for a clever decoding method for error-correcting codes (see §17). $[G : N]$ denotes the number of different cosets of G w.r.t. N, and is called the ***index*** of N in G.

Furthermore, the map $\varphi\colon N \to N \circ g;\ x \mapsto x \circ g$ is easily seen to be bijective (Exercise 7). So any two cosets are equipotent (to N), and we get:

10.17. Theorem (*Lagrange's Theorem*). *If $N \leq G$ and G is finite, then $|N|$ is a divisor of $|G|$. More precisely, $|N|[G : N] = |G|$.*

So a group of order 50 can only have subgroups of order 1, 2, 5, 10, 25, and 50.

10.18. Corollary. *A group of prime order cannot have any "proper" subgroup.*

For determining the structure of fields we need a characterization of cyclic and of finite abelian groups. Suppose $G = \langle g \rangle$, the group generated by a single element g. Then $g, g \circ g =: g^2, g^3, \ldots$, and also $g^0 = n, g^{-1}, g^{-2}, g^{-3}, \ldots$ must be in $\langle g \rangle$. These elements $\{g^z \mid z \in \mathbb{Z}\}$ do form a subgroup, so

$\langle g\rangle = \{g^z \mid z \in \mathbb{Z}\}$. The map $z \mapsto g^z$ yields an epimorphism from $(\mathbb{Z}, +)$ to $(\langle g\rangle, \circ)$ since $g^{z+z'} = g^z \circ g^{z'}$.

Case I: All powers g^z are distinct. Then $G = \langle g\rangle \cong (\mathbb{Z}, +)$.

Case II: Some $g^s = g^t$. Without loss of generality we might assume that $s, t \in \mathbb{N}_0$ and $s < t$. Then $g^{t-s} = g^t \circ (g^s)^{-1} = n$. Let m be the minimal number $\in \mathbb{N}$ such that $g^m = n$. Then $n = g^0, g^1, \ldots, g^{m-1}$ are distinct. If $z \in \mathbb{Z}$, then $z = mq + r$ with $0 \leq r < m$ for some $q \in \mathbb{Z}$, whence $g^z = (g^m)^q \circ g^r = n^q \circ g^r = g^r$. Hence $\langle g\rangle = \{n = g^0, g^1, \ldots, g^{m-1}\} \cong (\mathbb{Z}_m, +)$.

We have seen

10.19. Theorem. *All cyclic groups are, up to isomorphism, given by* $\mathbb{Z}_1, \mathbb{Z}_2, \mathbb{Z}_3, \ldots,$ *and* $\mathbb{Z}$ *(all w.r.t. addition).*

We show for later use:

10.20. Theorem. *Let* $G = \langle g\rangle$ *be cyclic of order* m*. Then all generators of* G *are given by* g^k *with* $\gcd(k, m) = 1$.

Proof. If $d = \gcd(k, m) > 1$, then the powers of g^d are g^0, g^d, g^{2d}, $\ldots$, $g^{m/d-1}$, so g^d cannot be a generator. On the other hand, if $d = 1$, every $g^i \in \langle g\rangle$ is a power of g^k, because there are $r, s \in \mathbb{Z}$ with $kr + ms = 1$, hence $g = g^{kr+ms} = (g^k)^r(g^m)^s = (g^k)^r \in \langle g^k\rangle$, so each $g^i \in \langle g^k\rangle$. □

10.21. Definition. The ***order*** $\text{ord}(g)$ of an element $g \in G$ is the order of $\langle g\rangle$.

Hence $\text{ord}(g)$ is the smallest natural number k such that $g^k = n$ (if such a k exists). If G is finite, then $\text{ord}(g)$ divides $|G|$ by 10.17.

10.22. Corollary (*Fermat's Little Theorem*). *If* $g \in G$, G *a finite group, then* $g^{|G|} = n$.

Clearly, each cyclic group is abelian, and so are direct sums of cyclic groups. It can be shown that we get all finite abelian groups in this way. The proofs of the following results can be found in most introductory books on group theory or on modern algebra.

10.23. Theorem. $\mathbb{Z}_{m_1} \times \mathbb{Z}_{m_2}$ *is cyclic* $\iff \gcd(m_1, m_2) = 1$. *In this case,* $\mathbb{Z}_{m_1} \times \mathbb{Z}_{m_2} \cong \mathbb{Z}_{m_1 m_2}$.

10.24. Example. $\mathbb{Z}_{60} \cong \mathbb{Z}_4 \times \mathbb{Z}_3 \times \mathbb{Z}_5$. (In §24, we shall see that these decompositions give rise to faster ways of calculating with integers.)

10.25. Theorem (*Principal Theorem on Finite Abelian Groups*). *Every finite abelian group is isomorphic to the direct product of groups of the type $\mathbb{Z}_{p^k}$ (p prime).*

By rearranging the direct factors properly this principal theorem can be written in a different form.

10.26. Theorem and Definition. *Let G be a finite abelian group $\neq \{n\}$. Then there are natural numbers $s, d_1, d_2, \ldots, d_s$ with $d_1 > 1$ and $d_i \mid d_{i+1}$ for $1 \leq i < s$ such that*

$$G \cong \mathbb{Z}_{d_1} \times \mathbb{Z}_{d_2} \times \cdots \times \mathbb{Z}_{d_s}.$$

These numbers are uniquely determined by G and are called the ***elementary divisors*** *of G.*

10.27. Example. We want to find all abelian groups with 100 elements. By 10.26, we have to find all possible $s, d_1, \ldots, d_s$ with $d_i \mid d_{i+1}$ and $d_1 d_2 \ldots d_s = 100$.

If $s = 1$, then $d_1 = 100$, and we get $\mathbb{Z}_{100}$.
If $s = 2$, then $100 = 2 \cdot 50$ gives $\mathbb{Z}_2 \times \mathbb{Z}_{50}$,
$100 = 5 \cdot 20$ gives $\mathbb{Z}_5 \times \mathbb{Z}_{20}$,
$100 = 10 \cdot 10$ gives $\mathbb{Z}_{10} \times \mathbb{Z}_{10}$.

$s \geq 3$ turns out to be impossible (why?), so there are precisely four abelian groups of order 100 (up to isomorphism).

10.28. Theorem. *Every group of order p (p a prime) is cyclic, hence abelian. Every group of order p^2 is abelian, but not necessarily cyclic (take $\mathbb{Z}_p \times \mathbb{Z}_p$).*

These results show that a complete list of all abelian groups (up to isomorphisms) of order ≤ 15 is given by $\mathbb{Z}_1, \mathbb{Z}_2, \ldots, \mathbb{Z}_{15}$, plus the noncyclic groups $\mathbb{Z}_2 \times \mathbb{Z}_2$ (order 4), $\mathbb{Z}_2 \times \mathbb{Z}_4$ and $\mathbb{Z}_2 \times \mathbb{Z}_2 \times \mathbb{Z}_2$ (order 8), $\mathbb{Z}_3 \times \mathbb{Z}_3$ (order 9), and $\mathbb{Z}_2 \times \mathbb{Z}_6$ (order 12). It is less easy to see that there are precisely eight nonabelian groups of order ≤ 15: S_3 (order 6), D_4 and Q (see 11.8) (order 8), D_5 (order 10), D_6, A_4, and a further group (order 12), and D_7 (order 14). There are 14 nonisomorphic groups of order 16.

Finally, we need some concepts which are only important for nonabelian groups.

10.29. Definition. Two elements g, h of a group G are ***conjugate*** if there exists $x \in G$ such that $g = xhx^{-1}$; we denote this by $g \sim h$. It is easy to see that $\sim$ is an equivalence relation. The number of equivalence classes is called the ***class number*** of G.

Clearly, g and h in an abelian group are conjugate iff they are equal. In $\mathrm{GL}(n, \mathbb{R})$, matrices are conjugate iff they are similar; they then have the same determinant and the same eigenvalues. In 25.8(ii), we shall determine which elements in S_X are conjugate. Generally, conjugate elements "behave very similarly." The bigger the equivalence classes, the smaller is the class number of G, the "more abelian" is G. Here is another measure of "how nonabelian" a group is:

10.30. Definition. The ***center*** $C(G)$ of a group is the set of all $c \in G$ such that $g \circ c = c \circ g$ for all $g \in G$.

$(G, \circ)$ is abelian iff $C(G) = G$. The center of $\mathrm{GL}(n, \mathbb{R})$ can be seen to be $\{\lambda \mathbf{I} \mid \lambda \in \mathbb{R}\}$, and the center of S_X will be determined in 25.8(v): for $|X| \geq 3$, $C(S_X) = \{\mathrm{id}\}$.

10.31. Definition. A group G which has no normal subgroups except $\{1\}$ and G is called ***simple***.

A simple abelian group is hence an abelian group without proper subgroups. By 10.18, groups of prime order p are of this type; they are isomorphic to $\mathbb{Z}_p$ by 10.28 and 10.19. Conversely, if G is abelian and simple, then each $\langle g \rangle$ is a normal subgroup for $g \in G$. From this, it is easy to see (Exercise 11) that G must be cyclic of prime order. So the collection of all $\mathbb{Z}_p$, $(p \in \mathbb{P})$ is a complete list of all simple abelian groups (up to isomorphism). The smallest nonabelian simple group has order 60; it is the "alternating group of degree 5," which we shall meet in 25.12. Apart from these "alternating groups," there are some series derived from matrix groups which yield finite simple nonabelian groups (see the lines after Theorem 37.12). Apart from those, 26 "exceptional" groups were discovered. The largest one (the "monster") has order

$$2^{46} \cdot 3^{20} \cdot 5^9 \cdot 7^6 \cdot 11^2 \cdot 13^3 \cdot 17 \cdot 19 \cdot 23 \cdot 29 \cdot 31 \cdot 41 \cdot 47 \cdot 59 \cdot 71 \approx 8 \cdot 10^{53}.$$

The proof that this gives the complete list of all finite simple groups (and hence all finite "automata"—see 29.16), was completed around 1981, and comprised about 15,000 pages in print. It was later reduced to less than 10,000 pages, but this is still far too much for a single mathematician to check it thoroughly. This Classification Theorem might be the largest combined scientific effort of mankind—and it was successful!

Exercises

1. Show that in a group each element has precisely one inverse.
2. Write down the group tables of $(S_3, \circ)$ and $(D_4, \circ)$.
3. Do the same for $(\mathbb{Z}_2, +)$, $(\mathbb{Z}_3, +)$, $(\mathbb{Z}_2, +) \times (\mathbb{Z}_2, +)$, and $(D_3, \circ)$.
4. Show that $(\mathcal{P}(\{a, b\}), \Delta) \cong (\mathbb{Z}_2, +) \times (\mathbb{Z}_2, +)$ and $(S_3, \circ) \cong (D_3, \circ)$.
5. Prove Theorem 10.8.
6. Prove Theorem 10.12.
7. Show that any two cosets of a group are equipotent.
8. Are subgroups, arbitrary products, and homomorphic images of abelian groups again abelian? Does this also hold for "finite" instead of "abelian"?
9. Compute the order of each element in D_3 and D_4, and convince yourself that their orders divide the order of the group.
10. Find all abelian groups with 1000 elements and all groups of order 5328562009.
11. Determine all abelian simple groups.
12. Find all conjugacy classes in $(S_3, \circ)$ and $(D_4, \circ)$, and the class numbers of these groups. Find the centers of these groups.
13. Show that the group $(\mathbb{R}^*, \cdot)$ of nonzero real numbers is both a homomorphic image and (isomorphic to) a subgroup of $GL(n, \mathbb{R})$ for each $n \in \mathbb{N}$.
14. Show that subgroups of cyclic groups are cyclic and find all subgroups of $(\mathbb{Z}_{20}, +)$.
15. Study $(\mathbb{G}_n, \cdot)$ where $n \in \mathbb{N}$ and $\mathbb{G}_n := \{x \in \mathbb{Z}_n \mid \text{there is some } y \in \mathbb{Z}_n \text{ with } xy = 1\}$. Show that $(\mathbb{G}_n, \cdot)$ is a group and that $x \in \mathbb{Z}_n$ belongs to $\mathbb{G}_n$ iff $\gcd(x, n) = 1$. Cf. 28.7. The proof of 12.1 shows how to find the inverse of $x \in \mathbb{G}_n$.
16. Determine all $(\mathbb{G}_n, \cdot)$ by their group tables for $n \leq 12$.
17. Prove the subgroup criterion 10.4.
18. How many of the 14 groups of order 16 are abelian? Find them and give three examples of nonabelian groups of this order.

§11 Rings and Polynomials

In an additive group $(G, +)$ we can add and subtract. Now we also want to multiply.

11.1. Definition. A ***ring*** is a set R together with two binary operations, $+$ and $\cdot$, called ***addition*** and ***multiplication***, such that:

(i) $(R, +)$ is an abelian group;

(ii) the product $r \cdot s$ of any two elements $r, s \in R$ is in R and multiplication is associative;

(iii) for all $r, s, t \in R$: $r \cdot (s + t) = r \cdot s + r \cdot t$ and $(r + s) \cdot t = r \cdot t + s \cdot t$ (***distributive laws***).

We will then write $(R, +, \cdot)$ or simply R. In general, the neutral element in $(R, +)$ will be denoted by 0 and called ***zero***, the additive inverse of $r \in R$ will be denoted by $-r$. Instead of $r \cdot s$ we shall write rs. Let $R^* := R \setminus \{0\}$. Rings according to 11.1 are also called "***associative rings***" in contrast to "***nonassociative rings***" (where associativity of multiplication is not assumed). The "prototype" of a ring is $(\mathbb{Z}, +, \cdot)$.

11.2. Definitions. Let R be a ring. R is said to be ***commutative*** if this applies to multiplication. If there is an element 1 in R such that $r \cdot 1 = 1 \cdot r = r$ for any $r \in R$, then 1 is called an ***identity*** (or ***unit***) element. As for groups (see the lines after 10.1), we easily see that the identity is unique if it exists. If $r \neq 0, s \neq 0$ but $rs = 0$, then r is called a ***left divisor*** and s a ***right divisor of zero***. If R has no zero divisors, i.e., if $rs = 0$ implies $r = 0$ or $s = 0$ for all $r, s \in R$, then R is called ***integral***. A commutative integral ring with identity $1 \neq 0$ is called an ***integral domain***. If $(R^*, \cdot)$ is a group, then R is called a ***skew field*** or a ***division ring***. If, moreover, R is commutative, we speak of a ***field***. Hence a field is a ring $(R, +, \cdot)$ in which both $(R, +)$ and $(R^*, \cdot)$ are abelian groups. The ***characteristic*** of R is the smallest natural number k with $kr := r + \cdots + r$ (k-times) equal to 0 for all $r \in R$. We then write $k = \operatorname{char} R$. If no such k exists, we put $\operatorname{char} R = 0$. So if $k = \operatorname{char} R$, all elements in the group $(R, +)$ have an order dividing k.

Now we list a series of examples of rings. The assertions contained in this list are partly obvious, some will be discussed in the sequel; the rest is left to the reader.

11.3. Examples. $\mathbb{Z}$, $\mathbb{Q}$, $\mathbb{R}$, $\mathbb{C}$, and $n\mathbb{Z}$ (cf. 10.2(i)): All of these are integral commutative rings w.r.t. addition, $+$, and multiplication, $\cdot$; all of them have 1 as the identity, except $n\mathbb{Z}$ ($n \geq 2$), which does not contain an identity. Every nonzero element x in $\mathbb{Q}$, $\mathbb{R}$, and $\mathbb{C}$ has its multiplicative inverse $x^{-1} = \frac{1}{x}$ in the same set, respectively, so $\mathbb{Q}$, $\mathbb{R}$, and $\mathbb{C}$ are fields. $\mathbb{Z}$ is an integral domain. All of these rings have characteristic 0.

11.4. Examples. ***Matrix ring*** $\mathbb{M}_n(R)$. Let $\mathbb{M}_n(R)$ be the $n \times n$-matrix ring over a ring R with identity. Zero element is the zero matrix, identity is the identity matrix $\mathbf{I}$. If $n > 1$, $\mathbb{M}_n(R)$ is neither commutative nor integral. We have $\operatorname{char} \mathbb{M}_n(R) = \operatorname{char} R$.

11.5. Example. ***Function rings*** R^X. Let R be a ring and X a nonempty set. Then all functions from X to R form a ring w.r.t. pointwise addition and multiplication. Zero element is the zero function $\mathbf{o}\colon X \to R;\ x \mapsto 0$. If R has an identity 1, then $\bar{1}\colon X \to R;\ x \mapsto 1$ is the identity in R^X. In general (what are the exceptions?), R^X has zero divisors and is not commutative. Again, $\operatorname{char} R^X = \operatorname{char} R$.

11.6. Example. Let $(G, +)$ be an abelian group with zero 0. Define a multiplication $*$ in G via $g_1 * g_2 := 0$. Then $(G, +, *)$ is a commutative ring without identity (except $G = \{0\}$), called the ***zero-ring*** on G. This shows that every abelian group can be turned into a commutative ring.

11.7. Example. Let X be a set. Then the set $\mathcal{P}(X)$ of all subsets of X is a commutative ring w.r.t. the symmetric difference $\triangle$ as addition and with intersection as multiplication. Zero element is $\emptyset$, identity element is X. Since $A \cap A' = \emptyset$ for all $A \in \mathcal{P}(X)$, where A' denotes the complement of A, this ring consists only of zero divisors (and $\emptyset$), and since $A \triangle A = \emptyset$ for all $A \in \mathcal{P}(X)$, it has characteristic 2 (except $X = \emptyset$).

11.8. Examples. Let $Q := (\mathbb{R}^4, +, \cdot)$ with componentwise addition and with multiplication defined by

$$\begin{aligned}(a, b, c, d) \cdot (a', b', c', d') := (&aa' - bb' - cc' - dd',\ ab' + ba' + cd' - dc',\\ &ac' - bd' + ca' + db',\ ad' + bc' - cb' + da').\end{aligned}$$

With $1 := (1, 0, 0, 0)$, $i := (0, 1, 0, 0)$, $j := (0, 0, 1, 0)$, $k := (0, 0, 0, 1)$ we get $i^2 = j^2 = k^2 = -1$, $ij = k$, $jk = i$, $ki = j$, $ji = -k$, $kj = -i$, $ik = -j$, and

$$Q = \{a1 + bi + cj + dk \mid a, b, c, d \in \mathbb{R}\}.$$

Hence Q can be considered as a four-dimensional vector space over $\mathbb{R}$. $(Q, +, \cdot)$ is a skew-field; it is not a field and can be viewed as an "extension" of $\mathbb{C}$ to a system with three "imaginary units" i, j, k. The structure $(\{\pm 1, \pm i, \pm j, \pm k)\}, \cdot)$ is a group of order 8, called the ***quaternion group***. $\mathbb{Q}$ is called the ***skew-field of quaternions***.

We state some less trivial connections between these concepts. For proofs of (i), (iii), and (iv), see Exercise 7. Part (ii) is harder (see, e.g., Rowen (1988) or Herstein (1975)).

11.9. Theorem.

(i) *Every finite integral domain is a field.*

(ii) *Every finite skew-field is a field (**Wedderburn's Theorem**).*

(iii) *If R is an integral domain, then* $\operatorname{char} R$ *is prime.*

(iv) *Every field is an integral domain.*

By 11.9(i) and (ii), finite fields hence coincide with finite integral domains and with finite skew-fields. They will be described explicitly in §13. An example of a proper skew-field was given in Example 11.8.

Let R be a ring and let $S \subseteq R$. S is called a ***subring*** of R (denoted by $S \leq R$) if S is a ring with respect to the operations of R, restricted to S.

We have seen in 10.12 that every congruence in a group comes from a certain subgroup (i.e., from a normal one). What is the counterpart of normal subgroups in ring theory? 10.12(i) provides a hint: look at the class $[0]$ of the zero element.

11.10. Theorem. *Let $(R, +, \cdot)$ be a ring.*

(i) *If $\sim$ is a congruence in $(R, +, \cdot)$, then $[0]$ is a subring I of R with the property that for all $r \in R$ and $i \in I$*

$$ri \in I \text{ and } ir \in I. \tag{11.1}$$

(ii) *Conversely, if $I \leq R$ fulfills (11.1), then*

$$r \sim_I s :\Longleftrightarrow r - s \in I$$

gives a congruence on $(R, +, \cdot)$.

Again, the reader is urged to work out the proof in detail. The "hierarchy" of rings is given in Figure 11.1.

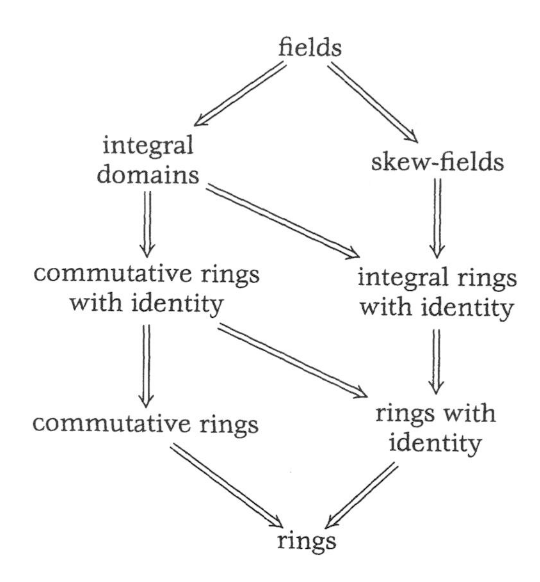

FIGURE 11.1

11.11. Definitions.

(i) A subring I of a ring R with (11.1) is called an ***ideal*** of R; we denote this by $I \trianglelefteq R$.

(ii) If the congruence $\sim$ "comes" from the ideal I, as $\sim = \sim_I$, then we write R/I instead of $R/\sim_I$.

The condition (11.1) is often written as "$IR \subseteq I$ and $RI \subseteq I$." We put $I \triangleleft R$ if $I \trianglelefteq R$ but $I \neq R$.

For instance, $n\mathbb{Z} \trianglelefteq \mathbb{Z}$ for all $n \in \mathbb{N}_0$. On the other hand, $\mathbb{Z}$ is a subring, but not an ideal of $\mathbb{Q}$. In fact, if an ideal I contains the identity 1 of R, then clearly $I = R$. Moreover, if F is a skew-field and $I \neq \{0\}$ is an ideal of F, take $i \in I^*$. Then $i^{-1}i \in I$, so $1 \in I$ and $I = F$. We see

11.12. Theorem and Definition. *A skew-field (and hence every field) F is **simple** in the sense that it does not contain any ideals (except $\{0\}$ and F).*

If F is a field, then $\mathbb{M}_n(F)$ can be shown to be simple for any $n \in \mathbb{N}$. So there do exist simple rings which are not (skew-)fields.

Similar to 10.10, every congruence relation $\sim$ in a ring R gives rise to a ***factor ring*** $R/\sim$, in which the equivalence classes $[r]$, $[s]$ are added and multiplied via

$$[r] \oplus [s] := [r + s],$$
$$[r] \odot [s] := [rs].$$

If no confusion is likely, we shall briefly write $(R/\sim, +, \cdot)$ or simply $R/\sim$ for this factor ring.

11.13. Example. Similar to 10.11, the relation $\equiv_n$ is also compatible with the multiplication in $\mathbb{Z}$. So we get factor rings $(\mathbb{Z}/\equiv_n, \oplus, \odot)$, again briefly denoted by $\mathbb{Z}_n$. We give addition and multiplication tables for $\mathbb{Z}_2$, $\mathbb{Z}_3$, and $\mathbb{Z}_4$; instead of $[a]$, we briefly write a.

$\mathbb{Z}_2$:

+	0	1
0	0	1
1	1	0

·	0	1
0	0	0
1	0	1

$\mathbb{Z}_3$:

+	0	1	2
0	0	1	2
1	1	2	0
2	2	0	1

·	0	1	2
0	0	0	0
1	0	1	2
2	0	2	1

$\mathbb{Z}_4$:

+	0	1	2	3
0	0	1	2	3
1	1	2	3	0
2	2	3	0	1
3	3	0	1	2

·	0	1	2	3
0	0	0	0	0
1	0	1	2	3
2	0	2	0	2
3	0	3	2	1

Observe that in $\mathbb{Z}_4$ we have nontrivial divisors of zero, since $2 \cdot 2 = 0$. Hence $\mathbb{Z}_4$ is not integral. In 12.1, we shall see which $\mathbb{Z}_n$ are integral.

As in 10.16, we can show

11.14. Theorem (*Homomorphism Theorem for Rings*). *Let $(R, +, \cdot)$ be a ring.*

(i) *If $I \trianglelefteq R$, then $f\colon R \to R/I$; $r \mapsto [r]$ is an epimorphism.*

(ii) *Conversely, if $e\colon R \to R'$ is an epimorphism, then $\operatorname{Ker} e \trianglelefteq R$ and $R/\operatorname{Ker} e \cong R'$.*

11.15. Theorem. *Every ring R can be embedded into a ring $R^{(1)}$ with identity.*

For $R^{(1)}$ we may take $(R \times \mathbb{Z}, +, \cdot)$, where addition of pairs is defined componentwise and multiplication is defined by $(r, z) \cdot (r', z') = (rr' + zr' + z'r, zz')$ for all $r, r' \in R$ and $z, z' \in \mathbb{Z}$; the identity is then given by $(0, 1)$.

It should be remarked that if R already has an identity 1, then 1 loses its role in this process of embedding R into $R^{(1)}$.

Intersections of ideals of R are again ideals. We can hence speak of the concept of a ***generated ideal*** (see the lines after 10.5). Ideals which are generated by a single element deserve special interest. Let R be a ring and $a \in R$. The ideal generated by a will be denoted by (a) and is called the ***principal ideal*** generated by a. If R is a commutative ring with identity, then for all $a \in R$ we get $(a) = aR = \{ar \mid r \in R\}$. In a ring with identity, $\{0\} = (0)$ and $R = (1)$ are principal ideals. In $\mathbb{Z}$, $n\mathbb{Z} = (n)$ is a principal ideal for every $n \in \mathbb{N}_0$. An integral domain in which every ideal is principal is called a ***principal ideal domain*** (**PID**). For example, $\mathbb{Z}$ is a PID, and the same can be shown for the polynomial ring $R[x]$ (see below) if R is a field (cf. Exercise 28).

An ideal I in R is called a ***maximal ideal***, if $I \neq R$ and there is no ideal strictly between I and R. In Exercise 11 we shall see that the ideal (n) is a maximal ideal in $\mathbb{Z}$ iff n is prime.

11.16. Theorem. *Let $I \lhd R$, and R a commutative ring with identity. Then I is maximal if and only if R/I is a field.*

The proof should be given in Exercise 29. Similar to 10.6, we define direct sums $R_1 \oplus R_2$ of rings R_1, R_2 and $\bigoplus_{i=1}^n R_i$ for rings $R_1, \ldots, R_n$.

Now we summarize some properties of "polynomials." Usually we think of polynomials as "formal expressions" $a_0 + a_1x + \cdots + a_nx^n$ in the "indeterminate" x. But, if x is "not determined" or "unknown," we might raise the question, how this "undetermined" x can be squared, added to some a_0, and so on. We will overcome this difficulty by introducing polynomials in a way which at first glance might look much too formal. But exactly this approach will prove very useful in applications. The whole matter starts by observing that a "polynomial $a_0 + a_1x + \cdots + a_nx^n$" is determined by the sequence $(a_0, a_1, \ldots, a_n)$ of its "coefficients." Again, see standard texts for the proofs of the following results.

11.17. Definitions. *For the rest of this section, let R be a commutative ring with identity.* All sequences of elements of R which have only finitely many nonzero elements are called ***polynomials*** over R, all sequences of elements of R are called ***formal power series*** over R. The set of polynomials over R is denoted by $R[x]$, the set of power series over R is $R[[x]]$. If $p = (a_0, a_1, \ldots, a_n, 0, 0, \ldots) \in R[x]$ we will also write $p = (a_0, a_1, \ldots, a_n)$.

If $a_n \neq 0$, then we call n the ***degree*** of p ($n = \deg p$); if $a_n = 1$ we call p ***monic***. We put $\deg(0, 0, 0, \ldots) := -\infty$. Polynomials of degree ≤ 0 are called ***constant***.

In $R[x]$ and $R[[x]]$ we define addition in terms of components as $(a_0, a_1, \ldots) + (b_0, b_1, \ldots) := (a_0 + b_0, a_1 + b_1, \ldots)$, and multiplication by $(a_0, a_1, \ldots) \cdot (b_0, b_1, \ldots) := (c_0, c_1, \ldots)$ with $c_k := \sum_{i+j=k} a_i b_j = \sum_{i=0}^{k} a_i b_{k-i}$, and note from Exercise 18 that $\deg pq = \deg p + \deg q$, for $p, q \in R[x]$, if R is integral.

With respect to the operations of addition and multiplication the sets $R[x]$ and $R[[x]]$ are commutative rings with identity $(1, 0, 0, \ldots)$. If R is integral, the same applies to $R[x]$ and $R[[x]]$ due to the formula for $\deg pq$ given above. Observe that the additive groups of $R[x]$ and $R[[x]]$ are just the direct sum (direct product, respectively) of countably many copies of $(R, +)$. The rings $(R[x], +, \cdot)$ and $(R[[x]], +, \cdot)$ are called the ***ring of polynomials over*** R and the ***ring of formal power series over*** R, respectively. In $R[x]$ and $R[[x]]$ we define $x := (0, 1, 0, 0, \ldots) = (0, 1)$. We then get $x \cdot x = x^2 = (0, 0, 1)$, $x^3 = (0, 0, 0, 1)$, and so on. With $x^0 := (1, 0, 0, 0, \ldots)$ and $a_i = (a_i, 0, 0, \ldots)$ we see that in $R[x]$ and $R[[x]]$ we can write

$$p = (a_0, a_1, a_2, \ldots) = a_0 + a_1 x + a_2 x^2 + \cdots =: \sum_{i \geq 0} a_i x^i.$$

This gives the familiar form of polynomials as $\sum_{i=0}^{n} a_i x^i$ and formal power series as $\sum_{i=0}^{\infty} a_i x^i$ (they are called "formal" since we are not concerned with questions of convergence). We see: x is *not* an "indeterminate" or a "symbol," it is just a special polynomial!

If $p, q \in R[x]$, $p = a_0 + a_1 x + \cdots + a_n x^n$, we also define their ***composition*** $p \circ q$ by $a_0 + a_1 q + \cdots + a_n q^n$. We get the rules $(p_1 + p_2) \circ q = p_1 \circ q + p_2 \circ q$ and $(p_1 p_2) \circ q = (p_1 \circ q)(p_2 \circ q)$ by easy computations. Cf. Exercises 18–20.

Let $p, q \in R[x]$. We say that p ***divides*** q (denoted by $p \mid q$) if $q = p \cdot r$ for some $r \in R[x]$. If $\deg q > \deg p > 0$, then p is called a ***proper divisor*** of q. A polynomial q with $\deg q \geq 1$ which has no proper divisors is called ***irreducible***.

Every polynomial of degree 1 is, of course, irreducible. If $R = \mathbb{C}$ there arc no more irreducible ones; if $R = \mathbb{R}$, "one-half" of the quadratic polynomials are irreducible as well. This is the content of the so-called "***Fundamental Theorem of Algebra***":

11.18. Theorem.

(i) *The irreducible polynomials in* $\mathbb{C}[x]$ *are precisely the ones of degree* 1.

(ii) *The irreducible polynomials in* $\mathbb{R}[x]$ *are the ones of degree* 1 *and the polynomials* $a_0 + a_1x + a_2x^2$ *with* $a_1^2 - 4a_0a_2 < 0$.

The best way to check if $g \mid f$ is to "divide" f by g and to see if the remainder is zero or not. This is possible by the following theorem.

11.19. Theorem (*Euclidean Division*). *Let* R *be a field and* $f, g \in R[x]$ *with* $g \neq 0$. *Then there exist uniquely determined* $q, r \in R[x]$ *with*

$$f = gq + r \quad \text{and} \quad \deg r < \deg g.$$

Here are some properties of polynomial rings over fields which will prove useful. See Exercise 32 for the proof.

11.20. Theorem. *Let* R *be a field and* $p, f, g \in R[x]$ *with* $p \neq 0$. *Then:*

(i) $R[x]$ *is a PID.*

(ii) (p) *is a maximal ideal* $\iff$ p *is irreducible.*

(iii) *If* p *is irreducible and* $p \mid fg$, *then* $p \mid f$ *or* $p \mid g$.

Theorem 11.20(i) has important consequences: we can define concepts like "greatest common divisors." If $f, g \in R[x]$, there is some $d \in R[x]$ such that the ideal generated by $\{f, g\}$ can be written as (d). From that we get:

11.21. Theorem. *Let* R *be a field and* $f, g \in R[x]$ *with* $f \neq 0$ *or* $g \neq 0$. *Then there exists exactly one* $d \in R[x]$ *which enjoys the following properties:*

(i) $d \mid f$ *and* $d \mid g$;

(ii) d *is monic;*

(iii) *if* $d' \mid f$ *and* $d' \mid g$, *then* $d' \mid d$.

For this d *there exist* $p, q \in R[x]$ *with* $d = pf + qg$.

The polynomial d in 11.21 is called the ***greatest common divisor*** of f and g, denoted by $\gcd(f, g)$. The polynomials f and g are called ***relatively prime*** (or ***coprime***) if $\gcd(f, g) = 1$; cf. Exercise 27.

The following method computes the $\gcd(f, g)$ of f and g and at the same time the polynomials p, q in 11.21.

11.22. Algorithm. Write a table with three entries in each row. The first two rows are (without loss of generality we assume $\deg f \geq \deg g$):

$$\begin{array}{rccc} \text{(i)} & f & 1 & 0; \\ \text{(ii)} & g & 0 & 1; \end{array}$$

divide f by g as $f = gq_3 + r_3$ and substract the q_3-fold of (ii) from line (i)

$$\begin{array}{rccc} \text{(iii)} & r_3 & 1 & -q_3. \end{array}$$

Now do the same with (ii) and (iii) instead of (i) and (ii), and so on:

$$\begin{array}{rccc} \text{(iv)} & r_4 & \alpha_4 & \beta_4; \\ \text{(v)} & r_5 & \alpha_5 & \beta_5; \\ \vdots & \vdots & \vdots & \vdots \end{array}$$

It is easy to see that $\deg f \geq \deg g > \deg r_3 > \deg r_4 > \dots$, so this process must eventually stop, say at

$$\begin{array}{rccc} (n) & r_n & \alpha_n & \beta_n; \\ (n+1) & 0 & \alpha_{n+1} & \beta_{n+1}. \end{array}$$

In each of the rows we have $r_i = \alpha_i f + \beta_i g$ and $\gcd(f, g) = \gcd(g, r_3) = \gcd(r_3, r_4) = \cdots = \gcd(r_n, 0) = r_n$. Hence $r_n = \gcd(f, g)$ and with $p := \alpha_n$, $q := \beta_n$ we get

$$\gcd(f, g) = pf + qg.$$

By working out the details of this algorithm, we get a proof of Theorem 11.21 (see Exercise 33).

11.23. Example. Let $f = 2x^5 + 2x^4 + x^2 + x$, $g = x^3 + x^2 + x + 1$ in $\mathbb{Z}_3[x]$.

$$\begin{array}{rccl} 2x^5 + 2x^4 + x^2 + x & 1 & 0 & \\ x^3 + x^2 + x + 1 & 0 & 1 & \mid \cdot x^2 \\ x^3 + 2x^2 + x & 1 & x^2 & \mid \cdot 2 \\ 2x^2 + 1 & 2 & 2x^2 + 1 & \mid \cdot x \\ 2x^2 + 2x & 2x + 1 & 2x^3 + x^2 + x & \mid \cdot 2 \\ x + 1 & x + 1 & x^3 + x^2 + 2x + 1 & \mid \cdot x \\ 0 & & & \end{array}$$

Hence $\gcd(f, g) = x + 1$ and

$$x + 1 = (x + 1)f + (x^3 + x^2 + 2x + 1)g.$$

11.24. Theorem (*Unique Factorization Theorem*). *Let R be a field. Then every $f \in R[x]$ has a representation (which is unique up to order) of the form $f = r \cdot p_1 \cdot p_2 \cdot \dots \cdot p_k$, with $r \in R$, and where $p_1, \dots, p_k$ are monic irreducible polynomials over R.*

We mention explicitly that the proof of this theorem (see Exercise 34) does not indicate how to construct this "prime factor decomposition." In general, there is no constructive way to do this. However, we will describe a (constructive) algorithm in §15, if R is a finite field.

In general, rings with a property analogous to 11.24 are called "unique factorization domains" (UFD). It can be shown that every PID is a UFD.

The penetrating similarity between $\mathbb{Z}$ and $R[x]$ (R a field) is enunciated by the following table. Let f, g be in $\mathbb{Z}$ or in $R[x]$, according to the left-hand side or right-hand side of the table. Cf. Exercise 28.

$\mathbb{Z}$	$R[x]$
"Norm": absolute value of f.	"Norm": degree of f.
Invertible: numbers with absolute value 1.	Invertible: polynomials of degree 0.
Every integer can be represented in the form $a_0 + a_1 10 + a_2 10^2 + \cdots + a_n 10^n$.	Every polynomial can be represented in the form $a_0 + a_1 x + a_2 x^2 + \cdots + a_n x^n$.
$g \mid f$ if $f = g \cdot q$ for some $q \in \mathbb{Z}$	$g \mid f$ if $f = g \cdot q$ for some $q \in R[x]$.
There are $q, r \in \mathbb{Z}$ with $f = gq + r$ and $0 \leq r < \lvert g \rvert$.	There are $q, r \in R[x]$ with $f = gq + r$ and $\deg r < \deg g$.
$\mathbb{Z}$ is an integral domain and a PID.	$R[x]$ is an integral domain and a PID.
f and g have a uniquely determined greatest common divisor d, which can be written as $d = pf + qg$ with $p, q \in \mathbb{Z}$.	f and g have a uniquely determined greatest common divisor d, which can be written as $d = pf + qg$ with $p, q \in R[x]$.
f is prime iff it has no proper divisors.	f is irreducible iff it has no proper divisors.
Every integer > 0 is a "unique" product of primes.	Every polynomial of degree > 0 is a "unique" product of irreducible polynomials.

Many people think of "functions" when polynomials are discussed. In fact, every polynomial induces a "polynomial function," but in general this is not a 1-1-correspondence:

If $p = (a_0, \dots, a_n) = a_0 + a_1x + \dots + a_nx^n \in R[x]$, then $\bar{p}\colon R \to R$; $r \mapsto a_0 + a_1r + \dots + a_nr^n$ is called the ***polynomial function*** induced by p. Let $P(R) := \{\bar{p} \mid p \in R[x]\}$ be the set of all polynomial functions induced by polynomials over R. If no confusion is to be expected, we will simply write p instead of $\bar{p}$. Note that often polynomials p are written as $p(x)$.

11.25. Theorem.

(i) *For all* $p, q \in R[x]$, $\overline{p+q} = \bar{p} + \bar{q}$, $\overline{p \cdot q} = \bar{p} \cdot \bar{q}$, *and* $\overline{p \circ q} = \bar{p} \circ \bar{q}$.

(ii) $P(R)$ *is a subring of* R^R.

11.26. Theorem.

(i) *The map* $h\colon R[x] \to P(R)$; $p \mapsto \bar{p}$, *is a (ring-) epimorphism.*

(ii) *If all nonzero elements of* $(R, +)$ *are of infinite order, then* h *is an isomorphism.*

Proof. (i) follows readily from 11.25. To show (ii), assume that $p = a_0 + a_1x + \dots + a_nx^n$ is a nonzero polynomial of minimal degree in $\operatorname{Ker} h$. So $\bar{p} = \bar{0}$. Then $q := p \circ (x + 1) - p$ also induces the zero function, but $q = na_{n-1}x^{n-1} + b_{n-2}x^{n-2} + \dots + b_0$ for some $b_{n-2}, \dots, b_0 \in \mathbb{R}$. By assumption, $q \neq 0$, so p cannot be of minimal degree in $\operatorname{Ker} h \setminus \{0\}$. □

If R is finite, $P(R)$ is also finite as a subring of R^R. On the other hand, $R[x]$ is infinite as soon as $|R| > 1$. Hence the map h of 11.26 cannot be an isomorphism in this case. So we get the first half of the next corollary. The proof of the other half is the content of Exercise 42.

11.27. Corollary.

(i) *If* R *is finite, then infinitely many polynomials in* $R[x]$ *induce the zero function.*

(ii) *Let* F *be a field. Then* $F[x] \cong P(F)$ *iff* F *is infinite.*

This result allows us to "identify" $R[x]$ and $P(R)$ if $R = \mathbb{R}$, for instance. This is, however, not allowed if R is a finite ring. In $\mathbb{Z}_n[x]$, the polynomial $p := x(x+1)\dots(x+(n-1))$ has degree n, but $\bar{p}$ is the zero function. Hence p is in $\operatorname{Ker} h$ of 11.26. An element $r \in R$ is called a ***root*** (or a ***zero***) of the polynomial $p \in R[x]$ if $\bar{p}(r) = 0$. There is an important link between roots and divisibility of polynomials:

11.28. Theorem. *An element r of a field R is a root of the polynomial $p \in R[x]$ iff $x - r$ divides p.*

Proof. If $x-r \mid p$, then $\overline{p}(r) = 0$. Conversely, if $\overline{p}(r) = 0$, write $p = (x-r)q+s$ with $\deg s < \deg(x-r) = 1$; so s must be constant. $\overline{p}(r) = 0$ shows $s = 0$. □

Let $r \in R$ be a root of $p \in R[x]$. If k is a positive integer such that p is divisible by $(x-r)^k$, but not by $(x-r)^{k+1}$, then k is called the ***multiplicity*** of r. If $k = 1$, then r is called a ***simple root*** of p. From 11.28 we can deduce

11.29. Corollary. *If F is a field and if $g \in F[x]$ has degree n, then g has at most n roots (even if we count the roots with their multiplicities).*

It follows from 11.18 that for $F = \mathbb{C}$, each $g \in \mathbb{C}[x]$ of degree n has precisely n roots (counted with their multiplicities). In 12.10, we will extend this notion of a root.

11.30. Theorem. *If F is a field, and if $(a_0, b_0), (a_1, b_1), \ldots, (a_k, b_k) \in F \times F$ are such that $a_i \neq a_j$ holds for $i \neq j$, then there is precisely one polynomial $p \in F[x]$ of degree $\leq k$ such that $p(a_0) = b_0, \ldots, p(a_k) = b_k$.*

This "interpolation" polynomial (function) can be found by the familiar interpolation formulas of Lagrange or Newton. For fast algorithms and a comparison between Lagrange and Newton interpolation, see Geddes, Czapor & Labahn (1993, Section 5.7). We recapitulate, for instance, Lagrange's formula:

11.31. Theorem (*Lagrange's Interpolation Formula*). *Let the notation be as in 11.30. Form polynomials p_i $(0 \leq i \leq k)$ via*

$$p_i := \frac{(x-a_0)(x-a_1)\ldots(x-a_{i-1})(x-a_{i+1})\ldots(x-a_k)}{(a_i-a_0)(a_i-a_1)\ldots(a_i-a_{i-1})(a_i-a_{i+1})\ldots(a_i-a_k)}.$$

Then $\overline{p}_i(a_i) = 1$ for all i and $\overline{p}_i(a_j) = 0$ for $i \neq j$. So $p := b_0p_0 + b_1p_1 + \cdots + b_kp_k$ fulfills the interpolation job in 11.30 and p has degree $\leq k$.

11.32. Example. To find a polynomial function passing through the points $(1, 3), (2, 5), (3, 10) \in \mathbb{R}^2$, we compute p_0, p_1, p_2 according to 11.31:

Solution:

$$p_0 = \frac{(x-2)(x-3)}{(1-2)(1-3)} = \tfrac{1}{2}(x^2 - 5x + 6),$$
$$p_1 = \frac{(x-1)(x-3)}{(2-1)(2-3)} = -1(x^2 - 4x + 3),$$
$$p_2 = \frac{(x-1)(x-2)}{(3-1)(3-2)} = \tfrac{1}{2}(x^2 - 3x + 2).$$

So $p = 3p_0 + 5p_1 + 10p_2 = \frac{3}{2}x^2 - \frac{5}{2}x + 4$ determines a polynomial function $\bar{p}$ which passes through the given points.

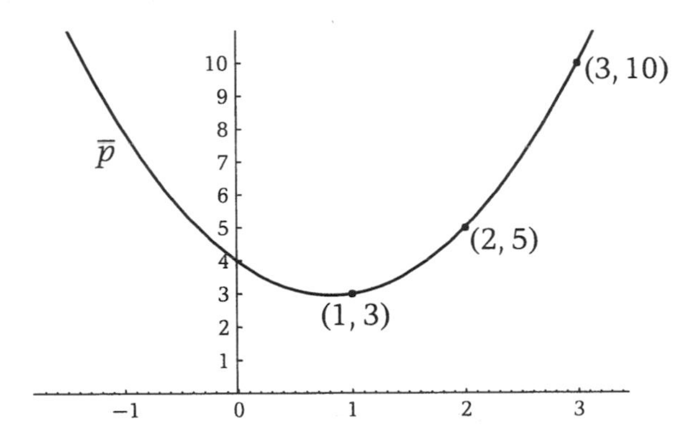

11.33. Corollary. *Let F be a field. Then every function from F to F is a polynomial function iff F is finite. Thus, the finite fields are precisely the* ***polynomially complete*** *fields (cf. the line after* 4.18).

We give another nice application of 11.31 due to Shamir (1979).

11.34 ("*How to share a secret*"). Suppose a central place in a company (the safe, for instance) should only be reachable if at least k of a total number of n employees are present at the same time. A possible solution is this: Choose a large prime number p, a "master key" $m \in \mathbb{Z}_p$, and a polynomial $q = m + a_1x + \cdots + a_{k-1}x^{k-1} \in \mathbb{Z}_p[x]$. The "privileged" employees get the subkeys $s_1 := \bar{q}(1), \ldots, s_n := \bar{q}(n)$ on a chip card. Now if at least k of these n employees are present, they put their subkeys in a computer which can uniquely reconstruct q by Theorems 11.30 and 11.31, hence also $\bar{q}(0) = m$. If this is the correct master key, the computer opens the door, and the employees can access the safe.

Note that the computer can generate p, m, and q at random by itself, compute $s_1, \dots, s_n$ automatically, and then it can throw away q, which really increases the security.

Finally, we remark that by $(R[x])[y] =: R[x, y]$, etc., we get polynomial rings in "more indeterminates."

Exercises

1. Determine all nonisomorphic rings with two and three elements.
2. How many possibilities are there to define a multiplication on $\mathbb{Z}_4$ in order to make $\mathbb{Z}_4$ into a ring?
3. Prove or disprove: A ring $R \neq \{0\}$ is a skew-field if and only if for every nonzero $a \in R$ there is a unique $x \in R$ such that $axa = a$.
4. Show that if a finite ring R has a nonzero element which is not a zero divisor, then R has an identity.
5. Give an example of a ring R with identity 1 and a subring R' of R with identity $1'$ such that $1 \neq 1'$.
6. Let R be a finite ring of characteristic p, a prime. Show that the order of R is a power of p.
7. Prove (i), (iii), and (iv) in Theorem 11.9.
8. Show that $\{\pm 1, \pm i, \pm j, \pm k\}$ in 11.8 yields a multiplicative group, and find its group table.
9. Are there integral rings of characteristic 6?
10. Let R be a commutative ring of characteristic p, a prime. Show that
$$(x + y)^p = x^p + y^p \quad \text{and} \quad (xy)^p = x^p y^p$$
hold for all $x, y \in R$.
11. Show that (n) is a maximal ideal of $\mathbb{Z}$ iff n is a prime.
12. Prove that an ideal $I \neq R$ of a ring R is maximal if and only if R/I is simple.
13. Let R be a ring with identity and let $I \lhd R$. Prove that I is contained in a maximal ideal.
14. Is the following true: "$n \leq m \iff \mathbb{Z}_m \leq \mathbb{Z}_n$"?
15. Find all ideals in $\mathbb{Z}_3$, $\mathbb{Z}_4$, and $\mathbb{Z}_6$.

16. Find all maximal ideals in the following rings: $\mathbb{R}$, $\mathbb{Q}$, $\mathbb{Z}_2$, $\mathbb{Z}_3$, $\mathbb{Z}_4$, $\mathbb{Z}_6$, $\mathbb{Z}_{24}$, $\mathbb{Z}_{p^e}$ (p a prime), $\mathbb{Z}_n$.
17. Are the following rings PIDs: $\mathbb{Z}_n$, $n\mathbb{Z}$, skew-fields, fields, simple rings?
18. Show that $\deg(pq) = \deg p + \deg q$ holds for $p, q \in R[x]$ if R is integral. Can we say something about $\deg(p+q)$? What about $\deg(p \circ q)$?
19. Do $(p_1 + p_2) \circ q = p_1 \circ q + p_2 \circ q$ and $(p_1 p_2) \circ q = (p_1 \circ q)(p_2 \circ q)$ hold in $R[x]$?
20. Is $(R[x], +, \circ)$ a ring?
21. Let $f = 1 + x + x^3 + x^6$ and $g = 1 + x + x^2 + x^4$ be polynomials over R. If $R = \mathbb{R}$, is f divisible by g? If $R = \mathbb{Z}_2$, is f divisible by g?
22. Show that $f_n := (x^n - 1)/(x - 1)$ is a polynomial over $\mathbb{Q}$ for all positive integers n. Prove that if f_n is irreducible, then n is prime. [Remark: The converse can also be shown.]
23. Let $f = x^6 + 3x^5 + 4x^2 - 3x + 2$, $g = x^2 + 2x - 3 \in \mathbb{Z}_7[x]$. Determine $q, r \in \mathbb{Z}_7[x]$ such that $f = gq + r$ and $\deg r < 2$.
24. If $f = x + 2x^3 + 3x^4 + 3x^5 + 2x^6 + 3x^7 + x^8$ and $g = x + 3x^2 + x^3$ are polynomials over R, compute polynomials q and r (with $\deg r < 3$) such that $f = gq + r$. Does $g \mid f$ hold? Answer these questions for $R = \mathbb{R}$, $R = \mathbb{Q}$, and finally $R = \mathbb{Z}_5$. Do you get the same answers?
25. With f, g of Exercise 24, compute $\gcd(f, g)$ again for $R = \mathbb{R}$, $R = \mathbb{Q}$, and $R = \mathbb{Z}_5$.
26. Decompose $x^5 + x^4 + 3x^3 + 3x^2 + x + 1 \in \mathbb{Z}_5[x]$ into irreducible factors over $\mathbb{Z}_5$.
27. Let $\{0\} \neq I \trianglelefteq R[x]$, R a field. By 11.20, $I = (d)$ for some $d \in R[x]$. Show that we can take d to be a nonzero polynomial of minimal degree in I.
28. Show that $\mathbb{Z}$ and $R[x]$ (R a field) are PIDs. Is $\mathbb{Z}[x]$ a PID?
29. Prove that R is a field if and only if R is a simple commutative ring with identity. Use this and Exercise 12 to prove Theorem 11.16.
30. Show that if I is an ideal of a ring R, then $I[x]$ is an ideal of $R[x]$.
31. Show that $(x^2 - 1)$ is not a maximal ideal of $\mathbb{R}[x]$.
32. Prove Theorem 11.20.
33. Work out the details in 11.22 and usc 11.22 to prove Theorem 11.21.
34. Prove the Unique Factorization Theorem 11.24. [Hint: Use induction on $\deg f$ and 11.20(iii).]
35. Find the inverse of $1 - x$ in $\mathbb{R}[[x]]$.
36. If R is a field, show that $a_0 + a_1 x + \cdots \in R[[x]]$ is invertible iff $a_0 \neq 0$.

37. Show that the ideals of $R[[x]]$, R a field, are precisely the members of the chain

$$R[[x]] = (1) \supset (x) \supset (x^2) \supset \cdots \supset (0).$$

Hence $R[[x]]$ is a PID with exactly one maximal ideal.

38. Determine all homomorphic images of $F[x]$, F a field.

39. Let $f(x) = 3 + 4x + 5x^2 + 6x^3 + x^6$ and $g(x) = 1 + x + x^2$ be two polynomials over $\mathbb{Z}_{11}$. Is f divisible by g? Determine $\bar{f}(3) + \bar{g}(3)$ and $\overline{f+g}(3)$.

40. Prove Theorem 11.25.

41. Show that $f, g \in F[x]$, F a field, have a common zero b iff b is a zero of $\gcd(f, g)$. Find the common zeros of $f = 2x^3 + x^2 - 4x + 1$ and $g = x^2 - 1$ in $\mathbb{R}$.

42. Use 11.29 to prove 11.27(ii).

43. Determine all roots and their multiplicities of the polynomial $x^6 + 3x^5 + x^4 + x^3 + 4x^2 + 3x + 2$ over $\mathbb{Z}_5$.

44. Show that $p \in R[x]$ (R a field and $2 \leq \deg p \leq 3$) is irreducible iff $\bar{p}$ has no zero.

45. Find infinitely many distinct polynomials in $\mathbb{Z}_3[x]$ which induce the zero function.

46. Show that $P(R) \leq \{f \in R^R \mid r - s \in I \Longrightarrow f(r) - f(s) \in I \text{ for each } I \trianglelefteq R\}$.

47. Let R be a commutative ring with identity. Use Exercise 46 to show that R is polynomially complete iff R is a finite field.

§12 Fields

Let us first find out when $(\mathbb{Z}_n, +, \cdot)$ is a field.

12.1. Theorem. *For every $n \in \mathbb{N}$, the following conditions are equivalent:*

(i) *$\mathbb{Z}_n$ is an integral domain;*

(ii) *$\mathbb{Z}_n$ is a field;*

(iii) *$n \in \mathbb{P}$.*

Proof. By 11.9(i), statements (i) and (ii) are equivalent. Suppose p is prime. If $a \in \mathbb{Z}_p^*$, then $\gcd(a, p) = 1$, so there are $r, s \in \mathbb{Z}$ with $ar + ps = 1$. In $\mathbb{Z}_p$ this means $[a][r] + [0][s] = 1$, so $[r] = [a]^{-1}$. If n is not prime, however, say $n = st$, then $[0] = [n] = [s][t]$, so by 11.9(iv), $\mathbb{Z}_n$ cannot be a field. □

So the fields $\mathbb{Z}_p$ (p prime) give us the first examples of finite fields. We shall devote §13 to finite fields.

Rings R which can be embedded in a field F (in symbols $R \hookrightarrow F$) obviously have to be commutative and integral, since F has these properties and every subring inherits them. We may assume (see 11.15) that R is an integral domain. Conversely, we get

12.2. Theorem. *For every integral domain $\neq \{0\}$ there is a field F with the following properties:*

(i) $R \hookrightarrow F$.

(ii) *If $R \hookrightarrow F'$ and F' is a field, then $F \hookrightarrow F'$.*

Sketch of Proof (the details are left as an exercise). Let $S := R \times R^*$. On S we define $(a, b) + (c, d) := (ad + bc, bd)$ and $(a, b) \cdot (c, d) := (ac, bd)$, as well as $(a, b) \sim (c, d) :\iff ad = bc$. We have to check that $(S, +, \cdot)$ is a ring, $\sim$ is a congruence relation in S, and $F := S/\sim$ is a field. The map h sending r into the equivalence class of $(r, 1)$ is an embedding. If F' is another field with an embedding $h'\colon R \to F'$, then the map $g\colon F \to F'$ sending the equivalence class of (a, b) into $h'(a)h'(b)^{-1}$ is well defined and is an embedding as well. The equivalence class of (a, b) is usually denoted by a/b.

Thus every integral domain can be embedded in a "minimal field." The field of 12.2 is called the ***quotient field*** of R (or ***field of quotients*** of R).

12.3. Theorem. *Let $h\colon R_1 \to R_2$ be an isomorphism between the integral domains R_1 and R_2. Then h can be uniquely extended to an isomorphism $\bar{h}$ between the quotient fields F_1 and F_2 of R_1 and R_2.*

Proof. $\bar{h}\colon F_1 \to F_2$; $u/b \mapsto h(u)/h(b)$ does the job. □

12.4. Corollary. *Any two quotient fields of an integral domain are isomorphic.*

Thus we can speak of *the* quotient field of an integral domain. Applying this construction to our three standard examples of integral domains yields the following fields:

12.5. Examples and Definitions.

(i) $\mathbb{Q}$ is the quotient field of $\mathbb{Z}$.

(ii) Let R be a field. The field of quotients of $R[x]$ is denoted by $R(x)$ and is called the ***field of rational functions*** over R. But the name is quite misleading: the elements in $R(x)$ are *not* functions at all. They consist of fractions p/q with $p, q \in R[x]$, $q \neq 0$ (in the sense of the proof of 12.2).

(iii) If R is a field, then the quotient field of $R[[x]]$ is denoted by $R\langle x\rangle$ and is called the ***field of formal Laurant series*** over R. In an isomorphic copy, $R\langle x\rangle$ consists of sequences $(a_{-n}, \ldots, a_0, a_1, \ldots)$ of elements of R with $n \in \mathbb{N}_0$.

12.6. Definitions. A subset U of a field F is called a ***subfield*** of F, in symbols $U \leq F$, if U is a subring of F and U is a field with respect to the operations in F. If also $U \neq F$, then $(U, +, \cdot)$ is called a ***proper subfield*** of $(F, +, \cdot)$, in symbols $U < F$. $(F, +, \cdot)$ is called an ***extension field*** (or ***extension***) of the field $(U, +, \cdot)$ if $(U, +, \cdot)$ is a subfield of $(F, +, \cdot)$. A field P is called a ***prime field*** if it has no proper subfields.

The following theorem characterizes all prime fields:

12.7. Theorem. *Up to isomorphism, all distinct prime fields are given by $\mathbb{Q}$ and $\mathbb{Z}_p$, p prime.*

Proof. Let P be a prime field and let 1 be its identity. It can be verified immediately that $C = \{z1 \mid z \in \mathbb{Z}\}$ is an integral domain in P. The mapping $\psi: \mathbb{Z} \to C;\ z \mapsto z1$ is an epimorphism of $\mathbb{Z}$ onto C. We distinguish between two cases:

(i) If $\operatorname{Ker}\psi = \{0\}$, then ψ is an isomorphism. The quotient field of C is the smallest field containing C and is isomorphic to the quotient field of $\mathbb{Z}$, which is $\mathbb{Q}$. Therefore $P \cong \mathbb{Q}$.

(ii) If $\operatorname{Ker}\psi \neq \{0\}$, then there is a $k \in \mathbb{N} \setminus \{1\}$ with $\operatorname{Ker}\psi = (k)$. The Homomorphism Theorem 11.14 implies $\mathbb{Z}_k = \mathbb{Z}/(k) \cong C$. Now C and $\mathbb{Z}_k$ are finite integral domains, so they are fields and k must be a prime. In this case, $P = C \cong \mathbb{Z}_k$. □

Hence every field is an extension of $\mathbb{Q}$ or of some $\mathbb{Z}_p$ (p prime).

12.8. Corollary. *Let P be the prime field of the field F. Then:*

(i) *If* $\operatorname{char} F = 0$, *then* $P \cong \mathbb{Q}$.

(ii) *If* $\operatorname{char} F = p \in \mathbb{P}$, *then* $P \cong \mathbb{Z}_p$.

A field with prime characteristic does not necessarily have to be finite, as the following example shows:

12.9. Example. The field $\mathbb{Z}_2(x)$ of all rational functions f/g, where $f, g \in \mathbb{Z}_2[x]$, $g \neq 0$, has as prime field $\mathbb{Z}_2$, i.e., its characteristic is 2, but it has infinitely many elements.

As mentioned earlier, we extend the notion of a root:

12.10. Definition. An element r lying in some extension field of a field F is called a ***root*** (or a ***zero***) of $p \in F[x]$ if $\bar{p}(r) = 0$.

We proceed to prove one of the basic theorems of modern algebra, Theorem 12.13, which is a result due to L. Kronecker (1821–1891). This theorem guarantees the existence of an extension containing a zero of f, for an arbitrary polynomial f over a field. The proof of the theorem also gives a method of construction of such an extension. Let F be a field. We already know that the ideals in $F[x]$ are principal ideals (see 11.20). Let $f \in F[x]$ be a polynomial of positive degree and let (f) denote the principal ideal generated by f.

Suppose f is a polynomial of degree k over F. Let $g + (f)$ be an arbitrary element in $F[x]/(f)$. By Euclidean division 11.19, we get $h, r \in F[x]$ with $g = hf + r$, where $\deg r < k$. Since $hf \in (f)$, it follows that $g + (f) = r + (f)$. Hence each element of $F[x]/(f)$ can be uniquely expressed in the form

$$a_0 + a_1x + \cdots + a_{k-1}x^{k-1} + (f), \qquad a_i \in F. \tag{12.1}$$

If we identify F with the subring $\{a + (f) \mid a \in F\}$ of $F[x]/(f)$, then the element in (12.1) can be written as $a_0 + a_1(x + (f)) + \cdots + a_{k-1}(x + (f))^{k-1}$. If $x + (f) =: \alpha$, we can write this uniquely as

$$a_0 + a_1\alpha + \cdots + a_{k-1}\alpha^{k-1} \tag{12.2}$$

and we may regard $F[x]/(f)$ as a vector space over F with basis $\{1, \alpha, \alpha^2, \ldots, \alpha^{k-1}\}$.

Since $0 + (f)$ is the zero element of $F[x]/(f)$, we have $\bar{f}(\alpha) = f + (f) = 0 + (f)$, i.e., α is a root of f. Clearly, α is an element in $F[x]/(f)$ but in general not in F. Thus the elements in $F[x]/(f)$ of the form (12.2) can be regarded so that α is an element with the property that $\bar{f}(\alpha) = 0$. From 11.16 and 11.20 we get altogether:

12.11. Theorem. *Let F be a field and $f \in F[x]$ with $\deg f = k$. Then $F[x]/(f) = \{a_0 + a_1x + \cdots + a_{k-1}x^{k-1} \mid a_i \in F\}$ is a k-dimensional vector space over F with basis $\{1, \alpha, \alpha^2, \ldots, \alpha^{k-1}\}$, where $\alpha = [x] = x + (f)$. We have $\bar{\alpha} = 0$ and $F[x]/(f)$ is a field iff f is irreducible.*

12.12. Examples.

(i) Let F be the field $\mathbb{Z}_2 = \{0, 1\}$; then $f = x^2 + x + 1$ is an irreducible polynomial of degree 2 over $\mathbb{Z}_2$. Hence $\mathbb{Z}_2[x]/(x^2 + x + 1)$ is a field whose elements can be represented in the form $a + b\alpha$, $a, b \in \mathbb{Z}_2$, where α satisfies $\bar{f}(\alpha) = 0$, i.e., $\alpha^2 + \alpha + 1 = 0$, which means that $\alpha^2 = \alpha + 1$, due to $-1 = 1$ in $\mathbb{Z}_2$. Hence $\mathbb{Z}_2[x]/(x^2 + x + 1)$ is a field with four elements:

$$\mathbb{Z}_2[x]/(x^2 + x + 1) = \{0, 1, \alpha, 1 + \alpha\}.$$

For instance, $\alpha \cdot (1 + \alpha) = \alpha + \alpha^2 = \alpha + \alpha + 1 = 1$. The addition and multiplication tables are given by

$+$	0	1	α	$1+\alpha$
0	0	1	α	$1+\alpha$
1	1	0	$1+\alpha$	α
α	α	$1+\alpha$	0	1
$1+\alpha$	$1+\alpha$	α	1	0

$\cdot$	0	1	α	$1+\alpha$
0	0	0	0	0
1	0	1	α	$1+\alpha$
α	0	α	$1+\alpha$	1
$1+\alpha$	0	$1+\alpha$	1	α

(ii) Similarly,

$$\mathbb{Z}_2[x]/(x^3 + x + 1) = \{0, 1, \alpha, 1 + \alpha, \alpha^2, 1 + \alpha^2, \alpha + \alpha^2, 1 + \alpha + \alpha^2\}$$

is a field with eight elements, subject to $\alpha^3 = \alpha + 1$.

(iii) $\mathbb{Z}_3[x]/(x^2 + 1) = \{0, 1, 2, \alpha, 1 + \alpha, 2 + \alpha, 2\alpha, 1 + 2\alpha, 2 + 2\alpha\}$ with $\alpha^2 = -1 = 2$ is a field with nine elements.

(iv) Later (in 13.6) we shall see that the fields in (i)–(iii) are the three smallest fields which are not of the $\mathbb{Z}_p$-type.

These examples indicate that calculations in $F[x]/(f)$ and $\mathbb{Z}_n$ can be performed in a "similar" way.

12.13. Theorem (*Kronecker's Theorem*). *Let F be a field and let g be an arbitrary polynomial of positive degree in $F[x]$. Then there is an extension field K of F such that g has a zero in K.*

Proof. If g has a zero in F, then the statement is trivial. If this is not the case, then there is a divisor f of g of degree at least 2 which is irreducible over F. Let $K := F[x]/(f)$ and consider g as a polynomial over K. Denoting the element $x + (f)$ of K by α we have $\bar{f}(\alpha) = 0 + (f)$, the zero in K, i.e., α is a zero of f and therefore also a zero of g. □

12.14. Example. Let $F = \mathbb{R}$ and $g = x^2 + 1 \in \mathbb{R}[x]$. Then g does not have a zero in $\mathbb{R}$. We form $K := \mathbb{R}[x]/(x^2 + 1)$ as in the proof of Kronecker's Theorem. We know from 12.11 that $K = \{a + b\alpha \mid a, b \in \mathbb{R}\}$ with $\alpha^2 = -1$, and α is a root of g; in fact, $g = (x - \alpha)(x + \alpha)$. Hence we have "invented" the complex numbers $\mathbb{C}$ as $\mathbb{C} = K = \mathbb{R}[x]/(x^2 + 1)$.

We now consider a field which is large enough to contain all zeros of a given polynomial.

12.15. Definition. A polynomial $f \in F[x]$ is said to ***split*** in an extension K of F if f can be expressed as a product of linear factors in $K[x]$. The field K is called a ***splitting field*** of f over F if f splits in K, but does not split in any proper subfield of K containing F.

12.16. Example. $\mathbb{C}$ is the splitting field of $x^2 + 1 \in \mathbb{R}$.

12.17. Corollary. *Let F be a field and let $g \in F[x]$ be of positive degree. Then there is an extension K of F such that g splits into linear factors over K.*

Proof. The polynomial g has $x - \alpha$ as a divisor in $K_1[x]$, where $K_1[x]$ is a suitable extension of F (see Theorem 12.13). If g does not split into linear factors over K_1, then we repeat the construction of 12.13 and construct extensions $K_2, K_3, \ldots$ until g splits completely into linear factors over K.

The following notation will prove useful. Let F be a subfield of a field M and let A be an arbitrary set of elements in M. Then $F(A)$ denotes the intersection of all subfields of M which contain both F and A. $F(A)$ is called the extension of F which is obtained by ***adjunction*** of the elements of A. If $A = \{a\}$, $a \notin F$, then $F(\{a\})$ is called a ***simple extension*** of F. We also write $F(a)$ in this case, $F(\{a_1, \ldots, a_n\})$ is abbreviated by $F(a_1, \ldots, a_n)$. Examples will come soon.

12.18. Theorem. *Let $f \in F[x]$ be of degree n and let K be an extension of F. If $f = c(x - a_1) \cdots (x - a_n)$ in $K[x]$, then $F(a_1, \ldots, a_n)$ is a splitting field of f over F.*

Proof. Clearly f splits over $F(a_1, \ldots, a_n)$. Suppose that L is a subfield of $F(a_1, \ldots, a_n)$ and f splits over L. Thus $f = c(x - b_1) \cdots (x - b_n)$ with $b_i \in L$ for $1 \leq i \leq n$. By the Unique Factorization Theorem 11.24, $\{a_1, \ldots, a_n\} = \{b_1, \ldots, b_n\}$, so $L \supseteq F(a_1, \ldots, a_n)$. □

Corollary 12.17 secures the existence of splitting fields. The proof of the uniqueness of the splitting field of a polynomial f over a field F is slightly more complicated and we omit it, see, e.g., Lidl & Niederreiter (1983).

12.19. Theorem. *For any field F and polynomial $f \in F[x]$ of degree ≥ 1 all splitting fields of f over F are isomorphic.*

We now introduce special types of extension fields of a given field F.

12.20. Definition. Let K be an extension of a field F. An element α of K is called ***algebraic*** over F if there is a nonzero polynomial g with coefficients in F such that $\bar{g}(\alpha) = 0$. If α is not algebraic over F, then α is said to be ***transcendental*** over F.

12.21. Examples. The real number $\sqrt{2}$ is algebraic over $\mathbb{Q}$, since $\sqrt{2}$ is a zero of $x^2 - 2$. If $\beta \in K$ is an element of F, then β is a zero of $x - \beta \in F[x]$. Thus any element in F is algebraic over F. Also $i = \sqrt{-1}$ is algebraic over $\mathbb{Q}$ and over $\mathbb{R}$ as a zero of x^2+1 by 12.14. Moreover, it will follow from 12.28 that *every* element of $\mathbb{C}$ is algebraic over $\mathbb{R}$. It can be shown, though with considerable difficulty, that π and e are transcendental over $\mathbb{Q}$.

Let K be an extension of F. Then K can be regarded as a vector space over F by considering the additive group $(K, +)$ together with scalar multiplication by elements in F. We can divide all extension fields into two classes:

12.22. Definitions. An extension K of a field F is called ***algebraic*** if each element of K is algebraic over F. If K contains at least one element which is transcendental over F, then K is a ***transcendental extension***. The ***degree*** of K over F, in symbols $[K : F]$, is the dimension of K as a vector space over F. K is called a ***finite extension*** of F if $[K : F]$ is finite. Otherwise K is called an ***infinite extension*** of F.

In particular, $\mathbb{R}$ is a transcendental extension of $\mathbb{Q}$. The following two theorems determine all simple extensions up to isomorphisms.

12.23. Theorem. *Let K be an extension of F, and let $\alpha \in K$ be transcendental over F. Then the extension $F(\alpha)$ is isomorphic to the field $F(x)$ of rational functions in x.*

Proof. $F(\alpha)$ must contain the set

$$R(\alpha) := \left\{ \frac{a_0 + a_1\alpha + \cdots + a_n\alpha^n}{b_0 + b_1\alpha + \cdots + b_m\alpha^m} \middle| n, m \in \mathbb{N}_0, a_0, \ldots, a_n, b_0, \ldots, b_m \in F, b_m \neq 0 \right\}$$

of all "rational expressions" in α. Since $b_0 + b_1\alpha + \cdots + b_m\alpha^m = 0$ iff $b_0 + b_1x + \cdots + b_mx^m = 0$ (otherwise α would be algebraic), the denominators in $R(\alpha)$ are never $= 0$. Since $R(\alpha)$ is a field, $R(\alpha) = F(\alpha)$. The obvious map $h\colon F(x) \to F(\alpha)$ extending $x \mapsto \alpha$ and acting identically on F is an epimorphism from $F(x)$ to $F(\alpha)$. Since α is transcendental, h is an isomorphism. □

For simplicity, we shall use the notation

$$F[\alpha] := \{a_0 + a_1\alpha + \cdots + a_n\alpha^n \mid n \in \mathbb{N}_0, a_i \in F\}.$$

12.24. Theorem. *Let K be an extension of F, and let $\alpha \in K$ be algebraic over F. Then:*

(i) *$F(\alpha) = F[\alpha] \cong F[x]/(f)$, where f is a uniquely determined, monic, irreducible polynomial in $F[x]$ with zero α in K.*

(ii) *α is a zero of a polynomial $g \in F[x]$ if and only if g is divisible by f.*

(iii) *If f in (i) is of degree n then, $1, \alpha, \ldots, \alpha^{n-1}$ is a basis of $F(\alpha)$ over F. We have $[F(\alpha) : F] = n$ and each element of $F(\alpha)$ can be uniquely expressed as $a_0 + a_1\alpha + \cdots + a_{n-1}\alpha^{n-1}$, $a_i \in F$.*

Proof. (i) We consider $\psi\colon F[x] \to F[\alpha]$ defined by $g \mapsto \bar{g}(\alpha)$. Then $F[x]/\operatorname{Ker}\psi \cong F[\alpha]$. Since α is algebraic over F, the kernel of ψ is not zero and not $F[x]$, i.e., it is a proper ideal. $\operatorname{Ker}\psi$ is a principal ideal, say $\operatorname{Ker}\psi = (f)$, where f is irreducible. We may assume f is monic, since F is a field. The uniqueness of f is clear. By the irreducibility of f, (f) is maximal and $F[x]/(f)$ is a field. Consequently $F[\alpha]$ is a field and we have $F[\alpha] = F(\alpha)$, since $F(\alpha)$ is the smallest field which contains $F[\alpha]$.

(ii) and (iii) follow from $\operatorname{Ker}\psi = (f)$ and 12.11. □

The polynomial f in Theorem 12.24(i) plays an important role in field extensions.

12.25. Definitions. Let $\alpha \in L$ be algebraic over a field F. The unique, monic, irreducible polynomial $f \in F[x]$ which has α as a zero is called the ***minimal polynomial*** of α over F. The ***degree*** of α over F is defined as the degree of f (or equivalently as $[F(\alpha) : F]$).

12.26. Example. The minimal polynomial of $\sqrt[3]{2} \in \mathbb{Q}(\sqrt[3]{2})$ over $\mathbb{Q}$ is $x^3 - 2$. We have $\mathbb{Q}(\sqrt[3]{2}) = \mathbb{Q}[\sqrt[3]{2}]$ and $[\mathbb{Q}(\sqrt[3]{2}) : \mathbb{Q}] = 3$. The elements 1,

$\sqrt[3]{2}$, $\sqrt[3]{4}$ form a basis of $\mathbb{Q}(\sqrt[3]{2})$ over $\mathbb{Q}$ so that every element of $\mathbb{Q}(\sqrt[3]{2})$ can be uniquely expressed in the form $a_0 + a_1\sqrt[3]{2} + a_2\sqrt[3]{4}$, with $a_i \in \mathbb{Q}$.

12.27. Theorem. *An element α in an extension K of F is algebraic over F if and only if it is a zero of an irreducible polynomial $f \in F[x]$.*

Proof. This follows from the fact that α is a zero of f iff it is a zero of an irreducible factor of f. □

Next we describe a relationship between extensions K of a field F and vector spaces over F. If $\alpha_1, \dots, \alpha_n$ is a basis of K over F, then

$$F(\alpha_1, \dots, \alpha_n) = \{c_1\alpha_1 + \cdots + c_n\alpha_n \mid c_i \in F\}.$$

12.28. Theorem. *Any finite extension K of F is an algebraic extension.*

Proof. If $n := [K : F]$, then any set of $n + 1$ elements in K is linearly dependent. Let $\alpha \in K$. Then $1, \alpha, \alpha^2, \dots, \alpha^n$ in K are linearly dependent over F, i.e., there are $c_i \in F$, not all zero, such that $c_0 + c_1\alpha + \cdots + c_n\alpha^n = 0$. Thus α is a zero of the polynomial $g = c_0 + \cdots + c_nx^n \in F[x]$ and therefore it is algebraic. □

12.29. Example. $\mathbb{Q}(\sqrt{2})$ is algebraic over $\mathbb{Q}$, and $\mathbb{C}$ is algebraic over $\mathbb{R}$. The fields with four, eight, and nine elements in 12.12 are algebraic over $\mathbb{Z}_2$, $\mathbb{Z}_2$, and $\mathbb{Z}_3$, respectively.

We briefly mention that there do exist algebraic extensions of a field which are not finite. An important example of an infinite algebraic extension is the field of all ***algebraic numbers***, which consists of all algebraic elements over $\mathbb{Q}$, like $\sqrt[3]{\sqrt{19} + \sqrt[10]{23}}$. Furthermore, it can be shown that if a field L is algebraic over K and K is algebraic over F, then L is algebraic over F. We shall show a special version of this fact now.

In a certain sense, the following theorem represents a generalization of the Theorem of Lagrange 10.17 to the case of finite extensions.

12.30. Theorem. *Let L be a finite extension of K and let K be a finite extension of F. Then $[L : K][K : F] = [L : F]$.*

Proof. Let $\{\alpha_i \mid i \in I\}$ be a basis of L over K and let $\{\beta_j \mid j \in J\}$ be a basis of K over F. It is not hard to verify that the $|I| \cdot |J|$ elements $\{\alpha_i\beta_j \mid i \in I, j \in J\}$ form a basis of L over F. □

12.31. Corollaries. Let K be a finite extension of F.

(i) The degree of an element of K over F divides $[K : F]$.

(ii) An element in K generates K over F if and only if its degree over F is $[K : F]$.

(iii) If $[K : F] = 2^m$ and f is an irreducible polynomial over F of degree 3, then f is irreducible over K.

Part (iii) of this corollary enables us to give proofs on the impossibility of certain classical Greek construction problems, i.e., constructions with the use of ruler and compass only. We mention the problem of doubling the cube. Given a cube of volume 1, then the construction of a cube of twice this volume makes it necessary to solve $x^3 - 2 = 0$. This polynomial is irreducible over $\mathbb{Q}$. In general, equations of circles are of degree 2 and equations of lines are of degree 1 so that their intersection leads to equations of degree 2^m. This implies that Greek construction methods lead to fields of degree 2^m over $\mathbb{Q}$. The irreducibility of $x^3 - 2$ over $\mathbb{Q}$ implies that $\mathbb{Q}(\sqrt[3]{2})$ has degree 3 over $\mathbb{Q}$ and so it is impossible to construct a side of a cube with twice the volume of a given cube by using ruler and compass alone.

The problem of trisecting an angle is similar. It is equivalent to determining the cosine of one-third of a given angle φ. By analytic geometry this cosine must be a zero of $4x^3 - 3x - c$ where $c = \cos(3\varphi)$. In general, this is an irreducible polynomial over $\mathbb{Q}(c)$, so this implies the impossibility of trisecting arbitrary angles using only ruler and compass constructions.

12.32. Examples. $\mathbb{Q}(\sqrt{2}, \sqrt{3})$ is a field of degree 4 over $\mathbb{Q}$. Since $(\sqrt{2} + \sqrt{3})^3 - 9(\sqrt{2} + \sqrt{3}) = 2\sqrt{2}$, we see that $\sqrt{2}$ is an element of $\mathbb{Q}(\sqrt{2} + \sqrt{3})$. Since $\sqrt{3} = (\sqrt{2} + \sqrt{3}) - \sqrt{2}$, we have $\sqrt{3} \in \mathbb{Q}(\sqrt{2} + \sqrt{3})$, and so $\mathbb{Q}(\sqrt{2}, \sqrt{3}) = \mathbb{Q}(\sqrt{2} + \sqrt{3})$.

This example shows that at least in special cases it is possible to regard an algebraic extension $F(\alpha_1, \ldots, \alpha_n)$ of F as a simple algebraic extension $F(\alpha)$ for suitable α.

An irreducible polynomial f in $F[x]$ is called ***separable*** if there are no multiple zeros of f in its splitting field. An arbitrary polynomial in $F[x]$ is called ***separable*** if each of its irreducible factors is separable. An algebraic element α over F is called ***separable*** if its minimal polynomial is separable over F. An algebraic extension E of F is ***separable*** if every element of E

is separable over F. For instance, $i = \sqrt{-1}$ is separable over $\mathbb{R}$, since its minimal polynomial is $x^2 + 1 \in \mathbb{R}[x]$, which has the two distinct roots $i, -i$ in its splitting field $\mathbb{C}$. Actually, it is quite "rare" not to be separable—see Corollary 12.37 and Theorem 12.38.

12.33. Theorem. *Let F be a field and let $\alpha_1, \ldots, \alpha_n$ be algebraic and separable over F. Then there is an element α in $F(\alpha_1, \ldots, \alpha_n)$ such that $F(\alpha) = F(\alpha_1, \ldots, \alpha_n)$.*

See Lidl & Niederreiter (1994) for a proof. A special case is shown in Theorem 13.4. This theorem is useful in the study of field extensions since simple extensions are more easily handled than multiple extensions. It is therefore important to be able to determine the separability or nonseparability of a polynomial. Here we need the concept of the ***formal derivative***. If $f = a_0 + a_1x + \cdots + a_nx^n \in F[x]$, let $f' := a_1 + \cdots + na_nx^{n-1}$. We may verify immediately that

$$(af + bg)' = af' + bg',\ a' = 0,\ (fg)' = f'g + fg', \text{ and } (f \circ g)' = (f' \circ g)g',$$

hold for all $a, b \in F$ and $f, g \in F[x]$. So the map $D\colon F[x] \to F[x];\ f \mapsto f'$ is a linear map of the vector space $F[x]$ over F, with kernel containing F. We call D the ***differential operator*** and f' the ***derivative*** of f. Derivatives are most helpful in deciding whether a given polynomial $f \in F[x]$ is ***square-free***, which means that there is no nonconstant $g \in F[x]$ with $g^2 \mid f$. This seems to be a hopeless task without knowing the factors of f (which can be impossible, e.g., for $F = \mathbb{R}$ and $\deg f \geq 5$). Hence the following result is surprising.

12.34. Theorem. *If $f \in F[x]$ has $\gcd(f, f') = 1$, then f is square-free.*

Proof. Suppose that f is not square-free. Then there are some $g, h \in F[x]$ with $f = g^2h$ and $\deg g \geq 1$. But then $f' = 2gg'h + g^2h' = g(2g'h + gh')$, so $g \mid \gcd(f, f')$, whence $\gcd(f, f') \neq 1$. □

Observe that 12.34 even shows that if $f \in F[x]$ is relatively prime to f', then f cannot have square factors in *any* possible extension field of F. So, for example, $f = x^{10000} + x^{200} + x^{10} + x + 1 \in \mathbb{Z}_5[x]$ can never have a square factor since $f' = 1$, so $\gcd(f, f') = 1$.

12.35. Theorem. *A polynomial f over F has no multiple zeros in its splitting field if and only if $\gcd(f, f') = 1$.*

Proof. Let S be the splitting field of f over F, so $f = c(x-s_1)\cdots(x-s_n)$ with $s_1, \ldots, s_n \in S$. Let $f_i := \prod_{j\neq i}(x-s_j)$. Then $f' = c(f_1+\cdots+f_n)$. If $\gcd(f, f') > 1$, then there must be some $k \in \{1, \ldots, n\}$ with $(x-s_k) \mid f'$. Now $x-s_k$ divides all f_i with $i \neq k$. So it also divides $f_k = \prod_{j\neq k}(x-s_j)$, which means there must be some $j \neq k$ with $s_j = s_k$. Hence f has multiple zeros. The converse is clear from the last result. □

12.36. Corollary. *An irreducible polynomial f over F is separable if and only if its derivative is nonzero.*

12.37. Corollary. *Every algebraic extension of a field F of characteristic 0 is separable.*

12.38. Theorem. *Every algebraic extension of a finite field is separable.*

The proof will follow later (see the lines after 14.1).

A field F is called ***algebraically closed*** if any nonconstant polynomial in $F[x]$ splits into linear factors in $F[x]$. A field $\bar{F}$ is called an ***algebraic closure*** of a field F, if $\bar{F}$ is algebraically closed and is an algebraic extension of F. We note that this is equivalent to saying that F does not have any algebraic extensions which properly contain $\bar{F}$. It is easy to see that a polynomial of degree $n > 0$ in $F[x]$, F algebraically closed, can be expressed as a product of n linear polynomials. In this context, we repeat, in other wording, the important Theorem 11.18, for which there are more than one hundred proofs, the first of which was given by C. F. Gauss in 1799.

12.39. Theorem (*Fundamental Theorem of Algebra*). *The field of complex numbers is algebraically closed.*

The field of algebraic numbers, mentioned after Example 12.29, can also be shown to be algebraically closed.

Most algebraists think that transcendental extensions have few applications to "practical" purposes. But recently, Binder (1996) has found a surprising application to "down-to-earth" problems. Let us take, for instance, $p = x^3 + 3x$ and $q = x^5 + x^4 \in \mathbb{R}[x]$. We know from 11.21 that there are polynomials $r, s \in \mathbb{R}[x]$ with $x = \gcd(p, q) = rp + sq$. Can x also be obtained from p, q as an "expression" with real coefficients (instead of the polynomials r, s)? We need a result, the proof of which can be found, for example, in van der Waerden (1970).

12.40. Theorem (*Lüroth's Theorem*). *Let F be a field. Then all fields between F and its simple transcendental extension $F(x)$ are simple, i.e., of the* *form $F(g)$ for some $g \in F(x)$. Furthermore, if $F(g)$ contains some nonconstant polynomial, g can be chosen as a polynomial.*

Note that the elements of $F(g)$ are of the form $h(g)$ for some $h \in F(x)$. In particular, if g is a polynomial, its degree divides the degrees of all of the polynomials in $F(g)$ (cf. Exercise 11.18).

So, returning to the example above, we see immediately from the degrees of x^3+3x and x^5+x^4, which are relatively prime, that $\mathbb{R}(x^3+3x, x^5+x^4) = \mathbb{R}(x)$; so there must be some $t \in \mathbb{R}(x, y)$ with $t(x^3+3x, x^5+x^4) = x$. To find such a t, however, is considerably more difficult than to prove its existence. Algorithms for this problem are discussed in Binder (1996). In our example, the shortest known solution is

$$x = \frac{p^3 - 3p^2 + pq + 36p + 9q}{7p^2 + pq - 9p - 3q + 108},$$

which we certainly could not find by guessing.

Exercises

1. Let $R = \{a + b\sqrt{2} \mid a, b \in \mathbb{Z}\}$. Define operations $+$ and $\cdot$ and show that $(R, +, \cdot)$ is a commutative ring with identity. Determine the quotient field of R in the field $\mathbb{R}$.
2. Let G denote the set of complex numbers $\{a + bi \mid a, b \in \mathbb{Z}\}$. With the usual addition and multiplication of complex numbers, G forms the domain of ***Gaussian integers***. Show that its quotient field is isomorphic to the subring of $\mathbb{C}$ consisting of $\{p + qi \mid p, q \in \mathbb{Q}\}$.
3. If F is a subfield of F', must F and F' have the same characteristic?
4. Show that $f = x^3 + x + 1$ is irreducible over $\mathbb{Z}_2$. Determine the elements of the field $\mathbb{Z}_2[x]/(f)$ and show that this field is the splitting field of f over $\mathbb{Z}_2$.
5. Show that $\alpha = \sqrt{2} + i$ is of degree 4 over $\mathbb{Q}$ and of degree 2 over $\mathbb{R}$. Determine the minimal polynomial of α in both cases.
6. Determine the multiplicative inverse of $1 + \sqrt[3]{2} + \sqrt[3]{4}$ in $\mathbb{Q}(\sqrt[3]{2})$.
7. Describe the elements of $\mathbb{Z}_2[x]/(x)$.
8. Find the degree and a basis for each of the given field extensions.

   (i) $\mathbb{Q}(\sqrt{2}, \sqrt{3}, \sqrt{5})$ over $\mathbb{Q}$.

   (ii) $\mathbb{Q}(\sqrt{2}, \sqrt{6})$ over $\mathbb{Q}(\sqrt{3})$.

(iii) $\mathbb{Q}(\sqrt{2}, \sqrt[3]{2})$ over $\mathbb{Q}$.

(iv) $\mathbb{Q}(\sqrt{2}+\sqrt{3})$ over $\mathbb{Q}(\sqrt{3})$.

9. (i) Find the splitting field of $(x^2-3)(x^3+1)$ over $\mathbb{Q}$.

   (ii) Find the splitting field of $(x^2-2x-2)(x^2+1)$ over $\mathbb{Q}$.

   (iii) Find the splitting field of x^2-3 and x^2-2x-2 over $\mathbb{Q}$.

   (iv) Find the splitting field of x^2+x+1 over $\mathbb{Z}_2$.

   (v) Find the splitting field of x^3+x+2 over $\mathbb{Q}$.

10. Show directly that every complex number is algebraic over $\mathbb{R}$ (cf. the last line of 12.21).

11. Show that an element α in an extension K of a field F is transcendental (algebraic) over F, if the map ψ: $F(x) \to F(\alpha)$; $f \mapsto \bar{f}(\alpha)$ is an isomorphism (not an isomorphism).

12. Let L be an extension of F, let $f \in F[x]$, and let φ be an automorphism of L that maps every element of F into itself. Prove that φ must map a root of f in L into a root of f in L.

13. Deduce algebraically that by repeated bisection it is possible to divide an arbitrary angle into four equal parts. (Use a relationship between $\cos 4\theta$ and $\cos\theta$.)

14. (i) Can the cube be "tripled"?

    (ii) Can the cube be "quadrupled"?

15. A regular n-gon is ***constructible*** for $n \geq 3$ if and only if the angle $2\pi/n$ is constructible. $2\pi/n$ is constructible if and only if a line segment of length $\cos(2\pi/n)$ is constructible. Prove: If the regular n-gon is constructible and if an odd prime p divides n, then p is of the form $2^{2^k}+1$.

16. Given a segment s, show that it is impossible to construct segments m and n such that $s : m = m : n = n : 2s$.

17. Determine whether the following polynomials have multiple roots:

    (i) $x^4-5x^3+6x^2+4x-8 \in \mathbb{Q}[x]$;

    (ii) $x^{32} \in \mathbb{R}[x]$;

    (iii) $x^{10}+x^9+x^8+x^7+x^6+x^5+x^4+x^3+x^2+x+1 \in \mathbb{Z}_2[x]$.

18. Is it possible to construct a regular 7-gon using ruler and compass constructions?

19. Show that an automorphism of a field maps every element of its prime field into itself.

20. Does it make sense to speak of a "transcendental closure"?
21. Suppose that for $f \in \mathbb{Z}_q[x]$ we have $\gcd(f, f') = f$, where $q = p^k$. Show that there exists some $g \in \mathbb{Z}_q[x]$ with $f = g \circ x^p$.

§13 Finite Fields

A field with m elements ($m \in \mathbb{N}$) is called a finite field of order m. One of the main aims of this section is to show that m must be the power of a prime and that, conversely, for any prime power p^n there is (up to isomorphism) exactly one finite field of this order. This field is the splitting field of $x^{p^n} - x$ over $\mathbb{Z}_p$. We know from §12 that a finite field F must be of prime characteristic p and that the prime field of F is isomorphic to $\mathbb{Z}_p$. We shall identify the prime field of a finite field with $\mathbb{Z}_p$, i.e., we shall regard any field of prime characteristic p as an extension field of the field $\mathbb{Z}_p$ of the integers mod p.

13.1. Theorem. *Let F be a finite field of characteristic p. Then F contains p^n elements, where $n = [F : \mathbb{Z}_p]$.*

Proof. The field F, considered as a vector space over its prime field $\mathbb{Z}_p$, contains a finite basis of n elements. Each element of F can be expressed as a unique linear combination of the n basis elements with coefficients in $\mathbb{Z}_p$. Therefore there are p^n elements in F. □

13.2. Theorem. *Let F be a finite field with q elements.*

(i) *The multiplicative group $(F^*, \cdot)$ of the nonzero elements of F is cyclic of order $q - 1$.*

(ii) *All elements a of F satisfy $a^q - a = 0$.*

Proof. We prove the theorem by using the Principal Theorem on finite abelian groups in the form 10.26. The multiplicative group G of nonzero elements of F is a group of order $q - 1$. G is a direct product of cyclic subgroups $U_1, \dots, U_m$, where $|U_i|$ divides $|U_{i+1}|$. This implies that the order of each element in G divides the order r of U_m. For any element a in G we therefore have $a^r - 1 = 0$. The polynomial $x^r - 1$ over F can have at most r zeros in F, hence $|G| = q - 1 \le r$. Since $|U_m|$ divides $|G|$, we have $r \le q - 1$, so $r = q - 1$, thus $G = U_m$, which proves (i). Part (ii) follows from the fact that any nonzero element is a zero of $x^{q-1} - 1$, as mentioned above. □

In the special case $F = \mathbb{Z}_p$ we have elements $a \in \mathbb{Z}_p$ such that the powers $a, a^2, \dots, a^{p-1}$ represent all nonzero elements of $\mathbb{Z}_p$, e.g.,

p	2	3	5	7	11	13	17	19	23	29	31	37	41	43	47	53	59
a	1	2	2	3	2	2	3	2	5	2	3	2	6	3	5	2	2

Such an element a is called a ***primitive root modulo*** p. A generator of the cyclic group of a finite field F is called a ***primitive element***.

13.3. Theorem. *Let α be a primitive element for the finite field F. If $|F| = q$, then*

$$F = \{0, 1, \alpha, \alpha^2, \dots, \alpha^{q-2}\},$$

where $\alpha^{q-1} = 1$. Moreover, α^k is also primitive iff $\gcd(k, q-1) = 1$.

The first statement is trivial, the rest follows from Theorem 10.20.

13.4. Theorem. *Let F be a finite field and let $\alpha_1, \dots, \alpha_k$ be algebraic over F. Then $F(\alpha_1, \dots, \alpha_k) = F(\alpha)$ for some α in $F(\alpha_1, \dots, \alpha_k)$.*

Proof. The extension field $F(\alpha_1, \dots, \alpha_k)$ is finite over F by 12.30, so it is an algebraic extension of F. Therefore it is a finite field with cyclic multiplicative group. If α is a generating element of this group, then the theorem follows. □

We therefore have shown Theorem 12.33 for this case.

13.5. Corollary. *Let F be a finite field of characteristic p and let $[F : \mathbb{Z}_p] = n$. Then there is an element α in F such that α is algebraic of degree n over $\mathbb{Z}_p$ and $F = \mathbb{Z}_p(\alpha)$.*

Theorem 13.2(ii) ensures that any finite field F consists of the roots of the polynomial $x^{p^n} - x$ for some n where $p = \operatorname{char} F$. The following theorem describes all finite fields and shows that there is precisely one finite field for any prime power p^n.

13.6. Theorem.

(i) *Any finite field F is of order p^n where $p = \operatorname{char} F$ is a prime and n is a positive integer.*

(ii) *For any prime p and any $n \in \mathbb{N}$ there is a field of order p^n.*

(iii) *Any field of order p^n is (up to isomorphism) the splitting field of $x^{p^n} - x$ and also of $x^{p^n-1} - 1 \in \mathbb{Z}_p[x]$.*

(iv) *Any two fields of order p^n are isomorphic.*

Proof. (i) follows from Theorem 13.1. Let K be the splitting field of $x^{p^n} - x =: f$ over $\mathbb{Z}_p$. Since $f' = -1$, Theorem 12.37 shows that all roots are simple. Now f has p^n roots in K. It is easily verified that sums, products, and inverses of roots of f in K are also roots of f. Thus the roots of f form a field with p^n elements, which must be the splitting field of f over $\mathbb{Z}_p$ by Theorem 12.18. This implies $[K : \mathbb{Z}_p] = n$, which proves (ii) and (iii). The uniqueness (iv) follows from the uniqueness of the splitting field (see 12.19). □

This implies a result which is in sharp contrast to the fields $\mathbb{R}$ and $\mathbb{C}$ (cf. 12.39).

13.7. Corollary. *For any positive integer n, the finite field of p^n elements is the splitting field of an irreducible polynomial of degree n in $\mathbb{Z}_p[x]$, and there is an irreducible polynomial of degree n in $\mathbb{Z}_p[x]$.*

Theorem 13.6 enables us to speak of *the* finite field with p^n elements. This field is also called the ***Galois field***, in honor of Evariste Galois (1811–1832) and is denoted by $GF(p^n)$ or $\mathbb{F}_{p^n}$. The multiplicative group of $\mathbb{F}_{p^n}$ is denoted by $\mathbb{F}_{p^n}^*$. The prime field of $\mathbb{F}_{p^n}$ is isomorphic to $\mathbb{Z}_p$. Note that $\mathbb{F}_p$ is the same as $\mathbb{Z}_p$.

The results obtained so far make it possible to determine the elements of a finite field. We know that $\mathbb{F}_{p^n}$ is a vector space of dimension n over $\mathbb{F}_p$. Moreover, it is a simple extension of the prime field $\mathbb{F}_p$, say $\mathbb{F}_{p^n} = \mathbb{F}_p(\alpha)$, and any $n + 1$ elements of $\mathbb{F}_{p^n}$ are linearly dependent, so that there are $a_0, a_1, \ldots, a_n \in \mathbb{F}_p$ with $a_0 + a_1\alpha + \cdots + a_n\alpha^n = 0$. This means that α is a root of the polynomial $a_0 + a_1x + \cdots + a_nx^n$ in $\mathbb{F}_p[x]$. Let f be the minimal polynomial of α, then $\mathbb{F}_{p^n} = \mathbb{F}_p(\alpha) \cong \mathbb{F}_p[x]/(f)$. In order to obtain the elements of $\mathbb{F}_{p^n}$ explicitly, we determine an irreducible monic polynomial of degree n over $\mathbb{F}_p$ (from tables, for instance) and form $\mathbb{F}_p[x]/(f)$. More generally, to obtain $\mathbb{F}_{q^m}$, $q = p^n$, we find an irreducible, monic polynomial g of degree m over $\mathbb{F}_q$ and form $\mathbb{F}_q[x]/(g)$, which is then isomorphic to $\mathbb{F}_{q^m}$.

13.8. Example. We determine the elements of $\mathbb{F}_{2^3}$. If we regard $\mathbb{F}_{2^3}$ as a simple extension of degree 3 of the prime field $\mathbb{F}_2$, then this extension is obtained by adjoining to $\mathbb{F}_2$ a root of an irreducible cubic polynomial over $\mathbb{F}_2$. It is easily verified that x^3+x+1 and x^3+x^2+1 are irreducible over $\mathbb{F}_2$. Therefore $\mathbb{F}_{2^3} \cong \mathbb{F}_2[x]/(x^3+x+1)$ and also $\mathbb{F}_{2^3} \cong \mathbb{F}_2[x]/(x^3+x^2+1)$. Let α be a root of $f = x^3+x+1$, then $1, \alpha, \alpha^2$ form a basis of $\mathbb{F}_{2^3}$ over $\mathbb{F}_2$. The elements of $\mathbb{F}_{2^3}$ are of the form

$$a + b\alpha + c\alpha^2 \quad \text{for all } a, b, c \in \mathbb{F}_2 \quad \text{with } \alpha^3 + \alpha + 1 = 0,$$

as we have seen in 12.12(iii). We can also use $g = x^3+x^2+1$ to determine the elements of $\mathbb{F}_{2^3}$. Let β be a root of g, so $\beta^3+\beta^2+1=0$. It can be easily verified that $\beta+1$ is a root of f in $\mathbb{F}_2[x]/(g)$. The two fields $\mathbb{F}_2[x]/(f)$ and $\mathbb{F}_2[x]/(g)$ are splitting fields of x^8-x and are thus isomorphic. Therefore there is an isomorphism ψ such that $\psi(\alpha) = \beta+1$ and ψ restricted to $\mathbb{F}_2$ is the identity mapping. The elements $1, \beta+1, (\beta+1)^2$ form a basis of $\mathbb{F}_2[x]/(g)$ over $\mathbb{F}_2$. Thus the isomorphism ψ is given by

$$\psi(a + b\alpha + c\alpha^2) = a + b(\beta+1) + c(\beta+1)^2 \qquad \text{with } a, b, c \in \mathbb{F}_2.$$

The multiplication table of the multiplicative group $\mathbb{F}_2[x]/(x^3+x^2+1)\backslash\{0\}$ is as follows (β is as above):

$\cdot$	1	β	$\beta+1$	β^2	$\beta^2+\beta$	β^2+1	$\beta^2+\beta+1$
1	1	β	$\beta+1$	β^2	$\beta^2+\beta$	β^2+1	$\beta^2+\beta+1$
β	β	β^2	$\beta^2+\beta$	β^2+1	1	$\beta^2+\beta+1$	$\beta+1$
$\beta+1$	$\beta+1$	$\beta^2+\beta$	β^2+1	1	$\beta^2+\beta+1$	β	β^2
β^2	β^2	β^2+1	1	$\beta^2+\beta+1$	β	$\beta+1$	$\beta^2+\beta$
$\beta^2+\beta$	$\beta^2+\beta$	1	$\beta^2+\beta+1$	β	$\beta+1$	β^2	$\beta+1$
β^2+1	β^2+1	$\beta^2+\beta+1$	β	$\beta+1$	β^2	$\beta^2+\beta$	1
$\beta^2+\beta+1$	$\beta^2+\beta+1$	$\beta+1$	β^2	$\beta^2+\beta$	β^2+1	1	β

Hence

$$\begin{aligned}\mathbb{F}_2[x]/(g) &= \{0, 1, \beta, \beta+1, \beta^2, \beta^2+\beta, \beta^2+1, \beta^2+\beta+1\} \\ &= \{0, 1, \beta, \beta^2, \beta^3, \beta^4, \beta^5, \beta^6\}\end{aligned}$$

with $\beta^3 = \beta^2+1$ and $\beta^7 = 1$.

So in every finite field $\mathbb{Z}_p[x]/(f)$ we have two possibilities to represent the elements: as "polynomials" in $\alpha = x+(f)$ and, except for 0, as powers of a primitive element. Later we shall see that this primitive element can be chosen to be $\alpha = x+(f)$, provided f is "maximally periodic." For addition and subtraction, the "polynomial representation" is very convenient; for multiplying, dividing, and raising powers, the "power representation" is much better. For example, in 13.8 we can see immediately that

$$(\beta^2)^{10} = \beta^{20} = \beta^6 \qquad \text{(since } \beta^7 = 1\text{)}$$

and

$$\sqrt[3]{\beta^6} = \beta^2, \quad \sqrt[3]{\beta^2} = \sqrt[3]{\beta^9} = \beta^3, \quad \text{etc.}$$

The conversion between these two representations is highly nontrivial: how can we see, for instance, for which k we have $\beta^k = 1 + \beta^2$? This number k is the ***discrete logarithm*** of $1+\beta^2$, and the problem of computing these logarithms (other than just trying) seems computationally so hard that it can be used to transmit secret messages (see §23). For small fields we can write down a conversion table, sometimes via the "Zech logarithm" (see Exercise 26). For the example in 13.8 we get

polynomial	power of β
1	0
β	1
$1+\beta$	5
β^2	2
$1+\beta^2$	3
$\beta+\beta^2$	6
$1+\beta+\beta^2$	4

so $1 + \beta = \beta^5$, etc., and we also write $\log_\beta(1 + \beta) = 5$. Sometimes the logarithm is also called the ***index*** and we write $\mathrm{ind}_\beta(1 + \beta) = 5$.

If F is a subfield of order p^m in $\mathbb{F}_{p^n}$, then F is the splitting field of $x^{p^m} - x$ in $\mathbb{F}_{p^n}$ over $\mathbb{F}_p$. We shall describe all subfields of a finite field. The following lemma can be proved as Exercise 12.

13.9. Lemma. *Let F be a field and p a prime. The following are equivalent:*

(i) $m \mid n$;

(ii) $p^m - 1 \mid p^n - 1$;

(iii) $x^m - 1 \mid x^n - 1$.

13.10. Theorem. *Let p be a prime and let m, n be natural numbers.*

(i) *If $\mathbb{F}_{p^m}$ is a subfield of $\mathbb{F}_{p^n}$, then $m \mid n$.*

(ii) *If $m \mid n$, then $\mathbb{F}_{p^m} \hookrightarrow \mathbb{F}_{p^n}$. There is exactly one subfield of $\mathbb{F}_{p^n}$ with p^m elements.*

Proof.

(i) Theorem 12.30 implies

$$[\mathbb{F}_{p^n} : \mathbb{F}_p] = [\mathbb{F}_{p^n} : \mathbb{F}_{p^m}][\mathbb{F}_{p^m} : \mathbb{F}_p].$$

Since the term on the left-hand side is n and the second factor on the right-hand side is m, we have $m \mid n$.

(ii) If $m \mid n$, we have $p^m - 1 \mid p^n - 1$, thus (by 13.9) $x^{p^m-1} - 1 \mid x^{p^n-1} - 1$ and $x^{p^m} - x \mid x^{p^n} - x$. The roots of $x^{p^m} - x$ form a subfield of $\mathbb{F}_{p^n}$ of order p^m, which is isomorphic to $\mathbb{F}_{p^m}$. There cannot be another subfield with p^m elements, because otherwise there would be more than p^m roots of $x^{p^m} - x$ in $\mathbb{F}_{p^n}$. □

13.11. Example. We draw a diagram of all subfields of $\mathbb{F}_{2^{30}}$.

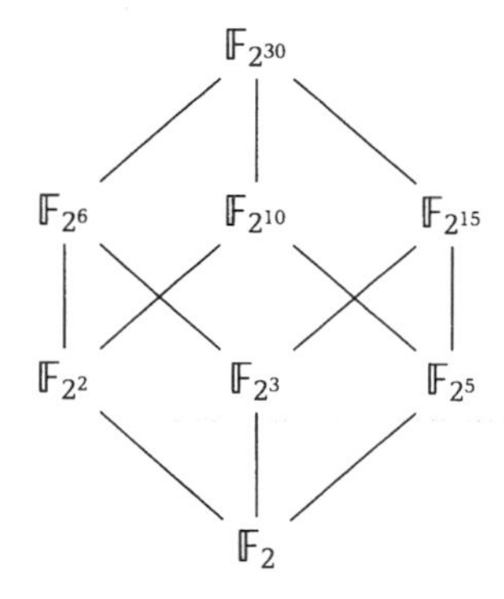

FIGURE 13.1

Because of Theorem 13.10 and the property $m! \mid n!$ for positive integers $m < n$, we have an ascending chain of fields

$$\mathbb{F}_p \subset \mathbb{F}_{p^{2!}} \subset \mathbb{F}_{p^{3!}} \subset \ldots.$$

We define $\mathbb{F}_{p^\infty}$ as $\bigcup_n \mathbb{F}_{p^{n!}}$ and note that $\mathbb{F}_{p^\infty}$ is a field which contains $\mathbb{F}_{p^n}$ as a subfield for any positive integer n. Each element in $\mathbb{F}_{p^\infty}$ is of finite multiplicative order, but $\mathbb{F}_{p^\infty}$ itself is infinite, with finite characteristic p. The field $\mathbb{F}_{p^\infty}$ can be shown to be algebraically closed; it is the algebraic closurc of $\mathbb{F}_p$.

Of importance in field theory is the set of automorphisms of an extension K of F which fix all elements of F. It is clear that this set G forms a group under composition of mappings. G is called the ***Galois group*** of

K over F. In the case of automorphisms of finite fields $\mathbb{F}_{p^n}$, all elements of $\mathbb{F}_p$ remain automatically fixed. Thus G consists of all automorphisms of $\mathbb{F}_{p^n}$.

13.12. Definition. Let $q = p^n$. The mapping $\theta\colon \mathbb{F}_q \to \mathbb{F}_q;\ a \mapsto a^p$, is called the ***Frobenius automorphism*** of $\mathbb{F}_q$.

It can be verified (see Exercise 11.10) that θ is an automorphism. If a is a generating element of $\mathbb{F}_q^*$ of order $q-1$, then $\theta^n(a) = a^{p^n} = a$. For $i = 1, 2, \ldots, n-1$ we have $\theta^i(a) = a^{p^i} \neq a$; therefore θ is an automorphism of order n. We state without proof:

13.13. Theorem. *The group G of automorphisms of $\mathbb{F}_{p^n}$ over $\mathbb{F}_p$ is cyclic of order n and G consists of the elements $\theta, \theta^2, \ldots, \theta^{n-1}, \theta^n = \mathrm{id}$.*

Finally, we consider a generalization of the well-known concept of complex roots of unity. In $\mathbb{C}$ the nth roots of unity are $z_k = e^{2\pi ik/n}$, $k = 0, 1, \ldots, n-1$. Geometrically these roots can be represented by the n vertices of a regular polygon in the unit circle in the complex plane. All z_k with $\gcd(k, n) = 1$ are generators by Theorem 10.20, so in particular $z_1 = e^{2\pi i/n}$ is a generator. They are called ***primitive*** nth ***roots of unity.*** For an arbitrary field F, we define:

13.14. Definitions. Let F be a field. A root of $x^n - 1$ in $F[x]$ is called an nth ***root of unity***. The ***order*** of an nth root α of unity is the least positive integer k such that $\alpha^k = 1$. An nth root of unity of order n is called ***primitive***. The splitting field S_n of $x^n - 1 \in F[x]$ is called the ***associated cyclotomic field***.

13.15. Definition. Let φ be ***Euler's phi-function*** (or the ***totient function***) where $\varphi(n)$ indicates the number of positive integers less than or equal to n that are relatively prime to n (see Exercise 6).

If $n = p_1^{t_1} \ldots p_k^{t_k}$, where p_i are distinct primes, then

$$\varphi(n) = n\left(1 - \frac{1}{p_1}\right) \cdots \left(1 - \frac{1}{p_k}\right).$$

13.16. Theorem. *Let n be a positive integer and let F be a field whose characteristic does not divide n.*

(i) *There is a finite extension K of F which contains a primitive nth root of unity.*

(ii) *If α is a primitive nth root of unity, then $F(\alpha)$ is the splitting field of $f = x^n - 1$ over F.*

(iii) *$x^n - 1$ has exactly n distinct roots in $F(\alpha)$. These roots form a cyclic group. The order of an nth root of unity α is just the order of α in this group. The primitive nth roots of unity in $F(\alpha)$ are precisely the generators of this group. There are $\varphi(n)$ primitive nth roots of unity, which can be obtained from one primitive root by raising it to the powers $k < n$ with* $\gcd(k, n) = 1$.

Proof. The proofs of (i) and (ii) are similar to that of 13.6 and are omitted. We show that the set of nth roots of unity is a cyclic group and leave the rest of (iii) to the reader. Let $\operatorname{char} F = p$. Then the extension $F(\alpha)$ of F contains the splitting field S_n of $x^n - 1$ over $\mathbb{F}_p$. This field S_n is finite and hence has a cyclic multiplicative group. The roots $\alpha_1, \ldots, \alpha_n$ of $x^n - 1$ in S_n form a subgroup S of S_n^*, which is cyclic. Since $\operatorname{char} F \nmid n$, all roots $\alpha_1, \ldots, \alpha_n$ are distinct by Theorem 12.35. The case where $\operatorname{char} F = 0$ must be treated separately (see Lidl & Niederreiter (1994)). □

In factoring $x^n - 1$ into irreducible factors, a task which will turn out to be essential in coding theory, the so-called cyclotomic polynomials are useful.

13.17. Definition. Let n and F be as in 13.16, and let α be a primitive nth root of unity. The polynomial

$$Q_n := (x - \alpha_1) \cdots (x - \alpha_{\varphi(n)}) \in F(\alpha)[x]$$

where $\alpha_1, \ldots, \alpha_{\varphi(n)}$ are the $\varphi(n)$ primitive nth roots of unity in $F(\alpha)$, is called the *nth **cyclotomic polynomial** over* F. We shall see below that Q_n does not depend on α.

13.18. Example. Let $n = 8$ and $F = \mathbb{F}_3$. Since $n = 3^2 - 1$, we might find a primitive eighth root of unity in $\mathbb{F}_9 = \mathbb{Z}_3[x]/(x^2 + x + 2)$ due to 13.6(iii). We might try $\alpha = [x] = x + (x^2 + x + 2)$ and succeed (in 14.24 we shall see why we succeed): α is a primitive eighth root of unity. The other primitive eighth roots unity are α^3, α^5, and α^7 (recall that $\varphi(8) = 4$); so we get $Q_8 = (x - \alpha)(x - \alpha^3)(x - \alpha^5)(x - \alpha^7)$. Expanding this and observing that $\alpha^2 = 2\alpha + 1$, we get $Q_8 = x^4 + 1$. Hence Q_8 is not only in $\mathbb{F}_9[x]$ but even in $\mathbb{F}_3[x]$. From Corollary 13.24 we shall see that this surprising fact is always true: Q_8 *always* has coefficients in $\mathbb{F}_p$.

Let α be a primitive nth root of unity. Then it follows from 13.16 that $Q_n = \prod_i (x - \alpha^i)$, where the product is formed over all i with $\gcd(i, n) = 1$. The polynomial Q_n is of degree $\varphi(n)$. Let $n = kd$ so that α^k is of order d and is a primitive dth root of unity. The dth cyclotomic polynomial is of the form

$$Q_d = \prod_{\gcd(i,d)=1} (x - \alpha^{ik}).$$

Any nth root of unity is a primitive dth root of unity for exactly one d. Therefore we can group the nth roots of unity together and obtain

13.19. Theorem.

$$x^n - 1 = \prod_{d|n} Q_d.$$

For example, $Q_2 = x + 1$ over any field F, since $x^2 - 1 = (x + 1)(x - 1)$ and -1 is the only primitive second root of unity. Right now, it seems to be hard to compute Q_n explicitly. A first result:

13.20. Theorem. *Let p be a prime number, F a field not of characteristic p, and m a positive integer. Then*

$$Q_{p^m} = 1 + x^{p^{m-1}} + \cdots + x^{(p-1)p^{m-1}}.$$

Proof. Theorem 13.19 shows

$$Q_{p^m} = \frac{x^{p^m} - 1}{Q_1 Q_p \cdots Q_{p^{m-1}}} = \frac{x^{p^m} - 1}{x^{p^{m-1}} - 1}$$

which yields the result. □

The decomposition of $x^n - 1$ in 13.19 does not necessarily give irreducible factors, and these factors seem to live in "large" extensions of F. So it will come as a pleasant surprise that cyclotomic polynomials have in fact coefficients in F. Theorem 13.19 is called the ***cyclotomic decomposition*** of $x^n - 1$. Using the so-called Möbius inversion formula we can derive a formula for cyclotomic polynomials.

13.21. Definition. The mapping $\mu\colon \mathbb{N} \to \{0, 1, -1\}$ defined by

$$\begin{aligned} \mu(1) &:= 1, \\ \mu(p_1 \dots p_t) &:= (-1)^t \qquad \text{if } p_i \text{ are distinct primes,} \\ \mu(n) &:= 0 \qquad\qquad \text{if } p^2 \mid n \text{ for some prime } p, \end{aligned}$$

is called the ***Möbius function*** or μ-function.

There is a very simple and useful property of the μ-function.

13.22. Lemma.

$$\sum_{d|n} \mu(d) = \begin{cases} 1 & \text{if } n = 1, \\ 0 & \text{if } n > 1. \end{cases}$$

To verify this, for $n > 1$, we have to take into account only those positive divisors d of n for which $\mu(d) \neq 0$, i.e., for which $d = 1$ or d is a product of distinct primes. Thus, if $p_1, p_2, \dots, p_k$ are the distinct prime divisors of n, we get

$$\begin{aligned} \sum_{d|n} \mu(d) &= \mu(1) + \sum_{i=1}^{k} \mu(p_i) + \sum_{1 \le i_1 < i_2 \le k} \mu(p_{i_1} p_{i_2}) + \cdots + \mu(p_1 p_2 \dots p_k) \\ &= 1 + \binom{k}{1}(-1) + \binom{k}{2}(-1)^2 + \cdots + \binom{k}{k}(-1)^k = (1 + (-1))^k = 0. \end{aligned}$$

The case $n = 1$ is trivial.

13.23. Theorem (*Möbius Inversion Formula*).

(i) (Additive Form) *Let $f, g\colon \mathbb{N} \to (A, +)$ be mappings from $\mathbb{N}$ into an additive abelian group A; then*

$$g(n) = \sum_{d|n} f(d) \iff f(n) = \sum_{d|n} \mu\left(\frac{n}{d}\right) g(d).$$

(ii) (Multiplicative Form) *Let $f, g\colon \mathbb{N} \to (A, \cdot)$ be mappings from $\mathbb{N}$ into a multiplicative abelian group A; then*

$$g(n) = \prod_{d|n} f(d) \iff f(n) = \prod_{d|n} g(d)^{\mu(\frac{n}{d})}.$$

Proof. We show the additive form of the inversion formula. Assuming $g(n) = \sum_{d|n} f(d)$ we get from 13.22

$$\sum_{d|n} \mu\left(\frac{n}{d}\right) g(d) = \sum_{d|n} \mu(d) g\left(\frac{n}{d}\right) = \sum_{d|n} \mu(d) \sum_{c|\frac{n}{d}} f(c) = \sum_{c|n} \sum_{d|\frac{n}{c}} \mu(d) f(c)$$

$$f = \sum c \mid nf(c) \sum_{d|\frac{n}{c}} \mu(d) = f(n).$$

The converse is derived by a similar calculation. The multiplicative form follows from the additive form by replacing sums with products and products with powers. □

13.24. Corollary.

$$Q_n = \prod_{d|n} (x^d - 1)^{\mu(\frac{n}{d})} = \prod_{d|n} (x^{\frac{n}{d}} - 1)^{\mu(d)}.$$

It can be verified that all cyclotomic polynomials over $\mathbb{Q}$ have integer coefficients, while in the case of fields with characteristic p the coefficients are to be taken $\bmod p$. So Q_n is largely independent of α, of F, and of char F, a surprising fact. Cyclotomic polynomials can be shown to be irreducible over $\mathbb{Q}$, but not necessarily over $\mathbb{Z}_p$. A curious property of Q_n is that the first 104 cyclotomic polynomials have coefficients in $\{0, 1, -1\}$ only. In Q_{105} we have 2 as one of the coefficients (see Exercise 17).

13.25. Examples.

n	Q_n
1	$x - 1$
2	$x + 1$
3	$x^2 + x + 1$
4	$x^2 + 1$
5	$x^4 + x^3 + x^2 + x + 1$
6	$x^2 - x + 1$
7	$x^6 + x^5 + x^4 + x^3 + x^2 + x + 1$
8	$x^4 + 1$
9	$x^6 + x^3 + 1$
10	$x^4 - x^3 + x^2 - x + 1$
11	$x^{10} + x^9 + x^8 + x^7 + x^6 + x^5 + x^4 + x^3 + x^2 + x + 1$
12	$x^4 - x^2 + 1$
13	$x^{12} + x^{11} + x^{10} + x^9 + x^8 + x^7 + x^6 + x^5 + x^4 + x^3 + x^2 + x + 1$
14	$x^6 - x^5 + x^4 - x^3 + x^2 - x + 1$
15	$x^8 - x^7 + x^5 - x^4 + x^3 - x + 1$

It follows from 13.4 and 13.16 that the cyclotomic field S_n can be constructed as a simple extension of $\mathbb{Z}_p$ by using a polynomial which divides Q_n. The finite field $\mathbb{F}_{p^n}$ is the cyclotomic field of the (p^n-1)th roots of unity. In $\mathbb{F}_{p^n}$ we have

$$x^{p^n-1}-1=\prod_{d|p^n-1} Q_d.$$

As mentioned above, the polynomial Q_d is in general not irreducible. For instance, $Q_4=x^2+1=(x+1)(x+1)$ over $\mathbb{Z}_2$. But Q_d has as roots all elements of order d in $\mathbb{F}_{p^n}$. An element of order d in an extension of $\mathbb{F}_p$ has a minimal polynomial of degree k over $\mathbb{F}_p$, where k is the smallest integer such that $d \mid p^k-1$. Since there are $\varphi(d)$ elements of order d, we have $\varphi(d)/k$ irreducible polynomials of degree k over $\mathbb{F}_p$ with this property. The product of these polynomials is equal to Q_d.

13.26. Example. We want to factor $x^{15}-1$ over $\mathbb{F}_2$. First we consider $x^{15}-1$ as a product of cyclotomic polynomials, namely

$$x^{15}-1=Q_{15}Q_5Q_3Q_1,$$

where over $\mathbb{F}_2$

$$\begin{aligned}
Q_1 &= x+1,\\
Q_3 &= x^2+x+1,\\
Q_5 &= x^4+x^3+x^2+x+1,\\
Q_{15} &= x^8+x^7+x^5+x^4+x^3+x+1.
\end{aligned}$$

Q_1, Q_3, and Q_5 are irreducible over $\mathbb{F}_2$. Since 15 divides 2^4-1 and $\varphi(15)=8$, we conclude that Q_{15} is a product of two irreducible polynomials of degree 4 over $\mathbb{F}_2$. Trying a few candidates yields

$$Q_{15}=(x^4+x+1)(x^4+x^3+1).$$

13.27. Example. We connect to Example 13.18 and describe the elements of $\mathbb{F}_{3^2}$. This is the eighth cyclotomic field. We factor Q_8 over $\mathbb{F}_3$

$$Q_8=x^4+1=(x^2+x-1)(x^2-x-1).$$

A root α of x^2+x-1 was seen to be a primitive eighth root of unity over $\mathbb{F}_3$. Now all nonzero elements of $\mathbb{F}_{2^3}$ can be represented in the form α^i, $0\le i\le 7$, so $\mathbb{F}_{3^2}=\{0,1,\alpha,\alpha^2,\alpha^3,\alpha^4,\alpha^5,\alpha^6,\alpha^7\}$.

Exercises

1. (i) Show that $x^2 + 1$ is irreducible in $\mathbb{Z}_3[x]$.

   (ii) Let α be a zero of $x^2 + 1$ in an extension of $\mathbb{Z}_3$. Give the addition and multiplication tables for the nine elements of $\mathbb{Z}_3(\alpha)$.

2. If we choose $\mathbb{F}_9$ as $\mathbb{F}_3[x]/(x^2+1)$ as in 12.12(iii), is $[x]$ a primitive eighth root of unity?

3. Find all fields of order ≤ 25 explicitly and find generators of their multiplicative group.

4. Show that $f = x^2+x+1$ is irreducible over $\mathbb{Z}_5$. Let $\alpha = x+(f)$ be in $\mathbb{Z}_5[x]/(f)$ and let β be another zero of f. Determine an isomorphism from $\mathbb{Z}_5(\alpha)$ onto $\mathbb{Z}_5(\beta)$.

5. Let α be a zero of x^2+x+1 and let β be a zero of x^2+4 over $\mathbb{Z}_5$. Determine an element γ such that $\mathbb{Z}_5(\gamma) = \mathbb{Z}_5(\alpha, \beta)$.

6. Show the formula for Euler's phi-function after 13.15.

7. For $\alpha \in \mathbb{F}_{q^m} = F$ and $K = \mathbb{F}_q$, the trace $\mathrm{Tr}_{F/K}(\alpha)$ of α over K is defined by

$$\operatorname*{Tr}_{F/K}(\alpha) = \alpha + \alpha^q + \cdots + \alpha^{q^{m-1}}.$$

   Prove:

   (i) $\mathrm{Tr}_{F/K}(\alpha + \beta) = \mathrm{Tr}_{F/K}(\alpha) + \mathrm{Tr}_{F/K}(\beta)$ for all $\alpha, \beta \in F$;

   (ii) $\mathrm{Tr}_{F/K}(c\alpha) = c\,\mathrm{Tr}_{F/K}(\alpha)$ for all $c \in K$, $\alpha \in F$;

   (iii) $\mathrm{Tr}_{F/K}$ is a linear transformation from F onto K, where both F and K are viewed as vector spaces over K;

   (iv) $\mathrm{Tr}_{F/K}(a) = ma$ for all $a \in K$;

   (v) $\mathrm{Tr}_{F/K}(\alpha^q) = \mathrm{Tr}_{F/K}(\alpha)$ for all $\alpha \in F$.

8. Show that the sum of all elements in a finite field $F \neq \mathbb{Z}_2$ and in a finite vector space of dimension ≥ 2 is zero.

9. Let $a_0, \ldots, a_{q-1}$ be the elements of $\mathbb{F}_q$. Prove

$$\sum_{i=0}^{q-1} a_i^t = \begin{cases} 0 & \text{for } 1 \leq t \leq q-2, \\ -1 & \text{for } t = q-1. \end{cases}$$

10. Show as consequences of properties of finite fields that for $a \in \mathbb{Z}$ and $p \in \mathbb{P}$:

    (i) $a^p \equiv a \bmod p$ (***Fermat's Little Theorem***);

    (ii) $(p-1)! \equiv -1 \bmod p$ (***Wilson's Theorem***).

11. Find all subfields of $\mathbb{F}_{15625}$.
12. Prove Lemma 13.9.
13. Determine all primitive elements of $\mathbb{F}_7$.
14. Prove $f^q = f \circ x^q$ for $f \in \mathbb{F}_q[x]$.
15. Let ζ be an nth root of unity over a field K. Prove that $1+\zeta+\zeta^2+\cdots+\zeta^{n-1} = 0$ or n, according as $\zeta \neq 1$ or $\zeta = 1$.
16. Prove the following properties of cyclotomic polynomials:

    (i) If p is prime and $p \nmid m$, then

    $$Q_{mp^k} = Q_{pm} \circ (x^{p^{k-1}}).$$

    (ii) If p is prime and $p \nmid m$, then

    $$Q_{pm} = \frac{Q_m \circ x^p}{Q_m}.$$

    (iii) If $n \geq 2$, $Q_n = \prod_{d|n}(1 - x^{n/d})^{\mu(d)}$.

    (iv) If $n \geq 3$, n odd, then $Q_{2n} = Q_n \circ (-x)$.

17. Find the cyclotomic polynomials Q_{36} and Q_{105}.
18. The ***companion matrix*** of a monic polynomial

    $$f = a_0 + a_1 x + \cdots + a_{n-1}x^{n-1} + x^n$$

    of degree $n \geq 1$ over a field is defined to be the $n \times n$ matrix

    $$\mathbf{A} = \begin{pmatrix} 0 & 0 & \cdots & 0 & -a_0 \\ 1 & 0 & \cdots & 0 & -a_1 \\ \cdots & \cdots & \cdots & \cdots & \cdots \\ 0 & 0 & \cdots & 1 & -a_{n-1} \end{pmatrix}.$$

    $\mathbf{A}$ satisfies $f(\mathbf{A}) = a_0\mathbf{I} + a_1\mathbf{A} + \cdots + a_{n-1}\mathbf{A}_{n-1} + \mathbf{A}^n = 0$. If f is irreducible over $\mathbb{F}_p$, then $\mathbf{A}$ can play the role of a root of f and the polynomials in $\mathbf{A}$ over $\mathbb{F}_p$ of degree less than n yield a representation of the elements of $\mathbb{F}_q$ where $q = p^n$.

(i) Let $f = x^2 + 1 \in \mathbb{F}_3[x]$. Find the companion matrix $\mathbf{A}$ of f and a representation of $\mathbb{F}_9$ using $\mathbf{A}$. Establish the multiplication table for the elements of $\mathbb{F}_9$ given in terms of $\mathbf{A}$.

(ii) Let $f = x^2 + x + 2 \in \mathbb{F}_3[x]$ be an irreducible factor of the cyclotomic polynomial $Q_8 \in \mathbb{F}_3[x]$. Find the companion matrix $\mathbf{A}$ of f and a representation of the elements of $\mathbb{F}_9$ in terms of $\mathbf{A}$.

19. Let $\mathbb{F}_q$ be a finite field and $\mathbb{F}_r$ a finite extension. Show that $\mathbb{F}_r$ is a simple algebraic extension of $\mathbb{F}_q$ and that every primitive element of $\mathbb{F}_r$ can be adjoined to $\mathbb{F}_q$ to give $\mathbb{F}_r$.

20. Show that a finite field $\mathbb{F}_q$ is the $(q-1)$th cyclotomic field over any one of its subfields.

21. Let $\mathbb{F}_{q^m}$ be an extension of $\mathbb{F}_q$ and let $\alpha \in \mathbb{F}_{q^m}$. The elements $\alpha, \alpha^q, \ldots, \alpha^{q^{m-1}}$ are called the ***conjugates*** of α with respect to $\mathbb{F}_q$. Take a root of $x^4 + x + 1$ in $\mathbb{F}_{16}$ and find its conjugates with respect to $\mathbb{F}_2$ and with respect to $\mathbb{F}_4$.

22. A basis of $\mathbb{F}_{q^m}$ over $\mathbb{F}_q$, which consists of a suitable element $\alpha \in \mathbb{F}_{q^m}$ and its conjugates with respect to $\mathbb{F}_q$, is called a ***normal basis*** of $\mathbb{F}_{q^m}$ over $\mathbb{F}_q$. Find a normal basis of $\mathbb{F}_8$ over $\mathbb{F}_2$ and show how easy it is to raise elements to the qth power using a normal basis.

23. Let $\{\alpha, \alpha^2, \ldots, \alpha^{2^{m-1}}\}$ be a normal basis of $\mathbb{F}_{2^m}$. Let, similar to Exercise 7, the trace of $\beta = b_0 + b_1\alpha + \cdots + b_{m-1}\alpha^{2^{m-1}}$ be $\mathrm{Tr}(\beta) = b_0 + b_1 + \cdots + b_{m-1}$. Show that the quadratic equation $x^2 + x + \beta = 0$ has two solutions in $\mathbb{F}_{2^m}$ if $\mathrm{Tr}(\beta) = 0$, namely $x_0 + x_1\alpha + \cdots + x_{m-1}\alpha^{2^{m-1}}$ with $x_0 = \delta$, $x_1 = \delta + b_1$, $x_2 = \delta + b_1 + b_2, \ldots, x_{m-1} = \delta + b_1 + b_2 + \cdots + b_{m-1}$ for $\delta \in \{0, 1\}$, and that $x^2 + x + \beta = 0$ is not solvable if $T(\beta) = 1$. Why can we not use the usual formula for solving quadratic equations over $\mathbb{F}_{2^m}$?

24. Show that every quadratic equation $x^2 + px + q = 0$ with $p \neq 0$ can be transformed into one of the form $x^2 + x + \beta = 0$ by changing x to px.

25. Let α and β be nonzero elements of $\mathbb{F}_q$. Show that there exist elements $a, b \in \mathbb{F}_q$ such that $1 + \alpha a^2 + \beta b^2 = 0$.

26. Let α be a primitive element of $\mathbb{F}_q$. Then ***Zech's Logarithm*** Z is a function which is defined on the integers, for $0 \leq n \leq q-1$, in such a way that $\alpha^{Z(n)} = \alpha^n + 1$. This can be used to add elements α^i and α^j in $\mathbb{F}_q$ by using the equation $\alpha^i + \alpha^j = \alpha^{j+Z(i-j)}$. Determine the Zech Logarithm in $\mathbb{F}_{2^4}$ and evaluate $\alpha^3 + \alpha^5$ and also $\alpha^4 + \alpha^{13}$ where $\alpha^4 + \alpha + 1 = 0$.

§14 Irreducible Polynomials over Finite Fields

We have seen in §12 and §13 that irreducible polynomials over a field are of fundamental importance in the theory of field extensions. In this section we consider polynomials over $\mathbb{F}_q$, which also have many applications in combinatorics, number theory, and algebraic coding theory.

We recall that the splitting field of an irreducible polynomial of degree k over $\mathbb{F}_q$ is $\mathbb{F}_{q^k}$.

From now on, we shall write the induced polynomial function $\bar{p}$ simply as p.

14.1. Theorem. *An irreducible polynomial f over $\mathbb{F}_q$ of degree k divides $x^{q^n} - x$ if and only if k divides n.*

Proof. Suppose $f \mid x^{q^n} - x$. Then f has its roots in $\mathbb{F}_{q^n}$ and its splitting field $\mathbb{F}_{q^k}$ must be contained in $\mathbb{F}_{q^n}$. Theorem 13.10 implies $k \mid n$. Conversely, let $k \mid n$, so that $\mathbb{F}_{q^k}$ is a subfield of $\mathbb{F}_{q^n}$. Since f and $x^{q^n} - x$ split into linear factors in $\mathbb{F}_{q^n}$, $f \mid x^{q^n} - x$ holds over $\mathbb{F}_{q^n}$ by 13.2(ii). □

By 13.16(iii), $x^{q^n} - x \in \mathbb{F}_q[x]$ has only simple roots. Theorem 14.1 implies that this is so for all irreducible polynomials in $\mathbb{F}_q[x]$. This shows Theorem 12.38 as well as (cf. Exercises 13.14 and 13.21):

14.2. Theorem. *If α is any root of an irreducible polynomial of degree k, all other roots are given by $\alpha^q, \alpha^{q^2}, \ldots, \alpha^{q^{k-1}}$, called the* ***conjugates*** *of α.*

14.3. Theorem. *$x^{q^n} - x = \prod_i f_i$, where the product is extended over all distinct monic irreducible polynomials over $\mathbb{F}_q$, with degree a divisor of n.*

Proof. If f_i and f_j are two distinct monic irreducible polynomials over $\mathbb{F}_q$ whose degrees divide n, then f_i and f_j are relatively prime and hence $f_i f_j \mid x^{q^n} - x$. The theorem follows from Theorem 14.1 and the fact that $x^{q^n} - x$ has only simple roots in its splitting field over $\mathbb{F}_q$. □

14.4. Example. x, $x+1$, x^2+x+1 are precisely the irreducible polynomials over $\mathbb{F}_2$ of degree ≤ 2. Hence $x(x+1)(x^2+x+1) = x^4 - x$.

Theorem 14.1 asserts that any element $\alpha \in \mathbb{F}_{q^n}$ is a root of an irreducible polynomial of degree $\leq n$ over $\mathbb{F}_q$ (in accordance with Theorem 12.28). The factorization of Theorem 14.3 can also be regarded as the product of all distinct minimal polynomials of elements of $\mathbb{F}_{q^n}$ over $\mathbb{F}_q$. If m_α is the minimal polynomial of α, then $\{f \in \mathbb{F}_q[x] \mid f(\alpha) = 0\}$ is a principal ideal of $\mathbb{F}_q[x]$ generated by m_α. So we see the following properties of minimal polynomials (see Exercise 2):

14.5. Theorem. *Let $\alpha \in \mathbb{F}_{q^n}$. Suppose the degree of α over $\mathbb{F}_q$ is d and let m_α be the minimal polynomial of α over $\mathbb{F}_q$. Then:*

(i) *m_α is irreducible over $\mathbb{F}_q$ and* $\deg m_\alpha = d$ *divides n.*

(ii) *$f \in \mathbb{F}_q[x]$ satisfies $f(\alpha) = 0$ if and only if $m_\alpha \mid f$.*

(iii) *If α is primitive, then* $\deg m_\alpha = n$.

(iv) *If f is a monic irreducible polynomial of $\mathbb{F}_q[x]$ with $f(\alpha) = 0$, then $f = m_\alpha$.*

(v) *m_α divides $x^{q^d} - x$ and $x^{q^n} - x$.*

(vi) *The roots of m_α are $\alpha, \alpha^q, \ldots, \alpha^{q^{d-1}}$, and m_α is the minimal polynomial over $\mathbb{F}_q$ of all these elements.*

As we saw before, it is often important to find the minimal polynomial of an element in a finite field. A straightforward method of determining minimal polynomials is the following one. Let ζ be a "defining element" of $\mathbb{F}_{q^n}$ over $\mathbb{F}_q$, so that $\{1, \zeta, \ldots, \zeta^{n-1}\}$ is a basis of $\mathbb{F}_{q^n}$ over $\mathbb{F}_q$. If we wish to find the minimal polynomial m_β of $\beta \in \mathbb{F}_{q^n}^*$ over $\mathbb{F}_q$, we represent $\beta^0, \beta^1, \ldots, \beta^n$ in terms of the basis elements. Let

$$\beta^{i-1} = \sum_{j=1}^{n} d_{ij}\zeta^{j-1} \qquad \text{for } 1 \leq i \leq n+1.$$

Let $m_\beta = c_n x^n + \cdots + c_1 x + c_0$; then $m_\beta(\beta) = c_n\beta^n + \cdots + c_1\beta + c_0 = 0$ leads to the homogeneous system of linear equations

$$\sum_{i=1}^{n+1} c_{i-1} d_{ij} = 0 \qquad \text{for } 1 \leq j \leq n, \qquad \square$$

with unknowns $c_0, c_1, \ldots, c_n$. Let $\mathbf{D}$ be the transposed matrix of coefficients of the system, i.e., $\mathbf{D}$ is the $(n+1) \times n$ matrix whose (i,j) entry is d_{ij}, and let r be the rank of $\mathbf{D}$. Then the dimension of the space of solutions of the system is $s = n+1-r$, and since $1 \leq r \leq n$, we have $1 \leq s \leq n$. Therefore we let s of the unknowns $c_0, c_1, \ldots, c_n$ take prescribed values and then the remaining ones are uniquely determined. If $s = 1$, we set $c_n = 1$, and if $s > 1$, we set $c_n = c_{n-1} = \cdots = c_{n-s+2} = 0$ and $c_{n-s+1} = 1$.

14.6. Example. Let $\zeta \in \mathbb{F}_{64}$ be a root of the irreducible polynomial $x^6 + x + 1$ in $\mathbb{F}_2[x]$. For $\beta = \zeta^3 + \zeta^4$ we have

$$\begin{aligned}
\beta^0 &= 1, \\
\beta^1 &= \qquad\qquad\qquad \zeta^3 + \zeta^4, \\
\beta^2 &= 1 + \zeta + \zeta^2 + \zeta^3, \\
\beta^3 &= \quad\;\; \zeta + \zeta^2 + \zeta^3, \\
\beta^4 &= \quad\;\; \zeta + \zeta^2 \qquad + \zeta^4, \\
\beta^5 &= 1 \qquad\qquad + \zeta^3 + \zeta^4, \\
\beta^6 &= 1 + \zeta + \zeta^2 \qquad + \zeta^4.
\end{aligned}$$

Therefore the matrix $\mathbf{D}$ is of the form

$$\mathbf{D} = \begin{pmatrix} 1 & 0 & 0 & 0 & 0 & 0 \\ 0 & 0 & 0 & 1 & 1 & 0 \\ 1 & 1 & 1 & 1 & 0 & 0 \\ 0 & 1 & 1 & 1 & 0 & 0 \\ 0 & 1 & 1 & 0 & 1 & 0 \\ 1 & 0 & 0 & 1 & 1 & 0 \\ 1 & 1 & 1 & 0 & 1 & 0 \end{pmatrix}$$

and its rank is $r = 3$. Hence $s = n+1-r = 4$, and we set $c_6 = c_5 = c_4 = 0$, $c_3 = 1$. The remaining coefficients are determined from (14.1), and this yields $c_2 = 1$, $c_1 = 0$, $c_0 = 1$. Therefore the minimal polynomial of β over $\mathbb{F}_2$ is $m_\beta(x) = x^3 + x^2 + 1$.

Another method of determining minimal polynomials is as follows. If we wish to find the minimal polynomial m_β of $\beta \in \mathbb{F}_{q^n}$ over $\mathbb{F}_q$, we compute the powers $\beta, \beta^q, \beta^{q^2}, \ldots$ until we find the least positive integer d for which $\beta^{q^d} = \beta$. This integer d is the degree of m_β, and m_β itself is given by $m_\beta = (x - \beta)(x - \beta^q) \cdots (x - \beta^{q^{d-1}})$. Observe that $m_\beta \in \mathbb{F}_q[x]$, although $\beta, \beta^q, \ldots$ are usually *not* in $\mathbb{F}_q$ but in $\mathbb{F}_{q^n}$.

14.7. Example. We compute the minimal polynomials over $\mathbb{F}_2$ of all elements of $\mathbb{F}_{16}$. Let $\zeta \in \mathbb{F}_{16}$ be a root of the polynomial $x^4 + x + 1$ over $\mathbb{F}_2$. Then ζ can be shown to bc primitive, so that every nonzero element of $\mathbb{F}_{16}$ can be written as a power of ζ. We have the following index table for $\mathbb{F}_{16}$:

i	ζ^i	i	ζ^i	i	ζ^i
0	1	5	$\zeta+\zeta^2$	10	$1+\zeta+\zeta^2$
1	ζ	6	$\zeta^2+\zeta^3$	11	$\zeta+\zeta^2+\zeta^3$
2	ζ^2	7	$1+\zeta+\zeta^3$	12	$1+\zeta+\zeta^2+\zeta^3$
3	ζ^3	8	$1+\zeta^2$	13	$1+\zeta^2+\zeta^3$
4	$1+\zeta$	9	$\zeta+\zeta^3$	14	$1+\zeta^3$

The minimal polynomials m_β of the elements β of $\mathbb{F}_{16}$ over $\mathbb{F}_2$ are:

$\beta = 0$: $m_0 = x$.

$\beta = 1$: $m_1 = x - 1$.

$\beta = \zeta$: The distinct conjugates of ζ with respect to $\mathbb{F}_2$ are $\zeta, \zeta^2, \zeta^4, \zeta^8$, and the minimal polynomials are

$$\begin{aligned} m_\zeta &= m_{\zeta^2} = m_{\zeta^4} = m_{\zeta^8} \\ &= (x-\zeta)(x-\zeta^2)(x-\zeta^4)(x-\zeta^8) = x^4 + x + 1. \end{aligned}$$

$\beta = \zeta^3$: The distinct conjugates of ζ^3 with respect to $\mathbb{F}_2$ are $\zeta^3, \zeta^6, \zeta^{12}, \zeta^{24} = \zeta^9$, and the minimal polynomials are

$$\begin{aligned} m_{\zeta^3} &= m_{\zeta^6} = m_{\zeta^{12}} = m_{\zeta^9} \\ &= (x-\zeta^3)(x-\zeta^6)(x-\zeta^9)(x-\zeta^{12}) = x^4 + x^3 + x^2 + x + 1. \end{aligned}$$

$\beta = \zeta^5$: Since $\beta^4 = \beta$, the distinct conjugates of this element with respect to $\mathbb{F}_2$ are ζ^5, ζ^{10}, and the minimal polynomials are

$$m_{\zeta^5} = m_{\zeta^{10}} = (x-\zeta^5)(x-\zeta^{10}) = x^2 + x + 1.$$

$\beta = \zeta^7$: The distinct conjugates of ζ^7 with respect to $\mathbb{F}_2$ are $\zeta^7, \zeta^{14}, \zeta^{28} = \zeta^{13}$, $\zeta^{56} = \zeta^{11}$, and the minimal polynomials are

$$\begin{aligned} m_{\zeta^7} &= m_{\zeta^{14}} = m_{\zeta^{13}} = m_{\zeta^{11}} \\ &= (x-\zeta^7)(x-\zeta^{11})(x-\zeta^{13})(x-\zeta^{14}) = x^4 + x^3 + 1. \end{aligned}$$

The following concept comes in very naturally.

14.8. Definition. Let $n \in \mathbb{N}$ divide some $p^m - 1$ and let $0 \le s < n$. The ***cyclotomic coset*** C_s of s modulo n is given by

$$C_s := \{s, ps, p^2 s, \dots, p^{r-1} s\},$$

where r is the smallest positive integer with $p^r s \equiv s \pmod{n}$ and each $p^i s$ is reduced modulo $p^m - 1$.

If $\gcd(s, p^m - 1) = 1$, then $r = m$ in 14.8, but if $\gcd(s, p^m - 1) \neq 1$, then r varies with s. By 14.5 we get

14.9. Theorem. *Let α be an element in $\mathbb{F}_{p^m}$ and let m_α be its minimal polynomial. Let ζ be a primitive element in $\mathbb{F}_{p^m}$ and let $\alpha = \zeta^i$. If i belongs to the cyclotomic coset C, then*

$$m_\alpha = \prod_{j \in C} (x - \zeta^j).$$

14.10. Example. In Example 14.7 we have the cyclotomic cosets $C_0 = \{0\}$, $C_1 = \{1, 2, 4, 8\}$, $C_3 = \{3, 6, 12, 9\}$, $C_5 = \{5, 10\}$, and $C_7 = \{7, 14, 13, 11\}$.

14.11. Example. Let $p^m - 1 = 63$. The cyclotomic cosets (mod 63) are

$$\begin{aligned} C_0 &= \{0\},\ C_1 = \{1,2,4,8,16,32\}, & C_3 &= \{3,6,12,24,48,33\},\\ C_5 &= \{5,10,20,40,17,34\}, & C_7 &= \{7,14,28,56,49,35\},\\ C_9 &= \{9,18,36\}, & C_{11} &= \{11,22,44,25,50,37\},\\ C_{13} &= \{13,26,52,41,19,38\}, & C_{15} &= \{15,30,60,57,51,39\},\\ C_{21} &= \{21,42\}, & C_{23} &= \{23,46,29,58,53,43\},\\ C_{27} &= \{27,54,45\}, & C_{31} &= \{31,62,61,59,55,47\}. \end{aligned}$$

Therefore we can conclude that there are nine irreducible factors of $x^{63} - 1$ of degree 6, two irreducible factors of degree 3, and one of each of degrees 2 and 1. To find the factors of degree 6 is not so easy.

That takes care of factoring $x^n - 1 \in \mathbb{F}_{p^m}[x]$ provided that $n = p^m - 1$. What about the other n? We still assume $\gcd(n, p) = 1$ (cf. 13.16), so that p has an inverse modulo n. Since $p \mid n$, we know that there is a smallest integer m such that n divides $p^m - 1$. This happens if and only if $x^n - 1$ divides $x^{p^m-1} - 1$, so that $\mathbb{F}_{p^m}$ is the smallest field of characteristic p that contains all the roots of $x^n - 1$. So we simply factor $x^n - 1$ over $\mathbb{F}_{p^n}[x]$. Note that this yields a factorization over $\mathbb{F}_p$.

14.12. Example. Let $p = 2$, $n = 9$. Then $m = 6$ and $9 \mid 2^m - 1$, and $x^9 - 1$ splits into linear factors over $\mathbb{F}_{2^6}$. The cyclotomic cosets modulo n are $C_0 = \{0\}$, $C_1 = \{1, 2, 4, 8, 7, 5\}$, $C_3 = \{3, 6\}$. We choose $\zeta \in \mathbb{F}_{2^6}$, where $\zeta^6 + \zeta^3 + 1 = 0$. Then the minimal polynomials according to Theorem 14.9 are

$$m_{\zeta^0} = x + 1, \qquad m_{\zeta^1} = x^6 + x^3 + 1, \qquad m_{\zeta^3} = x^2 + x + 1.$$

Therefore

$$x^9 - 1 = m_{\zeta^0} m_{\zeta^1} m_{\zeta^3}.$$

14.13. Example. To get some information about the factorization of $x^{23} - 1$ over $\mathbb{F}_2$ (which we shall need in 19.7(v)), we compute the cyclotomic cosets modulo 23:

$$C_0 = \{0\}, \qquad C_1 = \{1, 2, 4, 8, 16, 9, 18, 13, 3, 6, 12\},$$
$$C_5 = \{5, 10, 20, 17, 11, 22, 21, 19, 15, 7, 14\}.$$

This implies that $x^{23} - 1$ has two factors of degree 11 and one factor of degree 1, namely $x - 1$. Each is a minimal polynomial of a primitive twenty-third root of unity, which is contained in $\mathbb{F}_{2^{11}}$, since $r = 11$ is the smallest integer such that $23 \mid 2^r - 1$.

We now determine the number of irreducible polynomials.

14.14. Theorem. *The number of monic irreducible polynomials of degree k over $\mathbb{F}_q$ is given by*

$$I_q(k) = \frac{1}{k} \sum_{d \mid k} \mu\left(\frac{k}{d}\right) q^d.$$

Proof. Theorem 14.3 implies that the degree of the product of all monic irreducible polynomials over $\mathbb{F}_q$ whose degree divides k is equal to $\sum_{d \mid k} I_q(d) d = q^k$. The additive form of the Möbius inversion formula, Theorem 13.23(i), gives the desired result. □

The formula in Theorem 14.14 shows again that $I_q(k) \geq 1$ for any prime power q and any positive integer k (see Exercise 17).

14.15. Examples. For $q = 2$:

k	1	2	3	4	5	6	7	8	9	10	11	12	13	14
$I_q(k)$	2	1	2	3	6	9	18	30	56	99	186	335	630	1161

Explicit lists of irreducible polynomials can be found in Figure 14.1.

There is an interesting connection between minimal polynomials and primitive elements of a finite field. We introduce the order (also called the exponent or the period) of a nonzero polynomial over a finite field. The following result motivates the definition of an order.

14.16. Lemma. *Let $f \in \mathbb{F}_q[x]$ be a polynomial of degree $m \geq 1$ with $f(0) \neq 0$. Then there exists a positive integer $e \leq q^m - 1$ such that f divides $x^e - 1$.*

Proof. $\mathbb{F}_q[x]/(f)$ has $q^m - 1$ nonzero residue classes. Since the q^m residue classes $x^j + (f)$, $0 \leq j \leq q^m - 1$, are all nonzero, there exist integers s and t with $0 \leq s < t \leq q^m - 1$ such that $x^t \equiv x^s \bmod f$. Since $\gcd(x, f) = 1$, we have $f \mid (x^{t-s} - 1)$ and $0 < t - s \leq q^m - 1$. □

A method for determining e is to simply try if $f \mid x^e - 1$ for $e = m, m+1, \ldots$ until we succeed. Corollary 14.19 will help a lot (cf. 14.20).

14.17. Definition. Let $0 \neq f \in \mathbb{F}_q[x]$. If $f(0) \neq 0$, then the smallest natural number e with the property that $f \mid (x^e - 1)$ is called the ***order*** (or ***exponent*** or ***period***) of f. If $f(0) = 0$, then f is of the form $x^h g$ with $h \in \mathbb{N}$ and $g \in \mathbb{F}_q[x]$, $g(0) \neq 0$, for a unique polynomial g. The order $\operatorname{ord} g$ of f is then defined as the order of g.

The order of an irreducible polynomial can be characterized by its roots.

14.18. Theorem. *Let $f \in \mathbb{F}_q[x]$ be an irreducible polynomial over $\mathbb{F}_q$ of degree $m \geq 2$. Then $\operatorname{ord} f$ is equal to the order of any root of f in $\mathbb{F}_{q^m}^*$.*

Proof. Note that $f(0) \neq 0$, since otherwise x would be a proper factor. $\mathbb{F}_{q^m}$ is the splitting field of f over $\mathbb{F}_q$. The roots of f have the same order in $\mathbb{F}_{q^m}^*$; let $\alpha \in \mathbb{F}_{q^m}^*$ be a root of f. Then $\alpha^e = 1$ if and only if $f \mid x^e - 1$. The result follows from the definition of $\operatorname{ord} f$ and of the order of α in the group $\mathbb{F}_{q^m}^*$. □

14.19. Corollary. *If $f \in \mathbb{F}_q[x]$ is irreducible over $\mathbb{F}_q$ of degree k, then $\operatorname{ord} f$ divides $q^k - 1$.*

Proof. If $f = cx$ with $c \in \mathbb{F}_q^*$, then $\operatorname{ord} f = 1$. Otherwise the result follows from Theorem 14.18 and the fact that the order of $\mathbb{F}_{q^k}^*$ is $q^k - 1$. □

14.20. Examples.

(i) $f = x^3 + x + 1 \in \mathbb{F}_2[x]$ is irreducible. So $\operatorname{ord} f$ must be a divisor of $2^3 - 1 = 7$. Hence $\operatorname{ord}(f) = 7$.

(ii) If $f \in \mathbb{F}_2[x]$ is irreducible of degree 4, then $4 \leq \operatorname{ord} f \mid 2^4 - 1 = 15$. So $\operatorname{ord} f = 5$ or 15, which can be scanned much quicker than all candidates from 4 to 15. By 14.15, we have three choices of f:

$$f_1 = x^4 + x + 1,$$
$$f_2 = x^4 + x^3 + 1,$$
$$f_3 = x^4 + x^3 + x^2 + x + 1.$$

By inspection, we get $\operatorname{ord} f_1 = \operatorname{ord} f_2 = 15$ and $\operatorname{ord} f_3 = 5$ (since $x^5 - 1 = (x-1)(x^4 + x^3 + x^2 + x + 1)$).

(iii) Similarly to (i), every irreducible polynomial in $\mathbb{F}_2[x]$ of degree 5 must have order 31.

Note that f_3 of 14.20 is the only irreducible polynomial of degree < 6 in $\mathbb{F}_2[x]$ that has not the maximal possible order $2^k - 1$. We now take a closer look at those polynomials which have this maximal order. Recall that a primitive element of $\mathbb{F}_q$ is defined as a generator of the multiplicative group $(\mathbb{F}_q^*, \cdot)$.

14.21. Definition. A monic irreducible polynomial of degree m over $\mathbb{F}_q$ is called ***primitive*** if it is the minimal polynomial of a primitive element of $\mathbb{F}_{q^m}$.

How can we find out if a given polynomial is primitive?

14.22. Theorem. *A polynomial $f \in \mathbb{F}_q[x]$ of degree m is primitive if and only if f is monic, $f(0) \neq 0$, and the order of f is equal to $q^m - 1$.*

Proof. If f is primitive over $\mathbb{F}_q$, then f is monic and $f(0) \neq 0$. Now f is irreducible and has as root a primitive element over $\mathbb{F}_{q^m}$, so by Theorem 14.18, $\operatorname{ord} f = q^m - 1$. Conversely it suffices to show that f is irreducible. Suppose on the contrary that f is reducible. We have two cases to consider: Either $f = g_1 g_2$ where g_1 and g_2 are relatively prime polynomials of positive degrees k_1 and k_2, or $f = g^b$ where $g \in \mathbb{F}_q[x]$, $g(0) \neq 0$, is irreducible, and $b \in \mathbb{N}$.

In the first case let $e_i = \operatorname{ord} g_i$; then by Lemma 13.9, $\operatorname{ord} f \leq e_1 e_2$. By Theorem 14.1, $g \mid x^{q^{k_i}-1} - 1$ so $e_i \leq q^{k_i} - 1$. Hence $\operatorname{ord} f \leq e_1 e_2 \leq (q^{k_1} - 1)(q^{k_2} - 1) < q^{k_1+k_2} - 1 = q^m - 1$, a contradiction.

In the second case, let $e = \operatorname{ord} g$. By Theorem 14.1 and the fact that $g \mid f$ we have $e \mid q^m - 1$, so $p \nmid e$ where p is the characteristic of $\mathbb{F}_q$. By Lemma 13.9, we get $g \mid x^k - 1$ if $e \mid k$. So if $k = p^i j$ where $p \nmid j$ we have

$$x^k - 1 = x^{p^i j} - 1 = (x^j - 1)^{p^i}.$$

Since $x^j - 1$ has no repeated roots, every irreducible factor of $x^k - 1$ has multiplicity p^i. Let t be the unique integer with $p^{t-1} < b \leq p^t$, then

$\operatorname{ord} f = ep^t$. But $ep^t \le (q^n - 1)p^t$ where $n = m/b$, the degree of g. Moreover $ep^t < q^{n+t} - 1$. So $t \le p^{t-1} \le b - 1 \le (b-1)n$. Now combining these inequalities we have $\operatorname{ord} f < q^{n+t} - 1 \le q^{bn} - 1 = q^m - 1$, a contradiction. Therefore f is irreducible and by Theorem 14.18 the roots of f have order $q^m - 1$, so f is primitive. □

So the problem of the determination of primitive polynomials is "the same" as the one of determining all polynomials with maximal order. One approach is based on the fact that the product of all primitive polynomials over $\mathbb{F}_q$ of degree m is equal to the cyclotomic polynomial Q_e with $e = q^m - 1$. Therefore, all primitive polynomials over $\mathbb{F}_q$ of degree m can be determined by factoring the cyclotomic polynomial Q_e.

Another method depends on constructing a primitive element of $\mathbb{F}_{q^m}$ and then determining the minimal polynomial of this element over $\mathbb{F}_q$. To find a primitive element of $\mathbb{F}_{q^m}$, we start from the order $q^m - 1$ of such an element in the group $\mathbb{F}_{q^m}^*$ and factor it into the form $q^m - 1 = h_1 \dots h_k$, where the positive integers $h_1, \dots, h_k$ are pairwise relatively prime. If for each i, $1 \le i \le k$, we can find an element $\alpha_i \in \mathbb{F}_{q^m}^*$ of order h_i, then the product $\alpha_1 \dots \alpha_k$ can be shown to have order $q^m - 1$ and thus is a primitive element of $\mathbb{F}_{q^m}$.

14.23. Example. We determine a primitive polynomial over $\mathbb{F}_3$ of degree 4. Since $3^4 - 1 = 16 \cdot 5$, we first construct two elements of $\mathbb{F}_{81}^*$ of order 16 and 5, respectively. The elements of order 16 are the roots of the cyclotomic polynomial $Q_{16}(x) = x^8 + 1 \in \mathbb{F}_3[x]$. Since the multiplicative order of 3 modulo 16 is 4, Q_{16} factors into two monic irreducible polynomials in $\mathbb{F}_3[x]$ of degree 4. Now

$$x^8 + 1 = (x^4 - 1)^2 - x^4 = (x^4 - 1 + x^2)(x^4 - 1 - x^2),$$

and so $f(x) = x^4 - x^2 - 1$ is irreducible over $\mathbb{F}_3$ and with a root θ of f we have $\mathbb{F}_{81} = \mathbb{F}_3(\theta)$. Furthermore, θ is an element of $\mathbb{F}_{81}^*$ of order 16. It can be verified that $\alpha = \theta + \theta^2$ has order 5. Therefore $\zeta = \theta\alpha = \theta^2 + \theta^3$ has order 80 and is thus a primitive element of $\mathbb{F}_{81}$. The minimal polynomial g of ζ over $\mathbb{F}_3$ is

$$\begin{aligned} g(x) &= (x - \zeta)(x - \zeta^3)(x - \zeta^9)(x - \zeta^{27}) \\ &= (x - \theta^2 - \theta^3)(x - 1 + \theta + \theta^2)(x - \theta^2 + \theta^3)(x - 1 - \theta + \theta^2) \\ &= x^4 + x^3 + x^2 - x - 1, \end{aligned}$$

and we have thus obtained a primitive polynomial over $\mathbb{F}_3$ of degree 4.

We know that $\mathbb{F}_{q^m} = \mathbb{F}_q[x]/(f)$ where f is an irreducible polynomial of degree m in $\mathbb{F}_q[x]$. If f has maximal order, we get a nice reward from Theorem 14.22.

14.24. Corollary. *Let $\mathbb{F}_{q^m} = \mathbb{F}_q[x]/(f)$ where $\operatorname{ord}(f) = q^m - 1$. Then $\alpha = x + (f)$ is a primitive element in $\mathbb{F}_{q^m}$, so $\mathbb{F}_{q^m} = \{0, 1, \alpha, \alpha^2, \dots, \alpha^{q^m-2}\}$.*

Using the notation of Theorem 14.5 we observe the additional properties of minimal polynomials:

14.25. Theorem.

(i) *If $\alpha \neq 0$, then $\operatorname{ord} m_\alpha$ is equal to the order of α in $\mathbb{F}^*_{q^n}$.*

(ii) *m_α is primitive over $\mathbb{F}_q$ if and only if α is of order $q^d - 1$ in $\mathbb{F}^*_{q^n}$.*

Next we list some primitive polynomials over $\mathbb{F}_2$.

14.26. Table.

Degree	Polynomial	Degree	Polynomial
1	$x+1$	12	$x^{12}+x^7+x^4+x^3+1$
2	x^2+x+1	13	$x^{13}+x^4+x^3+x+1$
3	x^3+x+1	14	$x^{14}+x^{12}+x^{11}+x+1$
4	x^4+x+1	15	$x^{15}+x+1$
5	x^5+x^2+1	16	$x^{16}+x^5+x^3+x^2+1$
6	x^6+x+1	17	$x^{17}+x^3+1$
7	x^7+x+1	18	$x^{18}+x^7+1$
8	$x^8+x^6+x^5+x+1$	19	$x^{19}+x^6+x^5+x+1$
9	x^9+x^4+1	20	$x^{20}+x^3+1$
10	$x^{10}+x^3+1$	60	$x^{60}+x+1$
11	$x^{11}+x^2+1$	100	$x^{100}+x^8+x^7+x^2+1$

In Figure 14.1, we give a list of some irreducible polynomials $f = a_nx^n + a_{n-1}x^{n-1} + \cdots + a_0$ of degree n over $\mathbb{F}_p$. We abbreviate the polynomials by writing the coefficient vector as $a_na_{n-1}\dots a_0$. The column e indicates the order of f.

14.27. Theorem. *There are exactly $\varphi(e)/m$ monic irreducible polynomials of degree m and order $e \geq 2$ over $\mathbb{F}_q$.*

Proof. We have $e \mid (q^m - 1)$ and $e \nmid (q^k - 1)$ for $k < m$, since $\mathbb{F}_{q^m}$ is the smallest extension field of $\mathbb{F}_q$ which contains all zeros of an irreducible polynomial of degree m. Since $(x^e - 1) \mid (x^{q^{m-1}} - 1)$, $\mathbb{F}_{q^m}$ contains all eth roots

For $p = 2$:

$n = 1$	e
11	1

$n = 2$	e
111	3

$n = 3$	e
1011	7
1101	7

$n = 4$	e
10011	15
11001	15
11111	5

$n = 5$	e
100101	31
101001	31
101111	31
110111	31
111011	31
111101	31

$n = 6$	e
1000011	63
1001001	9
1010111	21
1011011	63
1100001	63
1100111	63
1101101	63
1110011	63
1110101	21

$n = 7$	e
10000011	127
10001001	127
10001111	127
10010001	127
10011101	127
10100111	127
10101011	127
10111001	127
10111111	127
11000001	127
11001011	127
11010011	127
11010101	127
11100101	127
11101111	127
11110001	127
11110111	127
11111101	127

For $p = 3$:

$n = 1$	e
11	2
12	1

$n = 2$	e
101	4
112	8
122	8

$n = 3$	e
1021	26
1022	13
1102	13
1112	13
1121	26
1201	26
1211	26
1222	13

$n = 4$	e
10012	80
10022	80
10102	16
10111	40
10121	40
10202	16
11002	80
11021	20
11101	40
11111	5
11122	80
11222	80
12002	80
12011	20
12101	40
12112	80
12121	10
12212	80

For $p = 5$:

$n = 1$	e
11	2
12	4
13	4
14	1

$n = 2$	e
102	8
103	8
111	3
112	24
123	24
124	12
133	24
134	12
141	6
142	24

For $p = 7$:

$n = 1$	e
11	2
12	6
13	3
14	6
15	3
16	1

$n = 2$	e
101	4
102	12
104	12
113	48
114	24
116	16
122	24
123	48
125	48
131	8
135	48
136	16
141	8
145	48
146	16
152	24
153	48
155	48
163	48
164	24
166	16

FIGURE 14.1 Irreducible polynomials of degree n over $\mathbb{F}_p$.

of unity, and thus contains all $\varphi(e)$ primitive eth roots of unity. Let α be such a root; then $\alpha, \alpha^q, \dots, \alpha^{q^{m-1}}$ are distinct and the minimal polynomial of a primitive eth root of unity has degree m, according to 14.5. □

14.28. Theorem. *For $e > 1$, $f = (x^e - 1)/(x-1) = x^{e-1} + x^{e-2} + \cdots + x + 1$ is irreducible over $\mathbb{F}_q$ if and only if e is a prime and q is a primitive $(e-1)$th root of unity modulo e.*

Proof. Left as Exercise 5. □

In Example 14.20(ii), $f_3 = (x^5 - 1)/(x-1)$ has $e = 5$ and 2 as a primitive fourth root of unity in $\mathbb{Z}_5$. In fact, f_3 is irreducible.

The following theorem gives a formula for the product of all monic, irreducible polynomials over $\mathbb{F}_q$.

14.29. Theorem. *The product $I(q, n)$ of all monic irreducible polynomials in $\mathbb{F}_q[x]$ of degree n is given by*

$$I(q, n) = \prod_{d|n} (x^{q^d} - x)^{\mu(n/d)} = \prod_{d|n} (x^{q^{n/d}} - x)^{\mu(d)}.$$

Proof. Theorem 14.3 implies $x^{q^n} - x = \prod_{d|n} I(q, d)$. Let $f(n) = I(q, n)$ and $g(n) = x^{q^n} - x$ for all $n \in \mathbb{N}$ in Theorem 13.23. The multiplicative Möbius inversion formula yields the result. □

14.30. Example. Let $q = 2$ and $n = 4$.

$$I(2, 4) = \frac{x^{16} - x}{x^4 - x} = x^{12} + x^9 + x^6 + x^3 + 1$$

and

$$\begin{aligned} x^{16} - x = x^{2^4} - x &= I(2,1)I(2,2)I(2,4) \\ &= (x^2 - x)(x^2 + x + 1)(x^{12} + x^9 + x^6 + x^3 + 1). \end{aligned}$$

It can be shown (see Exercise 6 or, e.g., Lidl & Niederreiter (1994)) that the following partial factorization of $I(q, n)$ holds.

14.31. Theorem. *For $n > 1$ we have $I(q, n) = \prod_m Q_m$, where Q_m is the mth cyclotomic polynomial over $\mathbb{F}_q$. The product is extended over all positive divisors m of $q^n - 1$ such that n is the multiplicative order of q modulo m.*

Exercises

1. Show that the three irreducible polynomials of degree 4 over $\mathbb{F}_2$ are x^4+x+1, x^4+x^3+1, $x^4+x^3+x^2+x+1$.
2. Prove Theorem 14.5.
3. Find the first prime p such that $(x^{19}-1)(x-1)^{-1}$ is irreducible over $\mathbb{F}_p$.
4. Prove that the polynomial $f \in \mathbb{F}_q[x]$, with $f(0) \neq 0$, divides $x^m - 1$ iff the order of f divides the positive integer m.
5. Prove Theorem 14.28.
6. Prove Theorem 14.31.
7. Compute $I(3,4)$.
8. Compute $I(2,6)$ from Theorem 14.29.
9. Compute $I(2,6)$ from Theorem 14.31.
10. How many irreducible polynomials of degree 32 do exist over $\mathbb{F}_2$? Over $\mathbb{F}_3$?
11. Let $g_1, \ldots, g_k$ be pairwise relatively prime nonzero polynomials over $\mathbb{F}_q$, and let $f = g_1 \ldots g_k$. Prove that $\operatorname{ord} f$ is equal to the least common multiple of $\operatorname{ord} g_1, \ldots, \operatorname{ord}_{g_k}$. (Use Exercise 4.)
12. Determine the order of $f = (x^2+x+1)^3(x^3+x+1)$ over $\mathbb{F}_2$.
13. Given that $f = x^4 + x^3 + x^2 + 2x + 2$ in $\mathbb{F}_3[x]$ is irreducible over $\mathbb{F}_3$ and $\operatorname{ord} f = 80$, show that f is primitive over $\mathbb{F}_3$.
14. Construct $\mathbb{F}_{2^4}$ by using $f = x^4+x^3+x^2+x+1$, which is irreducible over $\mathbb{F}_2$. Let $f(\alpha) = 0$ and show that α is not a primitive element, but that $\alpha+1$ is a primitive element. Find the minimal polynomial of $\alpha+1$.
15. Construct the finite field $\mathbb{F}_{2^{20}}$ by using the primitive polynomial $f = x^{20} + x^3 + 1$ over $\mathbb{F}_2$. Determine all subfields of $\mathbb{F}_{2^{20}}$.
16. Is $x^{31} + x^3 + 1$ a primitive polynomial over $\mathbb{F}_2$?
17. Show that Theorem 14.14 also implies that for every $n \in \mathbb{N}$ and every prime p, there is an irreducible polynomial of degree n over $\mathbb{F}_p$.
18. Determine the cyclotomic cosets mod 21 over $\mathbb{F}_2$ and find the factorization of $x^{21} - 1$ over $\mathbb{F}_2$.
19. What are the degrees of the irreducible factors of $x^{17} - 1$ over $\mathbb{F}_2$? What is the smallest field of characteristic 2 in which $x^{17} - 1$ splits into linear factors?

§15 Factorization of Polynomials over Finite Fields

A method for determining the complete factorization of a polynomial over $\mathbb{F}_q$ into a product of irreducible factors will be given. Such factorizations are useful within mathematics in the factorization of polynomials over $\mathbb{Z}$, in the determination of the Galois group of an equation over $\mathbb{Z}$, in factoring elements of finite algebraic number fields into prime elements, and in many other areas. Some other applications of factorizations of polynomials will be considered in §18. In §14, we have settled the case of factoring $x^n - 1$. Now we turn to general polynomials.

The problem of factoring a given polynomial into a product of irreducible polynomials over the integers is rather old. L. Kronecker gave an algorithm for the explicit determination of such a factorization. However, Kronecker's approach is not very economical, since the calculation time for this algorithm increases exponentially with the degree of the given polynomial. More recently so-called "homomorphism methods" have been developed to find the factorization of a given polynomial over $\mathbb{F}_q$. We shall describe an algorithm due to Berlekamp (1984) to express a given monic polynomial $f \in \mathbb{F}_q[x]$ in the form

$$f = f_1^{e_1} \dots f_r^{e_r},$$

where f_i are distinct monic irreducible polynomials over $\mathbb{F}_q$.

First we want to find out if f has multiple factors, i.e., if $e_i > 1$ for some i. By 12.35 we can proceed as follows: Let $d := \gcd(f, f')$.

Case 1: $d = 1$. Then f is square-free.

Case 2: $1 \neq d \neq f$. Then we already have a factor (namely d) of f and we factor d and f/d (which is easier than to factor f).

Case 3: $d = f$. By Exercise 12.21, f is of the form $f = g \circ x^p$, where $p = \operatorname{char} \mathbb{F}_q$. In this case, we factor g (which is much easier than to factor f); we can do this due to the lines before 11.18.

Therefore it suffices to consider the factorization of a monic square-free polynomial $f \in \mathbb{F}_q[x]$. Berlekamp's algorithm is based on some important properties: First we need a generalization of the Chinese Remainder Theorem for integers to the case of polynomials over $\mathbb{F}_q$. The

Chinese Remainder Theorem is one of the oldest results in number theory, and it has found applications in computer programming with fast adders (§24). The old result was used by the ancient Chinese to predict the common period of several astronomical cycles.

15.1. Theorem (*Chinese Remainder Theorem for Polynomials*). *Let $f_1, \dots, f_r$ be distinct irreducible polynomials over $\mathbb{F}_q$ and let $g_1, \dots, g_r$ be arbitrary polynomials over $\mathbb{F}_q$. Then the system of congruences $h \equiv g_i \pmod{f_i}$, $i = 1, 2, \dots, r$, has a unique solution h modulo $f_1 f_2 \dots f_r$.*

Proof. We use the Euclidean algorithm to determine polynomials m_i such that $m_i \prod_{j \neq i} f_j \equiv 1 \pmod{f_i}$. Then we put $h := \sum_i (g_i m_i \prod_{j \neq i} f_j)$; we get $h \equiv g_i \pmod{f_i}$ for all i. If $\hat{h}$ were also a solution of the system of congruences, then $\hat{h} - h$ would be divisible by each f_i and therefore $\hat{h} \equiv h \pmod{\prod_i f_i}$. □

Theorem 15.1 is equivalent to the statement:

$$\mathbb{F}_q[x]/(f_1 f_2 \dots f_r) \cong \bigoplus_{i=1}^{r} \mathbb{F}_q[x]/(f_i),$$

where $[g]_{f_1 \dots f_r} \mapsto ([g]_{f_1}, \dots, [g]_{f_r})$ establishes the isomorphism. This result is of great importance in itself. Instead of computing with polynomials in $\mathbb{F}_q[x]$, we might do this in $\mathbb{F}_q[x]/(f)$ if f has a sufficiently high degree. We shall be in a similar situation in §24, where we can do parallel computation. In general, decomposing a ring into a direct sum of other rings is an invitation to parallel computing.

In trying to factor a square-free polynomial f into distinct monic irreducible divisors f_i over $\mathbb{F}_q$, Theorem 15.1 says that we may expect information on f_i if we choose any constants $s_i \in \mathbb{F}_q$ and find h as in 15.1. Then

$$f_i \mid \gcd(h - s_i, f)$$

In fact, in Theorem 4.2 we shall see that we can find all f_i among these gcd's. For h we get

$$h^q \equiv s_i^q = s_i \equiv h \pmod{f_i}.$$

The multinomial theorem in $\mathbb{F}_q[x]$ yields $h^q = h \circ x^q$ (see Exercise 13.14) and thus

$$h \circ x^q - h \equiv 0 \pmod{f_i}.$$

Since f_i divides f, it is sufficient to find polynomials h with the property

$$h \circ x^q - h \equiv 0 \pmod{f}. \tag{15.1}$$

In order to find such polynomials we construct the $n \times n$ matrix ($n = \deg f$)

$$\mathbf{Q} = \begin{pmatrix} q_{00} & q_{01} & \cdots & q_{0,n-1} \\ \vdots & \vdots & \ddots & \vdots \\ q_{n-1,0} & q_{n-1,1} & \cdots & q_{n-1,n-1} \end{pmatrix}$$

such that the kth row is given as the coefficient vector of the congruence $x^{qk} \equiv q_{k,n-1}x^{n-1} + \cdots + q_{k1}x + q_{k0} \pmod{f}$.

A polynomial $h = \sum_{i=0}^{n-1} v_i x^i$ is a solution of (15.1) if and only if

$$(v_0, v_1, \ldots, v_{n-1})\mathbf{Q} = (v_0, v_1, \ldots, v_{n-1}).$$

This holds because

$$h = \sum_i v_i x^i = \sum_i \sum_k v_k q_{ki} x^i = \sum_k v_k x^{qk} = h \circ x^q \equiv h^q \pmod{f}.$$

The determination of a polynomial h satisfying (15.1) can be regarded as solving the system of linear equations $\mathbf{v}(\mathbf{Q} - \mathbf{I}) = \mathbf{0}$, where $\mathbf{v}$ is the coefficient vector of h, $\mathbf{Q} = (q_{ki})$, $\mathbf{I}$ is the $n \times n$ identity matrix, and $\mathbf{0}$ is the n-dimensional zero vector. Finding a suitable h is therefore equivalent to determining the null space of $\mathbf{Q} - \mathbf{I}$. We use the notation above to prove

15.2. Theorem.

(i) $f = \prod_{s \in \mathbb{F}_q} \gcd(f, h - s)$.

(ii) *The number of monic distinct irreducible factors f_i of f is equal to the dimension of the null space of the matrix $\mathbf{Q} - \mathbf{I}$.*

Proof.

(i) Since $h^q - h \equiv 0 \pmod{f}$, f divides $h^q - h = \prod_{s \in \mathbb{F}_q}(h - s)$. Therefore f divides $\prod_{s \in \mathbb{F}_q} \gcd(f, h - s)$. On the other hand, $\gcd(f, h - s)$ divides f. If $s \neq t \in \mathbb{F}_q$, then $h - s$ and $h - t$ are relatively prime, and so are $\gcd(f, h - s)$ and $\gcd(f, h - t)$. Thus $\prod_{s \in \mathbb{F}_q} \gcd(f, h - s)$ divides f. Since both polynomials are monic, they must be equal.

(ii) f divides $\prod_{s \in \mathbb{F}_q}(h - s)$ if and only if each f_i divides $h - s_i$ for some $s_i \in \mathbb{F}_q$. Given $s_1, \ldots, s_r \in \mathbb{F}_q$, Theorem 15.1 implies the existence of a unique polynomial $h \pmod{f}$, such that $h \equiv s_i \pmod{f_i}$. We have the choice of q^r elements s_i, therefore we have exactly q^r solutions of

$h^q - h \equiv 0 \pmod f$. We noted above that h is a solution of (15.1) if and only if

$$(v_0, v_1, \ldots, v_{n-1})\mathbf{Q} = (v_0, v_1, \ldots, v_{n-1}),$$

or equivalently

$$(v_0, v_1, \ldots, v_{n-1})(\mathbf{Q} - \mathbf{I}) = (0, 0, \ldots, 0).$$

This system has q^r solutions. Thus the dimension of the null space of the matrix $\mathbf{Q} - \mathbf{I}$ is r, the number of distinct monic irreducible factors of f, and the rank of $\mathbf{Q} - \mathbf{I}$ is $n - r$. □

Let k be the rank of the matrix $\mathbf{Q} - \mathbf{I}$. We have $k = n - r$, so that once the rank k is found, we know that the number of distinct monic irreducible factors of f is given by $n - k$. On the basis of this information we can already decide if f is irreducible or not. The rank of $\mathbf{Q} - \mathbf{I}$ can be determined by using row and column operations to reduce the matrix to echelon form. However, it is advisable to use only column operations because they leave the null space invariant. Thus, we are allowed to multiply any column of the matrix $\mathbf{Q} - \mathbf{I}$ by a nonzero element of $\mathbb{F}_q$ and to add any multiple of one of its columns to a different column. The rank k is the number of nonzero columns in the column echelon form.

After finding k, we form $r = n - k$. If $r = 1$, we know that f is irreducible over $\mathbb{F}_q$ and the procedure terminates. This is a surprisingly fast irreducibility test. In this case the only solutions of (15.1) are the constant polynomials, and the null space of $\mathbf{Q} - \mathbf{I}$ contains only the vectors of the form $(c, 0, \ldots, 0)$ with $c \in \mathbb{F}_q$. If $r \geq 2$, we take the polynomial h_2 and compute $\gcd(f, h_2 - s)$ for all $s \in \mathbb{F}_q$. If the use of h_2 does not succeed in splitting f into r factors, we compute $\gcd(g, h_3 - s)$ for all $s \in \mathbb{F}_q$ and all nontrivial factors g found so far. This procedure is continued until r factors of f are obtained. The process described above must eventually yield all the factors. If the null space has dimension r, then there exists a basis with r monic polynomials $h^{(1)}, \ldots, h^{(r)}$. We summarize these results.

15.3. Theorem (*Berlekamp's Algorithm*). *Let $f \in \mathbb{F}_q[x]$ be monic of degree n.*

Step 1. Check if f is square-free, i.e., if $\gcd(f, f') = 1$.

Step 2. If f is square-free, form the $n \times n$ matrix $\mathbf{Q} = (q_{ki})$, defined by

$$x^{qk} \equiv \sum_{i=0}^{n-1} q_{ki} x^i \pmod f, \qquad 0 \leq k \leq n - 1.$$

Step 3. Find the null space of $\mathbf{Q} - \mathbf{I}$*, determine its rank* $n - r$*, and find* r *linearly independent vectors* $\mathbf{v}^{(1)}, \dots, \mathbf{v}^{(r)}$ *such that* $\mathbf{v}^{(i)}(\mathbf{Q} - \mathbf{I}) = \mathbf{0}$ *for* $i = 1, 2, \dots, r$*. The integer* r *is the number of irreducible factors of* f*. If* $r = 1$*,* f *is irreducible; otherwise go to Step 4.*

Step 4. Compute $\gcd(f, h^{(2)} - s)$ *for all* $s \in \mathbb{F}_q$*, where* $h^{(2)} := \sum_i v_i^{(2)} x^i$*. This yields a nontrivial decomposition of* f *into a product of (not necessarily irreducible) factors. If* $h^{(2)}$ *does not give all* r *factors of* f*, then we can obtain further factors by computing* $\gcd(h^{(k)} - s, f)$ *for all* $s \in \mathbb{F}_q$ *and for* $k = 3, 4, \dots, r$*. Here* $h^{(k)} := \sum v_i^{(k)} x^i$*. The algorithm ends when all* r *irreducible factors of* f *are found.*

15.4. Example. We want to determine the complete factorization of

$$g = x^8 + x^6 + 2x^4 + 2x^3 + 3x^2 + 2x \qquad \text{over } \mathbb{F}_5.$$

We put $g = xf$ and factorize f. First we see that $\gcd(f, f') = 1$. Next we compute the matrix $\mathbf{Q}$, in this case a 7×7 matrix. We have to find $x^{5k} \pmod f$ for $k = 0, 1, 2, \dots, 6$. The first row of $\mathbf{Q}$ is $(1, 0, 0, 0, 0, 0, 0)$, since $x^0 = 1 \pmod f$. We give a systematic procedure for computing the rows of $\mathbf{Q}$. In general, let

$$f = x^n + a_{n-1}x^{n-1} + \cdots + a_0$$

and let

$$x^m \equiv a_{m0} + a_{m1}x + \cdots + a_{m,n-1}x^{n-1} \pmod f.$$

Then

$$\begin{aligned} x^{m+1} &\equiv a_{m0}x + a_{m1}x^2 + \cdots + a_{m,n-1}x^n \\ &\equiv a_{m0}x + a_{m1}x^2 + \cdots + a_{m,n-2}x^{n-1} \\ &\qquad + a_{m,n-1}(-a_{n-1}x^{n-1} - \cdots - a_1x - a_0) \\ &\equiv a_{m+1,0} + a_{m+1,1}x + \cdots + a_{m+1,n-1}x^{n-1} \pmod f, \end{aligned}$$

where

$$\begin{aligned} a_{m+1,j} &= a_{m,j-1} - a_{m,n-1}a_j, \\ a_{m,-1} &= 0. \end{aligned}$$

Thus we tabulate

m	a_{m0}	a_{m1}	a_{m2}	a_{m3}	a_{m4}	a_{m5}	a_{m6}	m	a_{m0}	a_{m1}	a_{m2}	a_{m3}	a_{m4}	a_{m5}	a_{m6}
0	1	0	0	0	0	0	0	0	1	0	0	0	0	0	0
1	0	1	0	0	0	0	0	16	3	2	4	2	0	3	2
2	0	0	1	0	0	0	0	17	1	2	3	0	2	3	3
3	0	0	0	1	0	0	0	18	4	2	1	2	0	4	3
4	0	0	0	0	1	0	0	19	4	0	1	0	2	2	4
5	0	0	0	0	0	1	0	20	2	2	2	3	0	3	2
6	0	0	0	0	0	0	1	21	1	1	3	3	3	3	3
7	3	2	3	3	0	4	0	22	4	2	0	2	3	0	3
8	0	3	2	3	3	0	4	23	4	0	1	4	2	0	0
9	2	3	0	4	3	4	0	24	0	4	0	1	4	2	0
10	0	2	3	0	4	3	4	25	0	0	4	0	1	4	2
11	2	3	4	0	0	0	3	26	1	4	1	0	0	4	4
12	4	3	2	3	0	2	0	27	2	4	1	3	0	1	4
13	0	4	3	2	3	0	2	28	2	0	1	3	3	1	1
14	1	4	0	4	2	1	0	29	3	4	3	4	3	2	1
15	0	1	4	0	4	2	1	30	3	0	2	1	4	2	2

Therefore the matrix $\mathbf{Q}$ is given as

$$\mathbf{Q} = \begin{pmatrix} 1 & 0 & 0 & 0 & 0 & 0 & 0 \\ 0 & 0 & 0 & 0 & 0 & 1 & 0 \\ 0 & 2 & 3 & 0 & 4 & 3 & 4 \\ 0 & 1 & 4 & 0 & 4 & 2 & 1 \\ 2 & 2 & 2 & 3 & 0 & 3 & 2 \\ 0 & 0 & 4 & 0 & 1 & 4 & 2 \\ 3 & 0 & 2 & 1 & 4 & 2 & 2 \end{pmatrix},$$

so that

$$\mathbf{Q} - \mathbf{I} = \begin{pmatrix} 0 & 0 & 0 & 0 & 0 & 0 & 0 \\ 0 & 4 & 0 & 0 & 0 & 1 & 0 \\ 0 & 2 & 2 & 0 & 4 & 3 & 4 \\ 0 & 1 & 4 & 4 & 4 & 2 & 1 \\ 2 & 2 & 2 & 3 & 4 & 3 & 2 \\ 0 & 0 & 4 & 0 & 1 & 3 & 2 \\ 3 & 0 & 2 & 1 & 4 & 2 & 1 \end{pmatrix}.$$

Step 3 requires that we find the null space of $\mathbf{Q} - \mathbf{I}$. By methods from linear algebra the null space is spanned by the two linearly independent vectors

$$\mathbf{v}^{(1)} = (1, 0, 0, 0, 0, 0, 0) \quad \text{and} \quad \mathbf{v}^{(2)} = (0, 3, 1, 4, 1, 2, 1).$$

This means that f has two irreducible factors.

In Step 4 we compute $\gcd(f, h^{(2)} - s)$ for all $s \in \mathbb{F}_5$, where

$$h^{(2)} = x^6 + 2x^5 + x^4 + 4x^3 + x^2 + 3x.$$

The first factor f_i is already found as $\gcd(f, h^{(i)} - 0)$:

$$f_1 = x^5 + 2x^4 + x^3 + 4x^2 + x + 3.$$

The second factor of f is obtained by division, and turns out to be x^2+3x+4. Thus the complete factorization of $g = xf$ over $\mathbb{F}_5$ is of the form

$$g = x(x^2 + 3x + 4)(x^5 + 2x^4 + x^3 + 4x^2 + x + 3).$$

15.5. Example. We want to find the factorization of

$$g = x^{16} + x^{12} + x^8 + x^6 + 1 \quad \text{over } \mathbb{F}_2.$$

Since $g' = 0$, $\gcd(g, g') = g$. In fact, $g = f \circ x^2 = f^2$ where $f = x^8 + x^6 + x^4 + x^3 + 1$. So we try to factor f. We verify that $\gcd(f, f') = 1$, thus f does not have any multiple factors. The 8×8 matrix $\mathbf{Q}$ can be obtained by means of a recursion. In general, for $\deg f = n$, we have

$$a_{m+1j} = a_{m,j-1} - a_{m,n-1}a_j$$

for the coefficients of $x^{2r} \pmod f$, where $f = \sum a_j x^j$ and $a_{m,-1} = 0$. Thus in our example

$$a_0 = a_3 = a_4 = a_6 = a_8 = 1, \qquad a_1 = a_2 = a_5 = a_7 = 0.$$

We obtain

m	a_{m0}	a_{m1}	a_{m2}	a_{m3}	a_{m4}	a_{m5}	a_{m6}	a_{m7}
0	1	0	0	0	0	0	0	0
1	0	1	0	0	0	0	0	0
2	0	0	1	0	0	0	0	0
3	0	0	0	1	0	0	0	0
4	0	0	0	0	1	0	0	0

m	a_{m0}	a_{m1}	a_{m2}	a_{m3}	a_{m4}	a_{m5}	a_{m6}	a_{m7}
5	0	0	0	0	0	1	0	0
6	0	0	0	0	0	0	1	0
7	0	0	0	0	0	0	0	1
8	1	0	0	1	1	0	1	0
9	0	1	0	0	1	1	0	1
10	1	0	1	1	1	1	0	0
11	0	1	0	1	1	1	1	0
12	0	0	1	0	1	1	1	1
13	1	0	0	0	1	1	0	1
14	1	1	0	1	1	1	0	0

Therefore the matrix $\mathbf{Q}$ consists of the rows for $m = 0, 2, 4, 6, 8, 10, 12, 14$ and is of the form

$$\mathbf{Q} = \begin{pmatrix} 1 & 0 & 0 & 0 & 0 & 0 & 0 & 0 \\ 0 & 0 & 1 & 0 & 0 & 0 & 0 & 0 \\ 0 & 0 & 0 & 0 & 1 & 0 & 0 & 0 \\ 0 & 0 & 0 & 0 & 0 & 0 & 1 & 0 \\ 1 & 0 & 0 & 1 & 1 & 0 & 1 & 0 \\ 1 & 0 & 1 & 1 & 1 & 1 & 0 & 0 \\ 0 & 0 & 1 & 0 & 1 & 1 & 1 & 1 \\ 1 & 1 & 0 & 1 & 1 & 1 & 0 & 0 \end{pmatrix};$$

$$\mathbf{Q} - \mathbf{I} = \begin{pmatrix} 0 & 0 & 0 & 0 & 0 & 0 & 0 & 0 \\ 0 & 1 & 1 & 0 & 0 & 0 & 0 & 0 \\ 0 & 0 & 1 & 0 & 1 & 0 & 0 & 0 \\ 0 & 0 & 0 & 1 & 0 & 0 & 1 & 0 \\ 1 & 0 & 0 & 1 & 0 & 0 & 1 & 0 \\ 1 & 0 & 1 & 1 & 1 & 0 & 0 & 0 \\ 0 & 0 & 1 & 0 & 1 & 1 & 0 & 1 \\ 1 & 1 & 0 & 1 & 1 & 1 & 0 & 1 \end{pmatrix}.$$

Computing the null space yields $r = 2$ and

$$h^{(2)} = x^7 + x^6 + x^5 + x^2 + x.$$

Now we can perform Step 4 in 15.3 and compute $\gcd(f, h^{(2)} - 1)$, which yields $x^2 + x + 1$ as an irreducible factor of f. The second factor is $x^6 + x^5 + x^4 + x + 1$. Hence $g = (x^2 + x + 1)^2(x^6 + x^5 + x^4 + x + 1)^2$.

Some additional remarks concerning Berlekamp's algorithm: It may be advantageous to find simple factors of the given polynomial by trial-and-error methods before applying more complicated methods such as 15.3.

If the order of the underlying finite field is large, then the following approach can be used to determine the matrix $\mathbf{Q}$ in Step 2 of 15.3. First we determine an auxiliary table consisting of all coefficients of $x^m \pmod f$ for $m = n, n+1, \dots, 2n-2$, where $n = \deg f$. If

$$x^k \equiv c_{n-1}x^{n-1} + \cdots + c_1x + c_0 \pmod f,$$

then

$$x^{2k} \equiv c_{n-1}^2 x^{2n-2} + \cdots + (c_1c_0 + c_0c_1)x + c_0^2 \pmod f,$$

where the powers $x^{2n-2}, \dots, x^n$ can be replaced by combinations of powers of lower degrees from the auxiliary table. In this way, $x^q \pmod f$ and thus the second row of $\mathbf{Q}$ can be obtained. The remaining rows of $\mathbf{Q}$ are computed by repeated multiplication of $x^q \pmod f$ by itself modulo f.

Step 4 in Berlekamp's algorithm clearly shows that practical calculations are very time consuming if q is large, since this step requires the computation of $\gcd(f, h - s)$ for all $s \in \mathbb{F}_q$. Berlekamp and Knuth have shown that some of the computations can be performed for all $s \in \mathbb{F}_q$ at once, but only for polynomials of degrees up to $\frac{n}{2}$. In this method, the element s remains in the calculations as a parameter. Factorization methods for "large" finite fields are based on the work by Berlekamp; see Lidl & Niederreiter (1983, Chapter 4).

Let us close this section with a brief look at an interesting problem related to factorization. Instead of factoring a polynomial $f \in F[x]$ into a product of irreducible factors, let us try to decompose f into a composition of "indecomposible" polynomials: $f = g_1 \circ g_2 \circ \cdots \circ g_n$. Is this always possible? Is the decomposition unique? How can this decomposition be achieved? What is it good for? Let us look at the last question first. Everybody knows that finding the roots of $f = x^4 - 5x^2 + 6 \in \mathbb{R}[x]$ can be done by the substitution $x^2 = y$, solving $y^2 - 5y + 6 = 0$, obtaining $y_1 = 2$, $y_2 = 3$, and hence $x_{1,2} = \pm\sqrt{2}$, $x_{3,4} = \pm\sqrt{3}$.

What we did here was to write f as $f = (x^2 - 5x + 6) \circ x^2$. This decomposition is obvious. It is less obvious to go a similar way to find the zeros of $f = x^4 - 12x^3 - 27x^2 - 54x + 20$. We can show that $f = (x^2 - 5x + 6) \circ (x^2 - 6x + 7)$, so we first find the roots of x^2-5x+6 as 2 and 3, then solving $x^2-6x+7=2$ ($=3$, respectively), which yields the zeros 1, 5, and $3 \pm \sqrt{5}$.

15.6. Definition. A polynomial $f \in F[x]$, F a field, is called ***decomposable*** if $f = g \circ h$ where g and h have smaller degrees than f; otherwise, f is called ***indecomposable***.

15.7. Theorem. *Let F be a field.*

(i) *Every $f \in F[x]$ can be written as a decomposition of finitely many indecomposable polynomials.*
(ii) *If f has prime degree, then f is indecomposable.*

Proof.

(i) Either f is indecomposable, or $f = g \circ h$, where g and h have smaller degrees than f. Similarly, g and h are either indecomposable or can be decomposed again, etc.. Since the degrees get smaller and smaller, this process must terminate after finitely many steps.
(ii) This follows from $\deg(g \circ h) = (\deg g)(\deg f)$, see Exercise 11.18. □

As to uniqueness of decompositions, observe that, e.g., $f = g \circ h$ implies $f = g \circ (x+1) \circ (x-1) \circ h$, so we have to discard linear "factors" in between. If we do so, the decomposition is "essentially unique", a theorem which is quite hard to prove (see Lausch & Nöbauer (1973) and Binder (1995)).

Recently, much work has been done in establishing fast algorithms to accomplish complete decompositions of polynomials, see, e.g., von zur Gathen (1990). Surprisingly, these algorithms are much faster than factorization algorithms. Observe that here we do not work in the polynomial ring $(F(x), +, \cdot)$ but rather in $(F(x), +, \circ)$, which is not a ring since $f \circ (g+h) = f \circ g + f \circ h$ does not hold in general (see Exercise 11.20). These "near-rings" will show up for completely different reasons in §36 and §39.

Exercises

1. How many distinct irreducible monic factors divide
$$f = x^5 + 2x^4 + x^3 + x^2 + 2 \qquad \text{in } \mathbb{F}_3[x]?$$
2. Determine the factorization of $x^{12} + x^8 + x^7 + x^6 + x^2 + x + 1$ over $\mathbb{F}_2$.

3. Determine the factorization of x^9+x+1 over $\mathbb{F}_2$.
4. Factor $x^{15}+x^{11}+x^7+x^3 \in \mathbb{Z}_2[x]$.
5. Determine the factorization of $x^7+x^6+x^5-x^3+x^2-x-1$ over $\mathbb{F}_3$.
6. Determine the number of factors of $f = x^4+1 \in \mathbb{F}_p[x]$ over $\mathbb{F}_p$ for all primes p, using Berlekamp's algorithm.
7. ***Kronecker's algorithm*** can be used to determine the divisors of degree $\leq s$ of a nonconstant polynomial f in $\mathbb{Q}[x]$; it works as follows:

   (i) Without loss of generality suppose $f \in \mathbb{Z}[x]$.

   (ii) Choose distinct integers $a_0, \ldots, a_s$ which are not zeros of f and then determine all divisors of $f(a_i)$, $0 \leq i \leq s$.

   (iii) For each $(s+1)$-tuple $(b_0, \ldots, b_s)$ for which $b_i \mid f(a_i)$, determine polynomials $g \in \mathbb{Q}[x]$ of degree $\leq s$ such that $g(a_i) = b_i$, $0 \leq i \leq s$.

   (iv) Determine which of these polynomials g divides f. If $\deg f = n \geq 1$, $\lfloor n/2 \rfloor = s$ and the only divisors of f are constants, then f is irreducible. In the case of nonconstant divisors g of f repeated application of (ii) and (iii) gives the complete factorization of f over $\mathbb{Q}$.

   Factor $\frac{1}{3}x^4 - x^3 + \frac{1}{2}x^2 - \frac{1}{2}x + \frac{1}{6} \in \mathbb{Q}[x]$.
8. Let $f_1 = x^2+x+1$, $f_2 = x^3+x+1$, $f_3 = x^4+x+1$, $g_1 = x+1$, $g_2 = x^2+1$, and $g_3 = x^3+1$ be polynomials over $\mathbb{F}_2$. Find a polynomial h such that the simultaneous congruences $h \equiv g_i \pmod{f_i}$, $i = 1, 2, 3$, are satisfied.
9. Show that x^9+x^4+1 is irreducible over $\mathbb{F}_2$.
10. Find x^{12} modulo x^5+2x^3+x+1 over $\mathbb{F}_3$.

Notes

Most of the definitions and theorems in §10 and §11 can be found in nearly any of the introductory books on modern algebra, e.g., Childs (1995), Herstein (1975), Lang (1984), MacLane & Birkhoff (1967), and van der Waerden (1970). More detailed accounts on groups can be found for instance in Rotman (1995), Mathiak & Stingl (1968), Robinson (1996), and Thomas & Wood (1980). An excellent source on ring theory is Rowen (1988). More on fields can be found in Lidl & Niederreiter (1994), Nagata (1977), and Winter (1974), to mention a few. Historically, the discovery of so-called p-adic fields by Hensel led Steinitz to the abstract formulation of the theory of fields, which was developed in the fundamental paper by Steinitz (1910), where he introduced prime fields, separable elements, etc.; earlier contributions in the area of field theory were made by Kronecker (1887) and Weber (1893), who designed the first axiomatic field theory. One of the most fundamental theorems in the theory of fields is Theorem 12.13, due to Kronecker (1887). Kronecker used the idea of Cauchy that the complex numbers can be defined as residue classes in $\mathbb{R}[x]/(x^2+1)$ (see Example 12.14). If we apply Kronecker's approach to $\mathbb{Z}/(p)$, we obtain "Galois imaginaries" due to Serret and Dedekind.

Richard Dedekind introduced the term "field" for the first time, although the concept and some properties of special types of fields were known and studied before him, e.g., by Gauss, Galois, and Abel.

Leibniz created the name "transcendentals" when he proved in 1682 that $\sin x$ cannot be an algebraic function of x. The existence of a transcendental number was verified by Liouville in 1844; in 1873, Hermite proved the transcendency of e, and in 1882, Lindemann showed that π is transcendental, which concluded the "squaring of the circle" problem of the ancient Greeks. The concepts (but not the modern terminology) of constructing an extension field by adjoining an element to a given field, and of irreducible polynomials and conjugate roots, were studied by Galois. Abel and Galois defined the elements of the underlying field as those quantities which can be expressed as rational functions of given quantities. In 1744, A. Legendre recognized that any root of an algebraic equation with rational coefficients is an algebraic number, and that numbers that are not algebraic are transcendental. E. E. Kummer, a pupil of Gauss and Dirichlet, introduced algebraic numbers of the form $a_0 + a_2\alpha + \cdots + a_{p-2}\alpha^{p-2}$, $a_i \in \mathbb{Z}$, α a complex pth root of unity, and then studied divisibility, primality, and other properties of these numbers. In order to obtain unique factorizations for his numbers he created the theory of "ideal numbers," which led to the concept of ideals. For instance, consider $\{a+b\sqrt{-5} \mid a, b \in \mathbb{Z}\}$; then $6 = 2\cdot 3 = (1+\sqrt{-5})(1-\sqrt{-5})$, where all factors are prime. Using the "ideal numbers"

$$\alpha = \sqrt{2}, \qquad \rho_1 = \frac{1+\sqrt{-5}}{\sqrt{2}}, \qquad \rho_2 = \frac{1-\sqrt{-5}}{\sqrt{2}},$$

we have $6 = \alpha^2\rho_1\rho_2$ uniquely expressed as a product of four factors. Beginning in 1837, and following work of C. F. Gauss, Kummer studied the arithmetic of cyclotomic fields for some 25 years in an effort to solve Fermat's Last Theorem (namely that $x^n + y^n = z^n$ is unsolvable in nonzero integers x, y, z for $n > 2$) in the case of n being an odd prime. (Kummer succeeded in proving the conjecture for primes $p < 100$.) Fermat's Last Theorem has been a celebrated conjecture for several centuries until A. Wiles provided a proof in 1995/1996, using deep algebraic methods.

Throughout the long history of mathematics over several millennia, the problem of solving equations has been an important one. We know from the clay tablets of the ancient Babylonians, from papyri of the ancient Egyptians, and from old Chinese and Hindu writings that the ancient civilizations dealt with the solution of equations of the first and second degree. Although the interest in such problems diminished during and after the Classical and Alexandrian Greek period because of the predominance of geometry, this interest greatly increased in the early sixteenth century due to Italian mathematicians, who solved equations of degrees 3 and 4 by radicals. From then on until the eighteenth century the solution of algebraic equations by radicals, i.e., "expressions involving roots," was a natural problem to consider. Since the middle of the eighteenth century the search was on for an *a priori* proof of the fundamental theorem of algebra, because all attempts at finding solutions by radicals failed for general equations of degree ≥ 5. After contributions by Tschirnhaus, Leibniz, de Moivre, Euler, d'Alembert, and others, Lagrange attempted a proof of the fundamental theorem in the early 1770s. C. F. Gauss gave several complete proofs, his first one in 1799. Today there are more than 100 different proofs known. Lagrange studied the relationship between the roots $\alpha_1, \ldots, \alpha_n$ and the coefficients $a_0, \ldots, a_{n-1}$ of a polynomial equation, namely

$$x^n + a_{n-1}x^{n-1} + \cdots + a_1x + a_0 = (x - \alpha_1) \ldots (x - \alpha_n),$$

where

$$\begin{aligned} -a_{n-1} &= \alpha_1 + \cdots + \alpha_n, \\ a_{n-2} &= 7\alpha_1\alpha_2 + a_1a_3 + \cdots + \alpha_{n-1}\alpha_n, \\ &\vdots \\ (-1)^n a_0 &= \alpha_1 \ldots \alpha_n. \end{aligned}$$

Using these relationships Lagrange devised a technique for solving third-

and fourth-degree equations by finding a related equation whose degree was smaller than the degree of the original equation. When he applied this approach to the general fifth-degree equation, he obtained a related equation which was of degree six. From this Lagrange deduced that the solution of the general fifth-degree equation by radicals might be impossible.

Work by Ruffini and Gauss led Abel and Galois to the complete solution of the problem. In 1824–1826, N. H. Abel showed that the solution by radicals of a general equation of fifth degree is impossible (nowadays called the Abel-Ruffini Theorem). E. Galois studied and extended work by Lagrange, Gauss, and Abel, and he gave the complete answer that the general equation of degree n is not solvable by radicals if $n \geq 5$. Galois' work (summarized under the name Galois theory) also supplied a criterion for constructibility that dispensed with some of the famous Greek construction problems. The solutions to the problems of the solvability of equations by radicals in the nineteenth century made way for a wider development of algebraic concepts in the decades following Galois' death in 1832.

The Greek school of sophists, which was the first school of philosophers in ancient Athens, singled out construction with ruler and compass alone, mostly for philosophical rather than mathematical reasons. Therefore Euclid in his *Elements* describes only construction problems of this type, now referred to as the classical Greek construction problems. The problems such as constructing a square equal in area to a given circle, trisecting any angle, constructing the side of a cube whose volume is double that of a cube with given edge, had to wait over 2000 years for their solution, namely a demonstration of the impossibility of certain such constructions. One version of the origin or the problem of doubling the cube, found in a work by Erathosthenes (approx. 284–192 B.C.), relates that the Delians, suffering from pestilence, consulted an oracle, who advised constructing an altar double the size of the existing one. The famous Plato was asked to help; he answered that the God of the oracle did not really need an altar of twice the size, but he censured the Greeks for their indifference to mathematics and their lack of respect for geometry.

Cotes, de Moivre, and Vandermonde studied solutions of $x^n - 1 = 0$ and divided the circle into n equal parts. Gauss in his fundamental work *Disquisitiones Arithmeticae* also studied this cyclotomic equation for n

an odd prime. He arrived at the famous result that a regular n-gon is constructible with ruler and compass if and only if all the odd primes dividing n are primes of the form $2^{2^k}+1$ whose squares do not divide n. Thus, for instance, the regular 60-gon is constructible, since $60 = 2^2 \cdot 3 \cdot 5$, where $3 = 2+1$ and $5 = 2^2+1$.

Although many textbooks on abstract or applied algebra treat the theory of finite fields in a few pages, more extensive treatments can be found, for instance, in Berlekamp (1984), Dornhoff & Hohn (1978), MacWilliams & Sloane (1977), McDonald (1974), and Ireland & Rosen (1982). A comprehensive book on finite fields (with an extensive bibliography) is Lidl & Niederreiter (1983); an abridged version is Lidl & Niederreiter (1994). See also a recent survey by Lidl & Niederreiter (1996). Other books on finite fields are, e.g., Cohen & Niederreiter (1996), McEliece (1977), McEliece (1987), Menezes, Blake, Gao, Mullin, Vanstone & Yaghoobian (1993), Mullen & Shiue (1993), Small (1991), Blake (1973), Jungnickel (1993), and Shparlinski (1992).

The history of finite fields up to 1915 is presented in Dickson (1901), Chapter 8. Some of the early results of Fermat, Euler, Lagrange, Legendre, and Gauss were concerned with many properties of the special finite field $\mathbb{Z}_p$. The concept of a finite field in general (that is, not only considered as the prime field $\mathbb{Z}_p$) first occurred in a paper by Galois (1897) in 1830. However, Gauss initiated a project of developing a theory of "higher congruences," as equations over finite fields were called. R. Dedekind observed, after following Dirichlet's lectures, that Gauss's project could be carried out by establishing an analogy with the theory of ordinary congruences and then emulating Dirichlet's approach to the latter topic. This resulted in an important paper by Dedekind (1857), which put the theory of higher congruences on a sound basis and also played a role in the conceptual development of algebraic number theory and abstract algebra. A posthumous paper by Gauss investigated the arithmetic in the polynomial rings $\mathbb{F}_p[x]$; in particular, he realized that in these rings the Euclidean algorithm and unique factorization are valid. The construction of extension fields of $\mathbb{F}_p$ was described by Galois in 1830, who used an "imaginary" root i of an irreducible polynomial over $\mathbb{F}_p$ of degree n and showed that the expressions $a_0 + a_1 i + \cdots + a_{n-1} i^{n-1}$ with $a_j \in \mathbb{F}_p$ form a field with p^n elements. During the nineteenth century further research on "higher congruences" was carried out by Dedekind, Schönemann, and Kummer.

Lidl & Niederreiter (1983) (in their Notes to Chapters 2–9) refer to a large number of varied contributions to the theory of finite fields and this book also includes further tables representing elements of finite fields.

Properties of polynomials over finite fields are described in Berlekamp (1984); see also MacWilliams & Sloane (1977), McDonald (1974), and Blake & Mullin (1975), to mention just a few books. In the older literature the word primitive was used for a polynomial whose coefficients had gcd equal to 1. Our primitive polynomials are sometimes also called indexing polynomials.

The term "irreducible polynomial" over $\mathbb{Q}$ goes back to the seventeenth century, when Newton and Leibniz gave methods for determining irreducible factors of a given polynomial. Gauss showed that $(x^n-1)/(x-1)$ is irreducible over $\mathbb{Q}$ if n is an odd prime, and he also verified that the roots of $(x^n-1)/(x-1) = 0$ are cyclic. He also found the formula for the number of irreducible polynomials over $\mathbb{F}_p$ of fixed degree; our proof of this formula and the Möbius multiplicative inversion formula is due to Dedekind.

The factorization algorithm based on the matrix $\mathbf{Q}$ (Berlekamp's algorithm) was first developed around 1965 by Berlekamp, see Berlekamp (1984) and it is also described in Knuth (1981) and Lidl & Niederreiter (1983), the latter book also contains other factorization algorithms for small and large finite fields. Rabin (1980) redesigned Berlekamp's deterministic algorithm into a probabilistic method for finding factorizations of polynomials over large finite fields. All these algorithms are probabilistic in the sense that the time of computation depends on random choices, but the validity of the results does not depend on them. The book by Zippel (1993) is a comprehensive reference on computations with polynomials in general, and on efficient modern algorithms for factorization in particular. Another excellent text is Kaltofen (1992).

Niederreiter (1993) describes an approach to factoring that is based on the $\mathbb{F}_p$-linear operator $y^{(p-1)} + y^{(p)}$: $\mathbb{F}_{p^n}[x] \to \mathbb{F}_{p^n}[x]$ which sends f into $f^{(p-1)} - f^{(p)}$, where $f^{(k)}$ denotes the kth derivative of f.

Cyclotomic cosets and applications to the factorization of $x^n - 1$ are studied in MacWilliams & Sloane (1977) and Pless (1989); Blake & Mullin (1975) call them q-chains. Factorization tables for the binomials $x^n - 1$ can be found in McEliccc (1977).

We indicated that applications of finite fields and polynomials over finite fields to areas outside algebra will be treated in §18 and in Chapter 7.

There are also other uses of finite fields which we cannot describe here due to space limitations. For instance, the analysis of switching circuits (see §7) can be based on arithmetic in finite fields. The calculation of switching functions or general logic functions can also be described in terms of finite fields. Other applications comprise finite-state algorithms and linear systems theory (§39).

There are a number of comprehensive or purpose-built computer systems available to facilitate algebraic computations. Some of these packages are particularly useful for polynomial computations over finite fields, e.g., AXIOM, GAP, MAGMA, Maple, or *Mathematica*. Some of the smaller specialized computer algebra packages for computations with finite fields are COSSLIB, CoCoA, GALOIS, and MACAULAY.

4 CHAPTER Coding Theory

In many ways, coding theory or the theory of error-correcting codes represents a beautiful example of the applicability of abstract algebra. Applications of codes range from enabling the clear transmission of pictures from distant planets to securing the enjoyment of listening to noise-free CDs. A variety of algebraic concepts can be used to describe codes and their properties, including matrices, polynomials and their roots, linear shift registers, and discrete Fourier transforms. The theory is still relatively young, having started in 1948 with an influential paper by Claude Shannon. This chapter provides the reader with an introduction to the basic concepts of (block) codes, beginning in §16 with general background, §17 deals with properties of linear codes, §18 introduces cyclic codes, and §19 and §20 contain material on special cyclic codes.

§16 Introduction to Coding

We consider the problem of transmitting a message over a channel which can be affected by "noise," and our task is to find out if and where this noise has distorted the message and what the original message was.

In order to achieve this, we must add some "control symbols" to our message. In the simplest case, we might repeat the message: Suppose **PILL** was the original message. We "encode" it to **PILLPILL**, but **PILLKILL** arrives due to some noise at the fifth position. The receiver (who must know the encoding convention) then knows that (at least) one error has occurred. But where? Was the original message **PILL** or **KILL** (or **WILL** or something completely different like **LIDL** or **PILZ**)?

So let's improve our idea, let's repeat the message twice. If **PILLKILL-PILL** arrives and if errors are "rare," the receiver will conclude that **PILL** was the original message (correct, in this case). But **KILLPILLKILL** could have been received, in which case we would "decode" to the wrong word **KILL**. Or **APABALBYLIPI** could have arrived, which would leave us in total confusion.

A few moments of thinking will convince the reader that no clever idea will work in all cases. Repeating a message ten times is also not a good solution: new errors will come in, and the message gets very long (often there are time constraints, for example, in flight corrections from Earth to a space ship).

But the general tracks for our investigations are laid: lengthen a message "skillfully" so that all single errors, like in the case of a single repetition, or even all double errors (achievable by three-fold repetition; why not by double repetition?), etc., can be detected. If all collections of $\leq t$ errors can be detected, we say that the encoding method is ***t-error-detecting***. If also all sets of up to s errors can be "automatically" corrected (like $s = 1$ for double repetitions), we call the encoding method ***s-error correcting***.

Of course, we can reach arbitrarily high values for t and s by repeating many times. But we also want the number of added symbols to be small. These are contradicting goals. Also, the decoding and encoding methods should often be performed quickly (see the space ship example above). This makes coding theory difficult and fascinating at the same time. And, surprisingly enough, deep algebraic ideas turn out to be most helpful.

Here are two examples which are already a lot better than "brutal" repetitions. Let us assume that the messages are already turned into 0-1-sequences, and let us consider 0, 1 as the elements of the field $\mathbb{F}_2$. So messages are finite sequences $(x_1, \ldots, x_n) \in (\mathbb{F}_2)^n$, which we briefly write as $x_1 \ldots x_n$.

16.1. Example (*Parity-Check*). Encode the message $x_1 \dots x_k$ by augmenting it by a single x_{k+1} such that $x_{k+1} = 0$ if there is an even number of 1's in $x_1 \dots x_n$, and $x_{k+1} = 1$ otherwise. More elegantly, this can be defined by $x_{k+1} = x_1 + \cdots + x_k$ (in $\mathbb{F}_2$). Obviously, adding this single "check symbol" x_{k+1} to the message gives us a way to detect all single errors. So it can "do the same" as the single repetition method, but the "codewords" have length $k+1$ instead of $2k$ there.

This simple, but most effective idea is used in many areas of daily life. The last digit of a bank account is usually a check digit which detects single typing errors when writing the account number. The last digit in the ISBN-code for books fulfills the same purpose (see Exercise 1).

In Example 16.1 we determined the check symbol via a linear equation in $x_1, \dots, x_k$. The same is true for a "repetition code": $x_{k+1} = x_1, \dots, x_{2k} = x_k$. We are on the right track.

16.2. Example. Let $k = 3$ and add three check symbols x_4, x_5, x_6 to a message x_1, x_2, x_3 as follows:

$$\begin{aligned} x_4 &= x_1 + x_2, \\ x_5 &= x_1 \phantom{{}+x_2} + x_3, \\ x_6 &= \phantom{x_1 + {}} x_2 + x_3; \end{aligned} \qquad \text{i.e.,} \qquad \begin{aligned} x_1 + x_2 \phantom{{}+x_3} + x_4 \phantom{{}+x_5+x_6} &= 0, \\ x_1 \phantom{{}+x_2} + x_3 \phantom{{}+x_4} + x_5 \phantom{{}+x_6} &= 0, \\ x_2 + x_3 \phantom{{}+x_4+x_5} + x_6 &= 0. \end{aligned}$$

We shall see shortly that this yields a 2-error detecting and 1-error correcting code. So it can do the same as the double-repetition code, but with (total) word length of 6 instead of 9. Observe that the possible codewords are just the solutions of a homogeneous system of equations and thus form a subspace of $(\mathbb{F}_2)^6$.

Before we treat the whole subject systematically, let us quickly calculate if we really need all that. Suppose we want to transmit $m = 1000$ signals (each 0 or 1). If the possibility of a correct transmission is 99%, we still have to expect 10 errors in our message. And they can be fatal. Yes, we need coding theory. In general the expected value of errors is $m(1-p)$. Those who know the Binomial Distribution will see immediately that the probability of t errors in a message of m binary signals is given by $\binom{m}{t} p^{m-t}(1-p)^t$. Here are some concrete values for $p = 99\%$ and $m = 50$ (we usually break up the message into smaller parts):

t	Probability of t errors
0	60.50%
1	30.56%
2	7.56%
3	1.22%
4	0.145%

In this example we can see (and easily prove rigorously, see Exercise 2) that these probabilities decrease rapidly if m is "small" ($m < \frac{p}{1-p}$). So we really may expect that "not too many errors have occurred."

In a computer there is a lot of data transfer, and it is not too rare that some error occurs in these transfer jobs. In order to avoid wrong computations, parity-checks were built in at an early stage. But similar to what we have seen with our "99%-example" above, this caused the computer to stop up to 100 times per hour. So an engineer and computer scientist, R. Hamming, by necessity created in some pioneering work the foundations of coding theory. We will meet his name several times along our discussions.

We begin by describing a simple model of a communication transmission system (see Figure 16.1). Messages go through the system starting from the sender (or source). We shall only consider senders with a finite number of discrete signals (e.g., telegraph) in contrast to continuous sources (e.g., radio). In most systems the signals emanating from the source cannot be transmitted directly by the channel. For instance, a binary channel cannot transmit words in the usual Latin alphabet. Therefore an encoder performs the important task of data reduction and suitably transforms the message into usable form. Accordingly we dis-

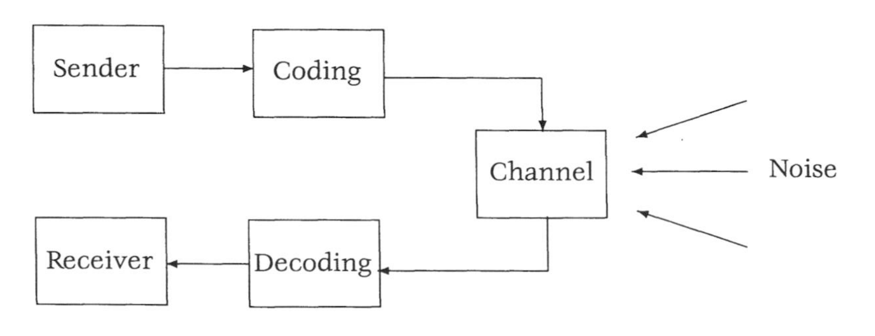

FIGURE 16.1

tinguish between source encoding and channel encoding. The former reduces the message to its essential (recognizable) parts, the latter adds redundant information to enable detection and correction of possible errors in the transmission. Similarly, on the receiving end we distinguish between channel decoding and source decoding, which invert the corresponding channel and source encoding besides detecting and correcting errors.

16.3. Example. We describe a simple model for a transmission channel, called the ***binary symmetric channel***. If p is the probability that a binary signal is received correctly, then $q := 1 - p$ is the probability of incorrect reception. Then the transmission probabilities for this channel are as shown in Figure 16.2. We assume that errors in transmission of successive signals are independent.

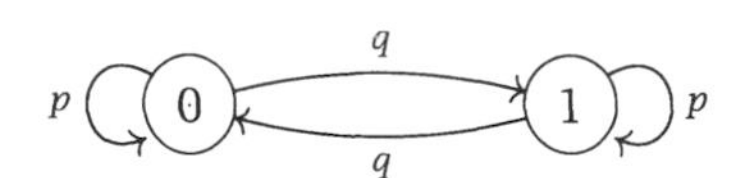

FIGURE 16.2

In the following we shall always assume that the elements of a finite field form the underlying alphabet for coding. Coding consists of transforming a block of k ***message symbols*** $a_1 \ldots a_k$, $a_i \in \mathbb{F}_q$, into a ***codeword*** $\mathbf{x} = x_1 \ldots x_n$, $x_i \in \mathbb{F}_q$, where $n \geq k$. Here we mostly assume that the first k symbols x_i are the message symbols, i.e., $x_i = a_i$, $1 \leq i \leq k$; the remaining $n - k$ elements $x_{k+1}, \ldots, x_n$ are ***check symbols*** (or ***control symbols***). Codewords will be written in one of the forms $\mathbf{x}$ or $x_1 \ldots x_n$ or $(x_1, \ldots, x_n)$ or $x_1, \ldots, x_n$.

If a message $\mathbf{a} = a_1 \ldots a_k$ is encoded as $\mathbf{x} = x_1 \ldots x_n$ and is transmitted through a "noisy" channel, the codeword $\mathbf{x}$ may be altered to a vector $\mathbf{y} = y_1 \ldots y_n$, where $\mathbf{y}$ may be different to $\mathbf{x}$.

16.4. Definition. A ***code*** over $\mathbb{F}_q$ of ***length*** n is a subset C of $\mathbb{F}_q^n$. If $\mathbf{x}$ is a transmitted codeword and $\mathbf{y}$ is received, then $\mathbf{e} = \mathbf{y} - \mathbf{x} = e_1 \ldots e_n$ is called the ***error word*** (or ***error vector***, or ***error***).

The decoder, on receiving $\mathbf{y}$, has to decide which codeword has been transmitted. According to our discussions this is the codeword of C which differs least from $\mathbf{y}$; this is called "maximum likelihood decoding."

16.5. Definition.

(i) The ***Hamming distance*** $d(\mathbf{x}, \mathbf{y})$ between two vectors $\mathbf{x} = x_1 \dots x_n$ and $\mathbf{y} = y_1 \dots y_n$ in $\mathbb{F}_q^n$ is the number of coordinates in which $\mathbf{x}$ and $\mathbf{y}$ differ.

(ii) The ***Hamming weight*** $w(\mathbf{x})$ of a vector $\mathbf{x} = x_1 \dots x_n$ in $\mathbb{F}_q^n$ is the number of nonzero coordinates x_i. So $w(\mathbf{x}) = d(\mathbf{x}, \mathbf{0})$.

We leave it as Exercise 4 to show:

16.6. Theorem. *The Hamming distance $d(C)$ is a metric on $\mathbb{F}_q^n$ and the Hamming weight w is a norm on $\mathbb{F}_2^n$.*

16.7. Definition. The ***minimum distance*** $d_{\min}(C)$, or briefly $d_{\min}$, of a linear code C is given as

$$d_{\min} := \min_{\substack{\mathbf{u},\mathbf{v} \in C \\ \mathbf{u} \neq \mathbf{v}}} d(\mathbf{u}, \mathbf{v}).$$

So, after receiving $\mathbf{y}$, we have to search through all q^k codewords and find the one which is closest to $\mathbf{y}$ w.r.t. the Hamming distance. This decoding method is called "***nearest neighbor decoding***." It is the maximum likelihood decoding if $n < \frac{p}{1-p}$. We shall assume this from now on. We also assume $p > \frac{1}{2}$ (if p is close to $\frac{1}{2}$, forget about transmitting messages!). Obviously this procedure is impossible for large k and one of the aims of coding theory is to find codes with faster decoding algorithms.

16.8. Definition. The set $S_r(\mathbf{x}) := \{\mathbf{y} \in \mathbb{F}_q^n \mid d(\mathbf{x}, \mathbf{y}) \leq r\}$ is called the ***sphere of radius*** r about $\mathbf{x} \in \mathbb{F}_q^n$.

For instance, in Example 16.2 we have

$$S_1(000000) = \{000000, 100000, 010000, 001000, 000100, 000010, 000001\}.$$

We might imagine these "spheres" around codewords $\mathbf{c}, \mathbf{c}', \mathbf{c}$,″ ... as in Figure 16.3. If $\mathbf{y}$ arrives and belongs to precisely one of the spheres about codewords, we thus decode y to the center of this sphere:

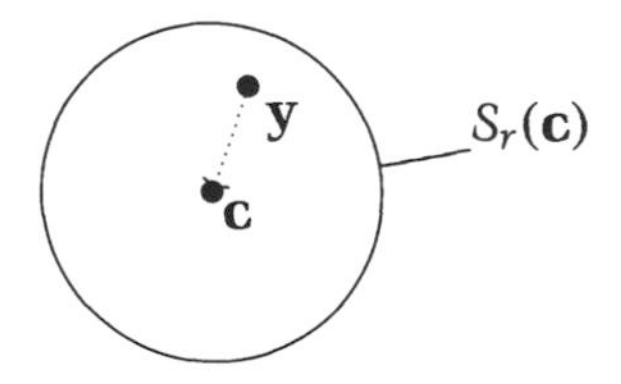

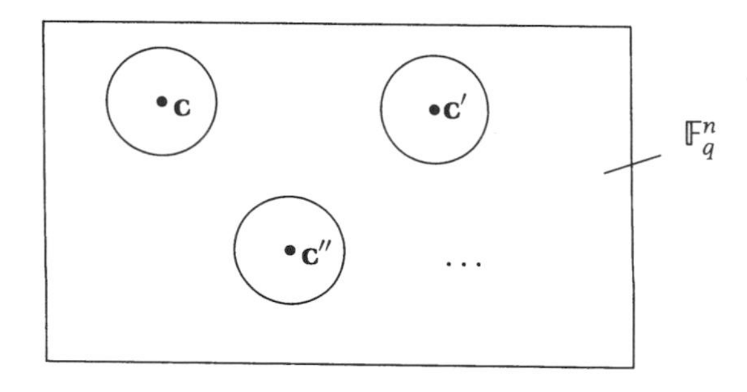

FIGURE 16.3

For "fairness reasons," all spheres should have the same radius (unless some codewords are more likely than others, which we exclude from our considerations). The radius r should be as large as possible, to be able to decode as many incoming words as possible. But on the other hand, r must be small enough so that two spheres do not intersect (they even must not "touch" each other):

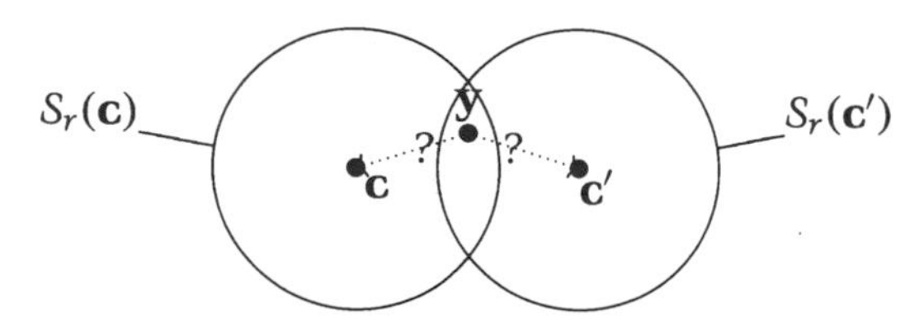

So r must be smaller than $\frac{1}{2}d_{\min}(C)$, and we get

16.9. Theorem. *A linear code with $d_{\min} = d$ can correct up to $\lfloor \frac{d-1}{2} \rfloor$ and detect up to $d - 1$ errors.*

In here, $\lfloor x \rfloor$ denotes the ***floor*** of $x \in \mathbb{R}$, i.e., the largest integer which is $\leq x$. So, we look for codes with a large minimum distance.

We shall see shortly that the code in 16.2 had minimum distance 3; hence it can detect up to two errors and correct one error.

One of the main problems of coding theory is not merely to minimize errors, but to do so without reducing the ***information rate*** $\frac{k}{n}$ unnecessarily. Errors can be corrected by lengthening the code blocks, but this reduces the number of message symbols that can be sent per second. To maximize the transmission rate we want enough code words to encode a given message alphabet, but at the same time the words should be no longer than necessary to achieve sufficient error detection and/or correction. In other words, a central problem in coding theory is:

Given $n, d \in \mathbb{N}$, find the maximum number $A(n, d)$ of vectors in $\mathbb{F}_2^n$ which are at distances $\geq d$ from each other, and, of course, find these vectors.

It is now clear that this can also be considered as a "discrete sphere packing problem" (cf. the end of §36 and, e.g., Conway & Sloane (1988)). The following table gives the first few values of $A(n, d)$ for $d = 3$, i.e., the maximum sizes of single-error-correcting codes, which equals the maximum number of nonintersecting spheres of radius 1 in $\mathbb{Z}_2^n$:

n	3	4	5	6	7	8	9	10
$A(n,3)$	2	2	4	8	16	20	40	Unknown, between 72 and 79

Let C be a code over $\mathbb{F}_q$ of length n with M codewords. Suppose C can correct t errors, i.e., C is t-error-correcting. There are $(q-1)^m\binom{n}{m}$ vectors of length n and weight m over $\mathbb{F}_q$. The spheres of radius t about the codewords of C are disjoint and each of the M spheres contains $1 + (q-1)\binom{n}{1} + \cdots + (q-1)^t\binom{n}{t}$ vectors. The total number of vectors in $\mathbb{F}_q^n$ is q^n. Thus we have

16.10. Theorem (*Hamming Bound*). *The parameters q, n, t, M of a t-error-correcting code C over $\mathbb{F}_q$ of length n with M codewords satisfy the inequality*

$$M\left(1 + (q-1)\binom{n}{1} + \cdots + (q-1)^t\binom{n}{t}\right) \leq q^n.$$

If all vectors of $\mathbb{F}_q^n$ are within or on spheres of radius t around codewords of a linear (n, k) code, then we obtain a special class of codes:

16.11. Definition. A t-error-correcting code over $\mathbb{F}_q$ is called ***perfect*** if equality holds in Theorem 16.10.

We shall come back to perfect codes later on (see Theorem 20.9).

If C is a code as in 16.10 with $d_{\min} = d = 2t + 1$, then deleting the last $d - 1$ symbols still gives a code in which all words are different. Since this code has length $n - d + 1$, we get

16.12. Theorem (*Singleton Bound*). *If $C \subseteq \mathbb{F}_q^n$ has minimum distance d, then $|C| \leq q^{n-d+1}$.*

16.13. Definition. If equality holds in 16.12, C is called ***maximum distance separable***, or briefly an ***MDS code***.

16.14. Example. In our Example 16.2 we have $M = 8$, $q = 2$, $n = 6$, $d = 3$, $t = 1$. The Hamming bound gives the inequality $8(1 + \binom{6}{1}) \leq 2^6$, i.e., $56 < 64$. This means that only eight 6-tuples of $\mathbb{F}_2^6$ lie outside of spheres and cannot be corrected properly (this is also clear from the fact that we have eight nonintersecting spheres with seven elements each. One example of these eight "very bad" incoming words is 100100; it has a distance ≥ 2 to all codewords. We shall see more of this in Example 17.17. The Singleton bound yields $8 \leq 2^4 = 16$; so this code is not MDS.

These bounds show up our limits. So it comes as a big surprise that there do exist "arbitrarily good codes": without proof (see, e.g., van Lint (1992)) we mention the following theorem, published in 1948, which many consider as the starting point of coding theory.

16.15. Theorem (*Shannon's Theorem*). *For every $\varepsilon > 0$, every $p \in [0, 1]$, and every r with $0 < r < 1 + p\log p + (1 - p)\log(1 - p)$, there is an $n \in \mathbb{N}$ and a code of length n and information rate r, for which each digit is correctly transmitted with probability p and in which the probability of incorrect decoding of words is $< \varepsilon$.*

The proof is a "pure existence proof" and does not give us a method at hand to actually construct such a code. So we have to continue our investigations, and we turn our attention to those codes which are subspaces of $\mathbb{F}_q^n$.

Exercises

1. Books nowadays have a (unique) ***ISBN*** ("International Standard Book Number"), like ISBN 0-387-96166-6. The "0" indicates the USA, 387 stands for Springer-Verlag, 96166 is an internal number, and 6 a check number so that a weighted sum of the ten digits $d_1, \ldots, d_{10}$, namely $d_{10}+2d_9+3d_8+\cdots+10d_1$ is $\equiv 0 \pmod{11}$. Here the possible "10" is replaced by X. Is the ISBN above possible? Complete 3-540-96035. Is the ISBN-system a kind of coding?

2. If p is the probability that a single binary signal is transmitted correctly and if m signals are sent, where $m < \frac{p}{1-p}$, prove that the probability for t errors is smaller than the probability of $t-1$ errors ($1 \leq t \leq m$).
3. Ten binary words of length 30 are sent off. The probability of a correct transmission of one digit is 99%. What is the probability of more than four errors in the whole message?
4. Prove Theorem 16.6.
5. Show that $A(n, 3) = 2, 2, 4$ if $n = 3, 4, 5$, respectively.
6. Give bounds for $A(n, 3)$ and $A(n, 4)$ with $n = 6, 7, 8, 9, 10$ using the Hamming bound and the singleton bound, and compare the results with the table on page 190.
7. Show that $A(n, d-1) = A(n+1, d)$ if d is even.
8. By dividing codewords of a code into two classes according if they start with 0 or with 1, show that $A(n-1, d) \geq \frac{1}{2}A(n, d)$.
9. Use Exercises 7 and 8 to improve the bounds you have found in Exercise 6.
10. Are parity-check codes perfect? Are they MDS-codes?

§17 Linear Codes

Suppose now that the check symbols can be obtained from the message symbols in such a way that the codewords $\mathbf{x}$ satisfy the system of linear equations

$$\mathbf{H}\mathbf{x}^{\mathrm{T}} = \mathbf{0},$$

where $\mathbf{H}$ is a given $(n-k) \times n$ matrix with elements in $\mathbb{F}_q$. A "standard form" for $\mathbf{H}$ is $(\mathbf{A}, \mathbf{I}_{n-k})$, with $\mathbf{A}$ an $(n-k) \times k$ matrix, and $\mathbf{I}_{n-k}$ the $(n-k) \times (n-k)$ identity matrix.

17.1. Example (*16.2 revisited*). Here we have $q = 2$, $n = 6$, $k = 3$. The check symbols x_4, x_5, x_6 are such that for

$$\mathbf{H} = \begin{pmatrix} 1 & 1 & 0 & 1 & 0 & 0 \\ 1 & 0 & 1 & 0 & 1 & 0 \\ 0 & 1 & 1 & 0 & 0 & 1 \end{pmatrix} =: (\mathbf{A}, \mathbf{I}_3)$$

we have $\mathbf{H}$ in standard form and $\mathbf{H}\mathbf{x}^{\mathrm{T}} = \mathbf{0}$ iff $\mathbf{x}$ is a codeword, i.e.,

$$a_1 + a_2 + a_4 = 0,$$
$$a_1 + a_3 + a_5 = 0,$$
$$a_2 + a_3 + a_6 = 0.$$

The equations in the system $\mathbf{H}\mathbf{x}^{\mathrm{T}} = \mathbf{0}$ are also called ***check equations***. If the message $\mathbf{a} = 011$ is transmitted, then the corresponding codeword is $\mathbf{x} = 011110$. Altogether there are 2^3 codewords:

$$000000, 001011, 010101, 011110, 100110, 101101, 110011, 111000.$$

In general, we define

17.2. Definition. Let $\mathbf{H}$ be an $(n-k) \times n$ matrix of rank $n-k$ with elements in $\mathbb{F}_q$. The set of all n-dimensional vectors $\mathbf{x}$ satisfying $\mathbf{H}\mathbf{x}^{\mathrm{T}} = \mathbf{0}$ over $\mathbb{F}_q$ is called a ***linear (block) code*** C over $\mathbb{F}_q$ of ***(block) length*** n. The matrix $\mathbf{H}$ is called the ***parity-check matrix*** of the code. C is also called a ***linear*** (n, k) ***code***. If $\mathbf{H}$ is of the form $(\mathbf{A}, \mathbf{I}_{n-k})$, then the first k symbols of the codeword $\mathbf{x}$ are the message symbols, and the last $n-k$ symbols in $\mathbf{x}$ are the check symbols. C is then also called a ***systematic*** linear (n, k) code; and $\mathbf{H}$ is said to be in ***standard form***. If $q = 2$, then C is a ***binary*** code.

17.3. Remark. The set C of solutions $\mathbf{x}$ of $\mathbf{H}\mathbf{x}^{\mathrm{T}} = \mathbf{0}$, i.e., the null space of $\mathbf{H}$, forms a subspace of $\mathbb{F}_q{}^n$ of dimension k. Since the codewords form an additive group, C is also called a ***group code***.

17.4. Example (*Repetition Code*). If each codeword of a code consists of only one message symbol $a_1 \in \mathbb{F}_2$ and the $n-1$ check symbols $x_2 = \cdots = x_n$ are all equal to a_1 (a_1 is repeated $n-1$ times), then we obtain a binary $(n, 1)$ code with parity-check matrix

$$\mathbf{H} = \begin{pmatrix} 1 & 1 & 0 & \cdots & 0 \\ 1 & 0 & 1 & \cdots & 0 \\ \vdots & \vdots & \vdots & \ddots & \vdots \\ 1 & 0 & 0 & \cdots & 1 \end{pmatrix}.$$

There are only two codewords in this code, namely $00\ldots0$ and $11\ldots1$.

In repetition codes we can, of course, also consider codewords with more than one message symbol. If we transmit a message of length k three times and compare corresponding "coordinates" x_i, x_{k+i}, x_{2k+i} of the codeword $x_1 \cdots x_i \cdots x_k x_{k+1} \cdots x_{k+i} \cdots x_{2k} x_{2k+1} \cdots x_{2k+i} \cdots x_{3k}$, then a "majority decision" decides which word has been sent, e.g., if $x_i = x_{k+i} \neq x_{2k+i}$, then most likely x_i has been transmitted. It is, however, often impracticable, impossible, or too expensive to send the original message more than once.

17.5. Example. The parity-check code of Example 16.1 is the binary $(n, n-1)$ code with parity-check matrix $\mathbf{H} = (11\dots1)$. Each codeword has one check symbol and all codewords are given by all binary vectors of length n with an even number of 1's.

We have seen that in a systematic code, a message $\mathbf{a} = a_1 \dots a_k$ is encoded as codeword $\mathbf{x} = x_1 \cdots x_k x_{k+1} \cdots x_n$, with $x_1 = a_1, \dots, x_k = a_k$. The check equations $(\mathbf{A}, \mathbf{I}_{n-k})\mathbf{x}^{\mathrm{T}} = \mathbf{0}$ yield

$$\begin{pmatrix} x_{k+1} \\ \vdots \\ x_n \end{pmatrix} = -\mathbf{A} \begin{pmatrix} x_1 \\ \vdots \\ x_k \end{pmatrix} = -\mathbf{A} \begin{pmatrix} a_1 \\ \vdots \\ a_k \end{pmatrix}.$$

Thus we obtain

$$\begin{pmatrix} x_1 \\ \vdots \\ x_n \end{pmatrix} = \begin{pmatrix} \mathbf{I}_k \\ -\mathbf{A} \end{pmatrix} \begin{pmatrix} a_1 \\ \vdots \\ a_k \end{pmatrix}.$$

We transpose and write this equation as

$$(x_1, \dots, x_n) = (a_1, \dots, a_k)(\mathbf{I}_k, -\mathbf{A}^{\mathrm{T}}).$$

17.6. Definition. The matrix $\mathbf{G} = (\mathbf{I}_k, -\mathbf{A}^{\mathrm{T}})$ is called a (***canonical***) ***generator matrix*** (or ***canonical basic matrix*** or ***encoding matrix***) of a linear (n, k) code with parity-check matrix $\mathbf{H} = (\mathbf{A}, \mathbf{I}_{n-k})$ in standard form.

We clearly have $\mathbf{G}\mathbf{H}^{\mathrm{T}} = \mathbf{0}$.

17.7. Example.

$$\mathbf{G} = \begin{pmatrix} 1 & 0 & 0 & 1 & 1 & 0 \\ 0 & 1 & 0 & 1 & 0 & 1 \\ 0 & 0 & 1 & 0 & 1 & 1 \end{pmatrix}$$

is a canonical generator matrix for the code of Example 16.2. The 2^3 codewords $\mathbf{x}$ of this binary code (see 16.2) can be obtained from $\mathbf{x} = \mathbf{aG}$ with $\mathbf{a} = a_1 a_2 a_3$, $a_i \in \mathbb{F}_2$.

17.8. Theorem. *Let* $\mathbf{G}$ *be a generator matrix of a linear code* C. *Then the rows of* $\mathbf{G}$ *form a basis of* C.

Proof. The k rows of $\mathbf{G}$ are clearly linearly independent by the definition of $\mathbf{G}$. If $\mathbf{r}$ is a row vector of $\mathbf{G}$, then $\mathbf{r}\mathbf{H}^{\mathrm{T}} = \mathbf{0}$, so $\mathbf{H}\mathbf{r}^{\mathrm{T}} = \mathbf{0}$, whence $\mathbf{r} \in C$. Now $\dim C$ is the dimension of the null space of $\mathbf{H}$, which is $n - \operatorname{Rk}\mathbf{H} = k$. Putting all together, we see that the rows of $\mathbf{G}$ are a basis for C. □

A code can have several parity-check matrices and generator matrices. Any $k \times n$ matrix whose row space is equal to C can be taken as a generator matrix of C.

What happens with a "generator matrix" if $\mathbf{H}$ is not in standard form? We might perform the "usual" row operations in $\mathbf{H}$ as known from Linear Algebra. This does not change C, the null space of $\mathbf{H}$, and then we permute the coordinates to get $\mathbf{H}$ finally into standard form $\mathbf{H}'$. The corresponding code C' is "equivalent" to C in the following sense:

17.9. Definition. Two codes C and C' of the same length n are called ***equivalent*** if there is a permutation π of $\{1, \ldots, n\}$ such that

$$(x_1, \ldots, x_n) \in C \iff (x_{\pi(1)}, \ldots, x_{\pi(n)}) \in C'.$$

Continuing the discussion above, we might form the generator matrix $\mathbf{G}'$ of C' (w.r.t. $\mathbf{H}'$) and then undo the permutation π of the coordinates.

One of the most important properties of linear codes is based on the fact that for linear codes we have $d(\mathbf{u}, \mathbf{v}) = d(\mathbf{u} - \mathbf{v}, \mathbf{0}) = w(\mathbf{u} - \mathbf{v})$. Therefore we get

17.10. Theorem. *The minimum distance of C is equal to the least weight of all nonzero codewords.*

For instance, for the code in 16.2 we have $d_{\min} = 3$ (by comparing all $\binom{8}{2} = 28$ pairs of different codewords or by finding the minimal weight of the seven nonzero codewords, which is much less work).

17.11. Definition. A linear code of length n, dimension k, and minimum distance d is called an ***(n, k, d) code***.

Let $\mathbf{u} = u_1 \ldots u_n$ and $\mathbf{v} = v_1 \ldots v_n$ be vectors in $\mathbb{F}_q^n$ and let $\mathbf{u} \cdot \mathbf{v} = u_1 v_1 + \cdots + u_n v_n$ denote the dot product of $\mathbf{u}$ and $\mathbf{v}$ over $\mathbb{F}_q^n$. If $\mathbf{u} \cdot \mathbf{v} = 0$, then u and v are called ***orthogonal***.

17.12. Definition. Let C be a linear (n, k) code over $\mathbb{F}_q$. The ***dual*** (or ***orthogonal***) ***code*** $C^{\perp}$ of C is defined by

$$C^{\perp} = \{\mathbf{u} \mid \mathbf{u} \cdot \mathbf{v} = 0 \text{ for all } \mathbf{v} \in C\}.$$

Since C is a k-dimensional subspace of the n-dimensional vector space $\mathbb{F}_q^n$, the orthogonal complement is of dimension $n-k$ and an $(n, n-k)$ code. It can be shown that if the code C has a generator matrix $\mathbf{G}$ and parity-check matrix $\mathbf{H}$, then $C^\perp$ has generator matrix $\mathbf{H}$ and parity-check matrix $\mathbf{G}$. Orthogonality of the two codes can be expressed by $\mathbf{GH}^\mathrm{T} = \mathbf{HG}^\mathrm{T} = \mathbf{0}$. We now summarize some of the simple properties of linear codes.

Let $\mathrm{mld}(\mathbf{H})$ be the minimal number of linearly dependent columns of $\mathbf{H}$. Obviously, $\mathrm{mld}(\mathbf{H}) \le \mathrm{Rk}(\mathbf{H}) + 1$ for every matrix $\mathbf{H}$.

17.13. Theorem. *Let* $\mathbf{H}$ *be a parity-check matrix of an* (n, k, d) *code* C *with* $n > k$. *Then the following hold:*

(i) $\dim C = k = n - \mathrm{Rk}\,\mathbf{H}$;

(ii) $d = \mathrm{mld}\,\mathbf{H}$;

(iii) $d \le n - k + 1$.

Proof. (i) is clear, and (iii) follows from (i) and (ii). In order to see (ii), suppose $\mathbf{H}$ has the columns $s_1, \ldots, s_n$. If $\mathbf{c} = (c_1, \ldots, c_n) \in C$, then $\mathbf{Hc}^\mathrm{T} = c_1 s_1 + \cdots + c_n s_n = 0$. Now apply 17.10. □

To verify the existence of linear (n, k) codes with minimum distance d over $\mathbb{F}_q$, it suffices to show that there exists an $(n-k) \times n$ matrix $\mathbf{H}$ with $\mathrm{mld}(\mathbf{H}) = d$, for some given n, k, d, q. We now construct such a matrix $\mathbf{H}$ and prove

17.14. Theorem (*Gilbert-Varshamov Bound*). *If*

$$q^{n-k} > \sum_{i=0}^{d-2} \binom{n-1}{i} (q-1)^i,$$

then we can construct a linear (n, k) *code over* $\mathbb{F}_q$ *with minimum distance* $\ge d$.

Proof. We construct an $(n-k) \times n$ parity-check matrix $\mathbf{H}$ for such a code. Let the first column of $\mathbf{H}$ be any nonzero $(n-k)$-tuple over $\mathbb{F}_q$. The second column is any $(n-k)$-tuple over $\mathbb{F}_q$ which is not a scalar multiple of the first column. Suppose $j-1$ columns have been chosen so that any $d-1$ of them are linearly independent. There are at most

$$\sum_{i=0}^{d-2} \binom{j-1}{i} (q-1)^i$$

vectors obtained by taking linear combinations of $d-2$ or fewer of these $j-1$ columns. If the inequality of the theorem holds, it will be possible to choose a jth column which is linearly independent of any $d-2$ of the first $j-1$ columns. This construction can be carried out in such a way that $\mathbf{H}$ has rank $n-k$. That the resulting code has minimum distance $d_{\min} \geq d$ follows from the fact that no $d-1$ columns of $\mathbf{H}$ are linearly dependent.

□

So if $q = 2$, $n = 15$, $k = 7$, we have $2^8 > \sum_{i=0}^{2} \binom{14}{i}$. But $2^8 < \sum_{i=0}^{3} \binom{14}{i}$, and we thus know that there must be a linear code of length 15 and dimension 7 with $d_{\min} = 4$, but we cannot be sure that there is also a linear $(15, 7)$ code with minimum distance 5. Nevertheless, we shall construct such a code in 19.1. The Hamming bound excludes a linear binary $(15, 7)$ code with $d_{\min} = 7$.

Let C be a linear (n, k) code over $\mathbb{F}_q$ and let $1 \leq i \leq n$ be such that C contains a codeword with nonzero ith component. Let D be the subspace of C containing all codewords with ith component zero. In C/D there are q elements, which correspond to q choices for the ith component of a codeword. If $|C| = M = q^k$ denotes the number of elements in C, then $M/|D| = |C/D|$, i.e., $|D| = q^{k-1}$. The sum of the weights of the codewords in C is $\leq nq^{k-1}(q-1)$. Since the total number of codewords of nonzero weight is $q^k - 1$, we get $nq^{k-1}(q-1) \geq$ sum of all weights $\geq d(q^k - 1)$, so

$$d \leq \frac{nq^{k-1}(q-1)}{q^k - 1} \leq \frac{nq^k(q-1)}{(q^k-1)q} = \frac{nM(q-1)}{(M-1)q}.$$

So the minimum distance d must satisfy the following inequality.

17.15. Theorem (*Plotkin Bound*). *If there is a linear code of length n with M codewords and minimum distance d over $\mathbb{F}_q$, then*

$$d \leq \frac{nM(q-1)}{(M-1)q}.$$

In our example above ($n = 15$, $q = 2$, $M = 2^7$) we get $d_{\min} \leq 7$.

In what follows, we shall describe a simple decoding algorithm for linear codes. Let C be a linear (n, k) code over $\mathbb{F}_q$, regarded as a subspace of $\mathbb{F}_q^n$. The factor space $\mathbb{F}_q^n/C$ consists of all cosets $\mathbf{a}+C = \{\mathbf{a}+\mathbf{x} \mid \mathbf{x} \in C\}$ for arbitrary $\mathbf{a} \in \mathbb{F}_q^n$. Each coset contains $|C| = q^k$ vectors (by the discussion before 10.17). There is a partition of $\mathbb{F}_q^n$ of the form

$$\mathbb{F}_q^n = C \cup (\mathbf{a}^{(1)} + C) \cup \cdots \cup (\mathbf{a}^{(t)} + C) \quad \text{for } t = q^{n-k} - 1.$$

If a vector $\mathbf{y}$ is received, then $\mathbf{y}$ must be an element of one of these cosets, say of $\mathbf{a}^{(i)} + C$. If the codeword $\mathbf{x}^{(1)}$ has been transmitted, then the error vector $\mathbf{e}$ is given as $\mathbf{e} = \mathbf{y} - \mathbf{x}^{(1)} \in \mathbf{a}^{(i)} + C - \mathbf{x}^{(1)} = \mathbf{a}^{(i)} + C$. Thus we have the following decoding rule.

17.16. Theorem. *If a vector* $\mathbf{y}$ *is received, then the possible error vectors* $\mathbf{e}$ *are precisely the vectors in the coset containing* $\mathbf{y}$. *The most likely error is the vector* $\bar{\mathbf{e}}$ *with minimum weight in this coset. Thus* $\mathbf{y}$ *is decoded to* $\bar{\mathbf{x}} = \mathbf{y} - \bar{\mathbf{e}}$.

Next we show how to find the coset of $\mathbf{y}$ and apply Theorem 17.16. The vector of minimum weight in a coset (if it is unique) is called the ***coset leader***. If there are several such vectors, then we arbitrarily choose one of them as coset leader. Let $\mathbf{a}^{(1)}, \ldots, \mathbf{a}^{(t)}$ be the coset leaders. We first establish the following table (due to Slepian):

$$\begin{array}{cccc|l}
\mathbf{x}^{(1)} = \mathbf{0} & \mathbf{x}^{(2)} & \cdots & \mathbf{x}^{(q^k)} & \} \text{ codewords in } C, \\
\mathbf{a}^{(1)} + \mathbf{x}^{(1)} & \mathbf{a}^{(1)} + \mathbf{x}^{(2)} & \cdots & \mathbf{a}^{(1)} + \mathbf{x}^{(q^k)} & \\
\cdots & \cdots & \cdots & \cdots & \text{other cosets.} \\
\underbrace{\mathbf{a}^{(t)} + \mathbf{x}^{(1)}}_{\text{coset leaders}} & \mathbf{a}^{(t)} + \mathbf{x}^{(2)} & \cdots & \mathbf{a}^{(t)} + \mathbf{x}^{(q^k)} &
\end{array}$$

If a vector $\mathbf{y}$ is received, then we have to find $\mathbf{y}$ in the table. Let $\mathbf{y} = \mathbf{a}^{(i)} + \mathbf{x}^{(j)}$; then the decoder decides that the error $\bar{\mathbf{e}}$ is the coset leader $\mathbf{a}^{(i)}$. Thus $\mathbf{y}$ is decoded to the codeword $\bar{\mathbf{x}} = \mathbf{y} - \bar{\mathbf{e}} = \mathbf{x}^{(j)}$. This codeword $\bar{\mathbf{x}}$ simply occurs as the first element in the column of $\mathbf{y}$.

17.17. Example. We take again the code in Example 16.2. The first row consists of the codewords (their first three digits always consist of the message itself). For getting the first coset other than C, we take any element $\notin C$, for instance 001111, and form $001111 + C = \{001111, 000001, 011010, 010100, 101100, 100010, 111001, 110111\}$. We choose one word with minimal weight as coset leader, namely 000001, and so on. We might arrive at the following table:

000000	001110	010101	011011	100011	101101	110110	111000
000001	001111	010100	011010	100010	101100	110111	111001
000010	001100	010111	011001	100001	101111	110100	111010
000100	001010	010001	011111	100111	101001	110010	111100
001000	000110	011101	010011	101011	100101	111110	110000
010000	011110	000101	001011	110011	111101	100110	101000
100000	101110	110101	111011	000011	001101	010110	011000
001001	000111	011100	010010	101010	100100	111111	110001

Observe that in the last row the coset leader is not unique: 001001, 010010, and 100100 have the minimal weight 2.

For larger codes it is very impracticable (or impossible) to store the huge resulting table and to search through this table to find the word which has arrived. The situation gets much better if we observe that for all words $\mathbf{w}$ in a given row of the table (i.e., in a given coset), $\mathbf{H}\mathbf{w}^{\mathrm{T}}$ has the same value.

17.18. Definition. Let $\mathbf{H}$ be the parity-check matrix of a linear (n, k) code. Then the vector $S(\mathbf{y}) := \mathbf{H}\mathbf{y}^{\mathrm{T}}$ of length $n-k$ is called the ***syndrome*** of $\mathbf{y}$.

17.19. Theorem.

(i) $S(\mathbf{y}) = \mathbf{0} \iff \mathbf{y} \in C$.

(ii) $S(\mathbf{y}) = S(\mathbf{w}) \iff \mathbf{y} + C = \mathbf{w} + C \iff$ $\mathbf{y}$ *and* $\mathbf{w}$ *are in the same coset.*

Proof.

(i) This follows from the definition of C with the parity-check matrix $\mathbf{H}$.

(ii)

$$\begin{aligned} S(\mathbf{y}) = S(\mathbf{w}) &\iff \mathbf{H}\mathbf{y}^{\mathrm{T}} = \mathbf{H}\mathbf{w}^{\mathrm{T}} \\ &\iff \mathbf{H}(\mathbf{y}^{\mathrm{T}} - \mathbf{w}^{\mathrm{T}}) = \mathbf{0} \\ &\iff \mathbf{y} - \mathbf{w} \in C \\ &\iff \mathbf{y} + C = \mathbf{w} + C. \end{aligned}$$

□

Thus we can define the cosets via syndromes. Let $\mathbf{e} = \mathbf{y} - \mathbf{x}$, $\mathbf{x} \in C$, $\mathbf{y} \in \mathbb{F}_q^n$; then

$$S(\mathbf{y}) = S(\mathbf{x} + \mathbf{e}) = S(\mathbf{x}) + S(\mathbf{e}) = S(\mathbf{e}),$$

i.e., $\mathbf{y}$ and $\mathbf{e}$ are in the same coset. We may now state a different formulation of the decoding algorithm 17.16.

17.16′. Theorem (*Decoding Algorithm*). *If $\mathbf{y} \in \mathbb{F}_q^n$ is a received vector, find $S(\mathbf{y})$ and the coset leader $\mathbf{e}$ with syndrome $S(\mathbf{y})$. Then the most likely transmitted codeword is $\mathbf{x} = \mathbf{y} - \mathbf{e}$ and we have $d(\mathbf{x}, \mathbf{y}) = min\{d(\mathbf{c}, \mathbf{y}) \mid \mathbf{c} \in C\}$.*

17.20. Example. In Examples 16.2 and 17.17 it is hence sufficient to store the coset leaders, together with their syndromes:

coset leader	000000	100000	010000	001000	000100	000010	000001	001001
syndrome	000	011	101	110	100	010	001	111

So if $\mathbf{y} = 101001$ arrives, we compute its syndrome $S(\mathbf{y}) = 100$, see that the most likely error word is the corresponding coset leader $\mathbf{e} = 000100$, conclude that $\mathbf{y} - \mathbf{e} = 101101$ is the most likely codeword which was sent off, and decode $\mathbf{y}$ to 101.

This is much better, but still, in large linear codes, finding explicitly the coset leaders as vectors in the cosets with least weights is practically impossible. (Any $(50, 20)$ code over $\mathbb{F}_2$ has approximately 10^9 cosets.) Therefore we construct codes which are more systematic, such as the following Hamming codes. But first we briefly look at the special case of codes over $\mathbb{F}_2$. In the binary case we see that $S(\mathbf{y}) = \mathbf{He}^{\mathrm{T}}$ for $\mathbf{y} = \mathbf{x} + \mathbf{e}$, $\mathbf{y} \in C$, $\mathbf{x} \in C$, so that we have

17.21. Theorem. *In a binary code the syndrome is equal to the sum of those columns of the parity-check matrix* $\mathbf{H}$ *in which errors occurred.*

So in our Example 17.20 the incoming word $\mathbf{y} = 101001$ has syndrome 100, which is the fourth column of $\mathbf{H}$. Note that we do not need any coset leaders, etc., for correcting single errors in binary codes.

In a single-error-correcting code the columns of $\mathbf{H}$ must be nonzero, otherwise we could not detect an error in the ith place of the message if the ith column is zero. All columns must be pairwise distinct, otherwise we could not distinguish between the errors. This is also clear since we must have $\operatorname{Rk} \mathbf{H} \geq 2$. So let's try *all* nonzero vectors as columns.

17.22. Definition. A binary code C_m of length $n = 2^m - 1$, $m \geq 2$, with an $m \times (2^m - 1)$ parity-check matrix $\mathbf{H}$ whose columns consist of all nonzero binary vectors of length m is called a binary ***Hamming code***.

Since any two, but not any three columns in $\mathbf{H}$ are linearly independent, $\operatorname{mld}(\mathbf{H}) = 3$. Hence C_m is a $(2^m - 1, 2^m - 1 - m, 3)$ code. The rank $\operatorname{Rk}(\mathbf{H})$ of $\mathbf{H}$ is m. Decoding in Hamming codes is particularly simple. Theorem 16.9 tells us that C_m can correct errors $\mathbf{e}$ with weight $w(\mathbf{e}) = 1$ and detect errors with $w(\mathbf{e}) \leq 2$. We choose the lexicographical order for the columns of $\mathbf{H}$, i.e., the ith column is the binary representation of i. If an error occurs in the ith column, then $S(\mathbf{y}) = \mathbf{Hy}^{\mathrm{T}} = \mathbf{He}^{\mathrm{T}}$ is the ith column of $\mathbf{H}$ (by 17.21) and hence the binary representation of i.

17.23. Example. Let C_3 be the $(7, 4, 3)$ Hamming code with parity-check matrix

$$\mathbf{H} = \begin{pmatrix} 0 & 0 & 0 & 1 & 1 & 1 & 1 \\ 0 & 1 & 1 & 0 & 0 & 1 & 1 \\ 1 & 0 & 1 & 0 & 1 & 0 & 1 \end{pmatrix}.$$

The first column is the binary representation of $1 = (001)_2$, the second column of $2 = (010)_2$, etc. If $S(\mathbf{y}) = (101)^{\mathrm{T}}$, then we know that an error occurred in the fifth column, since $(101)^{\mathrm{T}}$ is the binary representation of 5.

Interchanging columns of $\mathbf{H}$ gives that the parity-check matrix in the standard form $\mathbf{H}' = (\mathbf{A}, \mathbf{I}_m)$, where $\mathbf{A}$ consists of all columns with at least two 1's. We can easily see that C_m is uniquely determined "up to equivalence," i.e., any linear $(2^m - 1, 2^m - 1 - m, 3)$ code is equivalent to C_m. We summarize some further properties and remarks for Hamming codes.

17.24. Remarks.

(i) If we choose

$$\mathbf{H}'' = \begin{pmatrix} 1 & 1 & 1 & 0 & 1 & 0 & 0 \\ 0 & 1 & 1 & 1 & 0 & 1 & 0 \\ 0 & 0 & 1 & 1 & 1 & 0 & 1 \end{pmatrix}$$

as the parity-check matrix for the $(7, 4, 3)$ Hamming code in 17.23, then we see that the code is "cyclic," i.e., a cyclic shift, $x_1x_2 \ldots x_7 \mapsto x_7x_1 \ldots x_6$, of a codeword is also a codeword. All binary Hamming codes are cyclic in this sense. We shall study this property in §18.

(ii) The Hamming code C_m is a perfect 1-error-correcting linear code (see Exercise 16).

(iii) A ***generalized Hamming code*** over $\mathbb{F}_q$ can be defined by an $m \times (q^m - 1)/(q-1)$ parity-check matrix $\mathbf{H}$ such that no two columns of $\mathbf{H}$ are multiples of each other. This gives us a

$$((q^m - 1)/(q - 1), (q^m - 1)/(q - 1) - m, 3)$$

code over $\mathbb{F}_q$.

(iv) Hamming codes cannot correct any errors $\mathbf{e}$ with $w(\mathbf{e}) \geq 2$. A generalization of Hamming codes are the BCH codes, which will be studied in §19 and §20.

(v) We obtain a so-called ***extended code*** if we add a new element, the negative sum of the first n symbols, in an (n, k, d) code. The parity-check matrix $\overline{\mathbf{H}}$ of an extended code can be obtained from the parity-check matrix $\mathbf{H}$ of the original code by the addition of a column of zeros and then a row of ones. The code C_3 of Example 17.23 can be extended to an $(8, 4, 4)$ extended Hamming code with parity-check matrix

$$\overline{\mathbf{H}} = \begin{pmatrix} 1 & 1 & 1 & 1 & 1 & 1 & 1 & 1 \\ 0 & 0 & 0 & 1 & 1 & 1 & 1 & 0 \\ 0 & 1 & 1 & 0 & 0 & 1 & 1 & 0 \\ 1 & 0 & 1 & 0 & 1 & 0 & 1 & 0 \end{pmatrix}.$$

Any nonzero codeword in this extended code is of weight at least 4. By Theorem 16.10 this code can correct all errors of weight 1 and detect all errors of weight ≤ 3. Extending, however, works only once (why?), so we still have not seen a 2-error-correcting code, except four-fold repetition.

(vi) The dual code of the binary $(2^m - 1, 2^m - 1 - m, 3)$ Hamming code C_m is the ***binary simplex code*** with parameters $(2^m - 1, m, 2^{m-1})$.

(vii) The dual of an extended $(2^m, 2^m - 1 - m, 4)$ Hamming code is called a ***first-order Reed-Muller*** code. This code can also be obtained by "lengthening" a simplex code.

A code is lengthened by the addition of the codeword $11\ldots1$ and then is extended by adding an overall parity check. (This has the effect of adding one more message symbol.) The first-order Reed-Muller code has parameters $(2^m, m+1, 2^{m-1})$. The decoding algorithm for these codes is very fast and is based on the Fast Fourier Transform, which will be described in §34. This fast decoding is the main reason that all of NASA's Mariner deep-space probes launched between 1969 and 1977 were equipped with a $(32, 6, 16)$ Reed-Muller code for the transmission of pictures from the Moon and from Mars. Without coding theory, these pictures would have been completely unintelligible! Other deep-space probes have been equipped with convolutional codes (see McEliece (1977)).

Exercises

1. Let $\mathbf{G}$ be generator matrix of a binary $(5, 2)$ code

$$\mathbf{G} = \begin{pmatrix} 0 & 1 & 1 & 1 & 1 \\ 1 & 0 & 0 & 1 & 0 \end{pmatrix}.$$

   Determine a parity-check matrix, all syndromes, and coset leaders for this code.

2. A linear code $C \subseteq \mathbb{F}_2^5$ is defined by the generator matrix

$$\mathbf{G} = \begin{pmatrix} 0 & 1 & 0 & 0 & 1 \\ 0 & 0 & 1 & 0 & 1 \\ 1 & 0 & 0 & 1 & 1 \end{pmatrix}.$$

   Determine the rank of $\mathbf{G}$, the minimum distance of C, a parity-check matrix for C, and all the codewords.

3. Determine the dual code $C^\perp$ to the code C given in Exercise 2. Find the table of all cosets of $\mathbb{F}_2^5 \bmod C^\perp$, and determine the coset leaders and syndromes. If $\mathbf{y} = 01001$ is received, which word from $C^\perp$ is most likely to have been transmitted?

4. Show that the binary linear codes with generator matrices

$$\mathbf{G}_1 = \begin{pmatrix} 1 & 1 & 1 & 0 \\ 0 & 1 & 1 & 0 \\ 0 & 0 & 1 & 1 \end{pmatrix} \quad \text{and} \quad \mathbf{G}_2 = \begin{pmatrix} 1 & 0 & 1 & 1 \\ 0 & 1 & 1 & 1 \\ 1 & 0 & 0 & 1 \end{pmatrix},$$

   respectively, are equivalent.

5. Determine the generator matrix, all codewords, and the dual code of the $(4, 2)$ code over $\mathbb{F}_3$ with parity-check matrix

$$\mathbf{H} = \begin{pmatrix} 1 & 1 & 1 & 0 \\ 1 & 2 & 0 & 1 \end{pmatrix}.$$

6. Show that in a binary code either all vectors have even weight or half of them have even weight and half of them have odd weight.

7. Let $\mathbf{G}_1$ and $\mathbf{G}_2$ be generator matrices of a linear (n_1, k, d_1) code and (n_2, k, d_2) code, respectively. Show that the codes with generator matrices

$$\begin{pmatrix} \mathbf{G}_1 & \mathbf{0} \\ \mathbf{0} & \mathbf{G}_2 \end{pmatrix} \quad \text{and} \quad (\mathbf{G}_1 \quad \mathbf{G}_2)$$

   are $(n_1 + n_2, 2k, \min\{d_1, d_2\})$ and $(n_1 + n_2, k, d)$ codes, respectively, where $d \geq d_1 + d_2$.

8. Let $C_i^\perp$ be the dual code of a code C_i, $i = 1, 2$. Prove:

   (i) $(C_i^\perp)^\perp = C_i$.

   (ii) $(C_1 + C_2)^\perp = C_1^\perp \cap C_2^\perp$.

   (iii) If C_1 is the $(n, 1, n)$ binary repetition code, then $C_1^\perp$ is the $(n, n-1, 2)$ parity-check code.

9. A code C is ***self-orthogonal*** if $C \subseteq C^\perp$. Prove:

   (i) If the rows of a generator matrix $\mathbf{G}$ for a binary (n, k) code C have even weight and are orthogonal to each other, then C is self-orthogonal, and conversely.

   (ii) If the rows of a generator matrix $\mathbf{G}$ for a binary (n, k) code C have weights divisible by 4 and are orthogonal to each other, then C is self-orthogonal and all weights in C are divisible by 4.

10. Let C be a binary $(7, 3)$ code with generator matrix

$$\mathbf{G} = \begin{pmatrix} 0 & 0 & 0 & 1 & 1 & 1 & 1 \\ 0 & 1 & 1 & 0 & 0 & 1 & 1 \\ 1 & 0 & 1 & 0 & 1 & 0 & 1 \end{pmatrix}.$$

   Show that C is self-orthogonal. Find its dual code.

11. Let

$$\mathbf{H} = \begin{pmatrix} 1 & 0 & 0 & 1 & 1 & 0 & 1 \\ 0 & 1 & 0 & 1 & 0 & 1 & 1 \\ 0 & 0 & 1 & 0 & 1 & 1 & 1 \end{pmatrix}$$

   be a parity-check matrix of the $(7, 4)$ Hamming code. If $\mathbf{y} = 1110011$ is received, determine the codeword which was most likely sent.

12. For the $(15, 11)$ Hamming code with columns of the parity-check matrix in natural order using maximum likelihood decoding:

   (i) find the parity-check matrix;

   (ii) determine the information rate;

   (iii) decode the received word 111100101100010;

   (iv) find parity check equations (similar to Example 16.2).

13. Determine the probability of correct decoding when using a $(7, 4)$ Hamming code with maximum likelihood decoder over a binary symmetric channel.

14. Use the Hamming $(7, 4)$ code to encode the messages 0110 and 1011. Also encode these messages using the extended modified Hamming $(8, 4)$ code, which is obtained from the $(7, 4)$ Hamming parity-check matrix by adding a column of 0's and filling the new row with 1's. Decode 11001101 by this $(8, 4)$ code.
15. Find the parity-check matrix and generator matrix of the ternary Hamming code of length 13. Decode 0110100000011 using this code.
16. Prove that all binary Hamming codes and the repetition codes of odd block length are perfect.
17. Are Hamming codes and/or repetition codes MDS codes?
18. Use the method in the proof of the Gilbert-Varshamov Bound 17.14 to find a binary linear $(8, 3)$ code with minimum distance 3. Is there also a binary linear $(8, 3)$ code with $d_{\min} = 4$?
19. Find the length, dimension, minimum distance, basis, and parity-check matrix for the binary simplex code and the first-order Reed-Muller code with $m = 2$, 3, and 4.
20. Find three codewords in NASA's (32,6,16) Reed-Muller code.
21. The ***covering radius*** r for a binary code of length n is defined as the smallest number s so that spheres of radius s about codewords cover the vector space $\mathbb{F}_2^n$. Show that the covering radius is the weight of the coset of largest weight.

§18 Cyclic Codes

We now turn to special classes of linear codes C for which $(a_0, \ldots, a_{n-1}) \in C$ implies that $(a_{n-1}, a_0, \ldots, a_{n-2}) \in C$. Let again $\mathbb{F}_q^n$, $n \geq 2$, be the n-dimensional vector space of n-tuples $(a_0, \ldots, a_{n-1})$, with the usual operations of addition of n-tuples and scalar multiplication of n-tuples by elements in $\mathbb{F}_q$. The mapping

$$Z\colon \mathbb{F}_q^n \to \mathbb{F}_q^n; \qquad (a_0, \ldots, a_{n-1}) \mapsto (a_{n-1}, a_0, \ldots, a_{n-2}),$$

is a linear mapping, called a "***cyclic shift***."

We shall also consider the polynomial ring $\mathbb{F}_q[x]$. Now $\mathbb{F}_q[x]$ is not only a commutative ring with identity but also a vector space over $\mathbb{F}_q$ with countable basis $1, x, x^2, \ldots$ in the natural way. This is often expressed by calling $\mathbb{F}_q[x]$ an ***$\mathbb{F}_q$-algebra***. We define a subspace V_n of this vector space by

$$V_n := \{v \in \mathbb{F}_q[x] \mid \text{degree } v < n\}$$
$$= \{v_0 + v_1x + \cdots + v_{n-1}x^{n-1} \mid v_i \in \mathbb{F}_q, 0 \le i \le n-1\}.$$

We can identify the two spaces V_n and $\mathbb{F}_q^n$ by the isomorphism

$$\tau\colon \mathbb{F}_q \to V_n; \qquad (v_0, \ldots, v_{n-1}) \mapsto v_0 + v_1x + \cdots + v_{n-1}x^{n-1}.$$

From now on we shall use both terms "vector" and "polynomial" to denote

$$v = (v_0, \ldots, v_{n-1}) = v_0 + v_1x + \cdots + v_{n-1}x^{n-1}.$$

Consider the polynomial ring $R = (\mathbb{F}_q[x], +, \cdot)$ and its factor ring $R/(x^n - 1)$, modulo the principal ideal generated by $x^n - 1$ in R.

From Theorem 12.11, $\mathbb{F}_q[x]/(x^n - 1) = \{a_0 + a_1\alpha + \ldots + a_{n-1}\alpha^{n-1} \mid a_0, \ldots, a_{n-1} \in \mathbb{F}_q\}$, where $\alpha = [x] = x + (x^n - 1)$ satisfies $\alpha^n = 1$. Hence $\{c_0 + c_1x + \cdots + c_{n-1}x^{n-1} | c_i \in \mathbb{F}_q\}$ is a system of representatives and we might take these instead of the cosets. Addition is as usual, but for multiplication (which we denote by $*$) we first have to multiply as usual and then take the remainder after division by $x^n - 1$. In this way, $\mathbb{F}_q[x]/(x^n - 1)$ is again a ring and an n-dimensional vector space as well. Hence we also arrive at an $\mathbb{F}_q$-algebra, which we will briefly call W_n. The following result is easy to see:

18.1. Theorem. *The mapping*

$$\omega\colon \mathbb{F}_q^n \to W_n; \qquad (v_0, \ldots, v_{n-1}) \mapsto v_0 + v_1x + \cdots + v_{n-1}x^{n-1}$$

is an isomorphism between the $\mathbb{F}_q$-vector spaces $\mathbb{F}_q^n$ and W_n.

So, altogether, $\mathbb{F}_q^n$, $\mathbb{F}_q[x]/(x^n - 1) = W_n$, and V_n are all "the same," and from now on we shall identify them and use the notation V_n. Frequently we shall enjoy the rich structure which comes from the fact that we can interpret sequences as vectors and these in turn as polynomials. So we can add and multiply (via $*$) elements of V_n, we might consider inner products, scalar products, shifts, orthogonality, and so on.

18.2. Example. Let $n = 5$, $q = 7$. Then

$$\begin{aligned}(x^4 + x^3 + x^2 + x + 1) &* (x^3 + x^2 + 1)\\ &= x^7 + 2x^6 + 2x^5 + 3x^4 + 3x^3 + 2x^2 + x + 1\\ &\equiv x^2 + 2x + 2 + 3x^4 + 3x^3 + 2x^2 + x + 1\\ &\equiv 3x^4 + 3x^3 + 3x^2 + 3x + 3 \pmod{(x^5 - 1)}.\end{aligned}$$

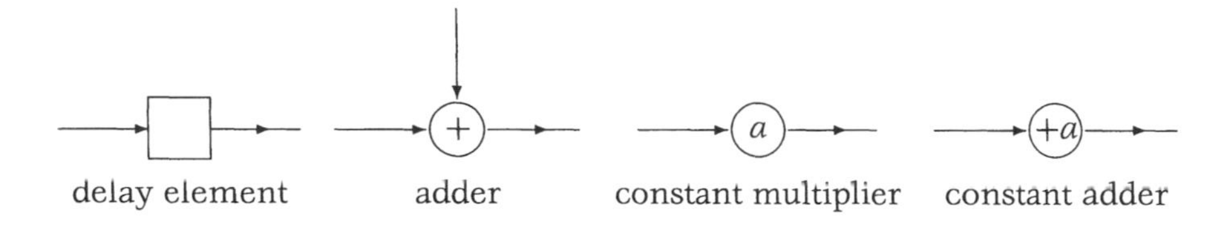

FIGURE 18.1

For the linear map Z we have $Z(v) = x * v$ and more generally $Z^i(v) = x^i * v$ for $v \in \mathbb{F}_q[x]$, $i \in \mathbb{N}$. Here $Z^i = Z \circ Z \circ \ldots \circ Z$ (i times), where $\circ$ denotes composition. Cyclic shifts can be implemented by using ***linear shift registers***. The basic building blocks of a shift register are ***delay*** (or storage) ***elements*** and ***adders***. The delay element (flip-flop) has one input and one output and is regulated by an external synchronous clock so that its input at a particular time appears as its output one unit of time later. The adder has two inputs and one output, the output being the sum in $\mathbb{F}_q$ of the two inputs. In addition a ***constant multiplier*** and a ***constant adder*** are used, which multiply or add constant elements of $\mathbb{F}_q$ to the input. Their representations are shown in Figure 18.1. A shift register is built by interconnecting a finite number of these devices in a suitable way along closed loops. As an example, we give a binary shift register with four delay elements a_0, a_1, a_2, and a_3 and two binary adders. At time 0, four binary elements, say 1, 1, 0, and 1 are placed in a_0, a_1, a_2, and a_3, respectively. These positions can be interpreted as message positions. After one time interval $a_0 = 1$ is output, $a_1 = 1$ is shifted into a_0, $a_2 = 0$ into a_1, $a_3 = 1$ into a_2, and the new element is entered into a_3. If the shift register is of the form shown in Figure 18.2 then this new element is $a_0 + a_2 + a_3$. To summarize the outputs for seven time intervals, we obtain

Outputs							a_0	a_1	a_2	a_3	Time
							1	1	0	1	0
						1	1	0	1	0	1
					1	1	0	1	0	0	2
				1	1	0	1	0	0	0	3
			1	1	0	1	0	0	0	1	4
		1	1	0	1	0	0	0	1	1	5
	1	1	0	1	0	0	0	1	1	0	6
1	1	0	1	0	0	0	1	1	0	1	7

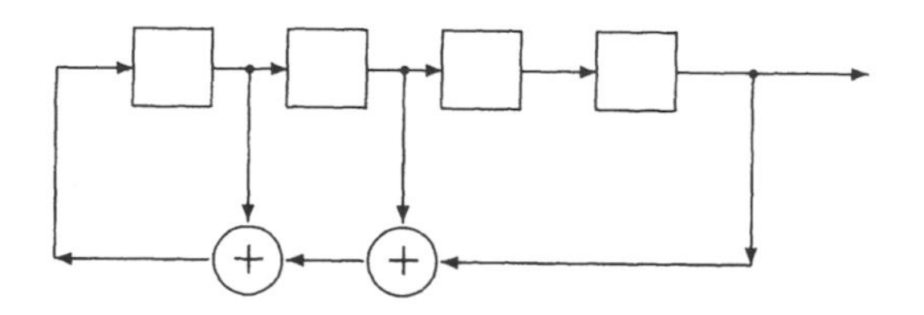

FIGURE 18.2

Continuing this process for time 8, 9, ... we see that the output vector 1101000 generated from the four initial entries 1101 will be repeated. The process can be interpreted as encoding the message 1101 into a code word 1101000. The code generated in this way is a linear code, with the property that whenever $(v_0, v_1, \dots, v_{n-1})$ is in the code then the cyclic shift $(v_{n-1}, v_0, v_1, \dots, v_{n-2})$ is also in the code.

18.3. Definition. A subspace C of $\mathbb{F}_q^n$ is called a ***cyclic code*** if $Z(v) \in C$ for all $v \in C$, that is,

$$v = (v_0, v_1, \dots, v_{n-1}) \in C \Longrightarrow (v_{n-1}, v_0, \dots, v_{n-2}) \in C.$$

If C is a cyclic, linear (n, k) code, we call C a cyclic (n, k) code, for brevity.

18.4. Example. Let $C \subseteq \mathbb{F}_2^7$ be defined by the generator matrix

$$\begin{pmatrix} 1 & 1 & 1 & 0 & 1 & 0 & 0 \\ 0 & 1 & 1 & 1 & 0 & 1 & 0 \\ 0 & 0 & 1 & 1 & 1 & 0 & 1 \end{pmatrix} = \begin{pmatrix} g^{(1)} \\ g^{(2)} \\ g^{(3)} \end{pmatrix}.$$

We show that C is cyclic. Each codeword of C is a linear combination of the linearly independent vectors $g^{(1)}, g^{(2)}, g^{(3)}$. The code C is cyclic if and only if $Z(g^{(i)}) \in C$ for $i = 1, 2, 3$. We have

$$Z(g^{(1)}) = g^{(2)}, \quad Z(g^{(2)}) = g^{(3)}, \quad Z(g^{(3)}) = g^{(1)} + g^{(2)}.$$

We now prove an elegant algebraic characterization of cyclic codes.

18.5. Theorem. *A linear code $C \subseteq V_n$ is cyclic if and only if C is an ideal in V_n.*

Proof. Let C be cyclic and let $f = \sum a_i x^i \in \mathbb{F}_q[x]$. Then for $v \in C$ we have $f * v = \sum a_i(x^i * v) = \sum a_i Z^i(v) \in C$, and moreover C is a subspace of V_n. So C is an ideal of V_n.

Conversely, let C be an ideal and $v \in C$. Since $Z(c) = x * c \in C$ for all $c \in C$, we see that C is cyclic. □

18.6. Theorem. *Let C be an ideal $\neq \{0\}$ of V_n. Then there exists a unique $g \in V_n$ with the following properties:*

(i) $g \mid x^n - 1$ *in* $\mathbb{F}_q[x]$*;*

(ii) $C = (g)$*;*

(iii) g *is monic.*

Hence every ideal of V_n is principal (note that V_n is not an integral domain, however).

Proof.

(i) and (iii) Let g be the nonzero monic polynomial of minimal degree in C. Then there are polynomials $s, r \in \mathbb{F}_q[x]$ with $x^n - 1 = gs + r$ and $\deg(r) < \deg g < n$. In V_n we have $0 = \tilde{g}\tilde{s} + \tilde{r}$ for $\tilde{s}, \tilde{r} \in V_n$. But then $r = \tilde{r} = -\tilde{g}\tilde{s} \in C$ so $r = 0$, and hence $g \mid x^n - 1$.

(ii) Since $g \in C$, $(g) \subseteq C$. Conversely, there are $f, t \in \mathbb{F}_q[x]$ such that $s = g \cdot f + t$ and $\deg t < \deg g$ holds in $\mathbb{F}_q[x]$. Similar to (i), $t = s - gf \in C$, so $C \subseteq (g)$.

Uniqueness: Suppose g' also fulfills (i)–(iii). By $(g) = (g') = C$, we get $g \mid g'$ and $g' \mid g$ in $\mathbb{F}_q[x]$. Since g and g' are monic, $g = g'$. □

18.7. Definition. The polynomial $g \in C$ in Theorem 18.6 is called the ***generator polynomial*** of C. The elements of C are called ***codewords***, ***code polynomials***, or ***code vectors***.

So we get all cyclic codes exactly once if we know all divisors of $x^n - 1 \in \mathbb{F}_q[x]$. And by 13.19, we do know them.

18.8. Remark. Let $g = g_0 + g_1 x + \cdots + g_m x^m \in V_n$, $g \mid (x^n - 1)$, and $\deg g = m < n$. Let C be a linear (n, k) code, with $k = n - m$, defined by the generator matrix

$$\mathbf{G} = \begin{pmatrix} g_0 & g_1 & \dots & g_m & 0 & \dots & 0 \\ 0 & g_0 & \dots & g_{m-1} & g_m & \dots & 0 \\ \dots & \dots & \dots & \dots & \dots & \dots & \dots \\ 0 & 0 & \dots & g_0 & g_1 & \dots & g_m \end{pmatrix} = \begin{pmatrix} g \\ xg \\ \vdots \\ x^{k-1}g \end{pmatrix}.$$

Then C is cyclic. The only thing to show is that the cyclic shift of the last row in $\mathbf{G}$ is a codeword. Let $x^n - 1 = gh$, where $h = h_0 + h_1 x + \dots + h_k x^k$. Then $x^k * g = -h_k^{-1}(h_0 g + h_1(xg) + \dots + h_{k-1}(x^{k-1}g))$ is a linear combination of rows in $\mathbf{G}$. So $x^k * g \in C = (g)$. The rows of $\mathbf{G}$ are linearly independent and $\operatorname{Rk} \mathbf{G} = k$, the dimension of C.

18.9. Encoding. If we use a cyclic code and g is its generator polynomial, encoding is done by $a_0 a_1 \dots a_{k-1} \mapsto (a_0 + a_1 x + \dots + a_{k-1} x^{k-1}) * g$.

18.10. Example. Suppose messages for transmission are elements of $\mathbb{F}_2^3$. Let $g = x^3 - 1 = 1 + x^3$, which trivially divides $x^3 - 1$, and hence g is the generator polynomial of a cyclic code. Then all possible messages can be encoded as codewords in the following way: We compute $a * g$ for each message polynomial a.

000	$\mapsto$ 000000	100	$\mapsto$ 100100
001	$\mapsto$ 001001	101	$\mapsto$ 101101
010	$\mapsto$ 010010	110	$\mapsto$ 110110
011	$\mapsto$ 011011	111	$\mapsto$ 111111

The corresponding generator matrix is of the form

$$\begin{pmatrix} 1 & 0 & 0 & 1 & 0 & 0 \\ 0 & 1 & 0 & 0 & 1 & 0 \\ 0 & 0 & 1 & 0 & 0 & 1 \end{pmatrix}.$$

From the decoding scheme and from the shape of the generator matrix, we see that this is just the repetition code.

Observe that $(g_1) \subseteq (g_2)$ iff $g_2 \mid g_1$, for $g_1, g_2 \in V_n$. Hence, if g_1 is irreducible, (g_1) must be "very large."

If F is a field $\mathbb{F}_{p^t}$, the polynomial $x^n - 1$ has derivative nx^{n-1}, so $\gcd(x^n - 1, nx^{n-1}) = 1$ iff $p \nmid n$. By 12.34, $x^n - 1$ is square-free in this case.

18.11. Convention. For the rest of this chapter, we require $p \nmid n$.

18.12. Definition. If $x^n - 1 = g_1 \dots g_t$ is the complete factorization of $x^n - 1$ into (different) irreducible polynomials over $\mathbb{F}_q$, then the cyclic codes (g_i) generated by polynomials g_i are called ***maximal cyclic codes***.

A maximal cyclic code is a maximal ideal in $\mathbb{F}_q[x]/(x^n - 1)$. If g is a generator polynomial of a cyclic code C, then $g \mid (x^n - 1)$. Therefore $h = (x^n - 1)/g$ is also a generator polynomial of a cyclic code. Let $\deg g = m = n - k$, so that $\deg h = k$.

18.13. Definition. Let g be a generator polynomial of a cyclic code C. Then $h = (x^n - 1)/g$ is called a ***check polynomial*** of C.

18.14. Theorem. *Let h be a check polynomial of a cyclic code $C \subseteq V_n$ with generator polynomial g and $v \in V_n$. Then*

$$v \in C \iff v * h = 0.$$

Proof. Let $v \in C$; then there is a polynomial a such that $v = a * g$. Since $g * h = 0$, we have $v * h = a * g * h = 0$. Conversely, if $v * h = 0$, then $vh = w(x^n - 1)$ for some $w \in \mathbb{F}_q[x]$. Moreover, $x^n - 1 = gh$ implies $vh = wgh$, so $v = wg$, i.e., $v \in C$. □

In Theorem 18.14, let $v = \sum v_i x^i$. Then the coefficient of x^j, $j = 0, 1, \dots, n-1$, in the product $v * h$ is given by

$$\sum_{i=0}^{n-1} v_i h_{j-i},$$

where $h = h_0 + h_1 x + \dots + h_k x^k$ and the indices are calculated mod n. If we form the matrix

$$\mathbf{H} = \begin{pmatrix} 0 & \cdots & 0 & h_k & & \cdots & h_1 & h_0 \\ 0 & \cdots & 0 & h_k & h_{k-1} & \cdots & h_0 & 0 \\ \dots & \dots & \dots & \dots & \dots & \dots & \dots & \dots \\ h_k & \cdots & h_1 & h_0 & 0 & \cdots & 0 & 0 \end{pmatrix}$$

and if $v \in V_n$ is in C, we get $\mathbf{H}\mathbf{v}^{\mathrm{T}} = (w_0, \dots, w_{n-1})^{\mathrm{T}}$, where, for $j \in \{0, \dots, n-1\}$, we have $w_j = \sum_{i=0}^{n-1} v_i h_{n-j-i} = 0$. So $\mathbf{H}\mathbf{v}^{\mathrm{T}} = \mathbf{0}$ holds for each $v \in C$. Now $\mathbf{H}$ has rank $n - k$, so thc null space of $\mathbf{H}$ has dimension $n - (n - k) = k$, from which we see that $\mathbf{H}$ is a parity-check matrix of $C = (g)$. Hence we have

18.15. Theorem. *Let C be a cyclic (n, k) code with generator polynomial g. Then the dual code $C^\perp$ of C is a cyclic $(n, n-k)$ code with generator polynomial*

$$h^\perp = x^{\deg h}(h \circ x^{-1}), \quad \textit{where } h = (x^n - 1)/g.$$

We note that a polynomial of the form $x^{\deg h}(h \circ x^{-1})$ is called the ***reciprocal polynomial*** of h. If $h = h_0 + h_1x + \cdots + h_kx^k$, then its reciprocal polynomial is $h_k + h_{k-1}x + \cdots + h_0x^k$. The code generated by h can be shown to be equivalent to the dual code $C^\perp$ of $C = (g)$.

18.16. Example. Let $g = x^3 + x + 1 \in \mathbb{F}_2[x]$ and $n = 7$. Then $h = (x^7 - 1)/g = x^4 + x^2 + x + 1$ and $h^\perp = x^4 + x^3 + x^2 + 1$. For this code,

$$\mathbf{G} = \begin{pmatrix} 1 & 1 & 0 & 1 & 0 & 0 & 0 \\ 0 & 1 & 1 & 0 & 1 & 0 & 0 \\ 0 & 0 & 1 & 1 & 0 & 1 & 0 \\ 0 & 0 & 0 & 1 & 1 & 0 & 1 \end{pmatrix} \quad \text{and} \quad \mathbf{H} = \begin{pmatrix} 0 & 0 & 1 & 0 & 1 & 1 & 1 \\ 0 & 1 & 0 & 1 & 1 & 1 & 0 \\ 1 & 0 & 1 & 1 & 1 & 0 & 0 \end{pmatrix}$$

are generator and check matrices, respectively. By inspecting $\mathbf{H}$, we see that (g) is precisely the binary Hamming code of length 7.

In fact, we might take a primitive element α of $\mathbb{F}_{2^m}$. Then $\mathbb{F}^*_{2^m} = \{1, \alpha, \alpha^2, \ldots, \alpha^{2^m-2}\}$. If we represent each $\alpha^i = a_0 + a_1x + \cdots + a_{m-1}x^{m-1}$ by $(a_0, \ldots, a_{m-1})^{\mathrm{T}}$, then the matrix $\mathbf{H} := (1, \alpha, \ldots, \alpha^{2^m-1})$, considered as an $m \times (2^m - 1)$ matrix over $\mathbb{F}_2$, consists of all binary column vectors $\neq \mathbf{0}$, and hence the resulting code C is just the Hamming code of length $2^m - 1$. A vector $\mathbf{v} = (v_0, \ldots, v_{n-1})$ belongs to C iff $\mathbf{H}\mathbf{v}^{\mathrm{T}} = \mathbf{0}$, which simply means that $v(\alpha) = 0$. This, in turn, holds iff the minimal polynomial m_α of α divides v. Hence $C = (m_\alpha)$, from which we see again that binary Hamming codes are cyclic. By 18.15 and 17.24(vi), simplex codes are cyclic, too. We shall return to these ideas shortly.

In §33 we shall consider linear recurrence relations of the form $\sum_{j=0}^{k} f_j a_{i+j} = 0$, $i = 0, 1, \ldots$, which are periodic of period n. Here $f_j \in \mathbb{F}_q$ are coefficients of $f = f_0 + f_1x + \cdots + f_kx^k$, $f_0 \neq 0$, $f_k = 1$. The set of the n-tuples of the first n terms of each possible solution, considered as polynomials modulo $x^n - 1$, can be shown to be the ideal generated by $g \in V_n$, where g is the reciprocal polynomial of $(x^n - 1)/f$ of degree $n - k$. Thus there is a close relationship between cyclic codes and linear recurrence relations. Moreover, this relationship facilitates the implementation of cyclic codes on feedback shift registers.

18.17. Example. Let again $g = x^3 + x + 1$; g divides $x^7 - 1$ over $\mathbb{F}_2$. Associated with g is the linear recurrence relation $a_{i+3} + a_{i+1} + a_i = 0$, which defines the (7, 3) cyclic code of the last example. This code encodes 111, say, as 1110010.

The following summarizes some results on principal ideals. Let J_1 and J_2 be ideals in V_n. Then the intersection $J_1 \cap J_2$ is an ideal while the (set-theoretic) union of J_1 and J_2 is not an ideal in general. The ideal $J_1 + J_2 = \{j_1 + j_2 \mid j_1 \in J_1, j_2 \in J_2\}$ is the smallest ideal containing J_1 and J_2. The product $J_1 J_2 = \{\sum a_{i_1} a_{i_2} \mid a_{i_1} \in J_1, a_{i_2} \in J_2\}$ is an ideal of V_n, too.

18.18. Theorem. *Let J_i, $i = 1, 2$, be two ideals in V_n with generating elements g_i. Then:*

(i) $J_1 J_2 \subseteq J_1 \cap J_2 \subseteq J_1, J_2 \subseteq J_1 + J_2$;

(ii) $J_1 + J_2$ *is generated by* $\gcd(g_1, g_2)$;

(iii) $J_1 \cap J_2$ *is generated by* $\operatorname{lcm}(g_1, g_2)$;

(iv) $J_1 J_2$ *is generated by* $\gcd(g_1 g_2, x^n - 1)$.

Here gcd computations are done in $\mathbb{F}_q[x]$.

While (i) is immediate, (ii)–(iv) will be proved in Exercise 9. For a better understanding of the following remarks we refer the reader to the relevant parts of §11. Let $M_i = ((x^n - 1)/g_i)$ be principal ideals of $\mathbb{F}_q[x]$, then $M_i \cap M_j = \{0\}$ for $i \neq j$ and $M_i + M_j = \{m_i + m_j \mid m_i \in M_i, m_j \in M_j\}$, where $\gcd(h_i, h_j) = (x^n - 1)/g_i g_j$ is a generating element of $M_i + M_j$ because of (iv) in 18.18.

For any ideal $J = (g)$ in V_n with $g = g_{i_1} \dots g_{i_s}$ we have

$$(x^n - 1)/g_{j_1} \dots g_{j_r} = g_{i_1} \dots g_{i_s},$$

if $\{j_1, \dots, j_r\} \cup \{i_1, \dots, i_s\} = \{1, \dots, k\}$ and J is the direct sum of the ideals $M_{j_1}, \dots, M_{j_r}$.

Let $x^n - 1 = g_1 \dots g_t$ be the factorization of $x^n - 1$ into irreducible polynomials over $\mathbb{F}_q$. The relation between a code C and its dual $C^\perp$ in Theorem 18.15 means that $(x^n - 1)/g_i$ is a generator polynomial of a so-called ***minimal*** or ***irreducible code*** and g_i is a generator polynomial of the corresponding maximal code.

Observe that altogether we have 2^t different cyclic codes of length n. Two of those are trivial: $(1) = V_n$ and $(x^n - 1) = \{\mathbf{0}\}$. Also, $x - 1$ is always a divisor of $x^n - 1$. The code $(x-1)$ can easily be seen to be $\{0\ldots 0, 1\ldots 1\}$, for $q = 2$, i.e., $(x-1)$ is the $(n-1)$-fold repetition code.

Irreducible codes have no divisors of zero. By 11.9(i) they are fields, and hence each cyclic code C is the direct sum of finite fields $F_1, \ldots, F_s$. If $e_1, \ldots, e_s$ are the unit elements in $F_1, \ldots, F_s$, respectively, then $e := e_1 + \cdots + e_s$ is an identity in the ring C, and the e_i fulfill $e_i^2 = e_i$ and $e_i e_j = 0$ for $i \neq j$. Hence they are called the ***orthogonal idempotents*** of a cyclic code.

Above, we have seen how to encode a message, but what about decoding? We start with an example.

18.19. Example.

(i) Let $g = 1 + x^2 + x^3$ be a generator polynomial of a binary $(7, 4)$ code. The message $a = 1010$ is encoded as $a * g = 1001110$.

(ii) Let $w = 1100001$ be a received word; then w/g is $0111 = x + x^2 + x^3$ with remainder $1 + x^2$. This shows that an error must have occurred. With the tools available so far we can, however, only use the lines after 17.16 for correcting errors in linear codes.

This is not very satisfactory. We need more theory, and start by dividing x^i by g. In this way, a canonical generator matrix of the form $\mathbf{G} = (\mathbf{I}_k, -\mathbf{A}_{k\times m})$ can be obtained by using the division algorithm for polynomials. Let $\deg g = m = n - k$; then there are unique polynomials $a^{(j)}$ and $r^{(j)}$ with $\deg r^{(j)} < m$, such that

$$x^j = a^{(j)} g + r^{(j)}, \qquad j \in \mathbb{N}.$$

Therefore

$$x^j - r^{(j)} \in C.$$

The k polynomials $g^{(j)} = x^k(x^j - r^{(j)})$ with $m \leq j \leq n-1$ and considered modulo $x^n - 1$ are linearly independent and form the rows of a matrix of the required form.

Similarly, a canonical parity-check matrix can be obtained by dividing $1, x, x^2, \ldots, x^{n-1}$ by g and by taking the remainders $r_0, \ldots, r_{n-1}$ as the columns of $\mathbf{H}$. This holds because the resulting rows are independent and for each $\mathbf{c} \in C$, $\mathbf{Hc} = \sum_{i=0}^{n-1} c_i r_i \equiv \sum_{i=0}^{n-1} c_i x^i \equiv 0 \pmod{g}$.

18.20. Example. Let $n = 7$, $q = 2$, $g = x^3 + x + 1$ as in Example 18.17. The remainders of $x^0, x^1, \ldots, x^7$ after division by g are $1, x, x^2, 1 + x, x +$

$x^2, 1+x+x^2$, and $1+x^2$, respectively. So we get $\mathbf{G}$ and $\mathbf{H}$ as

$$\mathbf{G} = \begin{pmatrix} 1 & 0 & 0 & 0 & 1 & 1 & 0 \\ 0 & 1 & 0 & 0 & 0 & 1 & 1 \\ 0 & 0 & 1 & 0 & 1 & 1 & 1 \\ 0 & 0 & 0 & 1 & 1 & 0 & 1 \end{pmatrix} \quad \text{and} \quad \mathbf{H} = \begin{pmatrix} 1 & 0 & 0 & 1 & 0 & 1 & 1 \\ 0 & 1 & 0 & 1 & 1 & 1 & 0 \\ 0 & 0 & 1 & 0 & 1 & 1 & 1 \end{pmatrix},$$

since $g^{(4)} = x^4(x^3 - (x+1)) = x^7 + x^5 + x^4 \equiv 1 + x^4 + x^5$, etc.

From this we can easily determine the syndrome. If we take $\mathbf{H}$ as above, and if $\mathbf{v} = (v_0, \ldots, v_{n-1}) \in V_n$, then $S(\mathbf{v}) = \mathbf{H}\mathbf{v}^{\mathrm{T}} = \sum_{i=0}^{n-1} v_i r_i \equiv \sum_{i=0}^{n-1} v_i x^i = v \pmod g$. This shows

18.21. Theorem. *Let $C = (g)$ be a cyclic code with canonial check matrix $\mathbf{H}$. The syndrome $S(\mathbf{v})$ of a received vector $\mathbf{v}$ fulfills $S(\mathbf{v}) \equiv v \pmod g$. So $\mathbf{v} = \mathbf{c}+\mathbf{e}$ with $\mathbf{c} \in C$ and $\deg e < \deg g$. $S(\mathbf{v}) = S(\mathbf{e})$ is just e itself.*

So we don't have to build up $\mathbf{H}$, we just have to divide v by g. Observe that if e is a coset leader, then we might take each $x^i e \in V_n$ again as different coset leaders ($1 \leq i \leq n-1$), because $x^i e$ is just a cyclic shift of e. Since $S(x^i v) \equiv x^i v \equiv x^i e$, we get

18.22. *Decoding Algorithm* *for t-error correcting cyclic codes $C = (g) \trianglelefteq V_n$.*

(i) *Divide an incoming word $v \in V_n$ by g to get the remainder r. We have $r \equiv S(v) = S(e)$, where e is the error vector (polynomial).*

(ii) *For $0 \leq i \leq n-1$, compute $S_i := x^i r \pmod g$ until some j is found such that S_j has weight $\leq t$. Then $x^{n-j} S_j \bmod (x^n - 1)$ is the most likely error.*

Observe that this algorithm only works if for some i, $\deg(x^i e) \leq \deg g$. In many cases this suffices, and in the next section, we shall see "complete" decoding algorithms for special cyclic codes.

18.23. Example. We take $q = 2$, $n = 15$, $g = x^8 + x^7 + x^6 + x^4 + 1 \in \mathbb{Z}_2[x]$, and $C = (g)$. In the next section, we shall see in Theorem 19.2 that $d_{\min}(C) = 5$, so C can correct up to $t = 2$ errors. Suppose $v = 1+x^3+x^4+x^5$ arrives. So $v = r = S(v)$. Now $v \equiv 1 + x^3 + x^4 + x^5 \pmod g$ has weight 4, $xv \equiv x + x^4 + x^5 + x^6 \pmod g$ also has weight 4, the same applies to $x^2 v$. Since $x^8 \equiv x^7 + x^6 + x^4 + 1 \pmod 8$; $x^3 v \equiv x^3 + x^6 + x^7 + x^8 \equiv 1 + x^3 + x^4$ has weight 3, as well as $x^4 v$, $x^5 v$, and $x^6 v$. Finally, $S_7 = x^7 v \equiv x^6 + 1$ has weight 2. So $x^{15-7} S_7 = x^8(x^6+1) = x^{14} + x^8$ is the most likely error, and v is decoded into the codeword $v + x^{14} + x^8 = 1 + x^3 + x^4 + x^5 + x^8 + x^{14}$. If we divide this by g, we get the original message $1 + x^3 + x^5 + x^6 = 1001011$.

Cyclic codes can be described in various other ways by using roots of the generator polynomial. The following result is easy to show.

18.24. Theorem. *Let g be a generator polynomial of a cyclic code C over* $\mathbb{F}_q$ *and let* $\alpha_1, \ldots, \alpha_m$ *be roots of g in an extension field of* $\mathbb{F}_q$*. All the roots are simple and (since* $g \mid x^n - 1$*) nth roots of unity. Also,*

$$v \in C \iff v(\alpha_1) = v(\alpha_2) = \cdots = v(\alpha_m) = 0.$$

As an example of the concurrence of the description of a cyclic code by a generator polynomial or by roots of code polynomials we prove the following result.

18.25. Theorem. *Let C be a cyclic* (n, k) *code with generator polynomial g and let* $\alpha_1, \ldots, \alpha_{n-k}$ *be the roots of g. Then* $v \in V_n$ *is a code polynomial if and only if the coefficient vector* $(v_0, \ldots, v_{n-1})$ *of v is in the null space of the matrix*

$$\mathbf{H} = \begin{pmatrix} 1 & \alpha_1 & \alpha_1^2 & \cdots & \alpha_1^{n-1} \\ \vdots & \vdots & \vdots & \ddots & \vdots \\ 1 & \alpha_{n-k} & \alpha_{n-k}^2 & \cdots & \alpha_{n-k}^{n-1} \end{pmatrix}.$$

Proof. Let $v = v_0 + v_1x + \cdots + v_{n-1}x^{n-1}$; then $v(\alpha_i) = 0$, i.e., $(1, \alpha_i, \ldots, \alpha_i^{n-1})(v_0, v_1, \ldots, v_{n-1})^{\mathrm{T}} = \mathbf{0}$, $1 \le i \le n-k$, if and only if $\mathbf{H}v^{\mathrm{T}} = \mathbf{0}$. □

In the case of cyclic codes, the syndrome can often be replaced by a simpler entity. Let α be a primitive nth root of unity in $\mathbb{F}_{q^m}$ and let the generator polynomial g be the minimal polynomial of α over $\mathbb{F}_q$. We know that g divides $v \in V_n$ if and only if $v(\alpha) = 0$. Therefore we can replace the parity-check matrix $\mathbf{H}$ in Theorem 18.25 by $(1, \alpha, \alpha^2, \ldots, \alpha^{n-1})$, and we can replace the syndrome by $v(\alpha)$.

Let v denote the transmitted codeword and w the received word. Suppose $e^{(j)}(x) = x^{j-1}$, $1 \le j \le n$, is an error polynomial with a single error, so $w = v + e^{(j)}$. Then

$$w(\alpha) = v(\alpha) + e^{(j)}(\alpha) = e^{(j)}(\alpha) = \alpha^{j-1}.$$

Here $e^{(j)}(\alpha)$ is called the ***error-location number***. $S(w) = \alpha^{j-1}$ indicates the error uniquely since $e^{(i)}(\alpha) \neq e^{(j)}(\alpha)$ for $i \neq j$, $1 \le i \le n$.

The zeros $\alpha_1, \ldots, \alpha_{n-k}$ of g in Theorem 18.24 are called the ***zeros of the code*** $C = (g)$. The following theorem shows we can choose generator polynomials other than the monic polynomial of least degree in C. Let $C = (g)$ be a cyclic (n, k) code as in Theorem 18.24. The proof of the following result should be done in Exercise 10.

18.26. Theorem. *The polynomials g and fg are generator polynomials of the same code if we have $f(\beta) \neq 0$ for all $\beta \notin \{\alpha_1, \dots, \alpha_m\}$, i.e., if f does not introduce new zeros of C.*

Suppose α is a primitive nth root of unity and $\operatorname{char} \mathbb{F}_q = p$. If α^s is a zero of $g = \sum_{i=0}^{n-1} g_i x^i$, then $g(\alpha^{ps}) = \sum_{i=0}^{n-1} g_i (\alpha^{si})^p = g(\alpha^s)^p = 0$, as well. So all α^t are zeros of g, where t runs through the cyclotomic coset C_s of s.

A cyclic code can hence be described in terms of a primitive nth root of unity α and a union K of cyclotomic cosets such that $\{\alpha^k \mid k \in K\}$ are the zeros of g. So

$$C = \{v \in V_n \mid v(\alpha^i) = 0 \text{ for all } i \in K\}.$$

The following theorem shows that the code is independent of the choice of α.

18.27. Theorem. *Let α and β be arbitrary primitive nth roots of unity in an extension field of $\mathbb{F}_q$ and let K be a union of cyclotomic cosets* mod n. *Then the polynomials*

$$g_\alpha = \prod_{i \in K} (x - \alpha^i) \quad \textit{and} \quad g_\beta = \prod_{i \in K} (x - \beta^i)$$

are generator polynomials of equivalent codes.

Proof. See, e.g., Blake & Mullin (1975). □

18.28. Definition. The ***Mattson-Solomon polynomial*** F_v of a polynomial $v \in V_n$ is

$$F_v = \frac{1}{n} \sum_{j=1}^{n} f_j x^{n-j}, \quad \text{where } f_j = v(\alpha^j) = \sum_{i=0}^{n-1} v_i \alpha^{ij}, \quad 0 \leq j \leq n.$$

If

$$D_n := \begin{pmatrix} 1 & 1 & 1 & 1 & 1 \\ 1 & \alpha & \alpha^2 & \dots & \alpha^{n-1} \\ 1 & \alpha^2 & \alpha^4 & \dots & \alpha^{2(n-1)} \\ \dots & \dots & \dots & \dots & \dots \\ 1 & \alpha^{n-1} & \alpha^{2(n-1)} & \dots & \alpha^{(n-1)(n-1)} \end{pmatrix},$$

then $(f_0, \dots, f_{n-1}) = D_n(v_0, \dots, v_{n-1})^{\mathrm{T}}$. The map $(v_0, \dots, v_{n-1}) \mapsto (f_0, \dots, f_{n-1})$ is called the ***Discrete Fourier Transform*** and will be studied in more detail in §34. The following result shows that we can "recover"

v from F_v. Recall from Exercise 13.15 that for an nth root $\beta \neq 1$ of unity, $\sum_{i=0}^{n-1} \beta^i = (\beta^n - 1)/(\beta - 1) = 0$. Observe also that $f_n = f_0$.

18.29. Theorem. *Let $v = (v_0, \ldots, v_{n-1})$ and let f_j be as in 18.28. Then $v_s = F_v(\alpha^s)$, $s = 0, 1, \ldots, n-1$.*

Proof.

$$F_v(\alpha^s) = \frac{1}{n} \sum_{j=1}^{n} \sum_{i=0}^{n-1} v_i \alpha^{ij} = \frac{1}{n} \sum_{i=0}^{n-1} v_i \sum_{j=1}^{n} \alpha^{(i-s)j} = v_s,$$

since $\sum_{j=1}^{n} \alpha^{(i-s)j} = 0$ for $i \neq s$. □

Theorem 18.29 enables us to describe a cyclic code C in the form

$$C = \{(F_v(\alpha^0), F_v(\alpha^1), \ldots, F_v(\alpha^{n-1})) \mid v \in C\}.$$

18.30. Theorem. *If α is a primitive nth root of unity and if α^i $(i \in K)$ are the zeros of g, then*

$$v = (v_0, \ldots, v_{n-1}) \in C \iff f_i = 0 \text{ for all } i \in K.$$

18.31. Lemma. *Let $g = \prod_{i \in K}(x - \alpha^i)$ be a generator polynomial of a cyclic code C and let $\{1, 2, \ldots, d-1\}$ be a subset of K, with K as above. Then $\deg F_v \leq n - d$ for $v \in C$.*

Proof. For all $j < d$, we have $v(\alpha^j) = 0$.

18.32. Theorem. *If there are exactly r nth roots of unity which are zeros of F_v, then the weight of v is $n - r$.*

Proof. The weight of v is the number of nonzero coordinates v_s. There are r values s, such that $v_s = 0$.

18.33. Example. From the lines after 18.16, we know that for $g = x^3 + x + 1 \in \mathbb{Z}_2[x]$ we have $C = (g) =$ the $(7, 4)$ Hamming code. Since g is irreducible, it is the minimal polynomial of some α, i.e., $\alpha^3 + \alpha + 1 = 0$, or $\alpha^3 = \alpha + 1$. Higher powers of α are given by $\alpha^4 = \alpha\alpha^3 = \alpha(\alpha + 1) = \alpha^2 + \alpha$, etc.,

$$\begin{aligned}
\alpha^3 &= \alpha + 1, \\
\alpha^4 &= \alpha^2 + \alpha, \\
\alpha^5 &= \alpha^2 + \alpha + 1, \\
\alpha^6 &= \alpha^2 + 1, \\
\alpha^7 &= 1.
\end{aligned}$$

So α is a primitive seventh root of unity. The Discrete Fourier Transform is given by the matrix

$$D_7 = \begin{pmatrix} 1 & 1 & 1 & 1 & 1 & 1 & 1 \\ 1 & \alpha & \alpha^2 & \alpha^3 & \alpha^4 & \alpha^5 & \alpha^6 \\ 1 & \alpha^2 & \alpha^4 & \alpha^6 & \alpha^8 & \alpha^{10} & \alpha^{12} \\ \cdots & \cdots & \cdots & \cdots & \cdots & \cdots & \cdots \\ 1 & \alpha^6 & \alpha^{12} & \alpha^{18} & \alpha^{24} & \alpha^{30} & \alpha^{36} \end{pmatrix} = \begin{pmatrix} 1 & 1 & 1 & 1 & 1 & 1 & 1 \\ 1 & \alpha & \alpha^2 & \alpha^3 & \alpha^4 & \alpha^5 & \alpha^6 \\ 1 & \alpha^2 & \alpha^4 & \alpha^6 & \alpha^1 & \alpha^3 & \alpha^5 \\ 1 & \alpha^3 & \alpha^6 & \alpha^2 & \alpha^5 & \alpha^1 & \alpha^4 \\ 1 & \alpha^4 & \alpha^1 & \alpha^5 & \alpha^2 & \alpha^6 & \alpha^3 \\ 1 & \alpha^5 & \alpha^3 & \alpha^1 & \alpha^6 & \alpha^4 & \alpha^2 \\ 1 & \alpha^6 & \alpha^5 & \alpha^4 & \alpha^3 & \alpha^2 & \alpha^1 \end{pmatrix}.$$

For example, the codeword $v = x * g = x^4 + x^2 + x = x + x^2 + x^4 = 0110100$ is transformed into $D_7 v^{\mathrm{T}} = (f_0, \ldots, f_6)^{\mathrm{T}}$. Now $f_0 = 1 + 1 + 1 = 1$, $f_1 = \alpha + \alpha^2 + \alpha^4 = \alpha(1 + \alpha + \alpha^3) = \alpha \cdot 0 = 0$, $f_2 = \alpha^2 + \alpha^4 + \alpha^6 = (\alpha + \alpha^2 + \alpha^3)^2 = 0$, $f_3 = \alpha^3 + \alpha^6 + \alpha^{12} = \alpha + 1 + \alpha^2 + 1 + \alpha^2 + \alpha + 1 = 1$, $f_4 = (\alpha + \alpha^2 + \alpha^4)^4 = 0$, $f_5 = \alpha^5 + \alpha^{10} + \alpha^{20} = \alpha^5 + \alpha^3 + \alpha^6 = 1$, $f_6 = \alpha^6 + \alpha^{12} + \alpha^{24} = (f_3)^2 = 1$. Hence $D_7 \mathbf{v}^{\mathrm{T}} = (1, 0, 0, 1, 0, 1, 1)$, so the Mattson-Solomon polynomial of $x * g$ is $\frac{1}{7}(x^4 + x^2 + x + 1)$. Similarly, $D_7(x^2 * g) = D_7(0011010) = (1, 0, 0, \alpha^3, 0, \alpha^5, \alpha^6)$, and so on. Observe that we don't need to know α itself. We could find it in $\mathbb{Z}_2[x]/(x^6 + x + 1)$ as $\alpha = [x]^9$.

It is customary to say that in considering the v_i $(0 \le i < n)$, we work in the "time domain" and i is called the ***time***. On the other hand, j in f_j is called the frequency, and with f_j we work in the "frequency domain"; $(f_0, \ldots, f_{n-1})$ is the ***spectrum*** of $(v_0, \ldots, v_{n-1})$. These names come from the applications of the "usual" (i.e., continuous) Fourier transforms.

18.34. Example. We consider again the binary $(7, 4)$ Hamming code as in 18.33, and we find the following table of codewords (at the top of next page) and their transforms.

The generator g can be found in the second line. So, by 18.30, C can be characterized as $\{(v_0, \ldots, v_6) \mid f_1 = f_2 = f_4 = 0\}$, hence the zeros of g are $\alpha^1, \alpha^2, \alpha^4$, which determine the cyclotomic coset $C_1 = \{1, 2, 4\} \bmod 7$ in a way which is easily visible.

Codewords (time domain codewords)							Codeword spectra (frequency domain codewords)						
v_0	v_1	v_2	v_3	v_4	v_5	v_6	f_0	f_1	f_2	f_3	f_4	f_5	f_6
0	0	0	0	0	0	0	0	0	0	0	0	0	0
1	1	0	1	0	0	0	1	0	0	α^4	0	α^2	α
0	1	1	0	1	0	0	1	0	0	1	0	1	1
0	0	1	1	0	1	0	1	0	0	α^3	0	α^5	α^6
0	0	0	1	1	0	1	1	0	0	α^6	0	α^3	α^5
1	0	1	1	1	0	0	0	0	0	α^5	0	α^6	α^3
1	1	1	0	0	1	0	0	0	0	α^6	0	α^3	α^5
1	1	0	0	1	0	1	0	0	0	α^3	0	α^5	α^6
0	1	0	1	1	1	0	0	0	0	α	0	α^4	α^2
0	1	1	1	0	0	1	0	0	0	α^2	0	α	α^4
0	0	1	0	1	1	1	0	0	0	α^4	0	α^2	α
1	0	0	0	1	1	0	1	0	0	α^2	0	α	α^4
1	0	1	0	0	0	1	1	0	0	α	0	α^4	α^2
1	1	1	1	1	1	1	1	0	0	0	0	0	0
0	1	0	0	0	1	1	1	0	0	α^5	0	α^6	α^3
1	0	0	1	0	1	1	0	0	0	1	0	1	1

Exercises

1. Determine all codewords of a code of length 7 with generator polynomial $g = 1+x+x^3$ over $\mathbb{F}_2$. Which of the following received words have detectable errors: 1000111, 0110011, 0100011?

2. Determine a table of coset leaders and syndromes for the binary (3, 1) code generated by $g = 1+x+x^2$. Do likewise for the binary (7, 3) code generated by $1 + x^2 + x^3 + x^4$.

3. How many binary cyclic codes of block length 15 are there?

4. What is the block length of the shortest cyclic binary code whose generator polynomial is $x^7 + x + 1$? Repeat the question for $x^9 + x + 1$.

5. Let $x^n - 1 = g_1 \ldots g_t$ be the factorization of $x^n - 1$ into irreducible polynomials over $\mathbb{F}_q$. Prove that (g_i), $1 \leq i \leq t$, is a maximal ideal in $\mathbb{F}_q[x]/(x^n - 1)$, and $(x^n - 1)/g_i$ generates a minimal ideal M_i, such that $M_i \cap M_j = \{0\}$

for $i \neq j$ and $M_i + M_j = \{m_i + m_j \mid m_i \in M_i, m_j \in M_j\}$ has generator $\gcd((x^n - 1)/g_i, (x^n - 1)/g_j) = (x^n - 1)/g_i g_j$.

6. Prove: The binary cyclic code with generator polynomial $1+x$ is the $(n-1)$-dimensional code C consisting of all even weight vectors of length n. A binary cyclic code $C = (g)$ contains only even weight vectors if and only if $1 + x$ divides g.

7. (i) Let $x^n - 1 = gh$ over $\mathbb{F}_q$. Prove that a cyclic code C with generator polynomial g is self-orthogonal (see Exercise 17.9) if and only if the reciprocal polynomial of h divides g.

   (ii) Consider a cyclic code of block length $n = 7$, with generator polynomial $1 + x^2 + x^3 + x^4$. Is this code self-orthogonal?

8. Let C be a binary (n, k) code generated by the irreducible polynomial g and having minimum distance $d_{\min}$. Show that $\operatorname{ord}(g) = n \iff d_{\min} = 3$. If $\operatorname{ord}(g) < 9$, find $d_{\min}$. Is this true for nonbinary codes?

9. Prove Theorem 18.18(ii)–(iv).

10. Prove Theorem 18.26.

11. Show that there are noncyclic Hamming codes.

12. A binary $(9, 3)$ code C is defined by

$$(v_0, v_1, \ldots, v_8) \in C \iff v_0 = v_1 = v_2,\ v_3 = v_4 = v_5,\ v_6 = v_7 = v_8.$$

    Show that C is equivalent to a cyclic code.

13. A binary cyclic code of length 63 has a generator polynomial $x^5 + x^4 + 1$. Show that the minimum distance of this code is ≤ 3 and encode the all-one-word.

14. Find the generator polynomials, dimensions, and idempotent generators for all binary cyclic codes of length $n = 7$. Identify dual codes and self-orthogonal codes.

15. Let $C = (f)$ over $\mathbb{F}_q$, $L = \{i \mid 0 \leq i \leq n-1,\ f(\alpha^i) = 0\}$, and let α be a primitive of $\mathbb{F}_q$. Prove $C = (g)$, with $g = \operatorname{lcm}\{m_{\alpha^i} \mid i \in L\}$.

16. Find all binary cyclic codes of length 10.

17. Find a canonical generator and a check matrix of a maximal binary code of length 10.

18. Let $C = (g)$ with $g = x^3 + x + 1 \in \mathbb{Z}_2[x]$. Decode (if possible) 1110010, 1110011, and 1110101.

19. Check the table in 18.34 by computing all codeword spectra. Also find the Mattson-Solomon polynomials for all codewords v.

§19 Special Cyclic Codes

In this section we shall consider some important examples of cyclic codes. After 18.16, we showed that the cyclic code of block length $2^m - 1$ over $\mathbb{F}_{2^m}$ is a Hamming code C_m if we choose the minimal polynomial of a primitive element ζ of $\mathbb{F}_{2^m}$ as a generating polynomial. The matrix $\mathbf{H} = (1, \zeta, \zeta^2, \ldots, \zeta^{n-1})$ is a parity-check matrix for C_m. This is a special case of Theorem 18.25. In the following example we shall generalize these codes and motivate the theory of these codes.

19.1. Example. Let $\zeta \in \mathbb{F}_{2^4}$ be a root of $x^4+x+1 \in \mathbb{F}_2[x]$. Since x^4+x+1 is irreducible, it is the minimal polynomial of ζ; it is also maximally periodic (due to 14.1, and so ζ is a primitive fifteenth root of unity. Let us take $\alpha_1 := \zeta$ and $\alpha_2 := \zeta^3$ in Theorem 18.24. Hence a cyclic code C over $\mathbb{F}_2$ is defined by the generator polynomial $g = M^{(1)}M^{(3)}$ and parity-check matrix

$$\mathbf{H} = \begin{pmatrix} 1 & \zeta & \cdots & \zeta^{13} & \zeta^{14} \\ 1 & \zeta^3 & \cdots & \zeta^{39} & \zeta^{42} \end{pmatrix}.$$

We shall see in 19.7(iii) that $d_{\min} \geq 5$ for this code. This means (by 16.9) that C can correct up to two errors and can detect up to four errors—our first double-error-correcting code. C is a cyclic $(15, 7)$ code since $n = 15$, $k = 15 - \deg g = 7$. We have

$$v \in C \iff S(v) = \mathbf{H}v^{\mathrm{T}} = 0 \iff S_1 = S_3 = 0,$$

where

$$S_1 := \sum_{i=0}^{14} v_i \zeta^i \quad \text{and} \quad S_3 := \sum_{i=0}^{14} v_i \zeta^{3i}$$

are the components of the syndrome $S(v) = (S_1, S_3)^{\mathrm{T}}$ of v with respect to $\mathbf{H}$. This explains why in studying S_1, S_3 we have omitted S_2: the syndrome $S_2 = \sum v_i \zeta^{2i}$ fulfills $S_2 = S_1^2$, hence $S_1 = 0 \iff S_2 = 0$, and we don't get any new information. If we use binary representation for elements of $\mathbb{F}_{2^4}$, then $\mathbf{H}$ is of the form

$$\mathbf{H} = \begin{pmatrix} 1 & 0 & 0 & 0 & 1 & 0 & 0 & 1 & 1 & 0 & 1 & 0 & 1 & 1 & 1 \\ 0 & 1 & 0 & 0 & 1 & 1 & 0 & 1 & 0 & 1 & 1 & 1 & 1 & 0 & 0 \\ 0 & 0 & 1 & 0 & 0 & 1 & 1 & 0 & 1 & 0 & 1 & 1 & 1 & 1 & 0 \\ 0 & 0 & 0 & 1 & 0 & 0 & 1 & 1 & 0 & 1 & 0 & 1 & 1 & 1 & 1 \\ 1 & 0 & 0 & 0 & 1 & 1 & 0 & 0 & 0 & 1 & 1 & 0 & 0 & 0 & 1 \\ 0 & 0 & 0 & 1 & 1 & 0 & 0 & 0 & 1 & 1 & 0 & 0 & 0 & 1 & 1 \\ 0 & 0 & 1 & 0 & 1 & 0 & 0 & 1 & 0 & 1 & 0 & 0 & 1 & 0 & 1 \\ 0 & 1 & 1 & 1 & 1 & 0 & 1 & 1 & 1 & 1 & 0 & 1 & 1 & 1 & 1 \end{pmatrix}.$$

Here we use the fact that $\zeta^4 + \zeta + 1 = 0$ in $\mathbb{F}_{2^4}$. The columns of $\mathbf{H}$ can be obtained as follows: The upper half of the first column is the coefficient vector $1\zeta^0 + 0\zeta^1 + 0\zeta^2 + 0\zeta^3$, i.e., $(1000)^T$. The upper half of the second column is the coefficient vector of $0\zeta^0 + 1\zeta^1 + 0\zeta^2 + 0\zeta^3$, i.e., $(0100)^T$, etc. The lower half of the second column is the coefficient vector of $0\zeta^0 + 0\zeta^1 + 0\zeta^2 + 1\zeta^3$, etc. Suppose the received vector $v = (v_0, v_1, \ldots, v_{14})$ is a vector with two errors. Let $e = x^{a_1} + x^{a_2}$, $0 \le a_1, a_2 \le 14$, $a_1 \ne a_2$, be the error vector. We have

$$S_1 = \zeta^{a_1} + \zeta^{a_2}, \qquad S_3 = \zeta^{3a_1} + \zeta^{3a_2},$$

We let

$$X_1 := \zeta^{a_1}, \qquad X_2 := \zeta^{a_2},$$

be the error-location numbers, then

$$S_1 = X_1 + X_2, \qquad S_3 = X_1^3 + X_2^3,$$

or

$$X_2 = S_1 + X_1, \qquad X_2^3 = S_1^3 + S_1^2 X_1 + S_1 X_1^2 + X_1^3.$$

Hence $S_1 X_1^2 + S_1^2 X_1 + (S_1^3 - S_3) = 0$. If two errors occurred, then this equation has two solutions. If only one error occurred, then $S_1 = X_1$, $S_3 = X_1^3$, and $S_1^3 + S_3 = 0$. Hence the quadratic equation reduces to $X_1 + S_1 = 0$. If no error occurred, then $S_1 = S_3 = 0$. To summarize, first we have to find the syndrome of the received vector v, which tells us if no, one, or at least two errors have occurred, then we find the solution(s) of the resulting equations. If these equations are not solvable, then we know that we have a detectable error e with more than two error locations, which is not correctable. See Exercise 13.23 for how to solve quadratic equations over fields of characteristic 2.

Suppose the vector $v = 100111000000000$ is the received word. Then the syndrome $S(v) = (S_1, S_3)^T$ is given by

$$S_1 = 1 + \zeta^3 + \zeta^4 + \zeta^5,$$
$$S_3 = 1 + \zeta^9 + \zeta^{12} + \zeta^{15}.$$

Here ζ is a primitive element of $\mathbb{F}_{2^4}$ with $\zeta^4 + \zeta + 1 = 0$. We use the following powers of ζ in our calculations:

$$\begin{aligned} \zeta^4 &= 1 + \zeta, & \zeta^{10} &= 1 + \zeta + \zeta^2, \\ \zeta^5 &= \zeta + \zeta^2, & \zeta^{11} &= \zeta + \zeta^2 + \zeta^3, \\ \zeta^6 &= \zeta^2 + \zeta^3, & \zeta^{12} &= 1 + \zeta + \zeta^2 + \zeta^3, \\ \zeta^7 &= 1 + \zeta + \zeta^3, & \zeta^{13} &= 1 + \zeta^2 + \zeta^3, \\ \zeta^8 &= 1 + \zeta^2, & \zeta^{14} &= 1 + \zeta^3, \\ \zeta^9 &= \zeta + \zeta^3, & \zeta^{15} &= 1. \end{aligned}$$

Then

$$S_1 = \zeta^2 + \zeta^3 \quad \text{and} \quad S_3 = 1 + \zeta^2.$$

give us the quadratic equation

$$(\zeta^2 + \zeta^3)X_1^2 + (\zeta^2 + \zeta^3)^2 X_1 + \left(1 + \zeta + \zeta^2 + \zeta^3 + \frac{1 + \zeta^2}{\zeta^2 + \zeta^3}\right) = 0.$$

We determine the solutions σ by trial-and-error and find that $X_1 = \zeta^8$ and $X_2 = \zeta^{14}$ in $\mathbb{F}_{2^4}$. Thus we know that errors must have occurred in the positions corresponding to x^8 and x^{14}. Observe that we started counting at 0. Therefore the corrected transmitted codeword is $w = 100111001000001$. Then w is decoded by dividing it by the generator polynomial $g = 1 + x^4 + x^6 + x^7 + x^8$. This gives $w/g = x^6 + x^5 + x^3 + 1$ with remainder 0. Therefore the original message was $a = 1001011$.

We describe a class of codes which has been introduced by Bose, Chaudhuri, and Hocquenghem, and which are therefore called BCH codes.

19.2. Definition and Theorem. Let $c, d, n, q \in \mathbb{N}$, where q is a prime power and $2 \leq d \leq n$. Also let m be the multiplicative order of q modulo n (recall that $\gcd(n, q) = 1$), and let ζ be a primitive nth root of unity in $\mathbb{F}_{q^m}$ and m_{ζ^i} the minimal polynomial of ζ^i. Finally let $I := \{c, c+1, \ldots, c+d-2\}$.

Then the ***BCH code*** $C \subseteq V_n$ of ***designed distance*** d is a cyclic code over $\mathbb{F}_q$ of length n defined by the following equivalent conditions:

(i) $v \in C \iff v(\zeta^i) = 0$ for all $i \in I$.

(ii) The polynomial $\operatorname{lcm}\{m_{\zeta^i} \mid i \in I\}$ is the generator polynomial of C.

(iii) A parity-check matrix of C is the matrix

$$\mathbf{H} = \begin{pmatrix} 1 & \zeta^{c} & \dots & \zeta^{c(n-1)} \\ 1 & \zeta^{c+1} & \dots & \zeta^{(c+1)(n-1)} \\ \dots & \dots & \dots & \dots \\ 1 & \zeta^{c+d-2} & \dots & \zeta^{(c+d-2)(n-1)} \end{pmatrix}.$$

Here C is the null space of $\mathbf{H}$.

19.3. Remark. If $c = 1$, then the code defined in 19.2 is called a ***narrow-sense BCH code***. If $n = q^m - 1$, the BCH code is called ***primitive***. The dual of a BCH code is, in general, not a BCH code. The dimension of C is $\geq n - m(d-1)$.

A practical example for the use of BCH codes is the European and trans-Atlantic information communication system, which has been using such codes for many years. The message symbols are of length 231 and the generator polynomial is of degree 24 such that $231 + 24 = 255 = 2^8 - 1$ is the length of the codewords. The code detects at least six errors, and its failure probability is one in sixteen million.

19.4. Theorem. *A BCH code of designed distance d defined by* 19.2 *has minimum distance $d_{\min} \geq d$.*

Proof. We show that no $d-1$ or fewer columns of $\mathbf{H}$ are linearly dependent over $\mathbb{F}_{q^m}$. We choose a set of $d-1$ columns with first elements $\zeta^{ci_1}, \zeta^{ci_2}, \dots, \zeta^{ci_{d-1}}$ and form the $(d-1) \times (d-1)$ determinant. We can factor out the divisors ζ^{ci_k}, $k = 1, 2, \dots, d-1$, and obtain

$$\zeta^{c(i_1+\dots+i_{d-1})} \begin{vmatrix} 1 & 1 & \dots & 1 \\ \zeta^{i_1} & \zeta^{i_2} & \dots & \zeta^{i_{d-1}} \\ \dots & \dots & \dots & \dots \\ \zeta^{(d-2)i_1} & \zeta^{(d-2)i_2} & \dots & \zeta^{(d-2)i_{d-1}} \end{vmatrix}$$

$$= \zeta^{c(i_1+\dots+i_{d-1})} \prod_{1 \leq k < j \leq d-1} (\zeta^{i_j} - \zeta^{i_k}) \neq 0.$$

The determinant is the Vandermonde determinant and is nonzero, which proves that the minimum distance of the code is at least d. □

We note that there is an even simpler proof of Theorem 19.4 in the case $c = 1$, by using Lemma 18.31. Let v be a codeword in the BCH code of 19.4, so that $\deg F_v \leq n - d$. Theorem 18.32 shows that the weight of v must be not less than $n - (n - d) = d$. In the spectral setting introduced at the end of §18, a primitive t-error-correcting BCH code of length $n = q^m - 1$ is the set of all vectors with n components in $\mathbb{F}_q$ whose spectrum is zero in a specified block of $2t$ consecutive components. For $n = q - 1$ we obtain the ***Reed-Solomon codes***. Encoding of these codes is simple. Since some set of $2t$ consecutive frequencies has been given zeros as components, the message symbols are loaded into the remaining $n - 2t$ symbols, and the result is inverse Fourier transformed to produce the (time-domain) codeword. The following definition introduces this special class of BCH codes.

19.5. Definition. A ***Reed-Solomon code*** (in short: ***RS code***) is a BCH code in the narrow sense of designed distance d and of length $n = q - 1$ over $\mathbb{F}_q$. Hence $m = 1$. The generator polynomial of an RS code is then given by

$$g = \prod_{i=1}^{d-1} (x - \zeta^i),$$

where ζ is a primitive element of $\mathbb{F}_q$.

19.6. Theorem. *The minimum distance of an RS code with generator polynomial* $g = \prod_{i=1}^{d-1} (x - \zeta^i)$ *is* d.

Proof. Theorem 19.4 shows that the minimum distance of the RS code is at least d. By Theorem 17.13, $d_{\min} \leq n - k + 1$. In our case $k = n - (d - 1)$, hence $d_{\min} \leq d$. Therefore $d_{\min} = d$. □

Reed-Solomon codes are used to obtain the high sound quality of compact discs. In such a disc there is, similar to the "old records," a track in spiral form, consisting of sequences of lower parts "pits" and "normal" ones ("lands"). The lengths of these pits make up the sound signals. When a CD player is in action, a laser beam is sent to this spiral track and then reflected from it. If the laser beam hits a pit, little of the light is reflected. Every time when a pit changes into a land (or conversely), the CD player stores a "1," otherwise a "0." So, for example,

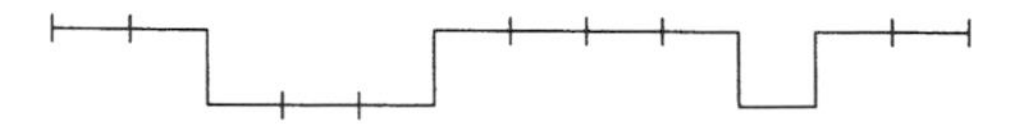

is recognized as the sequence 0010010001100. Per second, 44,100 of these bits are recorded.

These bits (i.e., the pits and the lands) were imprinted on the disc by using decoding with two RS codes over $\mathbb{F}_{256}$ and minimal distance 5 (more precisely, by "shortened" RS codes). In order to avoid burst errors (due to scratched discs, etc.), "interleaving" is done by "mixing" the 0–1-sequences properly to separate longer sequences of errors.

The CD player decodes the received 0–1-sequences. The probability of incorrect decoding is presently around $2 \cdot 10^{-6}$. If a larger error cannot be decoded, the CD player interpolates between the neighboring sounds. In order to give the CD player some time for decoding, the 0–1-sequences are sent into a data cache, which usually is about half full. So, also variations of the rotation speed of the disc have no influence on the quality. For protection, the disc is covered with a 1.2 mm transparent plastic cover. In this way, the high sound quality of CD's is obtained. Since the information for one sound is distributed on the whole disc, we can even drill a hole into the disc without recognizing this when the CD is played! For more information, see Pohlmann (1987).

19.7. More Examples of BCH Codes.

(i) Let $\zeta \in \mathbb{F}_{2^m}$ be a primitive element with minimal polynomial $m_\zeta \in \mathbb{F}_2[x]$. Let $n = 2^m - 1$, $\deg m_\zeta = m$, $c = 1$, $d = 3$, $I = \{1, 2\}$. Then according to 19.2, $C = (m_\zeta)$ is a BCH code and is also a binary Hamming code with m check symbols. Recall that ζ and ζ^2 have the same minimal polynomial.

(ii) Let $\zeta \in \mathbb{F}_{2^4}$ be a primitive element with minimal polynomial $m_\zeta = x^4 + x + 1$ over $\mathbb{F}_2$. The powers ζ^i, $0 \le i \le 14$, can be written as linear combinations of $1, \zeta, \zeta^2, \zeta^3$. Thus we obtain a parity-check matrix $\mathbf{H}$ of a code equivalent to the (15, 11) Hamming code,

$$\mathbf{H} = \begin{pmatrix} 1 & 0 & 0 & 0 & 1 & 0 & 0 & 1 & 1 & 0 & 1 & 0 & 1 & 1 & 1 \\ 0 & 1 & 0 & 0 & 1 & 1 & 0 & 1 & 0 & 1 & 1 & 1 & 1 & 0 & 0 \\ 0 & 0 & 1 & 0 & 0 & 1 & 1 & 0 & 1 & 0 & 1 & 1 & 1 & 1 & 0 \\ 0 & 0 & 0 & 1 & 0 & 0 & 1 & 1 & 0 & 1 & 0 & 1 & 1 & 1 & 1 \end{pmatrix}$$

$$= \begin{pmatrix} 1 & \zeta & \zeta^2 & \zeta^3 & \zeta^4 & \zeta^5 & \zeta^6 & \zeta^7 & \zeta^8 & \zeta^9 & \zeta^{10} & \zeta^{11} & \zeta^{12} & \zeta^{13} & \zeta^{14} \end{pmatrix}.$$

This code can be regarded as a narrow-sense BCH code of designed distance $d = 3$ over $\mathbb{F}_2$. Its minimum distance is also 3 and therefore this code can correct one error.

(iii) Let $q = 2$, $n = 15$, and $d = 4$. Then $x^4 + x + 1$ is irreducible over $\mathbb{F}_2$, and its roots, except 1, are primitive elements of $\mathbb{F}_{2^4}$. If ζ is such a root, then ζ^2 is a root, and ζ^3 is then a root of $x^4+x^3+x^2+x+1$. Thus

$$g = (x^4 + x + 1)(x^4 + x^3 + x^2 + x + 1)$$

is a generator polynomial of a narrow-sense BCH code with $d = 4$. This is also a generator for a BCH code with designed distance $d = 5$, since ζ^4 is a root of $x^4 + x + 1$. The dimension of this code is $15 - \deg g = 7$. This code was considered in detail in Example 19.1.

(iv) Let $q = 3$, $n = 8$, and $d = 4$, and let ζ be a primitive element of $\mathbb{F}_{3^2}$ with minimal polynomial $x^2 - x - 1$ over $\mathbb{F}_3$. Then ζ^2 has minimal polynomial x^2+1, and ζ^3 has x^2-x-1 as its minimal polynomial. The polynomial $g = (x^2 - x - 1)(x^2 + 1) = x^4 - x^3 - x - 1$ is the generating polynomial for a BCH code of length 8, dimension 4 over $\mathbb{F}_3$, and has minimum distance 4, since g is a polynomial of weight 4.

(v) Let $q = 2$, $n = 23$, and $d = 5$. The polynomial $x^{23} - 1$ has the following factorization into irreducibles over $\mathbb{F}_2$

$$\begin{aligned} x^{23} - 1 &= (x - 1)g_0 g_1 \\ &= (x-1)(x^{11}+x^9+x^7+x^6+x^5+x+1)(x^{11}+x^{10}+x^6+x^5+x^4+x^2+1). \end{aligned}$$

The roots of these polynomials are in $\mathbb{F}_{2^{11}}$ and they are the primitive twenty-third roots of unity over $\mathbb{F}_2$, since $2^{11}-1$ is a multiple of 23. If ζ is such a root then ζ^j, for $j = 1, 2, 4, 8, 16, 9, 18, 13, 3, 6, 12$, are its conjugates. Each of the cyclic codes in $\mathbb{F}_2^{23}$ generated by the irreducible factors g_0, g_1 of $x^{23} - 1$ of degree 11 is a BCH code of dimension 12 and of designed distance $d = 5$. These codes are equivalent versions of the so-called ***binary Golay code*** (see page 233). We state that its minimum distance is 7 and note that again $d_{\min} > d$.

BCH codes are very powerful since for any positive integer d we can construct a BCH code of minimum distance $\geq d$. To find a BCH code for a larger minimum distance, we have to increase the length n and hence increase the number m, i.e., the degree of $\mathbb{F}_{q^m}$ over $\mathbb{F}_q$. A BCH code of designed distance $d \geq 2t + 1$ will correct up to t errors, but at the same time, in order to achieve the desired minimum distance, we must use code words of great length.

Exercises

1. Determine the errors in Example 19.1 if the syndrome of a received word is given as $(10010110)^{\mathrm{T}}$.
2. Determine the dimension of a 5-error-correcting BCH code over $\mathbb{F}_3$ of length 80.
3. Find a generator polynomial of the 3-error-correcting BCH code of length 15 by using a primitive element ζ of $\mathbb{F}_{2^4}$, where $\zeta^4 = \zeta^3 + 1$.
4. Determine a generator polynomial g for a BCH code of length 31 with minimum distance $d = 9$.
5. If $d \mid n$, show that the binary BCH code of length n and designed distance d has $d_{\min} = d$.
6. Find all narrow sense primitive binary BCH codes of length $n = 15$.
7. Determine whether the dual of an arbitrary BCH code is a BCH code. Is the dual of a Reed-Solomon code again an RS code?
8. Find a generator polynomial of the RS code of length 4 over $\mathbb{F}_5$ with designed distance 3.
9. Let C be the $(n = q^m - 1, k)$ RS code of minimum distance d. Prove that the extended code, which is obtained by adding to each codeword an overall parity check, is a (q^m, k) code of minimum distance $d + 1$.
10. For any positive integer m and $t \leq 2^{m-1} - 1$, prove the existence of a binary BCH code of length $n = 2^m - 1$ that is t-error-correcting and has dimension $\geq n - mt$.
11. Find the zeros of the cyclic codes in 19.7.

§20 Decoding BCH Codes

We describe a general decoding algorithm for BCH codes. This will generalize the approach taken in Example 19.1. Let us denote by v, w, and e the transmitted code polynomial, the received polynomial, and the error polynomial, respectively, so that $w = v + e$. We view w, v, e as the corresponding coefficient vectors. As before, let ζ be a primitive root of unity. The syndrome of w is calculated as

$$S(w) = \mathbf{H}w^{\mathrm{T}} = (S_c, S_{c+1}, \ldots, S_{c+d-2})^{\mathrm{T}},$$

where

$$S_j = w(\zeta^j) = v(\zeta^j) + e(\zeta^j) \quad \text{for } c \le j \le c+d-2.$$

Let $e = \sum_{i \in I} e_i x^i$, where $I = \{i_1, \dots, i_r\}$ are the positions where errors have occurred. We have to assume that $r \le \lfloor \frac{d-1}{2} \rfloor$, i.e., that $2r < d$. Since $v(\zeta^j) = 0$ for $j \in J := \{c, c+1, \dots, c+d-2\}$, we get $S_j = e_i(\zeta^{ij}) = \sum_{i \in I} e_i \zeta^{ij} = 0$ for all $j \in J$. This is a first system of linear equations between the known S_j and the unknown entities I and e_i $(i \in I)$.

Next we define the polynomial $s := \prod_{i \in I}(x - \zeta^i) =: s_0 + s_1 x + \cdots + s_{r-1}x^{r-1} + x^r$. Since $s(\zeta^i) = 0$ for all $i \in I$, we get r equations (for each $i \in I$):

$$s_0 + s_1\zeta^i + \cdots + s_{r-1}\zeta^{i(r-1)} + \zeta^{ir} = 0.$$

If we multiply each equation by $e_i\zeta^{ij}$ (for $j \in J$) and add all these equations, we get

$$s_0 \sum_{i \in I} e_i\zeta^{ij} + \cdots + s_{r-1} \sum_{i \in I} e_i \zeta^{i(r+j-1)} + \sum_{i \in I} e_i \zeta^{i(r+j)} = 0$$

or, in a different notation,

$$s_0 S_j + \cdots + s_{r-1} S_{r+j-1} + S_{r+j} = 0 \qquad (j \in \{c, c+1, \dots, c+r-1\}),$$

which we might write as

$$\begin{aligned}
S_c s_0 + S_{c+1}s_1 + \cdots + S_{c+r-1}s_{r-1} &= -S_{c+r},\\
S_{c+1}s_0 + S_{c+2}s_1 + \cdots + S_{c+r}s_{r-1} &= -S_{c+r+1},\\
&\vdots\\
S_{c+r-1}s_0 + S_{c+r}s_1 + \cdots + S_{c+2r-2}s_{r-1} &= -S_{c+2r-1}
\end{aligned} \tag{20.1}$$

(observe that $2r - 1 \le d - 2$). The coefficient matrix of this system of linear equations in the unknowns $s_0, s_1, \dots, s_{r-1}$ is

$$\mathbf{S} := \begin{pmatrix} S_c & S_{c+1} & \dots & S_{c+r-1} \\ S_{c+1} & S_{c+2} & \dots & S_{c+r} \\ \cdots & \cdots & \cdots & \cdots \\ S_{c+r-1} & S_{c+r} & \dots & S_{c+2r-2} \end{pmatrix},$$

which can be written as $\mathbf{VDV}^{\mathrm{T}}$ with

$$\mathbf{V} = \begin{pmatrix} 1 & 1 & \dots & 1 \\ \zeta^{i_1} & \zeta^{i_2} & \dots & \zeta^{i_r} \\ \cdots & \cdots & \cdots & \cdots \\ \zeta^{i_1(r-1)} & \zeta^{i_2(r-1)} & \dots & \zeta^{i_r(r-1)} \end{pmatrix} \text{ and } \mathbf{D} = \begin{pmatrix} e_1\zeta^{i_1c} & 0 & \dots & 0 \\ 0 & e_2\zeta^{i_2c} & \dots & 0 \\ \vdots & \vdots & \ddots & \vdots \\ 0 & 0 & \dots & e_r\zeta^{i_rc} \end{pmatrix}.$$

Now $\mathbf{S}$ is nonsingular iff $\mathbf{V}$ is nonsingular, which is the case iff all $e_1, \ldots, e_r$ are distinct, since $\mathbf{V}$ is of the Vandermonde type. So we get

20.1. Lemma. *The system* (20.1) *is uniquely solvable iff at least r errors have occurred. The maximal number r such that* (20.1) *is uniquely solvable is precisely the number of positions where errors have occurred.*

If we solve (20.1), we get the polynomial S. From that we get the error positions by the definition of the ***error locator polynomial*** s.

20.2. Lemma. *The zeros of s are precisely those ζ^i such that an error has happened in position i.*

In the binary case, the finding of error locations is equivalent to correcting errors. We now summarize the BCH decoding algorithm.

20.3. BCH Decoding. Let v be a codeword and suppose that at most t errors occur in transmitting it by using a BCH code of designed distance $d \geq 2t+1$. For decoding a received word w, the following steps are executed:

Step 1. Determine the syndrome of w, where

$$S(\zeta) = (S_c, S_{c+1}, \ldots, S_{c+d-2})^{\mathrm{T}}.$$

Step 2. Determine the maximum number $r \leq t$ of equations of the form

$$S_j s_0 + S_{j+1} s_1 + \cdots + S_{j+r} = 0, \qquad c \leq j \leq c+r-1,$$

such that the coefficient matrix of the s_i is nonsingular and thus obtain the number r of errors that have occurred. Solve this maximal system and form the error-locator polynomial

$$s = \sum_{i=0}^{r} s_i x_i, \qquad \text{with } s_0 := 1.$$

Step 3. Find the zeros of s by trying $\zeta^0, \zeta^1, \ldots$.

Step 4. For binary codes, the zeros $\zeta^{i_1}, \ldots, \zeta^{i_r}$ determine e already. Otherwise, the e_i can be obtained from the equations $\sum_{i \in I} e_i \zeta^{ij} = 0$ for $j \in \{c, c+1, \ldots, c+d-2\}$ (since I is known now).

20.4. Remark. The most difficult step in this algorithm is Step 2. One way of facilitating the required calculations is to use an algorithm due to Berlekamp and Massey (see Berlekamp (1984) or Lidl & Niederreiter

(1983, Chapter 8)) to determine the unknown coefficients s_i in the system of Step 2. This system represents a linear recurrence relation for the S_j (see §33).

20.5. Example (*19.1 and 19.7(iii) revisited*). For $c = 1$, $n = 15$, $q = 2$ as parameters, we consider the BCH code with designed distance $d = 5$, which can correct up to two errors The generator polynomial of this BCH code was

$$g = 1 + x^4 + x^6 + x^7 + x^8.$$

Suppose now that the received word w is given by

$$100100110000100,$$

or written as a polynomial,

$$w = 1 + x^3 + x^6 + x^7 + x^{12}.$$

We calculate the syndrome according to Step 1;

$$\begin{aligned} S_1 &= e(\zeta) = w(\zeta) = 1, \\ S_2 &= e(\zeta^2) = w(\zeta^2) = 1, \\ S_3 &= e(\zeta^3) = w(\zeta^3) = \zeta^4, \\ S_4 &= e(\zeta^4) = w(\zeta^4) = 1. \end{aligned}$$

Then Step 2 yields the system of linear equations (recall that $r \leq 2$)

$$\begin{aligned} S_1 s_0 + S_2 s_1 &= S_3, \\ S_2 s_0 + S_3 s_1 &= S_4, \end{aligned}$$

or, after substituting for S_j,

$$\begin{aligned} s_0 + s_1 &= \zeta^4, \\ s_0 + \zeta^4 s_1 &= 1. \end{aligned}$$

Since these two equations are linearly independent, two errors must have occurred, i.e., $r = 2$. We solve the system and find $s_0 = \zeta$, $s_1 = 1$. So we get $s = \zeta + x + x^2$. As roots in $\mathbb{F}_{16}$, we find ζ^7 and ζ^9. Therefore we know that errors must have occurred in positions 7 and 9 of the code word (observe that we start the indices by 0). In order to correct these errors in the received polynomial, we form

$$\begin{aligned} w - e &= (1 + x^3 + x^6 + x^7 + x^{12}) - (x^7 + x^9) \\ &= 1 + x^3 + x^6 + x^9 + x^{12}, \end{aligned}$$

and thus obtain the transmitted code polynomial v. The corresponding codeword is 100100100100100. The initial message can be recovered by dividing the corrected polynomial v by g. This gives

$$v/g = 1 + x + x^4,$$

which yields the corresponding message word 1100100.

We conclude this chapter with some remarks on some other important types of codes and recall that a t-error-correcting code over $\mathbb{F}_q$ is perfect if equality holds in the Hamming bound (Theorem 16.10). In the binary case, if $t = 3$, a sphere about a codeword contains $1 + n + \binom{n}{2} + \binom{n}{3}$ points and this is a power of 2 for $n = 23$. Similarly, $1 + 2n + 4\binom{n}{2}$ is a power of 3 for $n = 11$, so for these values the Hamming bound is also satisfied as an equation. Probably this observation led Marcel Golay to find the corresponding codes and give the difficult existence proof for the important binary $(23, 12, 7)$ code and the ternary $(11, 6, 5)$ code. Both are perfect codes and are called the binary and ternary Golay codes, respectively. There are various ways of defining Golay codes, and one possibility for the binary Golay code was given in Example 19.7(v).

Golay codes are a special type of code, called quadratic residue codes. In the following let n be an odd prime and q a prime power such that $\gcd(n, q) = 1$, q is assumed to be a quadratic residue modulo n. U_0 denotes the set of quadratic residues mod n, that is the set of squares in $\mathbb{F}_n^*$, and U_1 denotes the set of nonsquares in $\mathbb{F}_n^*$. If ζ is a primitive nth root of unity in $\mathbb{F}_{q^{n-1}}$, then

$$g_0 = \prod_{i \in U_0} (x - \zeta^i), \qquad g_1 = \prod_{j \in U_1} (x - \zeta^j),$$

can be shown to be polynomials over $\mathbb{F}_q$ and we have $x^n - 1 = (x - 1)g_0 g_1$.

20.6. Definition. The ***quadratic residue codes*** (***QR codes***) C_0^+, C_0, C_1^+, C_1 are the cyclic codes with generator polynomials g_0, $(x - 1)g_0$, g_1, and $(x - 1)g_1$, respectively.

It can be easily verified that the codes C_0 and C_1 (and also C_0^+ and C_1^+) are equivalent and that $C_0^+ = C_0^\perp$, $C_1^+ = C_1^\perp$, if the prime n satisfies $n \equiv -1$

(mod 4). If $n \equiv 1 \pmod 4$, then $C_0^+ = C_1^\perp$, $C_1^+ = C_0^\perp$. Important examples will be given below.

20.7. Definition. The ***extended QR codes*** C_∞^+ are defined by adding an overall parity-check component c_∞ to C_0^+.

It can be shown that the binary QR code C_0 consists of all code words of even weight of the code C_0^+. If $\mathbf{G}_0$ is an $r \times n$ generator matrix for the binary code C_0, then $\begin{pmatrix} \mathbf{G}_0 \\ \mathbf{1} \end{pmatrix}$ is a generator matrix of C_0^+. A well-known result from number theory is that $q = 2$ is a quadratic residue mod n if and only if $n = \pm 1 \pmod 8$. A generator matrix for extended QR codes can be obtained by using idempotent generators (see Exercise 18.14). Let $q = 2$, n a prime $\equiv \pm 1 \pmod 8$, and $g = \sum_{i \in U_0} x^i$. Then it can be shown that g is an idempotent generator of C_0 if $n \equiv 1 \pmod 8$, and g is an idempotent generator of C_0^+ if $n \equiv -1 \pmod 8$. Now let $\mathbf{G}$ be a circulant $n \times n$ matrix over $\mathbb{F}_2$ with $g = (g_0, g_1, \ldots, g_{n-1})$ as its first row, that is a matrix of the form

$$\mathbf{G} = \begin{pmatrix} g_0 & g_1 & \cdots & g_{n-2} & g_{n-1} \\ g_1 & g_2 & \cdots & g_{n-1} & g_0 \\ \cdots & \cdots & \cdots & \cdots & \cdots \\ g_{n-1} & g_0 & \cdots & g_{n-3} & g_{n-2} \end{pmatrix}.$$

Let $\mathbf{a} = (0, \ldots, 0)$ for $n \equiv 1 \pmod 8$ and $\mathbf{a} = (1, \ldots, 1)$ for $n \equiv -1 \pmod 8$. Then

$$\mathbf{G}_\infty = \left(\begin{array}{c|c} 1 & 1 \ldots 1 \\ \hline \mathbf{a}^T & \mathbf{G} \end{array} \right)$$

is a generator matrix for the extended QR code C_∞^+. Observe that the rows of this matrix are clearly not independent. An important result on the minimum distance of a QR code is the following (see van Lint (1992)).

20.8. Theorem. *The minimum distance of the binary QR code C_0^+ of length n is the first odd number d, for which:*

(i) $d^2 > n$ *if* $n \equiv 1 \pmod 8$,

(ii) $d^2 - d + 1 \geq n$ *if* $n \equiv -1 \pmod 8$;

Now we observe that the binary $(23, 12)$ Golay code is a binary QR code of length $n = 23$. Its minimum distance is 7, because of Theorem 20.8(ii). If we extend this binary $(23, 12, 7)$ Golay code, we obtain the binary $(24, 12, 8)$ Golay code. Given an (n, k) code C for even n, we obtain a ***punctured*** code by removing a column of a generator matrix of C. Thus the binary $(23, 12, 7)$ Golay code can be obtained by puncturing the $(24, 12, 8)$ Golay code. The ***ternary*** $(11, 6, 5)$ ***Golay code*** can be defined as follows. Let

$$\mathbf{S}_5 = \begin{pmatrix} 0 & 1 & -1 & -1 & 1 \\ 1 & 0 & 1 & -1 & -1 \\ -1 & 1 & 0 & 1 & -1 \\ -1 & -1 & 1 & 0 & 1 \\ 1 & -1 & -1 & 1 & 0 \end{pmatrix}.$$

Then the matrix

$$\begin{pmatrix} 1 & \mathbf{0} & \mathbf{1} \\ \mathbf{0}^{\mathrm{T}} & \mathbf{I}_5 & \mathbf{S}_5 \end{pmatrix}$$

is a generator matrix for the ternary $(11, 6, 5)$ Golay code. This code is also a QR code. We consider the factorization

$$x^{11} - 1 = (x - 1)(x^5 - x^3 + x^2 - x - 1)(x^5 + x^4 - x^3 + x^2 - 1)$$

over $\mathbb{F}_3$. This factorization must be the factorization $x^{11} - 1 = (x - 1)g_0g_1$ in the sense of QR codes. Again the codes (g_0) and (g_1) are equivalent. We say that an (n, k, d) code with certain properties over $\mathbb{F}_q$ is ***unique*** if any two (n, k, d) codes over $\mathbb{F}_q$ with these properties are equivalent. It has been shown (see van Lint (1992), Pless (1989)) that the Hamming codes, the ternary Golay codes, the binary $(24, 12, 8)$ code, and the binary $(23, 12, 7)$ code are unique. We conclude with a remarkable result on perfect codes, which involves the Golay codes and classifies all perfect codes. Besides the binary repetition code of odd length with two codewords $11 1\dots 1$ and $00\dots 0$, the only trivial perfect codes are codes which consist of only one codeword or which contain the whole vector space (see Example 17.4). The 1-error-correcting Hamming codes over $\mathbb{F}_q$ of length $n = (q^m - 1)/(q - 1)$, dimension $k = (q^m - 1)/(q - 1) - m$, and minimum distance $d = 3$ are perfect. The binary $(23, 12, 7)$ Golay code and the $(11, 6, 5)$ Golay code over $\mathbb{F}_3$ are perfect (see 17.16). It had been shown in the mid-1970s by Tietäväinen and others that there are no other perfect codes, see van Lint (1994) or MacWilliams & Sloane (1977).

20.9. Theorem. *The only nontrivial multiple-error-correcting perfect codes are equivalent to either the binary* $(23, 12, 7)$ *Golay code or the ternary* $(11, 6, 5)$ *Golay code. The only nontrivial single-error-correcting perfect codes have the parameters of the Hamming codes.*

Exercises

1. A binary 2-error-correcting BCH code of length 31 is defined by a zero ζ of $x^5+x^2+1 \in \mathbb{F}_2[x]$. If a received word has syndrome $(1110011101)^T$, find the errors.
2. Let ζ be a primitive element of $\mathbb{F}_{2^4}$ with $\zeta^4+\zeta+1=0$, and let $g = x^{10}+x^8+x^5+x^4+x^2+x+1$ be a generator polynomial of a (15, 5) BCH code. Suppose the word $v = 110001001101000$ is received. Then determine the corrected codeword and decode it.
3. Find the generator polynomial of the double error correcting narrow sense ternary code of length 8 over $\mathbb{F}_3$. [Hint: $\zeta \in \mathbb{F}_{3^2}$ satisfying $\zeta^2+\zeta+2=0$ is primitive.] Decode $v = 22001001$.
4. Verify that the QR codes C_0^+ and C_1^+ are equivalent.
5. Show that the weight of all codewords of the ternary $(11, 6, 5)$ Golay code is not congruent to 1 (mod 3).
6. Show that the ternary Golay code has minimum distance 5 and that the code is perfect. [Hint: Use Exercise 5.]

Notes

Shannon's paper of 1948 marks the beginning of information theory as a mathematical discipline. The first nontrivial example of an error-correcting code over a finite field is contained in this paper; nowadays it would be referred to as the $(7, 4)$ Hamming code. A short history of the developments in algebraic coding theory is given in Blake (1973); this is a collection of important papers on the subject. There is quite a number of monographs and textbooks on algebraic coding theory; we refer the beginner who is interested in knowing somewhat more than we present, to the introductory texts by Pless (1989) or van Lint (1992). Some of the books on applied algebra or combinatorics also contain material on algebraic coding theory, which is easily accessible; see, e.g., Birkhoff & Bartee (1970), Dornhoff & Hohn (1978), and Lidl & Niederreiter (1983). Amongst the more advanced texts and monographs, we mention the very comprehensive volume by MacWilliams & Sloane (1977) (with an extensive bibliography on coding), Peterson & Weldon (1972), Berlekamp (1984), Blahut (1983), and Blake & Mullin (1975). The following books contain sections on coding in the general context of information theory: McEliece (1977), Guiasu (1977), and Gallager (1968). Cyclic codes were introduced by Prange, most of the books mentioned above contain the basics of cyclic codes (and some books much more). For connections between cyclic codes, shift registers, and linear recurring sequences (§33), see Berlekamp (1984, Chapter 5) and Zierler (1959). The engineering approach to coding was described by Blahut (1983) in order to bring coding closer to the subject of signal processing. BCH codes were introduced by Hocquenghem, Bose, and Ray-Chaudhuri in the binary case. Almost any standard book on coding includes a description of BCH codes and their properties, in particular, a presentation of decoding procedures. The interrelation between coding and combinatorics is of advantage to both disciplines and the books by Blake & Mullin (1975), Pless (1989), MacWilliams & Sloane (1977), and van Lint (1992) pay particular attention to this relationship. The vast material on algebraic coding makes it impossible to give details on references for other types of codes. It should suffice to refer the interested reader to MacWilliams & Sloane (1977), who treat some 30 kinds of codes, and give encoding and decoding procedures and nearly 1500 references; van Lint (1994) contains a carefully selected list of references.

We mention a few applications of codes. One of the important areas where codes are used is digital communication by satellite. The $(32, 6)$ Reed-Muller codes were used in all of NASA's Mariner class deep-space probes from 1969 to 1977. Block codes are used for satellite communication with mobile airborne terminals. The $(7, 2)$ Reed-Solomon (RS) code was used in the US Defense Department Tactical Satellite, whereas the $(31, 15)$ RS code has been adopted for tactical military communication links in the United States (see Berlekamp, 1984). As a code for high-speed data transmission with the INTELSAT V satellite the double-error-correcting and triple-error-detecting $(128, 112)$ BCH code has been selected. This is an extended version of a (127, 112) BCH code. The RS code of length 63 over $\mathbb{F}_{2^6}$ was used for correcting errors in binary data stored in a photodigital mass memory. Many thousands of lines of data are stored on a memory chip, each line representing one code word of the RS code. Dornhoff & Hohn (1978) describe a method based on cyclic codes, called the cyclic redundancy check, to correct errors in recorded data on magnetic tapes.

Bhargava, Haccoun, Matyas & Nuspl (1981) refer to some other examples of applications: Digital communication via satellite has been considered for the US Postal Service Electronic Message System, where binary error probabilities as low as 10^{-12} have been stated as goals to be achieved with appropriate error-control techniques. Long cyclic codes have been suggested for a UHF satellite system, where erasure bursts are due to radio-frequency interference with pulse lengths significantly longer than the bit transmission time. It was thought that codes of block length no greater than 128 should suffice for the US Navy's UHF demand-assignment multiple access system. In contrast to block codes, convolutional codes (which are not discussed in this text) are suitable when the message symbols to be transmitted are given serially in long sequences rather than in blocks, when block codes would be more appropriate. Several deep-space probes, such as Voyager, NASA's Pioneer missions, the INTELSAT SPADE system, and Germany's Helios have used convolutional codes. The INTELSAT SPADE system, for example, uses threshold decoding. Another decoding scheme was successfully used on a two-way digital video link with the Canadian experimental satellite Hermes, where the use of a Viterbi decoding scheme could be translated into a substantial reduction of antenna site and a smaller overall Earth station cost.

Recent texts on coding which are easy to read are, e.g., Mullen & Shiue (1993), van Lint & Wilson (1992), van Tilborg (1993), Vanstone & van Oorschot (1992), and the amusing book by Heise & Quattrocchi (1995). An excellent recent survey on Coding Theory is van Lint (1994). A common view of codes, cryptosystems (next chapter) and genetic "codes" (see §31) is given in Blakley & Borosh (1996).

5 CHAPTER Cryptology

The word cryptology stems from the Greek kryptos, "hidden," and logos, "word." Cryptology is the science of secure communications. Cryptology comprises cryptography and cryptanalysis, the former deals with methods to ensure the security, secrecy, or authenticity; the latter is concerned with methods of breaking secret messages or forging signals that will be accepted as authentic. In this chapter we will be concentrating mainly on those aspects of cryptography that rely on mathematical, in particular algebraic, techniques and tools.

The main problem areas in communication and cryptography can be highlighted by the questions: If a sender A (Alice) wants to send a message to a receiver B (Bob), how can both be certain that nobody else can receive the message? How can Alice ensure that Bob receives all of Alice's message? How can they ensure that Bob receives only Alice's message and no others? Answers to these questions are provided by studying the security, secrecy, and authenticity of message transfer.

Cryptographic work has increased rapidly during the past two decades and so has the importance of applications. We indicate a few examples of such applications in data security in computing. Often secret data are stored in computers or are exchanged between computers. Cryptographic devices can prevent that unauthorized personnel can obtain these data. Another example is represented by electronic mail and electronic transfers of money. Yet another example is designing secure phone systems. In most cases secrecy and authenticity are crucial. It is also essential to design security devices in many other instances, for example to prevent industrial espionage or sabotage of important installations such as power stations or military installations. Another application of cryptology lies in deciphering ancient secret documents written in ciphers. This branch is often called historical cryptanalysis; it is several hundred years old. Secret ciphers have existed for centuries to protect messages from being accessed by intelligent opponents. Thus the problem areas of secrecy and authenticity (i.e., protection from forgery) have very old beginnings. However, nowadays the most important aspects of cryptography are in the governmental as well as in the commercial and private domain of data security and information integrity in the widest sense. Contemporary cryptology has been characterized as a multimillion dollar problem awaiting solution and as an important question in applied mathematics.

In §21 we describe a few examples of classical (or conventional or secret key or single key) cryptography. §22 is devoted to basic concepts and examples of public key cryptography. §23 deals with discrete logarithms and cxamples of other cryptosystems.

§21 Classical Cryptosystems

All ciphers or cryptosystems prior to 1976 are sometimes referred to as conventional or classical cryptosystems. Since 1976 different types of ciphers have been developed, called public key ciphers, see §22. In this section we describe some of the pre-1976 ciphers, in particular those which are based on algebraic concepts.

Cryptography, or a cryptographic system (***cryptosystem*** or ***cipher***) consists of transforming a message (called the ***plaintext***) via an ***enciphering function*** (or ***encryption***) in such a way that only an authorized recipient can invert the transformation and restore the message. A message can be, for example, a readable text or a collection of computer data. The process of transforming the plaintext into a ciphertext (or ***cryptogram***) is called ***enciphering*** and each enciphering function E is defined by an ***enciphering algorithm***, which is common to every enciphering function of a given family, and a ***key*** k_e, which is a parameter particular to one transformation. The key is known to the sender and the (authorized) receiver but not to those from whom the plaintext is to be concealed. The key and ciphertext must determine the plaintext uniquely. Recovering the plaintext from a received ciphertext is called ***deciphering*** and is accomplished by a ***deciphering transformation*** (or function) D which is performed by a ***deciphering algorithm*** and a key k_d. For a given key, the corresponding deciphering function is the inverse of the enciphering function. An important feature of classical forms of cryptosystems is that sender and receiver use essentially the same key k. Such systems are referred to as ***single-key*** (or ***symmetric***) ***cryptosystems*** in order to distinguish them from systems using two different keys, the newer ***public-key cryptosystems*** or ***asymmetric ciphers***. The latter systems are well suited to protecting information transmitted over a computer network or for enciphering a user's private files. The security of cryptosystems lies in the large range of choice for the key k, which is known only to the sender and the receiver. It must be computationally impracticable, virtually impossible, or extremely difficult and time consuming (depending on the specific task) to find the value of the key, even if corresponding values of plaintext and cryptogram are known to an unauthorized interceptor. Thus in secure cryptosystems it should be computationally infeasible for an interceptor to determine the deciphering function, even if the plaintext corresponding to the intercepted ciphertext is known. It should also be computationally infeasible for an interceptor to systematically determine the plaintext from the intercepted ciphertext without knowing the deciphering function.

Figure 21.1 shows diagrammatically a classical cryptosystem. We note that the plaintext message usually has to be represented in a form that allows encryption via a (mathematical) algorithm or transformation. Several ways of representations will be described below. In Figure 21.1 the direct link between the encipher box and the decipher box is the insecure communication channel. However, the link between these same two boxes via the key source is assumed to be secure, i.e., no unauthorized interceptor is able to access k_e or k_d.

It is almost universally assumed in cryptography that the enemy cryptanalyst has full access to the ciphertext and knows the entire mechanism of encryption, except for the value of the secret key. In this case the cipher is designed to be secure against ***ciphertext-only attacks***. If in addition the cryptographer assumes that the cryptanalyst has access to

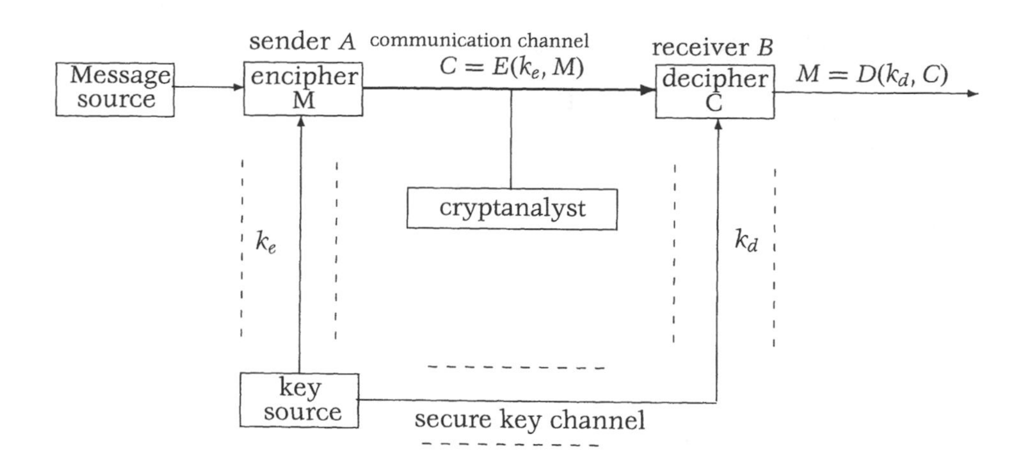

FIGURE 21.1 Classical cryptosystem.

some plaintext-ciphertext pairs formed with the secret key, then he is designing a cipher against a ***known-plaintext attack***. If the cryptographer assumes that the cryptanalyst can submit any plaintext message of her own and receive in return the correct ciphertext for the actual (still) secret key, then this is called a ***chosen-plaintext attack***. Most ciphers in use today are intended to be secure against at least a chosen-plaintext attack.

In classical cryptography two basic transformations of plaintext messages exist: ***transposition ciphers*** rearrange characters according to some rule (e.g., the plaintext is written into a matrix by rows and is sent by columns), whereas ***substitution ciphers*** replace plaintext letters (or characters) by corresponding characters of a ciphertext alphabet. If only one cipher alphabet is used, the cryptosystem is called ***monoalphabetic***. Cryptosystems in which a ciphertext letter may represent more than one plaintext letter are called ***polyalphabetic***. Such systems use a sequence of monoalphabetic cryptosystems. In deciphering unknown cryptosystems it was said by cryptanalysts that success is measured by four things in the order named: perseverance, careful methods of analysis, intuition, and luck. There are, however, some statistical and other techniques at the cryptanalyst's disposal. Often frequency analysis on the characters of a cryptogram, or on pairs or triples of characters lead to success. However, we shall not consider statistical methods in any detail.

In order to proceed from the message box to the enciphering box in Figure 21.1 we have to represent the particular cipher. Depending on the cipher, letters will be replaced by numbers as follows. In one possible representation we identify the letter A with 0, the letter B with 1, and so on until Z is being identified with 25.

Alternatively, we group together pairs of letters of a plaintext message and, say, represent the pair NO as $\mathsf{N} \cdot 26 + \mathsf{O}$, i.e., $13 \cdot 26 + 14 = 252$. In this way any two-letter block, any digraph, can be uniquely expressed as an integer $< 26^2$.

Let $\mathcal{A}$ denote the plaintext alphabet and $\mathcal{B}$ a cipher text alphabet. These alphabets may be letters of the English alphabet, or n-tuples of letters, or elements of $\mathbb{Z}_q$ (particularly $q = 2$, $q = 26$, or $q = 96$, as in some commercial systems), or elements of a finite field. We shall replace letters by numbers in $\mathbb{Z}_q$ and say that any one-to-one (or injective) mapping from $\mathcal{A}$ into $\mathcal{B}$ is called a ***key***.

The problem is to encipher a text consisting of letters in $\mathcal{A}$, i.e., a word $a_1a_2\dots a_n$, $a_i \in \mathcal{A}$. If we use the same mapping to encipher each a_i, then we use a fixed key, otherwise we use a variable key. In the latter case the key depends on some parameters, e.g., time. Time-dependent variable keys are called context free; text-dependent variable keys are called context sensitive.

A fixed key $f\colon \mathcal{A} \to \mathcal{B}$ can be extended to words with letters in $\mathcal{A}$ by considering $a_1a_2\dots a_n \mapsto f(a_1)f(a_2)\dots f(a_n)$. Similarly, a variable key $f_1, f_2, \dots$ can be extended by $a_1a_2\dots a_n \mapsto f_1(a_1)f_2(a_2)\dots f_n(a_n)$. Looking for the preimages is the process of deciphering. We shall describe several methods of enciphering: by using linear functions on $\mathbb{Z}_q$, by using matrices over $\mathbb{Z}_q$, by using permutations on $\mathbb{Z}_q$, by using scalar products, and by using big prime numbers. First we give an example of a simple substitution cipher. We verify easily that the mapping $a \mapsto an + k$ from $\mathbb{Z}_q$ into itself is injective if and only if $\gcd(n, q) = 1$.

21.1. Definition. The mapping $f\colon \mathbb{Z}_q \to \mathbb{Z}_q$; $a \mapsto an + k$, with fixed $n, k \in \mathbb{Z}_q$ and $\gcd(n, q) = 1$ is called a ***modular enciphering***; we simply speak of the key (n, k). For $n = 1$, we call the cipher a ***Caesar cipher***, for $n > 1$ and $\gcd(n, q) = 1$ the resulting cipher is called an ***affine cipher***.

The key with $n = 1$ and $k = 3$, i.e., cyclic shift of the alphabet by three letters, is of historical interest since it was used by the Roman emperor Gaius Julius Caesar. A Caesar cipher can be deciphered if only one association of an original letter with a letter in the cipher alphabet is known.

If, in an affine cipher, the ciphertext is $b = an+k$ in $\mathbb{Z}_q$, then deciphering yields the plaintext $a = (b-k)n^{-1}$ in $\mathbb{Z}_q$. An affine cipher with $k > 0$ can be deciphered as soon as two associations of plaintext and ciphertext letters are known.

21.2. Example. Let $\mathcal{A} = \mathbb{Z}_{26}$, and let the letters of the alphabet be represented as A by 0, B by 1, etc.

A	B	C	D	E	F	G	H	I	J	K	L	M	N	O	P	Q	R	S	T	U	V	W	X	Y	Z
0	1	2	3	4	5	6	7	8	9	10	11	12	13	14	15	16	17	18	19	20	21	22	23	24	25

For modular enciphering $a \mapsto 3a + 2$ in $\mathbb{Z}_{26}$ the word ALGEBRA is represented by 0 11 6 4 1 17 0, and is enciphered as 2 9 20 14 5 1 2, which in terms of letters is CJUOFBC.

21.3. Example. Suppose that the most frequently occurring letter in a (reasonably long) ciphertext of an affine cipher is the letter K and the second most frequent letter is D. Since in any (reasonably long) English text the letters E and T occur most frequently, in that order, we expect that the plaintext letter E corresponds to the ciphertext letter K, and T corresponds to D. In an affine cipher two such correspondences allow us to decipher any ciphertext letter b into a plaintext letter a. Namely,

$$a = bn' + k' \qquad \text{in } \mathbb{Z}_{26}.$$

Since

$$\begin{aligned} \mathsf{E} = \ 4 &= \mathsf{K}n' + k' = 10n' + k', \\ \mathsf{T} = 19 &= \mathsf{D}n' + k' = \ 3n' + k', \end{aligned}$$

solving these two equations in $\mathbb{Z}_{26}$ gives us $n' = 9$ and $k' = 18$.

To sum up modular enciphering or affine ciphers we note that for enciphering we use the key n, k since ciphertext b is obtained from plaintext a via $b = an + k \bmod q$. For deciphering the relevant formula for obtaining a plaintext a from a ciphertext b is $a = (b-k)n^{-1} \bmod q$, i.e., $a = bn^{-1} - kn^{-1} \bmod q$. Here the deciphering key is n^{-1} and $-kn^{-1}$, which mod q can be readily obtained from n and k. Thus we use essentially the same key for enciphering and deciphering and thus deal with a single key cryptosystem.

Polyalphabetic substitution ciphers conceal letter frequencies by using multiple substitutions.

21.4. Definition. A ***periodic substitution cipher*** with period p consists of p cipher alphabets $B_1, \ldots, B_p$ and the keys $f_i\colon A \to B_i$, $1 \leq i \leq p$. A plaintext message $m = m_1 \ldots m_p m_{p+1} \ldots m_{2p} \ldots$ is enciphered as $f_1(m_1) \cdots f_p(m_p) f_1(m_{p+1}) \ldots f_p(m_{2p}) \ldots$ by repeating the sequence of functions $f_1, \ldots, f_p$ every p characters.

A popular example of a periodic substitution cipher is the ***Vigenère cipher***, falsely ascribed to the sixteenth-century French cryptologist Claise de Vigenère (see Kahn (1967)). In this cipher the alphabets are $\mathbb{Z}_{26}$ and the key is given by a sequence of letters (often a key word $k = k_1 \ldots k_p$, where k_i indicates the cyclic shift in the ith alphabet). So $f_i(m) = m + k_i \bmod 26$. A Vigenère square can be used for enciphering, a special square with the letters A, B,..., Z of the alphabet as first row and column.

$$\begin{matrix} \mathsf{A} & \mathsf{B} & \cdots & \mathsf{Y} & \mathsf{Z} \\ \mathsf{B} & \mathsf{C} & \cdots & \mathsf{Z} & \mathsf{A} \\ \vdots & \vdots & \ddots & \vdots & \vdots \\ \mathsf{Y} & \mathsf{Z} & \cdots & \mathsf{W} & \mathsf{X} \\ \mathsf{Z} & \mathsf{A} & \cdots & \mathsf{X} & \mathsf{Y} \end{matrix}$$

A periodic column method of period p can be described by a simple key word of length p. Let GALOIS be a key word, thus $p = 6$, and then the keys f_i are given by the following simple tableau:

A	B	C	⋯	
G	H	I	⋯	$= f_1(\mathsf{A})f_1(\mathsf{B})f_1(\mathsf{C})\ldots,$
A	B	C	⋯	$= f_2(\mathsf{A})f_2(\mathsf{B})f_2(\mathsf{C})\ldots,$
L	M	N	⋯	$= f_3(\mathsf{A})f_3(\mathsf{B})f_3(\mathsf{C})\ldots,$
O	P	Q	⋯	$= f_4(\mathsf{A})f_4(\mathsf{B})f_4(\mathsf{C})\ldots,$
I	J	K	⋯	$= f_5(\mathsf{A})f_5(\mathsf{B})f_5(\mathsf{C})\ldots,$
S	T	U	⋯	$= f_6(\mathsf{A})f_6(\mathsf{B})f_6(\mathsf{C})\ldots.$

In this case the word ALGEBRA is enciphered as GLRSJJG $= f_1(\mathsf{A})f_2(\mathsf{L})f_3(\mathsf{G})f_4(\mathsf{E})f_5(\mathsf{B})f_6(\mathsf{R})f_1(\mathsf{A})$. We note that such a column method is essentially a sequence $f_1, f_2, \ldots$ of period 6 Caesar ciphers on $\mathbb{Z}_{26}$.

Two more examples of substitution ciphers on $\mathbb{Z}_q$ are obtained from the sequence (f_i), $i = 1, 2, \ldots$, where f_i is applied to the ith letter a_i in the plaintext and f_i is given as

$$f_i\colon \mathbb{Z}_q \to \mathbb{Z}_q; \qquad a \mapsto b_i := ka + d_i,$$

where $\gcd(k, q) = 1$, and

$$d_i = ca_{i-1} \quad \text{with } c, a_0 \text{ given,} \tag{21.1}$$

or

$$d_i = cb_{i-1} \quad \text{with } c, b_0 \text{ given.} \tag{21.2}$$

21.5. Example.

(i) If we number the letters from **A** to **Z** as in Example 21.2 then **ALGEBRA**, i.e., $0, 11, 6, 4, 1, 17, 0 = a_1a_2a_3a_4a_5a_6a_7$, is enciphered according to 21.1 by means of $f_i(a) = 3a + a_{i-1}$ and $a_0 := 4$ as follows:

$$\begin{aligned} f_1(0) &= f_1(a_1) = 3a_1 + 4 = 4, & f_5(1) &= 7,\\ f_2(11) &= f_2(a_2) = 3a_2 + 0 = 7, & f_6(17) &= 0,\\ f_3(6) &= f_3(a_3) = 3a_3 + 11 = 3, & f_7(0) &= 17.\\ f_4(4) &= f_4(a_4) = 3a_4 + 6 = 18, \end{aligned}$$

Thus **ALGEBRA** is enciphered as $4, 7, 3, 18, 7, 0, 17$, i.c., **EHDSHAR**.

(ii) If we use (21.2) with $f_i(a) = 3a + f_{i-1}(a_{i-1})$ and $f_0(a_0) := 4$, we obtain

$$\begin{aligned} f_1(0) &= f_1(a_1) = 3a_1 + f_0(a_0) = 0 + 4 = 4, & f_5(1) &= 18,\\ f_2(11) &= f_2(a_2) = 3a_2 + f_1(a_1) = 7 + 4 = 11, & f_6(17) &= 17,\\ f_3(6) &= f_3(a_3) = 3a_3 + f_2(a_2) = 18 + 11 = 3, & f_7(0) &= 17.\\ f_4(4) &= f_4(a_4) = 3a_4 + f_3(a_3) = 12 + 3 = 15, \end{aligned}$$

Thus **ALGEBRA** is enciphered as $4, 11, 3, 15, 18, 17, 17$, i.e., **ELDPSRR**.

The Vigenère cipher was considered unbreakable in its day, but it has been successfully cracked in the nineteenth century due to work by Kasiski. As in the Caesar cipher, a shift is applied to the alphabet, but the length of the shift varies, usually in a periodic way. For example, our opponent might decide to use shifts of lengths 1, 7, 4, 13, 5 over and over again as a running key. He then writes the sequence

$$1, 7, 4, 13, 5, 1, 7, 4, 13, 5, 1, 7, 4, 13, 5, \ldots$$

(call this the ***key*** sequence) for as long as necessary and "adds" it to the message, say

Message	S	E	N	D	M	O	R	E	M	E	N	A	N	D	M	O	R	E	A	R	M	S
Key Sequence	1	7	4	13	5	1	7	4	13	5	1	7	4	13	5	1	7	4	13	5	1	7
Cryptogram	T	L	R	Q	R	P	Y	I	Z	J	O	H	R	Q	R	P	Y	I	N	W	N	Z.

The changing shifts even out the overall letter frequencies, defeating the kind of analysis used to break Caesar ciphers, but the characteristic frequencies are retained in subsequences of the enciphered message corresponding to repetition in the key sequence (every five places in the above example). If we can find the length of the key's period, letters can be identified by frequency analysis.

The period can indeed be discovered, by looking for repeated blocks in the enciphered message. Some of these will be accidental, but a large proportion will result from matches between repeated words or subwords of the message and repeated blocks in the key sequence. When this happens, the distance between repetitions will be a multiple of the period. In our example, the block **RQRPYI** is undoubtedly a true repeat; the distance between its two occurrences is 10, indicating that the period length is 10 or 5. Examining all the repeats in a longer enciphered message we will find a majority at distances which are multiples of 5, at which time we will know that the period is 5.

The ultimate generalization of the Vigenère cipher was proposed by the American engineer G. S. Vernam. Let the key sequence be arbitrarily long and *random*, and use successive blocks of it for successive messages. If we assume an upper bound N on the length of all possible messages, then we take the number of keys to be at least as large as N. All keys are equally likely. A commonly used method is a modulo 26 adder (or a mod 2 adder for implementation in microelectronics). If the message is $\mathbf{m} = m_1 m_2 \dots m_n$ and each m_i is represented by one of the integers from 0 to 25 and each k_i is a number between 0 and 25, then the resulting cryptogram is $\mathbf{c} = c_1 c_2 \dots c_n$ where $c_i = m_i + k_i \pmod{26}$. This system is called a ***one-time pad*** and its name is derived from the fact that for enciphering we utilize written pads of random numbers to obtain the sequence $k_1 k_2 \dots k_n$ and we use each key tape only once. This is a cumbersome method, because both sender and receiver need to keep a copy of the long key sequence, but it is clearly unbreakable—the randomness of the key means that any two message sequences of the same length are equally likely to have produced the message, the cryptogram. It is said that a one-time pad was used for the hot line between Washington and Moscow.

The increase in security from Caesar cipher to one-time pad depends on increasing the length of the key. For a Caesar cipher the key is a single number between 1 and 26 (the length of shift), for a periodic Vigenère cipher it is a finite sequence of numbers, and for the one-time pad it is a potentially infinite sequence. The longer the key, the harder the cipher is to break, but for all the classical ciphers it is possible for an opponent to reconstruct the key by an amount of work which does not grow too exorbitantly, relative to key size.

Next we consider matrix methods based on investigations by Hill (1929). We note that a mapping from the space $(\mathbb{Z}_q)^n$ into itself, defined by

$$\mathbf{b} = \mathbf{Ka} + \mathbf{d} \quad \text{with } \mathbf{K} \in \mathbb{M}_n(\mathbb{Z}_q), \quad \mathbf{d} \in (\mathbb{Z}_q)^n,$$

is injective if and only if $\gcd(\det \mathbf{K}, q) = 1$. Therefore we can generalize modular enciphering as follows.

21.6. Definition. A ***Hill cipher*** with key matrix $\mathbf{K}$ is a mapping from $(\mathbb{Z}_q)^n$ into itself, defined by $\mathbf{a} \mapsto \mathbf{Ka} + \mathbf{d}$, such that $\gcd(\det \mathbf{K}, q) = 1$.

For enciphering we subdivide the plaintext into blocks of n letters each, replace each letter by its corresponding element in $\mathbb{Z}_q$, and write this in columns. Then we apply the given linear transformation to each block $\mathbf{a}$. In this context the so-called ***involutory*** (or self-inverse) matrices $\mathbf{K}$, defined by $\mathbf{K}^2 = \mathbf{I}$ or $\mathbf{K} = \mathbf{K}^{-1}$, are convenient to use since in deciphering we do not have to evaluate $\mathbf{K}^{-1}$ separately. A cryptogram which is constructed according to 21.6 can be deciphered if a plaintext word of sufficient minimal length (depending on n) is known. For $n = 3$ and $q = 26$ we can show that such a word of length ≥ 4 is sufficient. For $n = 2$ such a cryptogram can be deciphered even if no word of minimal length is known in the plaintext. In this case all possible involutory 2×2 matrices mod q are determined and tried. (There are 736 such matrices.) For larger n, deciphering by this trial-and-error method is lengthy, since, e.g., there are 22 involutory 3×3 matrices mod 2, and 66,856 such matrices mod 13, and therefore 1,470,832 involutory 3×3 matrices mod 26.

21.7. Example. Let the 26 letters of the alphabet be represented by the integers from 0 to 25. Using the involutory matrix

$$\mathbf{K} = \begin{pmatrix} 4 & 7 \\ 9 & 22 \end{pmatrix} \in (\mathbb{Z}_{26})_2^2$$

and taking $\mathbf{d} = \mathbf{0}$ in 21.6 we form

$$\begin{pmatrix} b_1 \\ b_2 \end{pmatrix} = \mathbf{K} \begin{pmatrix} a_1 \\ a_2 \end{pmatrix}. \tag{21.3}$$

Here a_1, a_2 are the numbers representing letters in the plaintext, which has been divided into blocks of two. If the plaintext is CRYPTOLOGY, we divide this into pairs, CR YP TO LO GY; in case of an odd number of letters we add an arbitrary letter. To each pair of letters corresponds a pair a_1, a_2 in (21.3) which determines b_1, b_2. Since CR $\leftrightarrow a_1a_2 = 2\ 17$, we have

$$\begin{pmatrix} b_1 \\ b_2 \end{pmatrix} = \begin{pmatrix} 4 & 7 \\ 9 & 22 \end{pmatrix} \begin{pmatrix} 2 \\ 17 \end{pmatrix} = \begin{pmatrix} 23 \\ 2 \end{pmatrix} = \begin{pmatrix} \mathsf{X} \\ \mathsf{C} \end{pmatrix},$$

i.e., CR is enciphered as XC, etc.

These methods are fixed transposition methods. Variable methods can be obtained which are similar to (21.1) and (21.2) (preceding 21.5) by subdividing the plaintext into blocks, each of which contains n letters, such that by substitution with elements in $\mathbb{Z}_q$, the plaintext is of the form

$$\underbrace{a_{11}a_{12}\dots a_{1n}}_{\mathbf{a}_1^{\mathrm{T}}}\underbrace{a_{21}a_{22}\dots a_{2n}}_{\mathbf{a}_2^{\mathrm{T}}}\underbrace{\dots}_{\dots} =: \mathbf{a}_1^{\mathrm{T}}\mathbf{a}_2^{\mathrm{T}}\dots.$$

A variable matrix method is obtained by letting $f_m(\mathbf{a}) := \mathbf{b}_m := \mathbf{K}\mathbf{a} + \mathbf{d}_m$, where $\gcd(\det \mathbf{K}, q) = 1$, and

$$\mathbf{d}_m = \mathbf{C}\mathbf{a}_{m-1} \quad \text{with given } \mathbf{C} \in \mathbb{M}_n(\mathbb{Z}_q) \text{ and } \mathbf{a}_0 \in (\mathbb{Z}_q)^n, \tag{21.4}$$

$$\mathbf{d}_m = \mathbf{C}\mathbf{b}_{m-1} \quad \text{with given } \mathbf{C} \in \mathbb{M}_n(\mathbb{Z}_q) \text{ and } \mathbf{b}_0 \in (\mathbb{Z}_q)^n. \tag{21.5}$$

21.8. Example. A variable key according to (21.4) for enciphering CRYPTOLOGY is given by

$$\mathbf{K} = \begin{pmatrix} 1 & 2 & 3 \\ 2 & 5 & 6 \\ 1 & 2 & 4 \end{pmatrix}, \quad \mathbf{C} = \begin{pmatrix} 4 & 1 & 1 \\ 2 & 0 & 3 \\ 1 & 2 & 0 \end{pmatrix}, \quad \mathbf{a}_0 = \begin{pmatrix} 1 \\ 2 \\ 3 \end{pmatrix},$$

$$\begin{pmatrix} b_{i1} \\ b_{i2} \\ b_{i3} \end{pmatrix} = \begin{pmatrix} 1 & 2 & 3 \\ 2 & 5 & 6 \\ 1 & 2 & 4 \end{pmatrix} \begin{pmatrix} a_{i1} \\ a_{i2} \\ a_{i3} \end{pmatrix} + \begin{pmatrix} 4 & 1 & 1 \\ 2 & 0 & 3 \\ 1 & 2 & 0 \end{pmatrix} \begin{pmatrix} a_{i-1,1} \\ a_{i-1,2} \\ a_{i-1,3} \end{pmatrix}. \tag{21.6}$$

Again, we use the simple representation A $\leftrightarrow 0$, B $\leftrightarrow 1$, etc., as in 21.2. For the first group of three letters we obtain

C $\leftrightarrow 2$, R $\leftrightarrow 17$, Y $\leftrightarrow 24$, i.e., CRY $= (a_{i1}, a_{i2}, a_{i3}) = (2, 17, 24)$. Thus

$$\begin{pmatrix} b_{i1} \\ b_{i2} \\ b_{i3} \end{pmatrix} = \begin{pmatrix} 1 & 2 & 3 \\ 2 & 5 & 6 \\ 1 & 2 & 4 \end{pmatrix} \begin{pmatrix} 2 \\ 17 \\ 24 \end{pmatrix} + \begin{pmatrix} 4 & 1 & 1 \\ 2 & 0 & 3 \\ 1 & 2 & 0 \end{pmatrix} \begin{pmatrix} 1 \\ 2 \\ 3 \end{pmatrix} = \begin{pmatrix} 13 \\ 10 \\ 7 \end{pmatrix} = \begin{pmatrix} \mathsf{N} \\ \mathsf{K} \\ \mathsf{H} \end{pmatrix}.$$

Deciphering means determining the inverse transformation of the transformation (21.6).

Digraphs, or two-letter blocks, xy over an alphabet with N letters can be represented in a one-to-one correspondence by the N^2 integers $xN + y$. The corresponding cipher of digraphs has an enciphering transformation

$$C \equiv a\mathcal{P} + b \bmod N^2$$

to obtain the ciphertext C from the plaintext $\mathcal{P}$ where $\mathcal{P}$ and C are digraphs. Similarly, the deciphering transformation is

$$\mathcal{P} \equiv a'C + b' \bmod N^2.$$

21.9. Example. Let $N = 26$, $a = 15$ and $b = 49$. The digraph GO corresponds to the integer $\mathsf{G} \cdot N + \mathsf{O} = 6 \cdot 26 + 14 = 170$. The ciphertext for this plaintext 170 is $a\mathcal{P} + b = 15 \cdot 170 + 49 \equiv 571 \bmod 676$, thus $C = 571$.

Breaking a digraph cipher is based on analyzing the frequency of two-letter blocks in (a reasonably long) ciphertext and comparing it with the number of two-letter blocks in an English text. In sufficiently long English texts the blocks TH and HE are the most frequently occurring digraphs.

21.10. Example. Suppose the cryptanalyst knows that a digraph cipher has been used and that the two most frequent digraphs in the ciphertext are EY and GY. Hence the cryptanalyst tests the hypothesis that the ciphertexts EY and GY correspond to the most frequent plaintext digraphs TH and HE, respectively. In order to find the deciphering key he solves $\mathcal{P} = a'C + b' \bmod 26^2$.

$$\mathsf{TH} = 19 \cdot 26 + 7 \equiv 501, \qquad \mathsf{HE} = 7 \cdot 26 + 4 \equiv 189,$$
$$\mathsf{EY} = 4 \cdot 26 + 24 \equiv 128, \qquad \mathsf{GY} = 6 \cdot 26 + 24 = 180.$$

Then

$$501 \equiv 128a' + b' \quad \text{and} \quad 189 \equiv 180a' + b' \bmod 676$$

has a solution $a' = 631$ and $b' = 177$.

Now suppose the cryptanalyst wants to decipher the digraphs WZ and VZ. Numerically they are $22 \cdot 26 + 25 \equiv 597$ and $21 \cdot 26 + 25 \equiv 571 \bmod 676$. Hence $a'C + b' = \mathcal{P}$ is $631 \cdot 597 + 177 \equiv 352$ and $631 \cdot 571 + 177 \equiv 170 \bmod 676$. We represent the plaintext numbers as digraphs: $352 = 13 \cdot 26 + 14$ and $170 = 6 \cdot 26 + 14$. Therefore the plaintext digraphs are NO and GO.

We now describe permutation keys. Let $\mathbb{F}_q$ be a finite field with $q = p^n$ elements and let ζ be a primitive element of $\mathbb{F}_q$. Each element of $\mathbb{F}_q$ can be expressed in the form

$$a = (c_0, \ldots, c_{n-1}) = c_0\zeta^{n-1} + c_1\zeta^{n-2} + \cdots + c_{n-1}, \qquad c_i \in \mathbb{F}_p.$$

$\mathbb{F}_q$ consists of $\{0, \zeta, \ldots, \zeta^{q-1}\}$. The primitive element is a zero of a primitive irreducible polynomial f over $\mathbb{F}_p$. We denote the integer

$$c_0 a^{n-1} + c_1 a^{n-2} + \cdots + c_{n-1}, \qquad c_i \in \mathbb{F}_p,$$

as the ***field-value*** of the element a in $\mathbb{F}_q$.

21.11. Example. Let $f = x^3 + 2x + 1$ be a primitive polynomial with zero ζ in $\mathbb{F}_{3^3}$. The elements of $\mathbb{F}_{3^3}$ are given in the following table. Column (1) will be used in the enciphering of plaintexts. Column (2) represents the field values of the elements in $\mathbb{F}_{3^3}$. In column (3) we describe the elements of $\mathbb{F}_{3^3}$ in vector form $c_0, c_1, \ldots, c_{n-1}$. Column (4) gives the exponents of the powers ζ^i of the elements of the cyclic group $\mathbb{F}_{3^3}^*$.

(1)	(2)	(3)	(4)	(1)	(2)	(3)	(4)
A	0	000	0	O	14	112	11
B	1	001	26	P	15	120	4
C	2	002	13	Q	16	121	18
D	3	010	1	R	17	122	7
E	4	011	9	S	18	200	15
F	5	012	3	T	19	201	25
G	6	020	14	U	20	202	8
H	7	021	16	V	21	210	17
I	8	022	22	W	22	211	20
J	9	100	2	X	23	212	5
K	10	101	21	Y	24	220	23
L	11	102	12	Z	25	221	24
M	12	110	10	␣	26	222	19
N	13	111	6				

21.12. Definition. A polynomial $g \in \mathbb{F}_q[x]$ is called a ***permutation polynomial*** of $\mathbb{F}_q$ if the polynomial function $\bar{g}$ induced by g is a permutation of $\mathbb{F}_q$.

If A is an alphabet with q letters, then a key can be defined as follows. First we determine a one-to-one correspondence between the q letters of A and the q letters of $\mathbb{F}_q$. If $P_1P_2P_3\ldots$ is a plaintext, then let $a_1a_2a_3\ldots$ be the corresponding sequence of elements in $\mathbb{F}_q$. This sequence will be mapped into $\bar{g}(a_1)\bar{g}(a_2)\bar{g}(a_3)\ldots$ by using a permutation polynomial g over $\mathbb{F}_q$. The elements $\bar{g}(a_i)$ are represented as letters of A and form a cryptogram. If $f\colon A \to \mathbb{F}_q$ is bijective and $P \in A$, then

$$P \mapsto f(P) \mapsto g(f(P)) \mapsto f^{-1}(g(f(P)))$$

is a key $f^{-1} \circ g \circ f$. If we use permutation polynomials g_i in enciphering, we obtain a variable key.

21.13. Example (*21.11 continued*). Columns (1) and (3) give the association between the letters of the natural alphabet and $\mathbb{F}_{3^3}$, and ⊔ denotes the twenty-seventh letter. Thus A = 000, H = 021, etc. Let $g_i = a_i x + b_i$, $a_i \neq 0$, and let a_i and b_i be determined by

$$a_{i+2} = a_{i+1}a_i \quad \text{and} \quad b_{i+2} = b_{i+1} + b_i, \qquad i = 1, 2, \ldots,$$

with initial values

$$a_1 = 021 = \zeta^{16}, \quad a_2 = 111 = \zeta^6, \quad b_1 = 002, \quad b_2 = 110.$$

The enciphering of the word GALOIS is given by

(a)	G	A	L	O	I	S
(b)	020	000	102	112	022	200
(c)	14	0	12	11	22	15
(d)	16	6	22	2	24	26
(e)	120	111	202	002	211	200
(f)	002	110	112	222	001	220
(g)	122	221	011	221	212	111
(h)	R	Z	E	Z	X	N

The rows can be calculated as follows: (a) plaintext; (b) vector form; (c) exponents of the elements $\alpha = \zeta^t$; (d) exponents t of the powers ζ^t of a_i; (e) vector form of the products $a_i\alpha_i$, i.e., (c) + (d) mod 26; (f) vector form of b_i; (g) vector form of $y_i = a_i\alpha_i + b_i$ of (e)+(f); (h) cipher description of (a).

Exercises

1. Decipher **AOPZ TLZZHNL PZ H MHRL**.
2. Use the word **ALGEBRA** as a key word (sequence) in a Vigenère cipher to encipher **CRYPTOGRAPHY**.
3. In a Hill cipher mod 26, suppose the plaintext **HELP** is enciphered as **HIAT**. Determine the 2×2 key matrix of the cipher.
4. How many pairs of letters P_1P_2 remain unchanged when the pairs are enciphered as C_1C_2 in the following way:

   (i) $C_1 \equiv 6P_1 + 5P_2 \pmod{26}$.
   $C_2 \equiv 5P_1 + P_2 \pmod{26}$.

   (ii) $C_1 \equiv 4P_1 + 17P_2 \pmod{26}$.
   $C_2 \equiv P_1 + 11P_2 \pmod{26}$.

   (iii) $C_1 \equiv 3P_1 + 5P_2 \pmod{26}$.
   $C_2 \equiv 6P_1 + 3P_2 \pmod{26}$.

5. Find the complete enciphering of the word **CRYPTOLOGY** in Example 21.8 and decipher the first three letters of that ciphertext.
6. The Playfair cipher utilizes a so-called ***Polybius square***. The alphabet is written into the square (with **I-J** always combined), usually led off by a key word (**PLAYF(A)IR** in this example):

P	L	A	Y	F
I	R	B	C	D
E	G	H	K	M
N	O	Q	S	T
U	V	W	X	Z

   To prepare the plaintext for enciphering, it is divided into digraphs. Where double letters could occur in the same digraph, a prearranged letter is inserted between the pair. Enciphering is performed for each digraph by using the Polybius square with the following rules:

   (i) if the two letters in the plaintext digraph appear in the same line, substitute for each one the letter immediately to its right;

   (ii) if the two letters appear in the same column, substitute for each the letter immediately below it; here the first column (row) is considered to be to the right (below) of the last column (row);

(iii) if the two letters are not in the same row or column, substitute for them the letters at the remaining two corners of the rectangle or square which they form, using first the letter on the same row as the first plaintext letter.

Encipher: CRYPTO ENIGMA.

(Deciphering simply uses the reverse of the enciphering steps.)

7. Decipher the following ciphertext assuming that it was obtained by a transformation of the form $x \to ax + b \pmod{26}$ from a plaintext in English:
R FX SNO FGYFRQ NG ONXNYYNZ GNY R WFEL TLLS PLTOLYQFP FSQ R CNEL ONQFP.

Note. The frequency of the letters of the alphabet in a sample of 1000 letters in an English article is as follows:
E 130, T 93, N 78, R 77, I 74, O 74, A 73, S 63, D 44, H 35, L35, C 30, F 28, P 27, U 27, M 25, Y 19, G 16, W 16, V 13, B 9, X 5, K 3, Q 3, J 2, Z 1.

8. Determine the values a which will make the matrix
$$\begin{pmatrix} 3 & 4 \\ a & 23 \end{pmatrix}$$
involutory over $\mathbb{Z}_{26}$.

9. Encipher ALGEBRA IS FUN by using mod 26 the 2×2 key matrix
$$\begin{pmatrix} 2 & 23 \\ 3 & 1 \end{pmatrix},$$
and decipher the resulting ciphertext.

10. Find the determinant of
$$\begin{pmatrix} 1 & 0 & 2 \\ 2 & 3 & 2 \\ 0 & 1 & 0 \end{pmatrix},$$
where entries are in $\mathbb{Z}_q$. Also find the inverse of this matrix for odd q.

11. (i) Prove that x^k is a permutation polynomial of $\mathbb{F}_q$ if and only if $\gcd(k, q-1) = 1$.

(ii) Given that $x^5 + x^3 + 2x$ is a permutation polynomial of $\mathbb{F}_{27}$, use this polynomial to encipher the word GALOIS according to the procedure outlined before and in Example 21.13.

12. A Hill cipher can also be constructed by using blocks of four message characters $abcd$ and forming a 2×2 matrix
$$\mathbf{M} = \begin{pmatrix} a & b \\ c & d \end{pmatrix},$$

which is enciphered as the matrix product **KM** with a Hill key **K**. Suppose letters A through Z are denoted by 0 through 25, a blank by 26, a period by 27, and a comma by 28. Suppose the key

$$\mathbf{K} = \begin{pmatrix} 1 & 2 \\ 3 & 4 \end{pmatrix}$$

was used for enciphering to obtain P␣X.AWDA. Decipher this cryptogram.

13. Suppose that for enciphering digraphs mod 26 the matrix $\begin{pmatrix} 1 & 7 \\ 5 & 8 \end{pmatrix}$ is used and the resulting digraph is enciphered again by using the matrix $\begin{pmatrix} 2 & 3 \\ 5 & 8 \end{pmatrix}$. Construct a single matrix that has the same effect as enciphering by two matrices.

§22 Public Key Cryptosystems

Unlike single key cryptosystems of the previous section, a different type of cipher, the public key cryptosystems, is based on the observation that the enciphering procedure and key do not have to be the same as the deciphering procedure and key. One of the advantages of public key cryptosystems is that they allow us to reduce considerably the difficulty that all key information must be kept secret and must be communicated or agreed to via a secure key channel between sender and receiver. In public key ciphers part of the information about keys (the public key) can be made public. This assists, for example, in the distribution of keys in large organizations for many users of the cipher.

Another advantage of public key ciphers is that several of these ciphers provide methods of message authentication and digital signatures.

We consider a diagram for public key ciphers, Figure 22.1, in comparison with Figure 21.1 for classical (single key) cryptosystems.

In Figure 22.1 the cryptanalyst has (unauthorized) access to the ciphertext C and the (public) key k_e, but not to the (private) key k_d for deciphering by the receiver.

In public key cryptosystems, the desirable features of the cipher in terms of an enciphering method E and a deciphering method D are:

1. Deciphering an enciphered message M yields M, i.e.,

$$D(E(M)) = M,$$

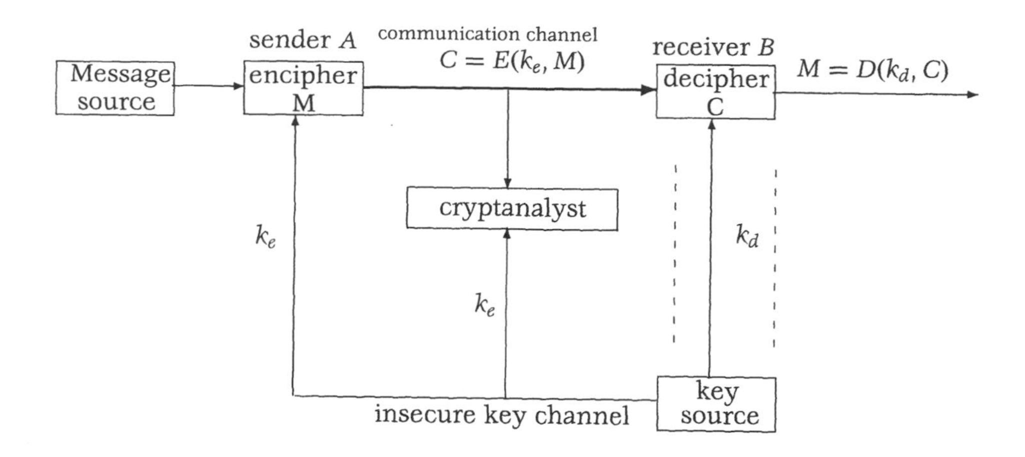

FIGURE 22.1 Public key cryptosystems.

but without knowledge of D it is practically impossible to decipher messages enciphered with E.

2. E and D can be easily computed.
3. E can be made public without revealing D, thus knowing E does not enable one to compute D.
4. Enciphering a deciphered message M yields M, i.e.,

$$E(D(M)) = M.$$

Public key cryptosystems thus eliminate the need for users of the system to agree beforehand on a common secret key, by using instead some initial interaction and trust. In 1976 Diffie & Hellman (1976) introduced these ciphers and called them public key cryptosystems. During the following years several examples of such ciphers were developed, some of them were found insecure, the security of others remains an open question.

For the transmission of messages with a public key cryptosystem the users have to adhere to a sequence of steps, called the protocol. Suppose a user B generates his enciphering and deciphering algorithms (and keys) E_B and D_B, respectively, and that a user A wants to send him a private message and tell him so. Then B places E_B in a public "directory" (like a telephone number, or an internet address). User A enciphers her message to B with E_B, which she can easily access. User B deciphers the cryptogram with D_B, which only he knows. Thus no one else can understand the message so exchanged. In this protocol A must trust the authenticity of B's public enciphering transformation E_B. If an impostor C substitutes B's public key E_B with his own, then A will end up telling C her private message, which was intended for B.

Therefore it is often preferable to use a trusted key management center that has responsibility (and trust) for updating and publicizing a directory of correct enciphering transformations/keys. A disadvantage of such a system is that it can be abused by criminals who can conduct their illegal business by exchanging information in secrecy and with great convenience. New technologies are under consideration to protect society by using "fair public key cryptosystems," see §23.

An advantage of public key ciphers is that several such ciphers allow us to use digital signatures for authentication purposes. User B "signs" messages he intends to send to another user A. To sign a message M, B first computes the digital signature $S = D_B(M)$ with his secret deciphering method D_B. Then B uses A's public enciphering method E_A, forms $E_A(S)$, and sends it to A together with a plaintext message of which he is the author. The receiver A will be able to check the authenticity, as we shall see in examples. Since A is the only user who knows the deciphering procedure D_A, she first computes $D_A(E_A(S)) = S$, thus obtaining the signature S, and then uses E_B from the public directory to obtain the message M as $E_B(S) = E_B(D_B(M)) = M$. Since only B can have created an S which deciphers to M by E_B, A knows that M can only have come from B.

(Here we are assuming that only a tiny fraction of symbol sequences actually are meaningful messages M, as is the case for sequences of letters in the English alphabet. Then any forgery S' is likely to be detected, because of the minuscule probability that $E_B(S')$ will be a meaningful message M'. This protection against forgery is analogous to the way error correcting codes give protection against random errors, namely, by having only a small fraction of possible binary blocks actually in the code.)

As a concrete example, suppose a test ban treaty for monitoring nuclear tests between country A and country B proposes that each nation places seismic instruments in each other's territory, to record any disturbances and hence detect underground tests. It is possible to protect the instruments (in the sense that they can be made to self-destruct if anyone tampers with them), but not the channel which sends their information (the host nation could cut the wires and send false information). Furthermore, if information is sent in ciphered form, the host nation may suspect that unauthorized information is also being sent, in addition to the agreed-on seismic data.

A digital signature system is the ideal solution to this problem. Nation B's seismic station contains a computer which converts the message M to $S = D_B(M)$. Nation A cannot substitute any S' for S, because of the overwhelming probability that $E_B(S') = M'$ will not be meaningful, and hence nation B will detect the forgery. However, nation B *can* supply A with the procedure E_B, which A can then use to recover $M = E_B(S)$, and thus be reassured that only an authorized message M is being sent.

The trapdoor one-way function is crucial for the construction of public key cryptosystems: An injective function $f(x)$ is called a ***one-way function*** if $f(x)$ can be computed for given x, but for almost all values of y in the range of f it is computationally infeasible or impracticable to find x such that $y = f(x)$, that is, the computation of the inverse of $f(x)$ can be said to be impracticable. f is a ***trapdoor*** one-way function if it is easy to compute its inverse given additional information, namely the secret deciphering key. One-way functions must lack all obvious structure in the table of $f(x)$ against x. Examples of such functions will be considered below.

We recall from elementary number theory that the reduced set of residues mod n is the subset of $\{0, 1, \dots, n-1\}$ consisting of integers relatively prime to n. Euler's φ-function $\varphi(n)$ gives the number of elements in the reduced set of residues mod n. For a prime p, $\varphi(p) = p - 1$, and for distinct primes p and q it can be easily verified (see 13.15) that $\varphi(pq) = (p-1)(q-1)$. Euler's Theorem says that $a^{\varphi(n)} \equiv 1 \pmod{n}$. This generalizes Fermat's Little Theorem: For p prime and a relatively prime to p we have $a^{p-1} \equiv 1 \pmod{p}$.

Rivest, Shamir & Adleman (1978) described a public key cryptosystem in 1978, the RSA cryptosystem, which is still the best known public key cryptosystem whose security seems to be intact. It is based on the power function as a one-way function. We note that if $\gcd(m, n) = 1$ and $st \equiv 1 \pmod{\varphi(n)}$, then $m \equiv c^t \pmod{n}$ is the inverse of $c \equiv m^s \pmod{n}$.

22.1. RSA Cryptosystem.

(i) Select two large primes p and q, and let $n = pq$.

(ii) Select a large random integer d, such that $\gcd(d, \varphi(n)) = 1$. Note that $\varphi(n) = (p-1)(q-1)$.

(iii) Determine the unique integer e, $1 \leq e < \varphi(n)$, such that $ed \equiv 1 \pmod{\varphi(n)}$. Note that $\gcd(e, \varphi(n)) = 1$.

(iv) Make e and n public, as the public key, keep d and the primes p and q secret, the private key.

(v) Represent the plaintext as a sequence of integers m, $0 \leq m \leq n-1$.

(vi) Enciphering: $c \equiv m^e \pmod{n}$.

(vii) Deciphering: $m \equiv c^d \pmod{n}$.

22.2. Theorem. *Given the public key $\{e, n\}$ and the private key d (and p, q) in an RSA cryptosystem, then for any plaintext message m, $0 \leq m \leq n-1$, and for the enciphering transformation $E(m) = c \equiv m^e \pmod{n}$ the deciphering transformation is $D(c) = c^d \equiv m \pmod{n}$ and $D(E(m)) \equiv m \pmod{n}$.*

Details of the easy proof are omitted, but we note that if $c \equiv m^e \pmod{n}$, then $c^d \equiv m^{ed} \pmod{n}$. Since $ed \equiv 1 \pmod{\varphi(n)}$, there is a k such that $ed = 1 + k\varphi(n)$. Then by Euler's Theorem

$$m^{ed} = m^{1+k\varphi(n)} = m(m^{\varphi(n)})^k \equiv m \pmod{n}.$$

The crucial aspect in the RSA cryptosystem is that n is the product of two large prime numbers of approximately equal size and that factoring large integers is infeasible. The message gets split up into blocks and each block, regarded as an integer, must lie between 0 and $n-1$, since we try to perform an encryption (mod n). The method is based on the fact that for large n it is virtually impossible to find d such that $ed \equiv 1 \pmod{\varphi(n)}$, without knowing the factorization of n. This is a generalization of Fermat's Theorem. Only the authorized receiver knows d and this factorization of n into p and q. The number d is the inverse of e, being in the prime residue class group modulo n of order $\varphi(n) = (p-1)(q-1)$. We use the extended form of the Euclidean algorithm to find e. This is easy to do if we know p and q, but it is extremely difficult or computationally infeasible if we only know n, for large n. The unauthorized decipherer thus has the problem of factoring a number into prime factors. If n has 200 digits and is as described above, even the fastest computer will normally need years in order to factor n. It is believed that the RSA cryptosystem will be secure in future if we use a well-chosen sufficiently large number n. Presently (1997) the "world record" in factoring an arbitrary integer is in the range $n \approx 10^{130}$ (see Pomerance (1996)).

We give two numerical examples of the RSA public key cryptosystem in order to demonstrate this method, but we have to mention that the values of the crucial parameters chosen here are too small to provide any security, i.e., too small to ensure that parts of the calculation are not feasible for an unauthorized interceptor. In these examples, the letters of the alphabet are represented as A $=$ 01, B $=$ 02,..., Z $=$ 26, and "space" $=$ 00.

22.3. Example. Let $n = pq = 3 \cdot 11 = 33$ and choose $d = 7$. Then $\varphi(n) = 20$, and $7e \equiv 1 \pmod{20}$ has the solution $e = 3$.

To encipher the word ALGEBRA we represent it in numerical form, namely $01, 12, 07, 05, 02, 18, 01$, i.e., $1, 12, 7, 5, 2, 18, 1$. By computing $1^3 \equiv 1 \pmod{33}$, $12^3 \equiv 12 \pmod{33}$, etc., we find the ciphertext $1, 12, 13, 26, 8, 24, 1$, which is the word ALMZIXA.

To decipher the ciphertext QZLLI or $17, 26, 12, 12, 9$ we calculate $17^7 \equiv 8 \pmod{33}$, etc., and find the message HELLO.

22.4. Example. Let $n = pq = 47 \cdot 59 = 2773$. Choose $d = 157$ and use the extended Euclidean Algorithm to solve $157e \equiv 1 \pmod{2668}$, where $\varphi(n) = (p-1)(q-1) = 2668$, and obtain $e = 17$. Because of (v) in 22.1 the message ITS ALL GO has to be broken at least into blocks of pairs of letters (or into single letters). Hence the message "blocks" m_i are $0920, 1900, 0112, 1200, 0715$, where $0 \le m_i \le n-1$. Computing $m_i^e \equiv c_i \pmod{n}$ gives the ciphertext c_i: $0948, 2342, 1084, 1444, \ldots$.

For deciphering the ciphertext numbers $948, 2342$, etc., we have $948^{157} \equiv 920 \pmod{2773}$, etc. Since message blocks have four digits we know that 0920 corresponds to the message IT.

22.5. Comments on RSA.

(i) The RSA cryptosystem can be implemented readily since there are well-known algorithms for finding large prime numbers, e can be obtained from d by using the extended Euclidean Algorithm (cf. 11.22), and exponentiation mod n can also be done in polynomial time.

(ii) Up to now, determining $\varphi(n)$ is not easier for the cryptanalyst than factoring n. But knowing the value $\varphi(n)$ gives the prime factors of n, since $\varphi(n) = (p-1)(q-1) = n - (p+q) + 1$. Thus if the numbers n and $\varphi(n)$ are known, then $p+q$ is known. Also, $(p-q)^2 = (p+q)^2 - 4pq = (p+q)^2 - 4n$, which shows that if $p+q$ and n are known, then $p-q$ is known. But then also p and q are known since $p = \frac{1}{2}((p+q)+(p-q))$, $q = \frac{1}{2}((p+q)-(p-q))$.

(iii) If $2^e > n$, then it is impossible to recover the message $m \neq 0, 1$ just by taking the eth root of c, i.e., every message $\neq 0, 1$ is enciphered by exponentiation followed by a reduction mod n. Also, $e-1$ must not be a multiple of $p-1$ or $q-1$, e.g., $e = \varphi(n)/2 + 1$ is a bad enciphering exponent since $m^e \equiv m \pmod{n}$. It can be shown that the number of unconcealable letters in an RSA cipher is

$$\big(1+\gcd(e-1, p-1)\big)\big(1+\gcd(e-1, q-1)\big).$$

(iv) At present it is not known how to break the RSA cryptosystem without factoring n with the exception of some special situations, see Simmons (1979). In general it is an open question whether RSA is as secure as factoring. Most of the fast factoring algorithms for the integers n that are of interest in cryptography have been shown to run in time $\exp((1+O(1))((\log n)(\log\log n))^{1/2})$ as $n \to \infty$. At present, the number field sieve algorithm is the fastest factoring method for the RSA modulus.

(v) It is not always true that making ciphers more complex implies improving security. If the primes p and q in the RSA cipher are replaced by secret irreducible polynomials p and q over $\mathbb{F}_2$ of degrees r and s, the public modulus is $n = pq$ and the enciphering exponent e is chosen to be relatively prime to $(2^r-1)(2^s-1)$. The resulting cryptosystem can be broken by factoring n, which is usually quite easy to do. It should be noted that replacing pq by $p \circ q$ yields a cryptosystem which is even easier to break.

(vi) The RSA cryptosystem requires "safe" primes p and q, otherwise iteration techniques can lead to a successful attack. Safe primes may be ***strong primes***; these are primes p, q such that $p-1$ and $q-1$ have large prime factors p_1 and q_1, respectively, and $p_1 - 1$ and $q_1 - 1$ have large prime factors p_2 and q_2, respectively, and both $p+1$ and $q+1$ should have large prime factors.

In the iteration technique, repeated enciphering with the exponent e may lead to a short cycle of iterations, i.e.,

$$c \equiv m^e \pmod{n},$$
$$c_1 \equiv c^e \pmod{n},$$

$$\vdots$$
$$c_{k+1} \equiv c_k^e \pmod{n},$$
$$c \equiv c_{k+1}^e \pmod{n}.$$

Therefore $c_{k+1} = m$. (Safe primes ensure long iteration cycles.) For example, let $n = 2773$ and $d = 157$. Then $e = 17$. Moreover, let $c = 1504$, then $c^{17} \equiv 2444 \pmod{n}$, $2444^{17} \equiv 470 \pmod{n}$, $470^{17} \equiv 2209 \pmod{n}$, $2209^{17} \equiv 1504 = c \pmod{n}$, hence $m = 2209$.

(vii) The primes p and q should be of similar size but with different numbers of digits, otherwise Fermat factoring may be a successful attack in finding the secret factors p and q from the given n. In Fermat factoring the aim is to find an integer $a \geq \sqrt{n}$ such that $n = a^2 - b^2 = (a+b)(a-b)$, i.e., a is such that $a^2 - n$ is a square. A trial-and-error approach starts with a approximately $\sqrt{n}$, if this does not work, use $a+1$, next $a+2$, etc., until $(a+k)^2 - n$ is a square.

(viii) To use the same modulus n for more than two users A and B is less secure. Suppose A sends the same message m to B_1 and B_2, whose RSA public keys are $\{e_1, n\}$ and $\{e_2, n\}$: A enciphers

$$c_i \equiv m^{e_i} \pmod{n}, \qquad i = 1, 2.$$

If $\gcd(e_1, e_2) = 1$, then the cipher can be broken, since in that case there exist integers a and b such that $ae_1 + be_2 = 1$. If $\gcd(c_1, n) > 1$, then c_1 is either p or q and the cipher is broken. If $\gcd(c_1, n) = 1$, then c_1^{-1} exists mod n. Without loss of generality, assume $a < 0$ and $b > 0$. Then

$$(c_1^{-1})^{|a|}(c_2)^b \equiv ((m^{e_1})^{-1})^{|a|}(m^{e_2})^b$$
$$\equiv m^{ae_1+be_2} \equiv m \pmod{n}.$$

In general the probability for $\gcd(e_1, e_2) = 1$ for random e_1 and e_2 is approximately 0.6079, for e_i in RSA ciphers it is approximately 0.8106.

(ix) Instead of the modulus $n = pq$ we can consider an arbitrary integer n, which is larger than any number representing a message block. It can be shown that for $a \in \mathbb{Z}$ and $\gcd(a, n) = 1$, we have $a^{\lambda(n)} \equiv 1 \pmod{n}$, where $\lambda(n)$ is the maximum order of elements of the multiplicative group $\mathbb{Z}_n$. Here $\lambda(n)$ is called the ***Carmichael***

function and is defined as follows: Let $n = p_1^{e_1} \dots p_r^{e_r}$ be a prime factorization of $n \in \mathbb{Z}$, and let $\varphi(p^t)$ be Euler's function, then

$$\lambda(n) = \operatorname{lcm}(\lambda(p_1^{e_1}), \dots, \lambda(p_r^{e_r})),$$

$$\lambda(p^t) = \begin{cases} \frac{1}{2}\varphi(2^t) & \text{for } p = 2 \text{ and } t > 2, \\ \varphi(p^t) & \text{for } p \text{ odd prime or } p = 2 \text{ and } t = 1 \text{ or } 2. \end{cases}$$

Moreover we can prove that for all $a \in \mathbb{Z}$: $a^{k+s} \equiv a^k \pmod{n}$ if and only if n is free of $(k+1)$th powers and $\lambda(n) \mid s$. From $a^{de} \equiv a \pmod{n}$ it follows that $de \equiv 1 \pmod{\lambda(n)}$.

Earlier we used $de \equiv 1 \pmod{\varphi(n)}$ for $n = pq$, which gives correct results, but could result in e being unnecessarily large, given d, or vice versa. In the case $n = pq$ the Carmichael function is $\lambda(n) = \operatorname{lcm}(p-1, q-1)$, which is less than $\varphi(n)$.

The RSA trapdoor one-way function is just the discrete exponentiation $c \equiv m^e \pmod{n}$ where the "trapdoor" is $\{p, q, e\}$. The inverse function is $m \equiv c^d \pmod{n}$, where $de \equiv 1 \pmod{\varphi(n)}$. The domain and range of this function both are the set of integers from 0 to $n-1$. Hence the RSA cipher can be used to form digital signatures in the manner described earlier on, below Figure 22.1. This digital signature capability is one of the important and useful features of the RSA.

Let n_A, d_A, e_A be the RSA parameters for sender A, D_A and E_A are A's deciphering and enciphering transformation, and similar notation is used for B. Suppose A wishes to send a signed message m to B. The following protocol describes the required steps.

A computes the signature $s = D_A(m) = m^{d_A} \pmod{n_A}$ and then enciphers it by using B's public key $\{e_B, n_B\}$. If $n_B > n_A$, A computes $c = E_B(s) \equiv s^{e_B} \pmod{n_B}$. A sends c to B.

B receives c, and applies his deciphering algorithm $D_B(c) = D_B(s^{e_B}) = s^{e_B d_B} \equiv s \pmod{n_B}$. Then B uses A's public key $\{e_A, n_A\}$ and forms $E_A(s) = E_A(D_A(m)) = m^{e_A d_A} \equiv m \pmod{n_A}$.

B is convinced that m came from A, since only A could have known (computed) s and correctly deciphers it to m by using the secret key d_B and the public key e_A.

If $n_B < n_A$, A "breaks" s into blocks of size less than n_B and enciphers each block by using B's public key. Alternatively, if $n_A > n_B$, the message can be signed by forming $D_A(E_B(m))$. Diagrammatically for $n_A < n_B$:

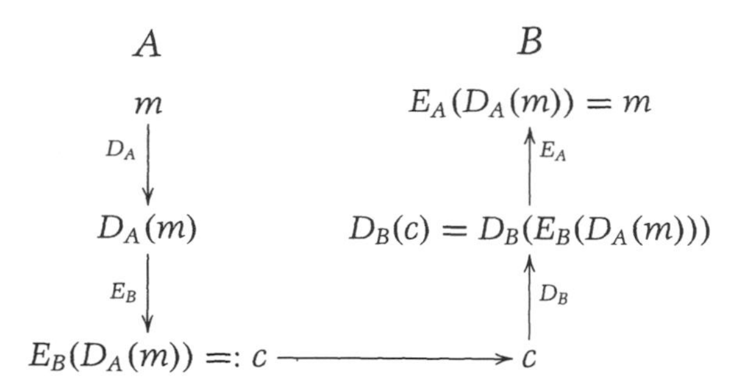

22.6. Example. Suppose A wants to send to B the plaintext message $m = 207$ signed by the RSA digital signature. The respective RSA keys are given by

$$A\colon n_A = 23 \cdot 47 = 1081, \quad d_A = 675,$$
$$B\colon n_B = 31 \cdot 59 = 1829, \quad e_B = 7.$$

We note that $n_B > n_A$. User A computes $s \equiv m^{d_A} \pmod{n_A}$, i.e., $276 \equiv 207^{675} \pmod{1081}$ and then $276^7 \equiv 386 \pmod{1829}$. Therefore the signed message is 386.

User B recovers the message by applying D_B and then E_A.

Exercises

1. Some exercises involving primes.

   (i) Show that if $n > 1$ and $a^n - 1$ is prime, then $a = 2$ and n is a prime.

   (ii) Prove that the number of digits of a prime must be prime if each digit is 1.

   (iii) A positive integer n is called ***perfect*** if it equals the sum of its divisors less than n. Prove that $2^{k-1}(2^k - 1)$ is perfect if $2^k - 1$ is a prime.

   (iv) Prove that there are infinitely many primes of the form $4k + 3$, $k = 0, 1, 2, \ldots$.

   (v) Estimate the number of digits in the primes $2^{127} - 1$ and $2^{859433} - 1$.

2. For $n = 3731$, are the requirements for an RSA cipher satisfied by n, apart from the requirement for the size of n? Does the cipher "work" with this n? Similarly for $n = 109871$.

3. In an RSA cipher with $n = 2773$, $e_A = 17$, and $e_B = 3$, a plaintext m is enciphered as ciphertext $c_A = 948$ and $c_B = 1870$ by users A and B, respectively. Determine the number m without factoring n.
4. How many of the messages $0, 1, \ldots, 14$ are unconcealable in the RSA cipher with $n = 15$ and $e = 3$? Similarly for $n = 2773$ and $e = 17$.
5. Suppose the letters **A** to **Z** are represented by the ASCII codes 65 to 90. Let $n = 9991$ and $e = 11$ be parameters of an RSA cipher. Decipher the ciphertext 9425, 9884 and represent the plaintext numbers as English plaintext.
6. Suppose the number $n = 536813567$ has been used in an RSA cipher with enciphering key $e = 3602561$. Assume that the plaintext $p_1 \ldots p_6$ consists of 6-letter blocks in the 26-letter alphabet **A** to **Z** (represented as 0 to 25), converted to an integer between 0 and $26^6 - 1$ in the usual way as $p_1 26^5 + p_2 26^4 + \cdots + p_6$. The ciphertext consists of 7-letter blocks in the same alphabet. Decipher the message **BNBPPKZBLGVPGX**.
7. In an RSA cipher let $p = 3336670033$ and $q = 9876543211$ and $e = 1031$. Determine d and decipher the ciphertext 899150261120482115 as a plaintext word in English. (Assume **A** $= 0, \ldots,$ **Z** $= 25$.)
8. Alice wants to send to Bob the plaintext 311 signed via an RSA digital signature. The respective RSA keys are $n_A = 1081$, $e_A = 3$ and $n_B = 1829$, $e_B = 7$. Find the signed ciphertext.
9. Use Fermat factoring to factor $n = 119143$.
10. Rabin's public key cryptosystem uses an enciphering function which is not bijective. It is defined by enciphering a message M as

$$C \equiv M(M + b) \pmod{n},$$

where n is the product of two approximately equally large primes p and q, M is the message to be enciphered, and b is an integer with $0 \leq b < n$. The integers n and b are public keys, the prime factors p and q of n are private. Show that for deciphering we have four solutions for M, given $C \pmod{n}$. Also show that it suffices to solve the two congruences

$$C \equiv M(M + b) \pmod{p}, \qquad C \equiv M(M + b) \pmod{q},$$

to determine a solution to the original congruence.
11. More on primes and pseudoprimes.

    (i) Give numerical examples for the following test: If for each prime p such that p divides $n-1$ there exists an a such that $a^{(n-1)/p} \not\equiv 1 \pmod{n}$ and $a^{n-1} \equiv 1 \pmod{n}$, then n is prime.

(ii) An integer N satisfying $b^{N-1} \equiv 1 \pmod{N}$ is called a ***base b pseudoprime***. Show that there are infinitely many composite base b pseudoprimes. Verify that 341 is a base 2 pseudoprime, also that 91 is a base 3 pseudoprime.

(iii) N is called a ***strong base b pseudoprime*** if $N = 2^s t + 1$ for odd t implies that either $b^t \equiv 1 \pmod{N}$ or $b^{t2^r} \equiv -1 \pmod{N}$ for some $0 \leq r < s$. Give an example of such a strong base b pseudoprime.

12. The ***Jacobi symbol*** $J(a, n)$ is defined as follows:

$$J(a,n) = \begin{cases} 1 & \text{if } a = 1, \\ J(a/2, n)(-1)^{(n^2-1)/8} & \text{if } a \text{ is even,} \\ J(n \bmod a, a)(-1)^{(a-1)(n-1)/4} & \text{if } a > 1 \text{ is odd.} \end{cases}$$

Evaluate $J(6, 13)$.

13. Use the following test due to Solovay and Strassen in a numerical example to show that a given number n is not prime. (This is an example of a probabilistic algorithm for checking primality. If the test has been passed t times, then the remaining probability that n is nonprime is less than 2^{-t}.) Choose an integer a at random from $1, \ldots, n-1$, and then test to see if $\gcd(a, n) = 1$ and $a^{(n-1)/2} \equiv J(a, n) \pmod{n}$. If n is prime, then it always passes the test. If n is not prime, it will pass the test for less than half of the values of a which are coprime to n.

§23 Discrete Logarithms and Other Ciphers

In §22 we showed how discrete exponentiation can be used as a candidate for a one-way function, but to date there is no proof that discrete exponentiation is truly one-way. In this section we use finite field theory to describe discrete logarithms. Let α be a primitive element of $\mathbb{F}_q$, let a be a nonzero element of $\mathbb{F}_q$, and compare the lines after 13.8.

We recall from §13 that the discrete logarithm of a with respect to the base α is the uniquely determined integer r, $0 \leq r < q - 1$, for which $a = \alpha^r$.

We use the notation $r = \log_\alpha a$ or simply $r = \log a$ if α is kept fixed. The discrete logarithm and the discrete exponential function $\exp_\alpha r = \exp r = \alpha^r$ form a pair of inverse functions of each other.

The discrete logarithm satisfies the following rules:

$$\log(ab) \equiv \log a + \log b \pmod{q-1},$$
$$\log(ab^{-1}) \equiv \log a - \log b \pmod{q-1}.$$

The discrete exponential function in $\mathbb{F}_q$ can be calculated by an analogue of the square and multiply technique. In order to compute $\exp r = \alpha^r$ in $\mathbb{F}_q$ we first perform repeated squaring on α to obtain $\alpha, \alpha^2, \alpha^4, \dots, \alpha^{2^t}$, where 2^t is the largest power of 2, that is $\leq r$. Then α^r is obtained by multiplying together an appropriate combination of these elements. So, for instance, α^{100} is not computed by 99 multiplications, but by α^2, $(\alpha^2)^2 = \alpha^4$, α^8, α^{16}, α^{32}, α^{64}, and $\alpha^{64} \cdot \alpha^{32} \cdot \alpha^4 = \alpha^{100}$, which needs only eight multiplications.

The apparent one-way nature of the discrete exponential function is the reason for its use in several classical cryptosystems and public key cryptosystems. Advances in computing discrete logarithms more efficiently have been important in the cryptanalysis of these ciphers. Particularly striking advances in algorithms for finding discrete logarithms in finite fields have been made, in particular for $\mathbb{F}_{2^n}$. At present, the complexity of finding discrete logarithms in a prime field $\mathbb{F}_p$ for a general prime p is essentially the same as the complexity of factoring an integer n of about the same size where n is the product of two approximately equal primes. When utilizing finite fields $\mathbb{F}_q$, where q is prime or $q = 2^k$, it is necessary to require $q-1$ to have a large prime factor, since otherwise it is easy to find discrete logarithms.

23.1. Example. Let m, e, and c denote the plaintext message, the secret enciphering key, and the ciphertext, respectively, where $m, c \in \mathbb{F}_q^*$, e is an integer with $1 \leq e \leq q-2$, $\gcd(e, q-1) = 1$, and q is a large prime power. Then $ed \equiv 1 \pmod{q-1}$ can be solved for d. For transmission of a message m in $\mathbb{F}_q$ we encipher it by computing $c = m^e$ and decipher it by $c^d = m$.

To find the secret enciphering key is as hard as finding discrete logarithms since $c = m^e$ is equivalent to $e \log m \equiv \log c \pmod{q-1}$. This is a classical cryptosystem with a wide choice for the key e since there are $\varphi(q-1)$ suitable integers e.

23.2. Example. The ***Diffie-Hellman scheme*** is a key exchange system for establishing a common key between A and B. Let α be a primitive element of $\mathbb{F}_q$. Users A and B choose random integers a and b, respectively, where

$2 \leq a, b \leq q-2$. Then A sends α^a to B, while B transmits α^b to A. Both take α^{ab} as their common key.

23.3. Example. We describe a classical cryptosystem, also called ***no-key*** or ***three-pass algorithm***, for message transmission. User A wishes to send a message $m \in \mathbb{F}_q^*$ to user B. The field $\mathbb{F}_q$ is publicly known.

Then A chooses a random integer a, where $1 \leq a \leq q-1$ and $\gcd(a, q-1) = 1$, and sends $x = m^a$ to B. User B chooses a random integer b, where $1 \leq b \leq q-1$ and $\gcd(b, q-1) = 1$, and sends $y = x^b = m^{ab}$ to A. Now A forms $z = y^{a'}$, where $aa' \equiv 1 \pmod{q-1}$, and sends z to B. Then B computes $z^{b'}$ to obtain m, where $bb' \equiv 1 \pmod{q-1}$, since $(((m^a)^b)^{a'})^{b'} = m$.

As a numerical example of this cipher with a very small finite field, $\mathbb{F}_{2^3}$, we consider transmission of the message $m = 011 = x + x^2$. Here $\mathbb{F}_{2^3}$ is defined by the polynomial $f(x) = x^3 + x + 1$. Suppose A chooses $a = 13$, computes $m^a = (x + x^2)^{13} = 1 + x$ in $\mathbb{F}_{2^3}$, and sends $1 + x$ to B. User B chooses $b = 11$, $(1 + x)^{11} = 1 + x + x^2$ and sends it to A. A finds $a' = 6$ from $13a' \equiv 1 \pmod 7$ and sends x^2, which she obtained from $(1 + x + x^2)^6 = x^2$, to B. Finally, B computes $b' = 2$ from $11b' \equiv 1 \pmod 7$ and obtains $(x^2)^2 = x + x^2 = m$.

23.4. Example. We describe the ***El Gamal public key cryptosystem*** for message transmission in $\mathbb{F}_q$. Let α be a primitive element of $\mathbb{F}_q$, and let α and q be public. User B selects a secret integer b at random and makes the element $\alpha^b \in \mathbb{F}_q$ his public key. If user A wants to send a message $m \in \mathbb{F}_q^*$ to B, then A selects a random integer a, $1 \leq a \leq q-1$, and sends the pair $(\alpha^a, m\alpha^{ab})$ to B. User B knows b and therefore can compute $\alpha^{ab} = (\alpha^a)^b$ and so recovers m.

In a very small numerical example that illustrates this cipher, let $q = 71$ and $\alpha = 7$. Let user B's public key be $\alpha^b = 3$ and let A select $a = 2$ as a random integer. In order for A to send the message $m = 20$, A sends the pair $(\alpha^a, m\alpha^{ab}) = (49, 38)$ to B, since $m\alpha^{ab} = 20 \cdot 9 = 38$ in $\mathbb{F}_{71}$.

If in the El Gamal cryptosystem m is a key chosen by A, then this system serves as a key exchange mechanism. However, it is not an authenticated key exchange, since an impostor could pretend to be A and send fake messages to B. The next example provides an ***El Gamal digital signature***.

23.5. Example. Let p be a prime and α a primitive element of $\mathbb{F}_p$, both p and α are public. Suppose A wishes to send a signed message to B. Then A chooses a random integer a, $1 \leq a \leq p-2$, and computes $y = \alpha^a$ in $\mathbb{F}_p$. For a given message m, $0 < m < p-1$, A chooses a secret k, $0 < k < p-1$, with $\gcd(k, p-1) = 1$, and computes $r = \alpha^k$ and $s \equiv k^{-1}(m - ra) \pmod{p-1}$. Now A sends m, r, and s to B. For authentication B computes $c = \alpha^m$ and $c' = y^r \cdot r^s$ in $\mathbb{F}_p$. We note that $m \equiv ks + ra \pmod{p-1}$, hence if m is authentic,

$$c = \alpha^m = (\alpha^a)^r \cdot (\alpha^k)^s = y^r \cdot r^s = c'.$$

We note that a different k must be chosen for each m, since using any k twice determines the secret a uniquely. Also, $p-1$ has to have a large prime factor, otherwise the discrete logarithm problem of finding a from α^a can be easily solved, see an algorithm for discrete logarithms given below.

Since public key cryptosystems may be open to abuse by criminals, recently the notion of ***fair public key cryptosystems*** was introduced by Micali. An informal description of such a system includes a guarantee for a special agreed-upon party, and only this party, under lawful circumstances to understand all messages enciphered by the system, even without the user's consent or knowledge. To be more concrete, the special agreed-upon party could be the Government and a court order could represent the lawful circumstances. Consider, as an example, the telephone system for the method of communication. Furthermore we assume the existence of a key management/distribution center for the system.

The essential ingredients of a fair public key cryptosystem are a fixed number of trustees, say five, and an arbitrary number of users. Each user chooses his own public and private keys independently in a given public key cryptosystem. Each user breaks the private deciphering key into five special pieces such that it can be reconstructed given knowledge of all five special pieces, it cannot be reconstructed if only any four, or fewer, special pieces are known (see 11.34), and each special piece can be individually verified to be correct.

The user submits to trustee i his own enciphered public key and the ith piece of this private key. The trustee approves (i.e., signs) the public key and safely stores the ith piece. The key management center approves any public key approved by all trustees, and these keys are the public keys used for private communication by users of the system.

When presented with a court order, the trustees will reveal the pieces of a given deciphering key in their possession. In this way the key management center (or the Government) can reconstruct the key from the correct pieces.

As a concrete example we describe a fair public key cryptosystem based on the Diffie-Hellman scheme, see Example 23.2, for $\mathbb{F}_p$, where p is a prime. Again α is a primitive element of $\mathbb{F}_p$.

23.6. Example. Each user A chooses five integers a_i, $i = 1, 2, \ldots, 5$, where $2 \leq a_i \leq p-2$, and lets a be their sum in $\mathbb{F}_p$. Then she computes $x_i = \alpha^{a_i}$ for $i = 1, 2, \ldots, 5$, and $x_A = \alpha^a$. Here x_A is A's public key and a her private key. The x_i are called the public pieces of x_A and the a_i are its private pieces. The product of the public pieces equals the public key. User A provides trustee T_i with x_A, x_i, and also, in enciphered form, a_i.

Trustee T_i verifies that $\alpha^{a_i} = x_i$ and, if this is the case, stores the pair (x_A, a_i), signs the pair (x_A, x_i), and gives the signed pair to the key management center. The center verifies that the product of the public pieces x_i, $i = 1, 2, \ldots, 5$, relative to a given public key x_A equals x_A. If that is the case the center approves x_A as a public key and thc remaining steps are exactly as in the Diffie and Hellman scheme.

It is left to the reader to establish why this scheme is "fair" in the sense described before the example.

Underlying most of the ciphers in this section is the difficulty of finding discrete logarithms in $\mathbb{F}_q$. In the case where $q-1$ has only small prime factors the ***Silver-Pohlig-Hellman algorithm*** allows us to compute discrete logarithms. First we sketch this algorithm and then demonstrate it in an example.

Solving the discrete logarithm problem means: given $a \in \mathbb{F}_q^*$, find the value r for which $a = \alpha^r$, where α is a primitive element of $\mathbb{F}_q$. Here $r = \log_\alpha a$. Let $q - 1 = \prod_{i=1}^k p_i^{e_i}$, where $p_1 < p_2 < \cdots < p_k$ are the distinct prime factors of $q-1$. The value of r will be determined modulo $p_i^{e_i}$, $i = 1, 2, \cdots, k$, and the result will then be combined by the Chinese Remainder Theorem for integers to obtain $r \bmod q-1$, which is unique since $0 \leq r < q-1$.

First we consider the representation of r in terms of p_i.

$$r \equiv \sum_{j=0}^{e_i-1} r_j p_i^j \pmod{p_i^{e_i}} \quad \text{with } 0 \leq r_j \leq p_i - 1.$$

We wish to determine r_j, $j = 0, \ldots, e_i - 1$, for each prime factor p_i and then combine these results by solving a system of congruences for r mod $p_i^{e_i}$.

First we determine r_0. We form

$$a^{(q-1)/p_i} = \alpha^{(q-1)r/p_i} = c_i^r = c_i^{r_0},$$

where $c_i = \alpha^{(q-1)/p_i}$ is a primitive p_ith root of unity. Therefore there are only p_i possible values for $a^{(q-1)/p_i}$ corresponding to $r_0 = 0, 1, \ldots, p_i - 1$ and evaluating these possibilities is feasible if p_i is small. It is convenient to perform some precalculations. We produce a table of $1, c_i, \ldots, c_i^{p_i-1}$. Then a simple table look-up of $a^{(q-1)/p_i}$ yields r_0. The resulting value uniquely determines r_0.

The next digit r_1 in the representation of r mod $p_i^{e_i}$ is obtained from

$$a\alpha^{-r_0} = \alpha^{t_1} \qquad \text{where } t_i = \sum_{j=1}^{e_i-1} r_j p_i^j.$$

Then

$$\alpha^{t_1(q-1)/p_i^2} = c_i^{t_1/p_i} = c_i^{r_1}$$

uniquely determines r_1, etc.

It can be shown that this algorithm has a running time of order at most $p_k^{1/2}(\log q)^2$, where p_k is the largest prime factor of $q - 1$. For more complex algorithms to compute discrete logarithms we refer to Odlyzko (1985).

23.7. Example. Let $q = 181$ and $\alpha = 2$. Then $q - 1 = 180 = 2^2 \cdot 3^2 \cdot 5$. We compute the discrete logarithm r of $a = 62$ to the base 2. Since $p_1 = 2$, $p_2 = 3$, and $p_3 = 5$ we consider three parts of the Silver-Pohlig-Hellman algorithm and combine the results by using the Chinese Remainder Theorem for integers.

First we determine $r = r_0 + 2r_1 \pmod 4$. From $c_1^{r_0} = 2^{90r_0} = 62^{90} \equiv 1$ we conclude that $r_0 = 0$. Next $62 \cdot 2^{-r_0} = 2^{t_1}$ and $62^{45} = 2^{r_1(q-1)/2} = 1$ if $r_1 = 0$ yields $r_1 = 0$. Thus $r \equiv 0 \pmod 4$.

Now we consider the factor 3^2 of $q - 1$ and determine $r = r_0 + 3r_1 \pmod 9$. Since $62^{60} = 48$ we deduce from $\alpha^{r_0(q-1)/p_2} = a^{(q-1)/p_2}$ that $r_0 = 1$. To determine r_1, consider $62 \cdot 2^{-1} = 31$ and $31^{180/9} = 1$, hence $r_1 = 0$. Altogether $r \equiv 1 \pmod 9$.

For the factor 5 of $q - 1$ we have to determine $r \equiv r_0 \pmod 5$. Since $62^{36} = 1$, we conclude that $r_0 = 0$ and $r \equiv 0 \pmod 5$.

Finally, solving the three congruences $r \equiv 0 \pmod 4$, $r \equiv 1 \pmod 9$, and $r \equiv 0 \pmod 5$ simultaneously yields $r \equiv 100 \pmod{180}$, thus $r = 100$.

In this example the look-up table has entries $\alpha^{k(q-1)/p} \bmod q$ for $0 \le k \le p-1$:

$k \backslash p$	2	3	5
0	1	1	1
1	180	48	59
2		132	42
3			125
4			135

Now we describe some other cryptosystems to illustrate attempts to link computationally hard problems, such as decoding linear codes or the knapsack problem, with the construction of one-way functions for public key cryptosystems. For linkages with coding we note

23.8. Lemma. *Let* $\mathbf{H}$ *be a parity-check matrix of an* (n, k) *linear t-error-correcting code* C. *Then the map* $\mathbf{H}\colon \mathbb{F}_q^n \to \mathbb{F}_q^{n-k}$ *is one-to-one when restricted to vectors of weight* $\le t$.

Proof. Let $\mathbf{y}, \mathbf{z} \in \mathbb{F}_q^n$ and $w(\mathbf{y}), w(\mathbf{z}) \le t$. Then $\mathbf{Hy} = \mathbf{Hz}$ implies $\mathbf{H}(\mathbf{y}-\mathbf{z}) = \mathbf{0}$, hence $\mathbf{y} - \mathbf{z} = \mathbf{c} \in C$. Since $d(\mathbf{y}, \mathbf{0}) = w(\mathbf{y}) \le t$, $d(\mathbf{y}, \mathbf{c}) = w(\mathbf{y} - \mathbf{c}) = w(\mathbf{z}) \le t$, we deduce that $\mathbf{c} = \mathbf{0}$, hence $\mathbf{y} = \mathbf{z}$ and the map is one-to-one.

23.9. Example. Let C be as in Lemma 23.8. A classical (or conventional) cryptosystem can be defined as follows:

Enciphering: A plaintext message $\mathbf{y} \in \mathbb{F}_q^n$ of weight $\le t$ is enciphered as ciphertext $\mathbf{Hy}^{\mathrm{T}}$.

Deciphering: Apply the decoding algorithm of C to the syndrome $\mathbf{Hy}$. Since $d(\mathbf{y}, \mathbf{0}) = w(\mathbf{y}) \le t$, view $\mathbf{y}$ as an error vector relative to $\mathbf{0}$.

23.10. Example. Let C be as before. Choose a random nonsingular $(n-k) \times (n-k)$ matrix $\mathbf{M}$, and a random $n \times n$ matrix $\mathbf{P}$ that is obtained by permuting the rows of a nonsingular diagonal matrix. Keep $\mathbf{M}$, $\mathbf{P}$, and $\mathbf{H}$ secret and compute $\mathbf{K} = \mathbf{MHP}$ as the public key, an $(n-k) \times n$ matrix. A vector $\in \mathbb{F}_q^n$ of weight $\le t$ is enciphered as $\mathbf{Ky}^{\mathrm{T}}$.

For deciphering, premultiply the ciphertext $\mathbf{Ky}^{\mathrm{T}}$ by $\mathbf{M}^{-1}$ to obtain $\mathbf{H}(\mathbf{yP}^{\mathrm{T}})^{\mathrm{T}}$. Then the decoding algorithm of C gives $\mathbf{yP}^{\mathrm{T}}$ and postmultiplication by $(\mathbf{P}^{\mathrm{T}})^{-1}$ recovers $\mathbf{y}$. It should be noted that factoring matrices, such as $\mathbf{K} = \mathbf{MHP}$, is not unique. A brute force attack, provided the parameters are sufficiently large, is not feasible since there are $\prod_{j=0}^{k-1}(q^{n-j}-1)/(q^{k-j}-1)$ k-dimensional subspaces of $\mathbb{F}_q^n$. There are $q^{(n-k)^2}\prod_{j=1}^{n-k}(1-q^{-j})$ nonsingular $(n-k)\times(n-k)$ matrices over $\mathbb{F}_q$, $n!(q-1)^n$ matrices $\mathbf{P}$, and $\sum_{j=0}^{t}\binom{n}{j}(q-1)^j$ vectors of weight $\le t$ in $\mathbb{F}_q^n$. However, there are several feasible cryptanalytic attacks, such as picking a submatrix $\mathbf{J}$ of $\mathbf{K}$ consisting of $(n-k)$ columns of $\mathbf{K}$ and computing $\mathbf{y}' = \mathbf{J}^{-1}(\mathbf{K}\mathbf{y}^{\mathrm{T}})$. Then $\mathbf{y}'$ will sometimes satisfy $\mathbf{K}\mathbf{y}' = \mathbf{K}\mathbf{y}^{\mathrm{T}}$.

23.11. Example. McEliece proposed a public key cryptosystem based on algebraic codes that seems to have withstood cryptanalytic attacks so far. Its disadvantage is that its keys are large and it is slow for transmission. It is based on so-called Goppa codes that can correct t errors. Let $\mathbf{G}'$ be a $k \times n$ generator matrix of such a code, let $\mathbf{P}$ be an $n \times n$ permutation matrix, and $\mathbf{S}$ a $k \times k$ nonsingular matrix. Here $\mathbf{G}'$, $\mathbf{P}$, and $\mathbf{S}$ are private. The $k \times n$ matrix $\mathbf{G} = \mathbf{SG}'\mathbf{P}$ is the public key, the messages are k-dimensional vectors over $\mathbb{F}_q$. We encipher $\mathbf{m}$ as $\mathbf{c} = \mathbf{mG} + \mathbf{z}$ ($\mathbf{z}$ a randomly chosen n-dimensional vector over $\mathbb{F}_q$ with weight $\le t$). For deciphering, let $\mathbf{c}' = \mathbf{cP}^{-1}$. The decoding algorithm finds $\mathbf{m}'$ such that the Hamming distance $d(\mathbf{m}'\mathbf{G}, \mathbf{c}') \le t$. Then $\mathbf{m} = \mathbf{m}'\mathbf{S}^{-1}$.

The last public key cryptosystem that we consider is based on the difficult problem of selecting from a collection of objects of various different sizes a set which will exactly fit a defined space. Expressed in one-dimensional terms this is equivalent to selecting rods of different lengths which, put end-to-end, will exactly equal a specified length K. This problem is called the ***knapsack problem***. First we express the knapsack problem in symbolic form. Let $\mathbf{y} = (y_1, \ldots, y_n)$ be the components of the knapsack written as a vector. In order to fill a specified knapsack a selection of these numbers is added together to form the total. This selection is a binary array $\mathbf{a} = (a_1, \ldots, a_n)$, in which $a_i = 0$ means that the ith value of $\mathbf{y}$ is not chosen, $a_i = 1$ means that it is chosen. If this selection of components fits the size of the knapsack, then

$$K = \mathbf{y} \cdot \mathbf{a},$$

where $\cdot$ denotes the scalar product of vectors. Usually it is not easy to find $\mathbf{a}$ when only K and $\mathbf{y}$ are given. In the application to cryptography $\mathbf{a}$ is the plaintext and K is the ciphertext. Cryptanalysis means solving the knapsack problem, that is, given K and $\mathbf{y}$, find $\mathbf{a}$. Deciphering means knowing a way to find $\mathbf{a}$ much more easily than by solving the general problem. It is known that determining a given knapsack problem with sum K and vector $\mathbf{y} = (y_1, \ldots, y_n)$, such that K is the sum of some y_i's, belongs to a class of problems for which no polynomial time algorithm is known (we say it is an ***NP-complete problem***). However, the complexity of a knapsack depends largely on the choice of $\mathbf{y}$. A ***simple knapsack*** (and as it turns out one that is easy to solve) is one where $\mathbf{y}$ is superincreasing in the sense that $y_i > y_1 + \cdots + y_{i-1}$ for $i = 2, \ldots, n$, and in this case it can be decided in linear time whether there is a solution and a simple algorithm will find it, if it exists. The basic idea of using a knapsack for a cryptosystem is to convert a simple knapsack into a complex one, a ***trapdoor knapsack***. This is done by choosing two large integers k and q, $\gcd(k, q) = 1$, and then forming a new trapdoor knapsack $\mathbf{z} = (z_1, \ldots, z_n)$ from a given knapsack vector $\mathbf{y}$ by letting $z_i \equiv ky_i \pmod q$. Knowing k and q makes it easy to compute $K \equiv k^{-1}L \pmod q$, where L is assumed to be the sum of some of the z_i, forming the given knapsack problem. Multiplying L by k^{-1} transforms L into K and reduces the given difficult knapsack with sum L and vector $\mathbf{z}$ into the easy knapsack with sum K and vector $\mathbf{y}$. This method can be used for a cryptosystem, where $\mathbf{z} = (z_1, \ldots, z_n)$ is the public key, the private keys are k and q in the calculation $z_i \equiv ky_i \pmod q$ for a given easily solvable knapsack with superincreasing vector $\mathbf{y}$. A message $\mathbf{a} = (a_1, \ldots, a_n)$ in binary form is enciphered by computing $L = a_1z_1 + \cdots + a_nz_n$, which is transmitted.

For applications it had been suggested that n be at least 200, and that other conditions on q, y_i, and k are imposed. However, from 1982 on, successful attacks on trapdoor knapsack systems have been outlined (see Simmons (1992)), whose security has been suspect already since its introduction in 1976 by Merkle and Hellman. Nevertheless, we describe the knapsack cryptosystem as an interesting example. First the letters of the alphabet A are replaced by the elements of $(\mathbb{Z}_2)^t$, for suitable values of t. Then we subdivide the resulting 0–1-sequences into blocks $(a_1, a_2, \ldots, a_n)$. Let $q \in \mathbb{N}$ and let $\mathbf{x} = (x_1, \ldots, x_n)$ be a fixed vector in $(\mathbb{Z}_q)^n$. A block $\mathbf{a} = (a_1, \ldots, a_n)$ is enciphered as an element $b \in \mathbb{Z}_q$ by means of

$$\mathbf{a} \mapsto b := \mathbf{a} \cdot \mathbf{x} = a_1x_1 + \cdots + a_nx_n.$$

Here we have a mapping from $A = (\mathbb{Z}_q)^n$ into $B = \mathbb{Z}_q$; this is not yet a key. In order to make this enciphering method into a key (to enable deciphering), the authorized receiver has to possess additional (trapdoor) information. There are several ways of doing this. We give an example.

23.12. Example. Let $t = 5$, i.e., we replace the letters of the alphabet by 5-tuples of elements in $\mathbb{Z}_2$.

A	B	C	D	E	F	G	H	I	J	K
00000	00001	00010	00011	00100	00101	00110	00111	01000	01001	01010

L	M	N	O	P	Q	R	S	T	U	V
01011	01100	01101	01110	01111	10000	10001	10010	10011	10100	10101

W	X	Y	Z	,	.	!	?	"	␣
10110	10111	11000	11001	11010	11011	11100	11101	11110	11111

Here ␣ represents a space. Now we choose $n = 10$, $q = 19999$, and a vector $\mathbf{y} = (4, 10, 64, 101, 200, 400, 800, 1980, 4000, 9000) \in (\mathbb{Z}_{19999})^{10}$. In $\mathbf{y}$ each coordinate is larger than the sum of the preceding coordinates if we regard the elements as integers, thus $\mathbf{y}$ is superincreasing, forming a simple knapsack. We form a new vector

$$\mathbf{x} = 200\mathbf{y} = (800, 2000, 12800, 201, 2, 4, 8, 16019, 40, 90) \in (\mathbb{Z}_{19999})^{10}.$$

If we wish to encipher the word CRYPTOLOGY we subdivide the message into groups of two: CR YP TO LO GY and encode:

$$\begin{aligned} \text{CR} &= 0001010001 \\ &\to (0,0,0,1,0,1,0,0,0,1) \cdot (800, 2000, 12800, 201, 2, 4, 8, 16019, 40, 90) \\ &= 295. \end{aligned}$$

Also:

$$\text{YP} \to 18957, \quad \text{TO} \to 17070, \quad \text{LO} \to 18270, \quad \text{GY} \to 13013.$$

Thus the enciphered plaintext is

$$(295, 18957, 17070, 18270, 13013) \in (\mathbb{Z}_{19999})^5.$$

If the receiver obtains the message

$$\mathbf{w} = (2130, 12800) \in (\mathbb{Z}_{19999})^3,$$

then we decipher it as follows. We evaluate $200^{-1}\mathbf{w} = 100\mathbf{w} \in (\mathbb{Z}_{19999})^3$. For the first component, say, we obtain $13010 = \mathbf{a} \cdot \mathbf{y} = 4a_1 + 10a_2 + \cdots +$

$9000a_{10}$, where $\mathbf{a} = (a_1, \ldots, a_{10})$, with $a_i \in \{0, 1\}$, represents the original first part of the message in numerical form.

13010 can only be represented in the form

$$1 \cdot 9000 + 1 \cdot 4000 + 1 \cdot 10 = 10a_2 + 4000a_9 + 9000a_{10},$$

if we subtract from 13010 the numbers 9000, 4000, etc. Thus we obtain $\mathbf{a} = (0, 1, 0, 0, 0, 0, 0, 0, 1, 1) \leftrightarrow 01000\,00011 \rightarrow$ ID. The receiver continues deciphering and reads the message EA, which gives the plaintext message IDEA.

We summarize these steps in the following remark.

23.13. A Trapdoor-Knapsack Cryptosystem. For sending a plaintext message, choose a correspondence between the letters of the alphabet and $\mathbb{Z}_2^t$, i.e., replace letters by t-tuples of zeros and ones. Then choose numbers q and n such that $t \mid n$ and a vector $(\mathbf{y} \in Z_q)^n$ such that $y_i > y_1 + \cdots + y_{i-1}$ for $2 \leq i \leq n$. Also choose a number k with $\gcd(k, q) = 1$. Then encipher and decipher as follows:

Enciphering: Form $\mathbf{x} := k\mathbf{y}$, divide the original plaintext message into blocks of n/t letters, and replace these by blocks of zeros and ones of length n. Multiply these blocks by the vector $\mathbf{x}$ mod q.

Deciphering: Multiply the received sequence by $k^{-1} \in \mathbb{Z}_q$ (which exists, since $\gcd(k, q) = 1$). Each term in the sequence corresponds to one $0-1$-sequence of length n. Interpret this as a group on n/t blocks of t letters each.

Here n, q, and $\mathbf{x}$ are openly published in a directory. The number k is kept secret between the participants. This method has the advantage that if we want to enable many participants to exchange information (e.g., between say 1000 branches of a banking group) we would need some half million arrangements between 1000 participants. A public key cryptosystem makes it possible for each participant to openly announce a vector $\mathbf{x}$ (similar to telephone numbers, e-mail addresses, or Internet addresses). The association letters $\rightarrow$ numbers and the numbers n and q can be openly agreed upon by the participants (e.g., the banking group). If participant A wants to transmit a message to participant B, she looks up B's vector $\mathbf{x}$ (e.g., given in a register) and multiplies (mod q) her message "piecewise" by $\mathbf{x}$. Then the enciphered message can be transmitted openly. It first was thought that only B is capable of deciphering the received message, because only he knows k and thus only he can restore the original message. However, this is not so, as we observed above, there is a feasible cryptanalytic attack of this system.

It could be thought that the security of some cryptosystems can be increased by repeating the enciphering procedure using new parameters. ***Iteration*** (or multiple encryption) does not always enhance security but in the case of the trapdoor knapsack it was initially thought to work by constructing doubly (or multiply) iterated and multiple knapsacks. But today successful attacks on iterated knapsacks are known. To obtain a doubly iterated knapsack, we choose large coprime integers k_1 and q_1 and construct a new knapsack vector $\mathbf{u} = (u_1, \ldots, u_n)$ where $u_i \equiv k_1 z_i \pmod{q_1}$. It is not clear to an unauthorized receiver whether one or more iterations are used, since all keys $k, q, k_1, q_1, \ldots$ are secret.

In conclusion to this chapter, we note that a widely used cryptosystem is the data encryption standard (DES), which is a conventional (classical) cryptosystem, see Denning (1983) or Seberry & Pieprzyk (1989). To break it is one of the challenges in cryptography. Another great challenge is proving the RSA cipher to be secure, or even to proving it equivalent to factoring. Yet another challenge for cryptographers, mathematicians, computer scientists, engineers, and others is producing new efficient and effective public key cryptosystems.

The principal advantage of the public key cryptosystem for privacy in communication is that it avoids the need to distribute keys. The keys which are distributed are public keys and they can be associated with the intended recipient of the message so that, assuming the corresponding secret key used in deciphering is secure, the message can be deciphered only by the intended recipient. Since anyone, with the aid of the public key, can produce an enciphered message, the cryptosystem in its simple form gives no information about the origin of the message and therefore message authentication is an important extra requirement.

The use of public key cryptosystems for signatures is effective only for those versions which introduce no redundancy in the ciphertext and therefore allow their inverses to be used as message transformations. For this subclass of public key systems a form of authentication can be employed which gives the message an unforgeable transformation that associates it with the sender, since only he could give the transformation which can be inverted to produce the message by means of his public key. The two transformations can be combined, one producing secrecy and the other producing authentication. For signatures, the RSA system is a very good technique.

Exercises

1. With modulus $p = 29$ and unknown enciphering key e, modular exponentiation produces the ciphertext $4, 19, 19, 11, 4, 24, 9, 15, 15$. Assuming that the ciphertext 24 corresponds to the plaintext 20 (i.e., the letter U), cryptanalyze the cipher.
2. Establish a common key for several parties.

   (i) For modulus $p = 101$, base $a = 5$, and individual keys $k_1 = 27$ and $k_2 = 31$.

   (ii) Same as (i) for $p = 1009$, $a = 3$, $k_1 = 11$, $k_2 = 12$, $k_3 = 17$, $k_4 = 19$.
3. In a no-key algorithm let $q = 2^7$ and $f = x^7 + x + 1$ defining $\mathbb{F}_q$ be publicly known. Let $a = 13$ and $b = 11$ be the secret random numbers chosen by A and B, respectively. Show how the message $x^6 + x^5 + x^4 + x^2 + 1$ in $\mathbb{F}_q$ can be securely transmitted from A to B in enciphered form using the no-key or three-pass algorithm.
4. In the El Gamal digital signature scheme, let $p = 11$, $\alpha = 2$, and $a = 8$. Also, let $k = 9$ and $m = 5$. Follow the steps for producing a digital signature.
5. Use the Silver-Pohlig-Hellman algorithm to compute the discrete logarithm of $a = 30$ for $q = 181$ and $\alpha = 2$.
6. Same as Exercise 5 for $a = 7531$, $q = 8101$, and $\alpha = 6$.
7. An example of a binary irreducible Goppa Code C is the linear $(8, 2)$ code C consisting of the four codewords

   $$00000000,\ 00111111,\ 11001011,\ 11110100$$

   of minimum distance 5. A generator matrix of this code is

   $$\mathbf{G} = \begin{pmatrix} 1 & 1 & 0 & 0 & 1 & 0 & 1 & 1 \\ 0 & 0 & 1 & 1 & 1 & 1 & 1 & 1 \end{pmatrix}.$$

   Let $\mathbf{S} = \begin{pmatrix} 1 & 1 \\ 0 & 1 \end{pmatrix}$ and

$$\mathbf{P} = \mathbf{P}^{-1} = \begin{pmatrix} 1 & 0 & 0 & 0 & 0 & 0 & 0 & 0 \\ 0 & 0 & 1 & 0 & 0 & 0 & 0 & 0 \\ 0 & 1 & 0 & 0 & 0 & 0 & 0 & 0 \\ 0 & 0 & 0 & 1 & 0 & 0 & 0 & 0 \\ 0 & 0 & 0 & 0 & 1 & 0 & 0 & 0 \\ 0 & 0 & 0 & 0 & 0 & 0 & 1 & 0 \\ 0 & 0 & 0 & 0 & 0 & 1 & 0 & 0 \\ 0 & 0 & 0 & 0 & 0 & 0 & 0 & 1 \end{pmatrix}$$

and let $\mathbf{z} = 10000000$. In a McEliece public key cryptosystem based on this code, determine the public key, encipher the plaintext $\mathbf{x} = 01$ as $\mathbf{y}$, and decipher $\mathbf{y}$ using nearest-neighbor decoding.

8. Encipher **ALGEBRA IS GREAT** by using the method described in 23.12. Decipher the message $(14800, 17152, 17167, 2044, 11913, 9718)$.

9. Use the knapsack scheme with $\mathbf{y} = (23, 57, 91, 179, 353)$, $q = 719$, and $k = 299$ to encipher the binary number 10110. Show how to decipher the cryptogram.

10. Let $\mathbf{y} = (14, 28, 56, 82, 90, 132, 197, 284, 341, 455)$ be a knapsack vector. How many binary vectors can you find such that $\mathbf{a} \cdot \mathbf{y} = 515$? Repeat the question for 516 instead of 515.

Notes

Cryptography is concerned with the protection of an established communication channel against (passive) eavesdropping and (active) forging, thus secrecy and authenticity being the two main concerns. Today cryptography draws heavily from mathematics and computer science. The importance of mathematics in cryptography was already recognized by the famous algebraist A. Adrian Albert, who said in 1941: "It would not be an exaggeration to state that abstract cryptography is identical with abstract mathematics."

Cryptography is a subject which has evolved rapidly recently, in particular in its applications in data security. The number of publications in cryptography has also grown rapidly. There are now several excellent books on cryptography available: Beker & Piper (1982) includes interesting applications of finite fields, Denning (1983) and Konheim

(1981) present a wealth of stimulating mathematical aspects of the subject. Denning (1983) includes an excellent introduction to encryption algorithms, cryptographic techniques, and chapters on access control, information flow control and inference control together with extensive bibliographical references. Myer & Matyas (1982) contains many references to applications. Whereas Galland (1970) and Sinkov (1978) cover the older literature and the mathematically more elementary aspects of cryptography, see Lidl & Niederreiter (1994). Other books, some of them of a survey nature, are Beth (1995), Beth, Frisch & Simmons (1992), Beth, Hess & Wirl (1983), Koblitz (1987), Leiss (1982), Merkle (1978), Patterson (1987), Salomaa (1990), Seberry & Pieprzyk (1989), Simmons (1979), Simmons (1992), van Tilborg (1988), and Welsh (1988). There are also a number of survey articles on various aspects of cryptography, see for instance Lidl & Niederreiter (1994). The NPL published a series of annotated bibliographies on data security and cryptology. There are also journals, *Cryptologia* and *Journal of Cryptology*, devoted entirely to this subject area. The number of annual conferences and workshops on cryptography gives an indication of the vigor and fast development of this discipline. Since this is a rather new and fascinating subject, we mention a few aspects in the history of cryptography with some further references to the bibliography. Kahn (1967) gives an exciting and mostly nonmathematical account of the historical development of cryptography. The first system of military cryptography was established by the Spartans in Ancient Greece in the fifth century B.C. They used a transposition cipher which was implemented on an apparatus called "skytale"—a staff of wood around which a strip of papyrus or leather was wrapped. The secret message is written on the papyrus down the length of the staff, the papyrus is then unwound and sent to the recipient. The Greek writer Polybius devised a cryptosystem which has as one of its characteristics the conversion of letters to numbers. Cryptology proper, which comprises both cryptography and cryptanalysis, was first discovered and described by the Arabs in 1412. Kahn traces the history of cryptography in detail over the following centuries, with emphasis on the important developments earlier in the twentieth century, particularly resulting from two World Wars.

It was recognized by Alberti (1466) and Viete (1589) many centuries ago that monoalphabetic substitution offers no security. F. W. Kasiski (1863) laid the foundation for breaking polyalphabetic substitutions with periodic or nonperiodic keys, if the same key is used several times. This lead to the development of running keys for one-time use. J. C. Mauborgne, of the US Army Signal Corps, introduced such random keys in 1918 based on Vernam ciphers, which had been invented for teletype machines by G. S. Vernam a year before. Claude Shannon studied the probabilistic nature of the one-time key and defined the concept of "perfect secrecy". The theoretical basis of this and other systems was established by information theory. One-time pads require extremely long keys and are therefore prohibitively expensive in most applications.

The period after World War I saw the development of special purpose machines for enciphering. Later the development of digital computers greatly facilitated the search for better cryptographic methods. World War II brought mathematicians in close contact with cryptology; for example, Alan Turing in England and A. A. Albert in the United States, who organized mathematicians for the work in cryptology. In 1940, the British were able to break the German ciphers produced by the ENIGMA machines, which are rotor machines for polyalphabetic enciphering. It is said that some 200,000 of these machines were in use. Another example of a cryptographic machine is the American Hagelin C-36 machine or M-209 converter. It was used by the US Army until the early 1950s.

United States cryptologists under the leadership of W. F. Friedman succeeded in cracking the Japanese PURPLE ciphers in 1941. German cryptanalysts under H. Rohrbach were able to break one of the early American M-138-A ciphers, which were then changed into new cipher machines, such as the M-134-C.

Only from the mid-1970s on was there also a wide-spread commercial interest in cryptography. Incentives for this development came from the need for protecting data, computer-supported message systems such as electronic mail or industrial espionage, to mention a few.

In the early 1970s IBM developed and implemented the system LUZIFER, which uses a transformation that alternatively applies substitution and transposition. This system was then modified into DES, the National Bureau of Standard's Data Encryption Standard. An interesting and rather new area of cryptography is the problem of secure speech transmission. There are, essentially, two techniques for encrypting speech: digital and analogue. Both techniques of voice scrambling are described in Beker & Piper (1982).

Simmons (1979) gives a survey of the state of cryptography, comparing classical methods (symmetric encryption) with newer developments of public-key cryptography (asymmetric encryption). Symmetric cryptosystems involve substitution and transposition as the primary operation; Caesar ciphers, Vigenères, Hill ciphers, and the DES fall into this category. One important advantage of the DES is that it is very fast and thus can be used for high-volume, high-speed enciphering.

The concept of a public-key cryptosystem is due to Diffie & Hellman (1976): as an important implementation the RSA cryptosystem was developed by Rivest, Shamir & Adleman (1978). The trapdoor knapsack is another implementation of a public key cryptosystem given by Merkle (1982) and Hellman (1979).

One of the public key cryptosystems that has recently received some popularity is PGP, which stands for "Pretty Good Privacy." It is a hybrid of the RSA and IDEA, the latter being a block cipher. PGP uses the RSA for key exchanges and IDEA for encryption (see Garfinkel, 1996).

The security of most cryptographic systems depends on the computational difficulty to discover the plaintext without knowledge of the key. The two modern disciplines which study the difficulty of solving computational problems are complexity theory and analysis of algorithms. The ultimate test of a system is the attack by a skilled cryptanalyst under very favorable conditions such as a *chosen plaintext attack*, where the cryptanalyst can submit an unlimited number of plaintext messages of his choice and examine the resulting cryptograms.

The material covered in Chapter 5 represents only a small part of the work in cryptology. Other parts include stream ciphers, see also §33, interactive proof systems and zero-knowledge smart cards, elliptic curve cryptosystems, threshold schemes, to mention a few.

In particular, considerable emphasis has been given to zero-knowledge proofs in recent years. A zero-knowledge proof permits A to demonstrate to B that she knows something (e.g., how to color a map with three colors), but gives him no way of conveying this assurance to anybody else, see Simmons (1992).

Today the need for governmental (diplomatic and military) cryptography and for private and commercial cryptography is recognized. The times are now over when the Secretary of State under President Hoover of the United States dismissed on ethical grounds the Black Chamber of the State Department, with the explanation "Gentlemen do not read each other's mail."

6 CHAPTER Applications of Groups

We now turn to some applications of group theory. The first application makes use of the observation that computing in $\mathbb{Z}$ can be replaced by computing in $\mathbb{Z}_n$, if n is sufficiently large; $\mathbb{Z}_n$ can be decomposed into a direct product of groups with prime power order, so we can do the computations in parallel in the smaller components. In §25, we look at permutation groups and apply these to combinatorial problems of finding the number of "essentially different" configurations, where configurations are considered as "essentially equal" if the second one can be obtained from the first one, e.g., by a rotation or reflection.

In §26, we study how to extract three-dimensional information out of two-dimensional images. Using representation theory we can determine the essential parameters and find suitable bases so that the relevant "3D-recovery equations" become very simple. Finally, we study all motions which fix a certain object (e.g., a molecule). The "symmetry operations" form a group, the symmetry group of the object under consideration. Objects with isomorphic symmetry groups enjoy similar properties. Again we employ representation theory to simplify eigenvalue problems for these objects. Group theory turns out to be the main tool to study symmetries.

§24 Fast Adding

Group and ring theory have interesting applications in the design of computer software. They result in computing techniques which speed up calculations considerably. If we consider two numbers such as $a = 37$ and $b = 56$, it makes no difference whether we add them as natural numbers or as members of some $\mathbb{Z}_n$ with n sufficiently large, say $n = 140$ in our case. The only requirement is that $a + b < n$.

We now decompose n canonically as

$$n = p_1^{t_1} p_2^{t_2} \cdots p_k^{t_k}.$$

The Principal Theorem on finite abelian groups (Theorem 10.25) shows that

$$\mathbb{Z}_n \cong \mathbb{Z}_{p_1^{t_1}} \oplus \cdots \oplus \mathbb{Z}_{p_k^{t_k}}.$$

An isomorphism is given by $h\colon [x]_n \mapsto ([x]_{p_1^{t_1}}, \dots, [x]_{p_k^{t_k}})$, where $[x]_m$ denotes the residue class of x modulo m. Surjectivity of h means that for all numbers $y_1, \dots, y_k \in \mathbb{Z}$ some $x \in \mathbb{Z}$ can be found with $x \equiv y_1 \pmod{p_1^{t_1}}, \dots, x \equiv y_k \pmod{p_k^{t_k}}$, a result which was already basically known in ancient China. It is therefore known as the ***Chinese Remainder Theorem*** (cf. Theorem 15.1). It is not hard to find this solution x explicitly. Similar to the proof of 15.1, form $q_i := n \cdot p_i^{-t_i}$. Because $\gcd(p_i^{t_i}, q_i) = 1$, q_i has a multiplicative inverse r_i in $\mathbb{Z}_{p_i^{t_i}}$. Thus

$$x := y_1 q_1 r_1 + \cdots + y_k q_k r_k,$$

and x is unique modulo n. A quick algorithm ("Garner's algorithm") can be found in Geddes, Czapor & Labahn (1993, p. 176).

The importance of this theorem lies in the fact that we can replace the addition of large natural numbers by parallel "small" simultaneous additions. We illustrate this by the example mentioned above:

$$n = 140 = 2^2 \cdot 5 \cdot 7,$$

$$\begin{array}{llr} a = 37 \to [37]_{140} \to ([37]_4, [37]_5, [37]_7) = & ([1]_4, [2]_5, [2]_7) & + \\ b = 56 \to [56]_{140} \to ([56]_4, [56]_5, [56]_7) = & ([0]_4, [1]_5, [0]_7) & \\ \hline a + b & ([1]_4, [3]_5, [2]_7) & \end{array}$$

Now we have to solve

$$x \equiv 1 \pmod 4),$$
$$x \equiv 3 \pmod 5),$$
$$x \equiv 2 \pmod 7),$$

by the method mentioned above. We get $x = 93$, hence $37 + 56 = 93$.

Of course, using this method does not make sense if we just have to add two numbers. If, however, some numbers are entered in a computer which has to work with them a great number of times (computing determinants, for instance), it definitely does make sense to transform the numbers to residue classes and to calculate in parallel in small $\mathbb{Z}_{p^t}$'s. This avoids an exponential growth of the coefficients.

Before we really adopt this method, we estimate the time we save by using it. The following gives a rough estimate. Adding devices in computers consist of a great number of "gates." Each gate has a small number r of inlets ($r \leq 4$ is typical), and one outlet. Each gate requires a certain unit time (10^{-8} seconds, say) to produce the output. We follow Dornhoff & Hohn (1978).

24.1. Notation. For $x \in \mathbb{R}$, let $\lceil x \rceil$ be the smallest integer $\geq x$.

24.2. Theorem. *The time required to produce a single output essentially depending on m inputs by means of r-input gates is* $\lceil \log_r m \rceil$.

Proof. In one time unit r inputs can be processed, in two time units r^2 inputs can be processed, and so on. Hence in t time units r^t inputs can be processed. Since we have to handle m inputs, we must have $r^t \geq m = r^{\log_r m}$, whence $t \geq \log_r m$, so $t \geq \lceil \log_r m \rceil$. □

Here is a sketch for $r = 3$ and $m = 8$:

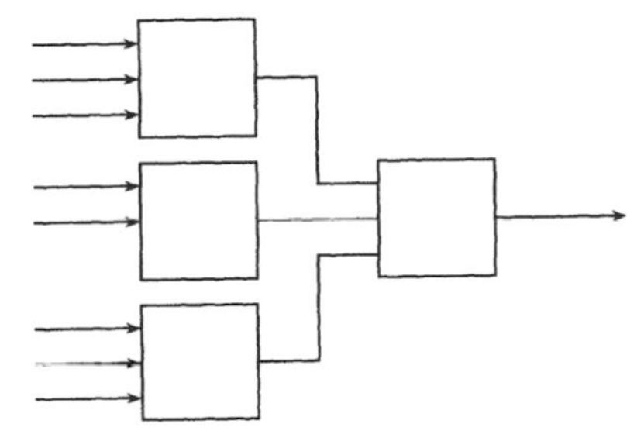

We need $2 = \lceil \log_3 8 \rceil$ time units to produce the single output.

If we add two m-digit binary numbers, we get $m+1$ outputs; one can easily see that the last output (the "carry-over") depends on all inputs. So we get:

24.3. Theorem. *Usual addition of two m-digit binary numbers needs* $\lceil \log_r 2m \rceil$ *time units.*

Proof. This follows from Theorem 24.2 by noting that the sum of two m-digit numbers $a_1 a_2 \ldots a_m$ and $b_1 b_2 \ldots b_m$ depends on the $2m$ inputs $a_1, \ldots, a_m, b_1, \ldots, b_m$. □

24.4. Theorem. *Addition modulo n (i.e., addition in* $\mathbb{Z}_n$*) in binary form consumes* $\lceil \log_r (2 \lceil \log_2 n \rceil) \rceil$ *time units.*

Proof. In order to write $0, 1, \ldots, n-1$ in binary form we need m-digit binary numbers with $2^m \geq n$, whence $m \geq \log_2 n$, so $m \geq \lceil \log_2 n \rceil$. Now apply 24.3. □

Similarly, we get the result for the proposed method of addition.

24.5. Theorem. *Addition of two (binary) numbers in* $\mathbb{Z}_n$ *by the method described at the beginning of this section needs* $\lceil \log_r (2 \lceil \log_2 n' \rceil) \rceil$ *time units. Here, n′ is the greatest prime power in the decomposition of n.*

Hence we will choose n in such a way that n' is as small as possible. It is wise to fix n' and look for a large n.

24.6. Example. We want $n' \leq 50$ and can choose n in an optimal way as

$$n := 2^5 \cdot 3^3 \cdot 5^2 \cdot 7^2 \cdot 11 \cdot 13 \cdot 17 \cdot 19 \cdot 23 \cdot 29 \cdot 31 \cdot 37 \cdot 41 \cdot 43 \cdot 47.$$

Now $n > 3 \cdot 10^{21}$ and $n' = 49$. So we can add numbers with 21 decimal digits. If we choose $r = 3$, the method in Theorem 24.4 needs

$$\lceil \log_3 (2 \lceil \log_2 (3 \cdot 10^{21}) \rceil) \rceil = \lceil 4.52 \rceil = 5 \text{ time units.}$$

With 24.5 we get (again with $r = 3$)

$$\lceil \log_3 (2 \lceil \log_2 49 \rceil) \rceil = \lceil 2.26 \rceil = 3 \text{ time units.}$$

Thus we can add nearly twice as fast.

24.7. Remark. Since the Chinese Remainder Theorem holds for the rings $\mathbb{Z}_n$ as well as for their additive groups, we get a method for fast multiplication as well. In general, this method is applicable to all unique factorization domains such as $F[x]$ (F a field). We shall return to this idea in §34.

Exercises

1. Compute "quickly" $8 \cdot (6 \cdot 43 + 7 \cdot 17 - 15^2)$ modulo 2520.
2. What is the minimal choice for n' if we want $n > 10^{12}$? Estimate the computing times in both cases as in 24.6 (with $r = 4$).
3. Add 516 and 1687 in two different ways, using the Chinese Remainder Theorem or performing usual addition in $\mathbb{Z}$, and compare the times required to obtain the sum.
4. Explain a method of "fast multiplication" in detail.
5. Try to compute $(3x+4)(x^2-1)^2+x^3+x+1$ modulo $x^2(x+1)(x+2)(x+3)(x+4)$ instead of in $\mathbb{R}[x]$.

§25 Pólya's Theory of Enumeration

Groups of permutations (i.e., subgroups of the symmetric group S_M) were the first groups which were studied. The notion of an abstract group came about half a century later by Cayley in 1854. N. H. Abel and E. Galois, for instance, studied groups of permutations of zeros of polynomials. Moreover, it is a big understatement to say that one is "only" studying permutation groups. The following classical theorem tells us that all groups are—up to isomorphism—permutation groups, more precisely, every group can be embedded into a permutation group.

25.1. Theorem (*Cayley's Theorem*). *If G is a group, then*

$$G \hookrightarrow S_G.$$

Proof. The map $h: G \to S_G;\ g \mapsto \varphi_g$ with $\varphi_g: G \to G;\ x \mapsto gx$ does the embedding job: $\varphi_{gg'} = \varphi_g \circ \varphi_{g'}$ ensures that h is a homomorphism, and $\operatorname{Ker} h = \{g \in G \mid \varphi_g = \mathrm{id}\} = \{1\}$, so h is a monomorphism. □

It follows that every group of order $n \in \mathbb{N}$ can be embedded in S_n. In writing products $\pi\sigma$ of permutations $\pi, \sigma \in S_n$ we consider π and σ to be functions. Hence in $\pi\sigma$ we first perform σ, then π.

25.2. Definition. $\pi \in S_n$ is called a ***cycle of length*** r if there is a subset $\{i_1, \ldots, i_r\}$ of $\{1, \ldots, n\}$ with $\pi(i_1) = i_2$, $\pi(i_2) = i_3, \ldots, \pi(i_r) = i_1$ and $\pi(j) = j$ for all $j \notin \{i_1, \ldots, i_r\}$. We will then write $\pi = (i_1, i_2, \ldots, i_r)$. Cycles of length 2 are called ***transpositions***. Cycles of length 1 are equal to the identity permutation, so they are often omitted (however, we shall need them in 25.20).

25.3. Examples.

(i) $\begin{pmatrix} 1 & 2 & 3 & 4 & 5 \\ 1 & 4 & 3 & 2 & 5 \end{pmatrix} = (2, 4) = (4, 2) = (1)(2, 4)(3)$ is a transposition in S_5.

(ii) $\begin{pmatrix} 1 & 2 & 3 & 4 & 5 \\ 1 & 5 & 4 & 2 & 3 \end{pmatrix} = (2, 5, 3, 4)$ is a 4-cycle in S_5.

(iii) $S_3 = \{\text{id}, (1, 2), (1, 3), (2, 3), (1, 2, 3), (1, 3, 2)\}$.

So every $\pi \in S_3$ is a cycle. This is not true any more from S_4 upward. But we shall see in 25.6 that every permutation in S_n is a product of cycles.

25.4. Definition. For $\pi \in S_n$ let $W_\pi := \{i \in \{1, \ldots, n\} \mid \pi(i) \neq i\}$ be the ***domain of action*** of π.

The following result shows that permutations with disjoint domains of action commute. The easy proof is left to the reader.

25.5. Theorem. *If $\pi, \sigma \in S_n$ with $W_\pi \cap W_\sigma = \emptyset$, then $\pi\sigma = \sigma\pi$.*

25.6. Theorem. *Every $\pi \in S_n$ can be written as a product of cycles with disjoint domains of action. This decomposition is unique up to the arrangement of the cycles.*

The decomposition of 25.6 is called "***canonical***." We shall give an example; the proof of 25.6 is just a formalization of this example and is omitted.

25.7. Example. Let

$$\pi = \begin{pmatrix} 1 & 2 & 3 & 4 & 5 & 6 & 7 & 8 \\ 2 & 1 & 5 & 6 & 7 & 4 & 3 & 8 \end{pmatrix} \in S_8.$$

π moves 1 into 2 and 2 into 1; this gives the first cycle $(1, 2)$. Next we observe that 3 is transferred into 5, 5 into 7, 7 into 3; second cycle: $(3, 5, 7)$. 4 and 6 are transposed, which yields $(4, 6)$ as the third cycle. Finally, 8 is left fixed. Hence $\pi = (1, 2)(3, 5, 7)(4, 6)(8)$. By 25.5, there is no reason to worry about the order of these cycles.

Without loss of generality we assume that the cycles in a canonical decomposition are ordered in such a way that their length is not decreasing. This gives, for instance, $\pi = (8)(1, 2)(4, 6)(3, 5, 7)$ in 25.7. Two canonical decompositions are called ***similar*** if the sequences of the lengths of the cycles involved are equal. Hence $\pi = (8)(1, 2)(4, 6)(3, 5, 7)$ and $\sigma = (6)(1, 3)(2, 4)(5, 8, 7)$ are different elements of S_8 having similar decompositions. These considerations prove very useful in looking at the structure of S_n: Let $n \in \mathbb{N}$. A ***partition*** of n is a sequence $(a_1, a_2, \dots, a_s) \in \mathbb{N}^s$ with $s \in \mathbb{N}$, $a_1 \leq a_2 \leq \cdots \leq a_s$ and $a_1 + a_2 + \cdots + a_s = n$. Let $P(n)$ be the number of all different partitions of n. We give a few values:

n	1	2	3	4	5	6	7	8	9	10	11	12
$P(n)$	1	2	3	5	7	11	15	22	30	42	56	77

Let $C(G) := \{c \in G \mid cg = gc \text{ for all } g \in G\}$ denote the ***center*** of G, as in 10.30.

25.8. Theorem.

(i) *If $\pi, \sigma \in S_n$, then $\sigma\pi\sigma^{-1}$ can be obtained from the canonical decomposition of π by replacing every i in its cycles by $\sigma(i)$.*

(ii) *Two cycles are conjugate if and only if they are of the same length.*

(iii) *$\pi_1, \pi_2 \in S_n$ are conjugate if and only if they have similar canonical decompositions.*

(iv) *$P(n)$ is the class number of S_n.*

(v) *For all $n \geq 3$, $C(S_n) = \{\text{id}\}$ and $C(S_n) = S_n$ for $n = 1, 2$.*

Proof.

(i) If $\pi = \xi_1\xi_2 \dots \xi_m$ is the canonical decomposition of π into cycles ξ_i, then $\sigma\pi\sigma^{-1} = (\sigma\xi_1\sigma^{-1})(\sigma\xi_2\sigma^{-1}) \dots (\sigma\xi_m\sigma^{-1})$, so that it suffices to look at a cycle ξ. Let $\xi = (i_1, \dots, i_r)$. If $1 \leq k \leq r - 1$, then $\xi(i_k) = i_{k+1}$, hence $(\sigma\xi\sigma^{-1})(\sigma(i_k)) = \sigma(i_{k+1})$. If $k = r$, then $\xi(i_k) = i_1$, whence $(\sigma\xi\sigma^{-1})(\sigma(i_k)) = \sigma(i_1)$. If $i \notin W_\xi$, then $\xi(i) = i$, and so $(\sigma\xi\sigma^{-1})(\sigma(i)) = \sigma(i)$. Thus $\sigma\xi\sigma^{-1} = (\sigma(i_1), \dots, \sigma(i_r))$.

(ii), (iii), and (iv) now follow from (i).

(v) Let $\pi \in S_n$, $\pi \neq \mathrm{id}$. Then there is some i with $\pi(i) \neq i$. If $n \geq 3$ then there is some $k \in \{1, \dots, n\}$ with $i \neq k \neq \pi(i)$. By (i) we get $\pi(i,k)\pi^{-1} = (\pi(i), \pi(k)) \neq (i,k)$, whence $\pi \notin C(S_n)$. Hence $C(S_n) = \{\mathrm{id}\}$ for $n \geq 3$. Since S_1 and S_2 are abelian, $C(S_n) = S_n$ in these cases. □

Since $(i_1, i_2, \dots, i_r) = (i_1, i_r)(i_1, i_{r-1}) \dots (i_1, i_2)$ we get from 25.6:

25.9. Theorem. *Every permutation is a product of transpositions.*

In contrast to 25.6, this type of decomposition is not unique.

25.10. Definition. For $\pi \in S_n$,

$$\mathrm{sign}(\pi) := \prod_{i>j} \frac{\pi(i) - \pi(j)}{i - j}$$

is called the ***signature*** of π.

It is easy to compute $\mathrm{sign}(\pi)$ by means of 25.6 or 25.7:

25.11. Theorem. *Let $n > 1$.*

(i) *sign: $S_n \to (\{1, -1\}, \cdot)$ is an epimorphism.*

(ii) *If $\pi = \tau_1 \tau_2 \dots \tau_s$, where all τ_i are transpositions, then* $\mathrm{sign}(\pi) = (-1)^s$.

(iii) *If ξ is a cycle of length k, then* $\mathrm{sign}(\xi) = (-1)^{k-1}$.

(iv) *If $\pi = \xi_1 \xi_2 \dots \xi_r$ is a canonical decomposition of π into cycles of length $k_1, \dots, k_r$, respectively, then* $\mathrm{sign}(\pi) = (-1)^{k_1 + k_2 + \dots + k_r - r}$.

(v) $A_n := \mathrm{Ker\,sign} = \{\pi \in S_n \mid \mathrm{sign}(\pi) = 1\}$ *is a normal subgroup of S_n.*

(vi) $[S_n : A_n] = 2$, *so* $|A_n| = n!/2$.

Proof.

(i) For $\pi, \sigma \in S_n$ we get

$$\begin{aligned}
\mathrm{sign}(\pi\sigma) &= \prod_{i>j} \frac{\pi(\sigma(i)) - \pi(\sigma(j))}{i - j} \\
&= \prod_{i>j} \frac{\pi(\sigma(i)) - \pi(\sigma(j))}{\sigma(i) - \sigma(j)} \cdot \frac{\sigma(i) - \sigma(j)}{i - j} \\
&= \left(\prod_{\sigma(i)>\sigma(j)} \frac{\pi(\sigma(i)) - \pi(\sigma(j))}{\sigma(i) - \sigma(j)} \right) \cdot \left(\prod_{i>j} \frac{\sigma(i) - \sigma(j)}{i - j} \right) \\
&= \mathrm{sign}(\pi)\,\mathrm{sign}(\sigma).
\end{aligned}$$

For the last equation but one observe that for $i < j$ and $\sigma(i) > \sigma(j)$ we have

$$\frac{\pi(\sigma(i)) - \pi(\sigma(j))}{\sigma(i) - \sigma(j)} = \frac{\pi(\sigma(j)) - \pi(\sigma(i))}{\sigma(j) - \sigma(i)}.$$

Obviously, the signature can take values in $\{1, -1\}$ only. Because $\operatorname{sign}(\tau) = -1$ for every transposition, $\operatorname{Im}\operatorname{sign} = \{1, -1\}$.

(ii) follows from (i) since $\operatorname{sign}(\tau) = -1$ for every transposition (Exercise 1).

(iii) is a special case of (ii) using the decomposition before 25.9.

(iv) If $\pi = \xi_1 \dots \xi_r$, then by (ii)

$$\begin{aligned}\operatorname{sign}(\pi) &= (\operatorname{sign}\xi_1)\dots(\operatorname{sign}\xi_r)\\ &= (-1)^{k_1-1}\dots(-1)^{k_r-1} = (-1)^{k_1+\dots+k_r-r}.\end{aligned}$$

(v) The kernel of every homomorphism is a normal subgroup due to 10.14(ii).

(vi) follows from 10.17. □

25.12. Definition. $A_n = \{\pi \in S_n \mid \operatorname{sign}(\pi) = 1\}$ is called the ***alternating group***. Permutations in A_n are also called ***even***.

Note that A_3 is abelian because $|A_3| = 3!/2 = 3$, whence $A_3 \cong \mathbb{Z}_3$. But A_n is nonabelian if $n \geq 4$.

We list several important properties of these alternating groups (without proofs).

25.13. Theorem.

(i) *A_n is the subgroup of S_n generated by the 3-cycles.*

(ii) *A_4 is not simple, since it has a normal subgroup isomorphic to $\mathbb{Z}_2 \times \mathbb{Z}_2$. But A_4, a group of order 12, has no subgroup of order 6.*

(iii) *For $n \geq 5$, A_n is simple and nonabelian. (This is the deep reason for the fact that there cannot exist "solution formulas" for equations of degree $\geq$ 5.) Also, A_5 has order 60 and is the smallest nonabelian simple group.*

(iv) *$C(A_n) = \{\mathrm{id}\}$ for $n \geq 4$; $C(A_n) = A_n$ for $n \in \{1, 2, 3\}$.*

(v) *$\bigcup_{n\in\mathbb{N}} A_n =: A$ is an infinite (nonabelian) simple group.*

Now suppose that X is a set and $G \leq S_X$. Then every $\pi \in G$ can be thought of as being an operator acting on X by sending $x \in X$ into $\pi(x)$.

We are interested in what happens to a fixed $x \in X$ under all $\pi \in G$. The easy proofs of the following assertions are left to the reader.

25.14. Theorem and Definition. *Let $G \leq S_X$ and $x, y \in X$. Then:*

(i) *x and y are called **G-equivalent** (denoted by $x \sim_G y$) if there is a $\pi \in G$ with $\pi(x) = y$.*

(ii) *$\sim_G$ is an equivalence relation on X.*

(iii) *The equivalence classes $\operatorname{Orb}(x) := \{y \in X \mid x \sim_G y\}$ are called **orbits** (of G on X).*

(iv) *For every $x \in X$, $\operatorname{Stab}(x) := \{\pi \in G \mid \pi(x) = x\}$ is a subgroup of G, called the **stabilizer** of x. As in 10.12(ii), we have for all $\pi, \tau \in G$:*

$$\pi \sim_{\operatorname{Stab}(x)} \tau \iff \pi\tau^{-1} \in \operatorname{Stab}(x).$$

25.15. Examples.

(i) Let G be a group, $S \leq G$ and $X = G$. By 25.1, S can be considered as a subgroup of $S_G = S_X$. If $g \in G$ the orbit $\operatorname{Orb}(g) = Sg$ (the right coset of g with respect to S) and $\operatorname{Stab}(g) = \{s \in S \mid sg = g\} = \{1\}$.

(ii) Let G be a group, $X = G$ and $\operatorname{Inn} G \leq S_G$, where $\operatorname{Inn} G = \{\varphi_x \mid x \in G\}$ is the set of all ***inner automorphisms*** $\varphi_x \colon G \to G$; $g \mapsto xgx^{-1}$. Then for each g in G we get $\operatorname{Orb}(g) = \{\varphi_x(g) \mid \varphi_x \in \operatorname{Inn} G\} =$ conjugacy class of g, and $\operatorname{Stab}(g) = \{\varphi_x \in \operatorname{Inn} G \mid xgx^{-1} = \varphi_x(g) = g\} = \{\varphi_x \in \operatorname{Inn} G \mid xg = gx\}$.

(iii) Let Y be a set. $\mathbb{Z}_n$ operates on $X := Y^n$ by "shifting the components cyclically": if $k \in \mathbb{Z}_n$, let $\pi_k \colon (y_1, \dots, y_n) \mapsto (y_k, y_{k+1}, \dots, y_{k-1})$. For $n = 3$, for instance, the orbit of (y_1, y_2, y_3) is $\{(y_1, y_2, y_3), (y_2, y_3, y_1), (y_3, y_1, y_2)\}$, and $\operatorname{Stab}(y_1, y_2, y_3) = \mathbb{Z}_3$ if $y_1 = y_2 = y_3$, and $= \{0\}$, otherwise.

These concepts turn out to be very useful in several applications. As an example, we mention an application to chemistry, due to G. Pólya.

25.16. Example. From the carbon ring (a) we can obtain a number of molecules by attaching hydrogen (H-) atoms or CH_3-groups in the places ① - ⑥ in (b). For instance, we can obtain xylole (c) and benzene (d) as shown in Figure 25.1.

Obviously, (c′) gives xylole as well, as shown in Figure 25.2.

The following problem arises: How many chemically different molecules can be obtained in this way? Altogether, there are 2^6 possi-

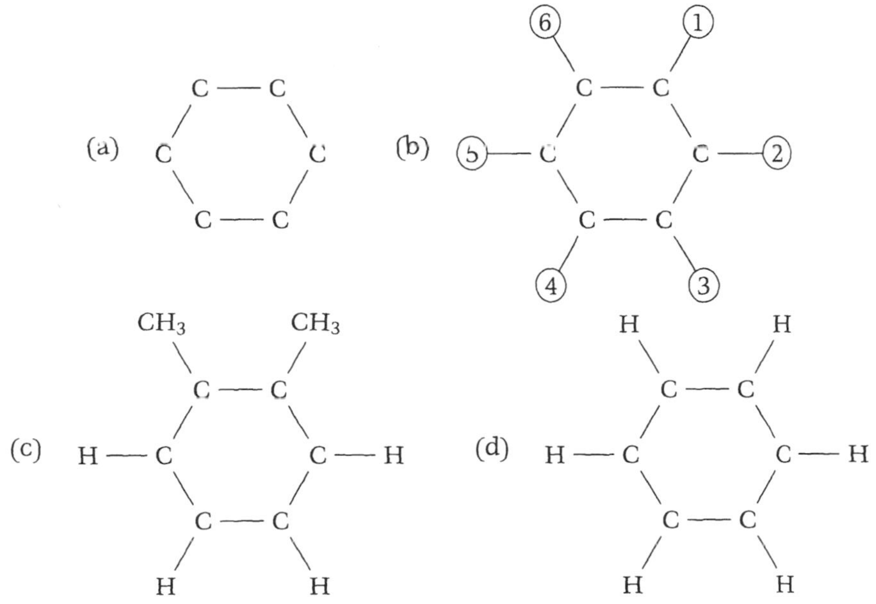

```
                                   (6)          (1)
                                      \        /
        C — C                          C — C
       /     \                        /     \
(a)   C       C          (b)  (5) — C         C — (2)
       \     /                        \     /
        C — C                          C — C
                                      /     \
                                   (4)       (3)

       CH3       CH3                    H         H
          \     /                        \       /
           C — C                          C — C
          /     \                        /     \
(c)  H — C       C — H          (d)  H — C       C — H
          \     /                        \     /
           C — C                          C — C
          /     \                        /     \
         H       H                      H       H
```

FIGURE 25.1

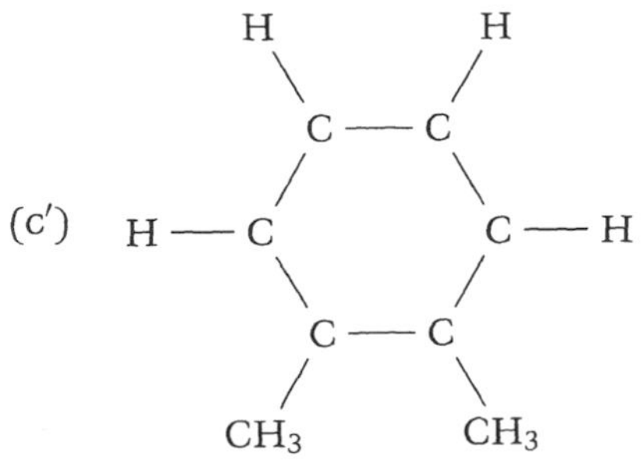

```
              H       H
               \     /
                C — C
               /     \
(c′)      H — C       C — H
               \     /
                C — C
               /     \
            CH3       CH3
```

FIGURE 25.2

bilities to attach either H or CH_3 in ① - ⑥. But how many attachments coincide chemically?

In order to solve this problem we can employ the following result of Burnside, Cauchy, and Frobenius.

25.17. Theorem (*Burnside's Lemma*). *Let X be finite and $G \leq S_X$. For every $x \in X$, $|\mathrm{Orb}(x)|$ divides $|G|$, and the number n of different orbits of X*

under G is given by

$$n = \frac{1}{|G|}\sum_{g\in G}|\mathrm{Fix}(g)| = \frac{1}{|G|}\sum_{x\in X}|\mathrm{Stab}(x)|,$$

where $\mathrm{Fix}(g) := \{x \in X \mid g(x) = x\}$. *Also* $[G : \mathrm{Stab}(x)] = |\mathrm{Orb}(x)|$, *and hence* $|\mathrm{Stab}(x)||\mathrm{Orb}(x)| = |G|$.

Proof. First we compute $|\mathrm{Orb}(x)|$. Let $x \in X$ and

$$f\colon \mathrm{Orb}(x) \to G/{\sim_{\mathrm{Stab}(x)}}; \qquad g(x) \mapsto g\,\mathrm{Stab}(x).$$

Since

$$\begin{aligned} g_1(x) = g_2(x) &\iff g_2^{-1}g_1(x) = x \iff g_2^{-1}g_1 \in \mathrm{Stab}(x)\\ &\iff g_1 \sim_{\mathrm{Stab}(x)} g_2 \iff g_1\,\mathrm{Stab}(x) = g_2\,\mathrm{Stab}(x), \end{aligned}$$

f is well defined. Obviously f is bijective and hence $|\mathrm{Orb}(x)| = |G/{\sim_{\mathrm{Stab}(x)}}| = [G : \mathrm{Stab}(x)]$, so $|\mathrm{Orb}(x)|$ divides $|G|$.

Now we compute the cardinality of $\{(g,x) \mid g \in G, x \in X, g(x) = x\}$ in two different ways:

$$\sum_{x\in X}|\mathrm{Stab}(x)| = |\{(g,x) \mid g \in G, x \in X, g(x) = x\}| = \sum_{g\in G}|\mathrm{Fix}(g)|.$$

Let us choose representatives $x_1, \dots, x_n$ from the n orbits. Since $\mathrm{Orb}(x) = \mathrm{Orb}(y)$ implies

$$\begin{aligned} |\mathrm{Stab}(x)| &= \frac{|G|}{[G : \mathrm{Stab}(x)]} = \frac{|G|}{|\mathrm{Orb}(x)|} = \frac{|G|}{|\mathrm{Orb}(y)|} = \frac{|G|}{[G : \mathrm{Stab}(y)]}\\ &= |\mathrm{Stab}(y)|, \end{aligned}$$

we get

$$\begin{aligned} \sum_{x\in X}|\mathrm{Stab}(x)| &= \sum_{i=1}^{n}|\mathrm{Orb}(x_i)|\cdot|\mathrm{Stab}(x_i)| = \sum_{i=1}^{n}\frac{|G|}{|\mathrm{Stab}(x_i)|}\cdot|\mathrm{Stab}(x_i)|\\ &= n\cdot|G|. \end{aligned}$$

□

Now we use 25.17 for our problem 25.16.

25.16. Example (*continued*). Let us denote the said $2^6 = 64$ attachments by $\{x_1, \dots, x_{64}\} =: X$. Attaching x_i and x_j will yield the same molecule if and only if x_j can be obtained from x_i by means of a symmetry operation of the hexagon ① - ⑥, i.e., by means of an element of D_6, the dihedral group of order 12 (see 10.2(vi)). Hence the number n of different possible molecules we are looking for is just the number of different orbits of X under D_6. From 25.17 we get

$$n = \frac{1}{|D_6|} \sum_{g \in D_6} |\mathrm{Fix}(g)| = \frac{1}{12} \sum_{g \in D_6} |\mathrm{Fix}(g)|.$$

Now id fixes all elements, whence $|\mathrm{Fix(id)}| = 64$. A reflection r on the axis ① - ④ in (b) fixes the four attachments possible in ① and ④, and also the four other possible attachments in ② and ③ if they are the same as those in ⑥ and ⑤, respectively. Hence $|\mathrm{Fix}(r)| = 4 \cdot 4 = 16$, and so on. Altogether we get $n = \frac{1}{12} \cdot 156 = 13$ different molecules.

Observe that reflections in space usually yield molecules with different chemical properties.

We see that this enumeration can be applied to situations where we look for the number of possible "attachments." The result in 25.17 shows that n is the arithmetic mean of the $|\mathrm{Fix}(g)|$'s, and also of the $|\mathrm{Stab}(x)|$'s in G.

We can improve the formula in 25.17 by the remark that if g_1 and g_2 are conjugate, then $|\mathrm{Fix}(g_1)| = |\mathrm{Fix}(g_2)|$. Of course, this only helps in the nonabelian case. So we get

25.18. Theorem. *Let X be finite and $G \leq S_X$. Let $g_1, \ldots, g_r$ be a complete set of representatives for the conjugacy classes in $G/\sim$ and let k_i be the number of elements conjugate to g_i. Then the number n of orbits of X under G is given by*

$$n = \frac{1}{|G|}(k_1|\mathrm{Fix}(g_1)| + \cdots + k_r|\mathrm{Fix}(g_r)|).$$

We give a simple example in which we can use our knowledge about the conjugacy classes of S_3.

25.19. Example. Find the number n of essentially different possibilities for placing three elements from the set {A, B, C, D, E} at the three corners 1, 2, 3 of an equilateral triangle such that at least two letters are distinct.

Solution. $G = S_3$ acts on $\{1, 2, 3\}$ as the group of symmetries. Take, in the language of Theorem 25.18, $g_1 = \mathrm{id}$, $g_2 = (1, 2)$, and $g_3 = (1, 2, 3)$. Then $k_1 = 1$, $k_2 = 3$, $k_3 = 2$. Now X contains all triples (a, b, c) such that at least two of the three elements a, b, c are distinct. Hence:

$|X| = 5(4 + 4 \cdot 5) = 120$,

$|\text{Fix}(g_1)| = 120$,

$|\text{Fix}(g_2)| = 20$, since g_2 fixes precisely all (a, a, b),

$|\text{Fix}(g_3)| = 0$, since g_3 fixes exactly all (a, a, a); these combinations are not allowed.

Hence Theorem 25.18 gives us

$$n = \frac{1}{6}(120 + 60 + 0) = 30. \qquad \square$$

The results 25.17 and 25.18 are indeed useful for Example 25.19. A direct treatment of 25.19 would require an examination of all $\binom{120}{2} = 7140$ pairs of attachments with respect to being essentially different. There might, however, remain quite a bit to do in 25.17 and 25.18, especially if G is big and if there are many conjugacy classes. So we might still be dissatisfied with what we have accomplished so far. Also, we still have no tool for finding a representative in each class of essentially equal attachments.

25.20. Definition. Suppose that $\pi \in S_n$ decomposes into j_1 cycles of length 1, j_2 cycles of length 2, ..., j_n cycles of length n according to 25.6 (we then have $1j_1 + 2j_2 + \cdots + nj_n = n$). We then call $(j_1, \ldots, j_n)$ the ***cycle index*** of π. If $G \leq S_n$, then

$$Z(G) := \frac{1}{|G|} \sum_{\pi \in G} x_1^{j_1} x_2^{j_2} \cdots x_n^{j_n} \in \mathbb{Q}[x_1, \ldots, x_n]$$

(where $(j_1, \ldots, j_n)$ is the cycle index of π) is called the ***cycle index polynomial*** of G.

25.21. Example. In S_3, we have one permutation (namely id) with cycle index $(3, 0, 0)$, three permutations $((1, 2), (1, 3), (2, 3))$ with cycle index $(1, 1, 0)$ (since $(1, 2) = (3)(1, 2)$, and so on), and two permutations with cycle index $(0, 0, 1)$. Hence

$$Z(S_3) = \frac{1}{6}(x_1^3 + 3x_1x_2 + 2x_3).$$

It is not very hard, but it is lengthy, to determine the cycle index polynomial of the following groups which often appear as groups of symmetries. We give the following list without proof. Let $\mathbb{Z}$, D_n, S_n, and A_n act on $\{1, \ldots, n\}$ in the "natural way."

25.22. List of Some Cycle Index Polynomials.

(i)
$$Z(S_n) = \sum \frac{1}{(1^{j_1}j_1!)(2^{j_2}j_2!)\cdots(n^{j_n}j_n!)} x_1^{j_1}x_2^{j_2}\cdots x_n^{j_n},$$

where the summation goes over all $(j_1, \ldots, j_n) \in \mathbb{N}_0^n$ with $1j_1 + 2j_2 + \cdots + nj_n = n$. In particular, we get

$$\begin{aligned} Z(S_1) &= x_1, \\ Z(S_2) &= \frac{1}{2}(x_1^2 + x_2), \\ Z(S_3) &= \frac{1}{6}(x_1^3 + 3x_1x_2 + 2x_3), \\ Z(S_4) &= \frac{1}{24}(x_1^4 + 6x_1^2x_2 + 3x_2^2 + 8x_1x_3 + 6x_4). \end{aligned}$$

(ii)
$$Z(A_n) = \sum \frac{1 + (-1)^{j_2+j_4+\cdots}}{(1^{j_1}j_1!)(2^{j_2}j_2!)\cdots(n^{j_n}j_n!)} x_1^{j_1}x_2^{j_2}\cdots x_n^{j_n},$$

where the summation is as in (i). In particular,

$$\begin{aligned} Z(A_1) &= x_1, \\ Z(A_2) &= x_1^2, \\ Z(A_3) &= \frac{1}{3}(x_1^3 + 2x_3), \\ Z(A_4) &= \frac{1}{12}(x_1^4 + 3x_2^2 + 8x_1x_3). \end{aligned}$$

(iii)
$$Z(\mathbb{Z}_n) = \frac{1}{n}\sum_{d|n} \varphi(d)x_d^{n/d} \qquad (\varphi = \text{Euler's phi-function, see 13.15}).$$

For instance,

$$\begin{aligned} Z(\mathbb{Z}_7) &= \frac{1}{7}(x_1^7 + 6x_7), \\ Z(\mathbb{Z}_8) &= \frac{1}{8}(x_1^8 + x_2^4 + 2x_4^2 + 4x_8), \\ Z(\mathbb{Z}_9) &= \frac{1}{9}(x_1^9 + 2x_3^3 + 6x_9). \end{aligned}$$

(iv)
$$Z(D_n) = \begin{cases} \frac{1}{2}Z(\mathbb{Z}_n) + \frac{1}{2}x_1x_2^{(n-1)/2} & \text{if } n \text{ is odd,} \\ \frac{1}{2}Z(\mathbb{Z}_n) + \frac{1}{4}(x_2^{n/2} + x_1^2x_2^{n/2-1}) & \text{if } n \text{ is even.} \end{cases}$$

Here D_n is the dihedral group (see 10.2(vi)).

For instance,

$$Z(D_4) = \frac{1}{8}(x_1^4 + 2x_1^2x_2 + 3x_2^2 + 2x_4),$$

$$Z(D_5) = \frac{1}{10}(x_1^5 + 5x_1x_2^2 + 4x_5),$$

$$Z(D_6) = \frac{1}{12}(x_1^6 + 3x_1^2x_2^2 + 4x_2^3 + 2x_3^2 + 2x_6).$$

(v) Let G be the group of all rotations mapping a given cube (or a given octahedron) into itself. Then $|G| = 24$. Consider G as a group C_c of permutations on the eight corners of the cube. Then

$$Z(C_c) = \frac{1}{24}(x_1^8 + 9x_2^4 + 6x_4^2 + 8x_1^2x_3^2).$$

(vi) Let G be as in (v). Now let G act as a group C_e on the twelve edges of the cube. So G is considered as a subgroup of S_{12}. Then

$$Z(C_e) = \frac{1}{24}(x_1^{12} + 3x_2^6 + 8x_3^4 + 6x_1^2x_2^5 + 6x_4^3).$$

(vii) Again, let G be as above. Now let G act as a permutation group $C_f < S_6$ on the six faces of the cube. Then

$$Z(C_f) = \frac{1}{24}(x_1^6 + 3x_1^2x_2^2 + 6x_1^2x_4 + 6x_2^3 + 8x_3^2).$$

The reader might be wondering what these cycle index polynomials might be good for. Recall our examples, in which we wanted to assign certain "figures" $f_1, \ldots, f_r$ (H- or CH_3-molecules in 25.16 or letters A, ..., E in 25.19) to a number m of "places" $1, 2, \ldots, m$ (free places in the carbon ring, corners of a triangle,...). We build up a mathematical model for these situations.

25.23. Definition. Let F be a set of r figures $f_1, \ldots, f_r$. Let $m \in \mathbb{N}$. If $G \leq S_m$, then G can be thought of as a permutation group on F^m via $g(f_1, \ldots, f_m) := (f_{g(1)}, \ldots, f_{g(m)})$; so G is considered as a subgroup of S_{r^m}.

25.24. Theorem (*Pólya's Theorem*). *In the situation of 25.23, the number n of different orbits on $X = F^m$ under G is given by*

$$n = Z(G)(r, r, \ldots, r).$$

(This equals the value of the induced polynomial function of $Z(G)$ at $x_1 = r, \ldots, x_m = r$.)

Proof. If $g \in G$, then $(f_1, \ldots, f_m) \in \mathrm{Fix}(g)$ if and only if all f_i, where i runs through the elements of a cycle of g, are equal. Hence $|\mathrm{Fix}(g)| =$

$r^{j_1+j_2+\cdots+j_m}$, where $(j_1, \ldots, j_m)$ is the cycle index of g. Now Burnside's Theorem 25.17 gives us the desired result

$$n = \frac{1}{|G|} \sum_{g \in G} |\mathrm{Fix}(g)| = \frac{1}{|G|} \sum_{g \in G} r^{j_1} r^{j_2} \cdots r^{j_m}. \qquad \square$$

25.16. Example (*revisited*). We have $F = \{\mathrm{H}, \mathrm{CH}_3\}$, $r = 2$, $m = 6$, and $G = D_6$. From 25.22(iv) we get

$$Z(D_6) = \frac{1}{12}(x_1^6 + 3x_1^2 x_2^2 + 4x_2^3 + 2x_3^2 + 2x_6),$$

hence

$$n = Z(D_6)(2,2,2,2,2,2) = \frac{1}{12}(64 + 48 + 32 + 8 + 4) = \frac{1}{12} \cdot 156 = 13.$$

25.25. Example. We want to color the faces of a cube with two colors. How many essentially different colorings can we get?

Solution. From 25.22(vii) we get $Z(C_f)$, in which we simply have to plug in $p = 2$. Hence $n = \frac{1}{24}(64 + 48 + 48 + 48 + 32) = 10$ different colorings. $\square$

If we actually want to find a representative in each orbit, we can simply try to find one by brute force. Otherwise, more theory is needed. For a proof and a detailed account of the following, see, e.g., Stone (1973) or Kerber (1991).

25.26. Theorem (*Redfield-Pólya Theorem*). *Recall the situation in 25.23. Let us "invent" formal products of the figures $f_1, \ldots, f_r$ and write f^2 for $f \cdot f$, etc. If we now substitute $f_1 + \cdots + f_r$ for x_1, $f_1^2 + \cdots + f_r^2$ for x_2, and so on, in $Z(G)$, then by expanding the products we get sums of the form $n_{i_1,\ldots,i_r} f_1^{i_1} f_2^{i_2} \ldots f_r^{i_r}$ with $i_1 + \cdots + i_r = m$. This means that there are precisely $n_{i_1,\ldots,i_r}$ orbits in $X = F^n$ under G such that each orbit-tuple consists of precisely i_1 figures f_1, i_2 of the figures f_2, etc.*

25.16. Example (*revisited*). Let f_1 be the H-groups and f_2 the CH_3-groups. If we expand as in 25.26, we get $\frac{1}{12}((f_1 + f_2)^6 + 3(f_1 + f_2)^2(f_1^2 + f_2^2)^2 + 4(f_1^2 + f_2^2)^3 + 2(f_1^3 + f_2^3)^2 + 2(f_1^6 + f_2^6)) = f_1^6 + f_1^5 f_2 + 3f_1^4 f_2^2 + 3f_1^3 f_2^3 + 3f_1^2 f_2^4 + f_1 f_2^5 + f_2^6$. Hence there are:

- one possibility to give only H-atoms;
- one possibility to give five H-atoms and one CH_3-molecule;
- three possibilities to take four H-atoms and two CH_3-molecules;

and so on. In order to find a complete set of representatives, we have to collect instances for each of these (altogether 13) possibilities.

Of course, if we replace all f_i by 1, we get Pólya's Theorem 25.24 as a corollary of 25.26. Even after the discovery of these two powerful results, they were not widely known. Let us consider a final example.

25.27. Example. Let us call two truth functions (or switching functions) $f_1, f_2\colon \{0,1\}^n \to \{0,1\}$ ***essentially similar*** if, after a suitable relabeling $(i_1, \dots, i_n)$ of $(1, \dots, n)$, we have

$$f_1(b_1, \dots, b_n) = f_2(b_{i_1}, \dots, b_{i_n})$$

for all $(b_1, \dots, b_n) \in \{0,1\}^n$. For switching theory, this means that the corresponding switching circuits of f_1 and f_2 "work identically" after a suitable permutation of the input wires.

Problem. How many essentially different such functions exist?

History. This problem was explicitly carried out and solved by means of a gigantic computer program in 1951 for $n = 4$. The total number of these functions is $2^{2^4} = 65\,536$, and 3984 essentially different functions were found.

Solution. Our solution is rather immediate. The group G is basically S_n. However, care must be taken, since G acts as described above on $\{0,1\}^n$ and *not* on $\{1, \dots, n\}$. If we take $n = 4$, for instance, the effect of $g = (1,2)(3,4)$ on the quadruple $(a,b,c,d) \in \{0,1\}^4$ is given by (b,a,d,c). Obviously, Fix(g) consists of precisely those functions which are constant on each cycle of g. In our case for $g = (1,2)(3,4)$ we get $|\mathrm{Fix}(g)| = 2\cdot 2 = 4$. Now S_4 decomposes into the following conjugacy classes (see 25.8):

(i) id;

(ii) six 2-cycles;

(iii) three products of two 2-cycles;

(iv) eight 3-cycles;

(v) six 4-cycles.

Now (i) fixes all 16 combinations (a,b,c,d), yielding x_1^{16} in the cycle index polynomial. Also, (ii) contributes $6x_1^8x_2^4$, since for instance $(1,2)$ yields the four 2-cycles $((0,1,c,d),(1,0,c,d))$ and fixes all $(0,0,c,d)$ and $(1,1,c,d)$, thus producing eight 1-cycles and four transpositions, and so on. The cycle index polynomial for G acting on $\{0,1\}^4$ is then given by

$$Z(G) = \frac{1}{24}(x_1^{16} + 6x_1^8x_2^4 + 3x_1^4x_2^6 + 8x_1^4x_3^4 + 6x_1^2x_2x_4^3).$$

Hence $x_1 = x_2 = x_3 = x_4 = 2$ gives 3984 equivalence classes of functions from $\{0,1\}^4$ to $\{0,1\}$. □

25.28. Remark. If we add complementation of inputs and outputs to G as a symmetry operation (i.e., $f_1(b_1,\dots,b_n) = f_2(b_1',\dots,b_n') \Longrightarrow f_1 \sim f_2$, etc.), then this number reduces further from 3984 to just 222. See Gilbert (1976, p. 144) and cf. Exercise 12.

Exercises

1. Show that every transposition has sign -1.
2. Show that each of the following three subsets of S_n $(n \geq 2)$ generates S_n:

   (i) $\{(1,n),(1,n-1),\dots,(1,2)\}$ (cf. 25.9);

   (ii) $\{(1,2),(2,3),\dots,(n-1,n)\}$;

   (iii) $\{(1,2),(1,2,\dots,n)\}$.

   Remark. You can consider this as a "game" in which you put the numbers $1,2,\dots,n$ to the vertices of a regular n-gon on the unit circle. In (i), you can interchange 1 with any other number, in (ii) you can transpose every number with the one following it, and in (iii) you can interchange 1 and 2 and turn the whole circle by $360/n$ degrees. The question is always: Can you get every arbitrary order of $1,\dots,n$ on the circle by means of these "allowed" manipulations?
3. If $|G| = n$, we know from 25.1 that $G \hookrightarrow S_n$. By (iii) of Exercise 2, S_n can be generated by two elements. Consider the conclusion that therefore every finite group can be generated by two elements. Is this correct, or where is the error?
4. Find the number of (up to rotations) distinct bracelets of five beads made up of red, blue, and white beads.
5. Three big and six small pearls are connected onto a circular chain. The chain can be rotated and turned. How many different chains can be obtained assuming that pearls of the same size are indistinguishable?

6. Find the number of ways of painting the four faces a, b, c, and d of a tetrahedron with two colors of paints.
7. If n colors are available, in how many different ways is it possible to color the vertices of a cube?
8. Let G be the group of all orthogonal transformations in $\mathbb{R}^3$ which move the edges of a cube (with center in the origin) into edges. Determine $|G|$ via 25.17.
9. Consider the class of organic molecules of the form

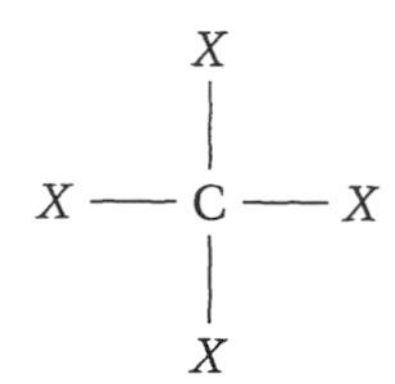

where C is a carbon atom, and each X can be any one of the following: CH_3, C_2H_5, H, or Cl. Model each molecule of this class as a regular tetrahedron with the carbon atom at the center and the X-components at its vertices. Find the number of different molecules of this class. (Hint: Find the number of equivalence classes of maps from the four vertices of the tetrahedron as the domain to the range $\{CH_3, C_2H_5, H, Cl\}$. A suitable permutation group is the group of the permutations which correspond to all possible rotations of the tetrahedron.)
10. Determine the number of switching functions (distinct up to permutations of entries) using three nonequivalent switches and find the corresponding cycle index polynomials and number of non-equivalent switching functions for two and for three switches. Suppose we also allow complementation of the outputs. What would then be a suitable group action?
11. Prove the formula 25.22(iii) for the cycle index polynomial of $\mathbb{Z}_5$.

§26 Image Understanding

Consider the following problems:

- A camera observes a traffic intersection. The goal is to get automatically (by a computer) information about the traffic flow ("How many cars turn right?"...). The problem is difficult: How can the computer know that certain data belong to the same car three seconds later? See Sung & Zimmermann (1986). How can the computer distinguish between cars and their shadows? (Nagel 1995, pp. 27–30.)

- In Computer Tomography, a rotating camera sends its (digitized) pictures to a computer which comes up with clear pictures of organs, blood vessels, etc., inside a human body. The problems are similar to the above: how can a computer know which input signals refer to the same organ? Cf. Beth (1995, pp. 118–120).
- A certain object should be identified by a robot or its shape should be controlled in a production plant (in order to take it off the production line if it is "seriously" out of shape). See Nagel (1995).
- Determine the type of an aircraft from its silhouettes in certain positions.

Moments of thinking will reveal that these problems are just special cases of one and the same general problem: ***image understanding***, the extraction of knowledge about a three-dimensional (3D-) scene from 2D-images. A similar problem is pattern recognition: obtain knowledge about a 2D-scene (e.g., hand-written letters) from 2D-images.

The term "knowledge" is most important here. It is vital in these problems not only to have a collection of data, but to be in a position to interpret the data as real-world objects. Let us establish a suitable notation.

26.1. Notation. Suppose we have a map from a 3D-scene to a 2D-(camera) image.

(i) ***Image parameters*** are numerical data extracted from the observed 3D-scene. The grey value or the value of light reflection at a certain point are image parameters.

(ii) ***Object parameters*** are "properties" of the 3D-scene, like the speed of a certain object, or the coordinates of a corner point.

(iii) (3D-)***recovery equations*** are equations relating the image and the object parameters.

If $\alpha_1, \dots, \alpha_n$ are the object and $c_1, \dots, c_m$ the image parameters, then the recovery equations may be written in the form

$$c_1 = F_1(\alpha_1, \dots, \alpha_n),$$
$$\vdots$$
$$c_m = F_m(\alpha_1, \dots, \alpha_n).$$

These recovery equations are usually obtained from laws of geometry, optics, and so on, and can also be implicit. The c_i are obtained through the camera and sent to a computer, and may be considered to be known. We want to obtain $\alpha_1, \dots, \alpha_n$ from these equations. The equations might be very complicated, and trying to solve them by brute force is usually impossible. So, the usual approach is to solve them numerically, for instance, via a Newton method. This has serious disadvantages: we have to "guess" initial solutions, and only one solution will be produced, although there might be more (even infinitely many). Furthermore, in this case, we might have "free parameters" involved, the meaning of which for the 3D-scene we often do not "understand."

26.2. Example (*Optical Flow*). In this example, as well as in the whole section, we shall closely follow Kanatani (1990), a fascinating text, which is highly recommended to the interested reader. Suppose a plane P with equation $z = px + qy + r$ is "traveling" through 3D-space in a motion which is composed of a translation and a rotation. A camera is fixed on the z-axis taking pictures of this plane; we assume that the pictures are taken orthogonally to the xy-plane (so we assume that the camera is high enough so that perspectives can be neglected, see Figure 26.1). Here, $p, q, r, a, b, c, w_1, w_2, w_3$ are suitable object parameters which describe completely the journey of the plane P. Hence we have nine parameters; (a, b, c) indicates translation, while (w_1, w_2, w_3) describes rotation. What we can measure via the camera, for instance, is the velocity field $\dot{x} = u(x, y)$, $\dot{y} = v(x, y)$. Elementary physics gives us the equations

$$u(x, y) = a + pw_2x + (qw_2 - w_3)y,$$
$$v(x, y) = b - (pw_1 - w_3)x - qw_1y,$$

which can be written as

$$\begin{pmatrix} u(x,y) \\ v(x,y) \end{pmatrix} = \begin{pmatrix} a \\ b \end{pmatrix} + \begin{pmatrix} A & B \\ C & D \end{pmatrix} \begin{pmatrix} x \\ y \end{pmatrix}$$

with $A := pw_2$, $B := qw_2 - w_3$, $C := -pw_1 + w_3$, $D := -qw_1$. It is advisable to measure u and v at "many" points and to fit a, b, A, B, C, D by means of the method of least squares.

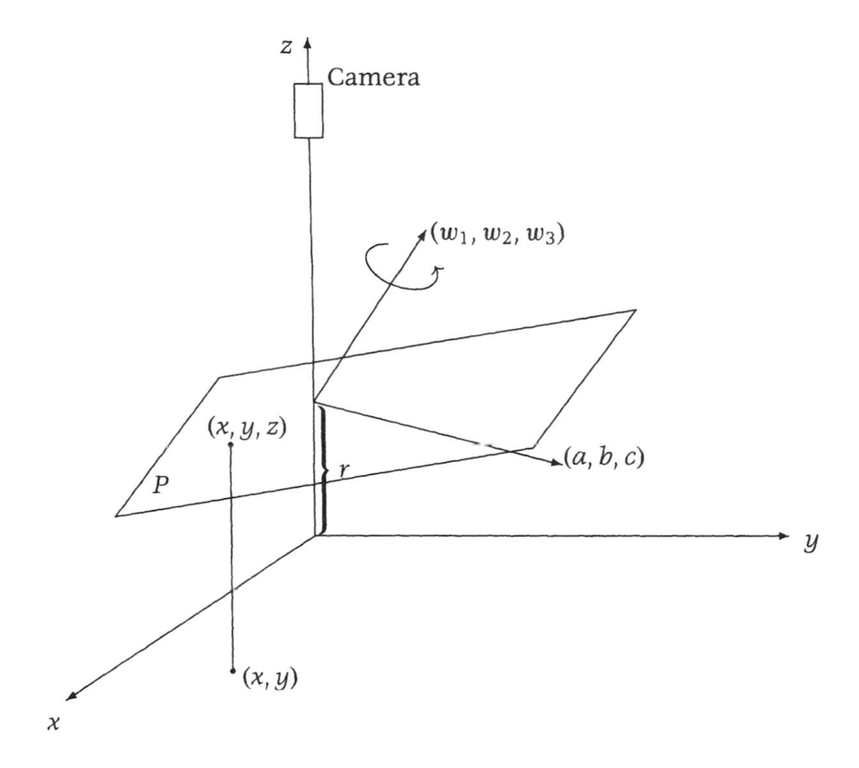

FIGURE 26.1 Translation and rotation of a plane.

So, a, b, A, B, C, D are "intermediate" parameters, and since they can be obtained from $u(x, y)$ and $v(x, y)$ by elementary methods, we might as well view them as image parameters, and

$$\begin{aligned}
a &= a, \\
b &= b, \\
A &= pw_2, \\
B &= qw_2 - w_3, \\
C &= -pw_1 + w_3, \\
D &= -qw_1,
\end{aligned}$$

are then the 3D-recovery equations. The first two equations are trivial; the remaining equations are four equations for the five unknowns p, q, w_1, w_2, w_3, and the object parameters r and c cannot be recovered at all (unless there is also equipment to measure r and c).

We now could solve the A, B, C, D equations by introducing a free parameter. But we would not get the "3D-meaning" of this parameter and of the solutions in general.

Let us interrupt this example for a short while. Not every image parameter has its own "geometrical meaning," independent of the other parameters. In the example above, a has no "meaning," nor has, say, $a + D$. But the pair (a, b) has a "meaning." Thus, not every collection $\{c_1, c_2, \ldots\}$ of image parameters is "meaningful." How can we suitably define "meaning"?

In our example, we have chosen the x- and the y-axis arbitrarily (the z-axis should pass through the lens of the camera). Let us look therefore at the group SO(2) of all rotations $\mathbf{R}_\theta$, where $\mathbf{R}_\theta$ is the rotation of the xy-plane around the origin of angle θ. Points $\begin{pmatrix} x \\ y \end{pmatrix}$ in the plane certainly will have a "meaning." If we take such a rotation $\mathbf{R}_\theta$ of the plane, the point $\begin{pmatrix} x \\ y \end{pmatrix}$ will have the new coordinates

$$\begin{pmatrix} x' \\ y' \end{pmatrix} = \begin{pmatrix} \cos\theta & -\sin\theta \\ \sin\theta & \cos\theta \end{pmatrix} \begin{pmatrix} x \\ y \end{pmatrix} = \mathbf{R}_\theta \begin{pmatrix} x \\ y \end{pmatrix}.$$

So we might say that a pair (c_1, c_2) of image characteristics has ***meaning*** if, after the rotation $\mathbf{R}_\theta$, these image characteristics have the values (c_1', c_2') such that

$$\begin{pmatrix} c_1' \\ c_2' \end{pmatrix} = \mathbf{R}_\theta \begin{pmatrix} c_1 \\ c_2 \end{pmatrix}.$$

Similarly, a single image parameter i is ***meaningful*** if it does not change its value after a coordinate rotation. For instance, the grey value of the darkest point(s) is a meaningful parameter, while just the x-coordinate of a point is not.

A little less obvious is the important case of four image parameters (c_1, c_2, c_3, c_4). We can consider this as a matrix $\mathbf{M} = \begin{pmatrix} c_1 & c_2 \\ c_3 & c_4 \end{pmatrix}$, i.e., as a linear transformation from $\mathbb{R}^2$ to $\mathbb{R}^2$. We call $\mathbf{M}$ ***meaningful*** if its entries are transformed into c_1', c_2', c_3', c_4' like a matrix representation of a linear transformation w.r.t. different bases, i.e., if

$$\begin{pmatrix} c_1' & c_2' \\ c_3' & c_4' \end{pmatrix} = \mathbf{R}_\theta \mathbf{M} \mathbf{R}_\theta^{-1} = \mathbf{R}_\theta \mathbf{M} \mathbf{R}_{-\theta} \quad \text{for all } \theta \in [0, 2\pi].$$

This again can be easily written as

$$\begin{pmatrix} c_1' \\ c_2' \\ c_3' \\ c_4' \end{pmatrix} = \begin{pmatrix} \cos^2\theta & -\cos\theta\,\sin\theta & -\cos\theta\,\sin\theta & \sin^2\theta \\ \cos\theta\,\sin\theta & \cos^2\theta & -\sin^2\theta & -\cos\theta\,\sin\theta \\ \cos\theta\,\sin\theta & -\sin^2\theta & \cos^2\theta & -\cos\theta\,\sin\theta \\ \sin^2\theta & \cos\theta\,\sin\theta & \cos\theta\,\sin\theta & \cos^2\theta \end{pmatrix} \begin{pmatrix} c_1 \\ c_2 \\ c_3 \\ c_4 \end{pmatrix}$$

$$=: \mathbf{S}_\theta \begin{pmatrix} c_1 \\ c_2 \\ c_3 \\ c_4 \end{pmatrix}. \tag{26.1}$$

26.3. Example (*26.2 continued*). Easy calculations (see Exercise 4) show that the pair (a, b) and the collection $\begin{pmatrix} A & B \\ C & D \end{pmatrix}$ are meaningful.

We still do not know the advantage of these considerations, so we need another step in our discussion. Suppose that (c_1, c_2) is meaningful, describing a certain "property" of the object under consideration. Then $(c_1 + c_2, c_1 - c_2)$ also describes this property since we can easily recover (c_1, c_2) from $(c_1 + c_2, c_1 - c_2)$. More generally, if $(c_1, \ldots, c_k)$ describes properties, then $C_1, \ldots, C_k$ with

$$\begin{pmatrix} C_1 \\ \vdots \\ C_k \end{pmatrix} = \mathbf{P} \begin{pmatrix} c_1 \\ \vdots \\ c_k \end{pmatrix}$$

describe the same properties if $\mathbf{P}$ is nonsingular. The new parameters $C_1, \ldots, C_k$ are then just linear combinations of the old image parameters $c_1, \ldots, c_k$.

Suppose we can find an invertible matrix $\mathbf{P}$ such that $\mathbf{P}\mathbf{R}_\theta\mathbf{P}^{-1}$ has the block diagonal form $\begin{pmatrix} \mathbf{Q} & \mathbf{0} \\ \mathbf{0} & \mathbf{R} \end{pmatrix}$ for all θ. If $\mathbf{R}_\theta$ is a $q \times q$ matrix, then the new image parameters $(C_1, \ldots, C_k)$ are transformed into $(C_1', \ldots, C_k')$ after a rotation of the xy-coordinate system such that

$$\begin{pmatrix} C_1' \\ \vdots \\ C_k' \end{pmatrix} = \mathbf{P} \begin{pmatrix} c_1' \\ \vdots \\ c_k' \end{pmatrix} = \mathbf{P}\mathbf{R}_\theta \begin{pmatrix} c_1 \\ \vdots \\ c_k \end{pmatrix} = \mathbf{P}\mathbf{R}_\theta\mathbf{P}^{-1} \begin{pmatrix} C_1 \\ \vdots \\ C_k \end{pmatrix}$$

$$= \begin{pmatrix} \mathbf{Q} & \mathbf{0} \\ \mathbf{0} & \mathbf{R} \end{pmatrix} \begin{pmatrix} C_1 \\ \vdots \\ C_k \end{pmatrix} = \begin{pmatrix} \mathbf{Q} \begin{pmatrix} C_1 \\ \vdots \\ C_q \end{pmatrix} & \mathbf{0} \\ \mathbf{0} & \mathbf{R} \begin{pmatrix} C_{q+1} \\ \vdots \\ C_k \end{pmatrix} \end{pmatrix},$$

since $c_1, \ldots, c_k$ are supposed to be meaningful. This shows that $(C_1, \ldots, C_q)$ and $(C_{q+1}, \ldots, C_k)$ are transformed independently of each other, and it makes sense to say that $(C_1, \ldots, C_q)$ and $(C_{q+1}, \ldots, C_k)$ indicate *different* properties.

We can try to apply this idea of decomposing components to $\mathbf{Q}$ and $\mathbf{R}$, and so on. If we find a nonsingular matrix $\mathbf{P}$, independent of θ, such that

$$\mathbf{P}\mathbf{R}_\theta\mathbf{P}^{-1} = \begin{pmatrix} \mathbf{Q}_1 & & \mathbf{0} \\ & \ddots & \\ \mathbf{0} & & \mathbf{Q}_r \end{pmatrix} \quad \text{for all } \theta,$$

where no further decompositions of the blocks $\mathbf{Q}_i$ are possible, we can say that $\mathbf{Q}_1 \begin{pmatrix} C_1 \\ \vdots \\ C_{q_1} \end{pmatrix}, \ldots$ indicate *single properties*. These vectors are called ***invariants***.

That is where group theory is useful. In our examples, we have the group $G = \mathrm{SO}(2)$ of all rotations of the xy-plane around the origin. On the points of this plane (i.e., on the meaningful pairs (c_1, c_2) of image parameters), G acts via the 2×2 matrices $\mathbf{R}_\theta = \begin{pmatrix} \cos\theta & -\sin\theta \\ \sin\theta & \cos\theta \end{pmatrix}$, and we have a natural group homomorphism $h\colon G \to \{\mathbf{R}_\theta \mid \theta \in [0, 2\pi)\}$.

26.4. Definition. A ***matrix representation*** h of dimension n of a group G is a homomorphism h from G into the group $\mathrm{GL}(n, \mathbb{F})$ of all invertible $n \times n$ matrices over a field $\mathbb{F}$. If there is some $\mathbf{P} \in \mathrm{GL}(n, \mathbb{F})$ with $\mathbf{P}\,h(g)\,\mathbf{P}^{-1} = \begin{pmatrix} h_1(g) & \mathbf{0} \\ \mathbf{0} & h_2(g) \end{pmatrix}$ for all $g \in G$, where h_1, h_2 are representations of G into $\mathrm{GL}(n_1, \mathbb{F})$ and $\mathrm{GL}(n_2, \mathbb{F})$, respectively, with $n = n_1 + n_2$, then the representation h is said to be the ***sum*** $h = h_1 + h_2$ of the representations h_1, h_2 or to be ***fully reducible*** to h_1 and h_2. A representation which cannot be written as the sum of two other nonzero representations is called ***irreducible***. A representation h is called ***faithful*** if h is injective; in this case, $G \cong h(G) \leq \mathrm{GL}(n, \mathbb{F})$.

For certain reasons (see below), we usually take $\mathbb{F}$ to be the field $\mathbb{C}$ of complex numbers. Any finite group G has a faithful representation in some $\mathrm{GL}(n, \mathbb{F})$: embed G into the symmetric group S_n and represent each $\pi \in S_n$ by its corresponding permutation matrix. If G has a representation in $\mathrm{GL}(n, \mathbb{F})$, then n is not unique. For example, $h\colon \mathrm{SO}(2) \to \{\mathbf{R}_\theta \,|\, \theta \in [0, 2\pi)\}$

is a faithful representation of SO(2) in GL(2, $\mathbb{R}$) (see Exercise 5), while the 4×4 matrices S_θ in (26.1) define a representation h' of SO(2) in GL(4, $\mathbb{R}$) (again faithful), and h'': SO(2) $\to$ GL(1, $\mathbb{R}$), sending every rotation to the 1×1 identity matrix is a nonfaithful representation.

So, for our purposes, we might say:

- Reducing a representation means the separation into independent properties.
- Irreducible representations correspond to single properties.

These statements cannot be proved (since we have not defined a "property" rigorously); they are often attributed to Hermann Weyl (1888–1955) as "Weyl's Thesis."

If a representation h is decomposed into $h = h_1 + h_2$ as in 26.4, we can find a basis B which is the disjoint union of B_1 and B_2 such that each $h(g)$ maps (by matrix multiplication) the linear hull L_1 of B_1 into L_1 and the linear hull L_2 of B_2 into L_2. So $\mathbb{F}^n = L_1 \oplus L_2$ and L_1, L_2 are $h(g)$-invariant for all $g \in G$. Sometimes, however, we can only find some $h(g)$-invariant subspace U of $\mathbb{F}^n$, but no complementary subspace W which is also $h(g)$-invariant. Then the matrix representation of each $h(g)$ can be written in the form $\begin{pmatrix} \mathbf{A} & \mathbf{B} \\ \mathbf{0} & \mathbf{C} \end{pmatrix}$. This means that each $h(g)$ maps U into (in fact, onto) itself, but the result $h(g)w$, for $w \in W$, can have components in U as well. Hence we do not get a "full reduction," but a "cascade-type" reduction, and in this case we say h is ***reducible*** (as opposed to "fully reducible") in 26.4. If h is not reducible, h is called ***simple***. Clearly a simple representation is irreducible. In many cases, however, we do get full reductions; in the following theorem, a ***compact group*** is a group with a topology on it such that $(g_1, g_2) \mapsto g_1 g_2$ and $g \mapsto g^{-1}$ are continuous maps from G^2 (with the product topology) to G (from G to G, respectively), and such that G is a compact topological space (see any text on topology if these concepts are not familiar).

26.5. Theorem. *Any irreducible representation of a group G is simple in the following cases:*

(i) *All matrices $h(g)$ are orthogonal w.r.t. some inner product of $\mathbb{F}^n$.*

(ii) *The same holds if "orthogonal" in (i) is replaced by "unitary" or "symmetric" or "hermitian."*

(iii) *G is finite;*

(iv) *G is a compact (topological) group;*

(v) *h is one-dimensional.*

Proof. Suppose each $h(g)$ is orthogonal, i.e., $\langle h(g)\mathbf{v}, \mathbf{w}\rangle = \langle \mathbf{v}, h(g)^{-1}\mathbf{w}\rangle$ holds for all $\mathbf{v}, \mathbf{w} \in V$. Let U be an $h(g)$-invariant subspace of $\mathbb{F}^n$ for each $g \in G$. If $U^{\perp}$ is the orthogonal complement, then

$$\begin{aligned} h(g)U^{\perp} &= \{h(g)\mathbf{w} \mid \langle \mathbf{w}, \mathbf{u}\rangle = 0 \text{ for all } \mathbf{u} \in U\} \\ &= \{\mathbf{v} \mid \langle h(g)^{-1}\mathbf{v}, \mathbf{u}\rangle = 0 \text{ for all } \mathbf{u} \in U\} \\ &= \{\mathbf{v}' \mid \langle \mathbf{v}', h(g)\mathbf{u}\rangle = 0 \text{ for all } \mathbf{u} \in U\} \\ &= \{\mathbf{v}' \mid \langle \mathbf{v}', \mathbf{u}'\rangle = 0 \text{ for all } \mathbf{u}' \in U\} \\ &= U^{\perp}. \end{aligned}$$

The cases in (ii) run along similar lines. If G is finite, introduce the inner product (cf. Exercise 6)

$$\langle \mathbf{v}, \mathbf{w}\rangle := \frac{1}{|G|} \sum_{g \in G} (h(g)\mathbf{v})(h(g)\mathbf{w})^{\mathrm{T}},$$

then $\langle h(g)\mathbf{v}, h(g)\mathbf{w}\rangle = \langle \mathbf{v}, \mathbf{w}\rangle$ for all $g \in G$ and $\mathbf{v}, \mathbf{w} \in \mathbb{F}^n$, so we can again use the argument above. If G is compact, we proceed similarly by setting

$$\langle \mathbf{v}, \mathbf{w}\rangle := \frac{1}{\mu(G)} \int_G (h(g)\mathbf{v})(h(g)\mathbf{w})^{\mathrm{T}}\, d\mu(g),$$

where μ denotes the translation-invariant Haar measure, which exists in any (locally) compact group, and $\mu(G) := \int_G d\mu(g)$. Finally, (v) is trivial. □

26.6. Corollary. *In the cases* (i)–(iv) *of Theorem* 26.5, *every representation is the sum of simple representations.*

The sophisticated theory of group representations gives explicit methods to reduce a representation into its irreducible or simple parts. For our "easy" group $G = \mathrm{SO}(2)$, we can derive the following results either from this theory or by a skillful application of methods of linear algebra (the latter is usually not possible for more complicated groups).

The representation $\{\mathbf{R}_\theta \mid \theta \in [0, 2\pi)\}$ is not reducible over $\mathbb{F} = \mathbb{R}$, but going from $\mathbb{R}$ to $\mathbb{C}$ makes things easier. In Exercise 7 we shall show:

26.7. Theorem. *The matrix* $\mathbf{P} := \begin{pmatrix} 1 & i \\ 1 & -i \end{pmatrix}$ *satisfies*

$$\mathbf{P}\mathbf{R}_\theta\mathbf{P}^{-1} = \begin{pmatrix} e^{-i\theta} & 0 \\ 0 & e^{i\theta} \end{pmatrix}.$$

So $h = h_1 + h_2$, *where the* $h_j\colon \mathrm{SO}(2) \to \mathrm{GL}(1, \mathbb{C})$ *send rotations of angle* θ *to* $e^{-i\theta}$ *or* $e^{i\theta}$, *respectively. Hence, the representation* h *is the sum of two irreducible one-dimensional representations* h_1, h_2. *If* (c_1, c_2) *is a meaningful set of image parameters,*

$$C_1 = c_1 + ic_2,$$
$$C_2 = c_1 - ic_2,$$

also gives a meaningful set of parameters (C_1, C_2), *which are transformed by a coordinate rotation of angle* θ *to* $C_1' = e^{-i\theta}C_1$ *and to* $C_2' = e^{i\theta}C_2$.

26.8. Theorem. *The matrix* $\mathbf{P} = \begin{pmatrix} 1 & 0 & 0 & 1 \\ 0 & 1 & -1 & 0 \\ 1 & i & i & -1 \\ 1 & -i & -i & -1 \end{pmatrix}$ *satisfies*

$$\mathbf{P}\mathbf{S}_\theta\mathbf{P}^{-1} = \begin{pmatrix} 1 & 0 & 0 & 0 \\ 0 & 1 & 0 & 0 \\ 0 & 0 & e^{-2i\theta} & 0 \\ 0 & 0 & 0 & e^{2i\theta} \end{pmatrix}$$

for all θ. *So* h' *is the sum of four irreducible representations, and*

$$\mathbf{P}\begin{pmatrix} c_1 \\ c_2 \\ c_3 \\ c_4 \end{pmatrix} = \begin{pmatrix} C_1 \\ C_2 \\ C_3 \\ C_4 \end{pmatrix}$$

gives the four invariants $C_1 = A + D$, $C_2 = B - C$, $C_3 = (A - D) + i(B + C)$, $C_4 = (A - D) - i(B + C)$ *for meaningful image parameters* $(A, B, C, D) = (c_1, \ldots, c_4)$. *A rotation of the xy-plane of angle* θ *turns* $C_1, \ldots, C_4$ *onto* $C_1' = C_1, C_2' = C_2, C_3' = e^{-2i\theta}C_3, C_4' = e^{2i\theta}C_4$.

26.9. Example (*26.2 revisited*). The last two theorems tell us that in the example of optical flow we get the following invariants: $a \pm bi$, $A + D$, $C - D$, $(A - D) \pm (B + C)i$. From that, we get a classification of the resulting flows: If A, B, C, D are all zero, the flow is of the form $\begin{pmatrix} u \\ v \end{pmatrix} = \begin{pmatrix} a \\ b \end{pmatrix}$, hence it is just a "translational flow." If, however, all parameters except $B - C$ are zero, we get a "rotational flow." If all parameters except $A + D$ are zero, we

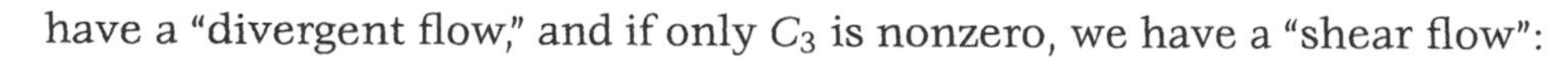
have a "divergent flow," and if only C_3 is nonzero, we have a "shear flow":

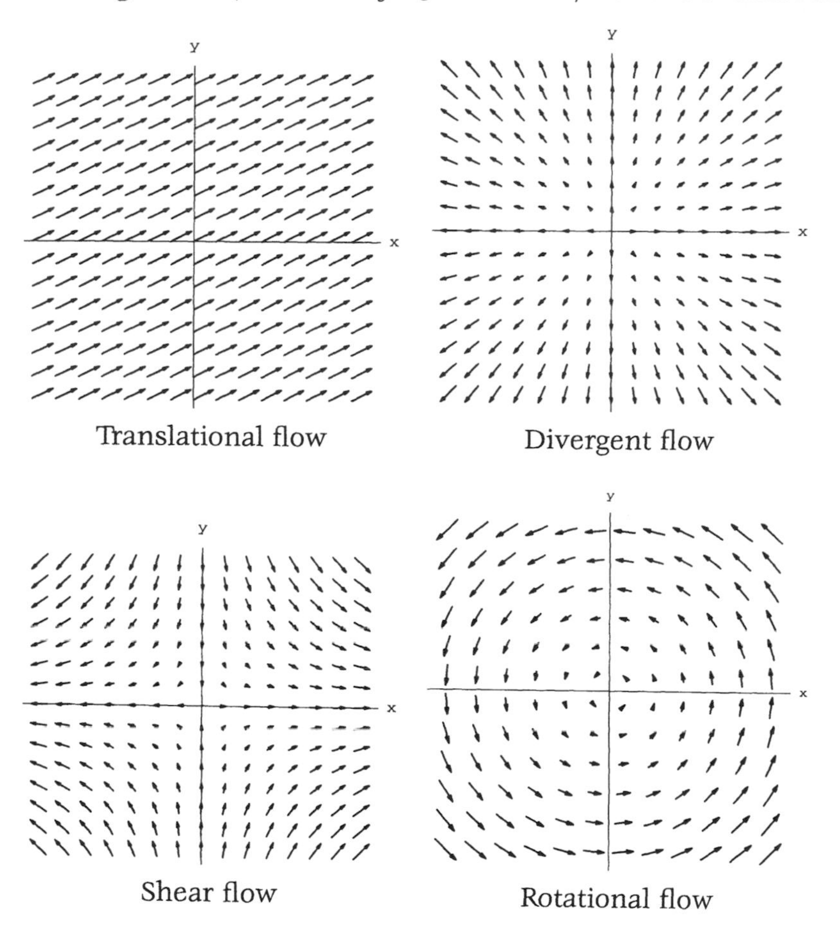

Translational flow   Divergent flow

Shear flow   Rotational flow

So, the invariant parameters characterize the kind of the flow. The 3*D*-recovery equations reduce to just two equations:

$$PW^* = (2w_3 + B - C) - (A + D)i,$$
$$PW = (-B - C) + (A - D)i,$$

where $P = p + qi$, $W = w_1 + w_2 i$, $W^* = w_1 - w_2 i$, which can be solved (see Exercise 3). The resulting free parameter is just a scaling factor. Much more can be said once these invariants are known.

In both cases, h and h' are sums of one-dimensional representations. This is true for all abelian groups, as we shall see now.

26.10. Theorem (*Schur's Lemma*). *Let h: $G \to \mathrm{GL}(n, \mathbb{C})$ be a simple representation of G. If a matrix* $\mathbf{P}$ *satisfies* $\mathbf{P}h(g) = h(g)\mathbf{P}$ *for all $g \in G$, then* $\mathbf{P}$ *is a scalar multiple of the identity matrix* $\mathbf{I}$.

Proof. Since the field $\mathbb{F} = \mathbb{C}$ is algebraically closed, $\mathbf{P}$ has an eigenvector $\mathbf{v} \in \mathbb{C}^n$ and a corresponding eigenvalue $\lambda \in \mathbb{C}$, so $\mathbf{Pv} = \lambda\mathbf{v}$. Now, $\mathbf{P}(h(g)\mathbf{v}) = h(g)\mathbf{Pv} = \lambda h(g)\mathbf{v}$ shows that, for all $g \in G$, $h(g)$ is also an eigenvector to λ. Hence, the eigenspace E of $\mathbf{P}$ w.r.t. λ is invariant under multiplication with all matrices $h(g)$, and so E must be trivial. Since $E \neq \{\mathbf{0}\}$, $E = \mathbb{C}^n$, and $(\mathbf{P} - \lambda\mathbf{I})\mathbf{v} = \mathbf{0}$ for all $\mathbf{v}$, whence $\mathbf{P} = \lambda\mathbf{I}$. □

26.11. Corollary. *All simple representations h of an abelian group G in* $\mathrm{GL}(n, \mathbb{C})$ *are one-dimensional.*

Proof. We can take $\mathbf{P} = h(g_0)$ in 26.10 for each $g_0 \in G$. Hence all $h(g_0)$ are of the form $\lambda\mathbf{I}$. This is a reducible representation unless $n = 1$. □

Conversely, if G has a faithful representation h in some $\mathrm{GL}(n, \mathbb{C})$ such that h is the sum of one-dimensional representations, then G clearly must be abelian (see Exercise 8).

To summarize, *reducing representations means decoupling of overlapping influences.*

Exercises

1. In Example 26.2, we measure $u(x, y)$ and $v(x, y)$ at the following points:

x	y	$u(x,y)$	$v(x,y)$
0	0	6.5	3.9
0	1	10.0	8.1
1	0	8.1	6.9
0	2	10.9	12.2
2	0	9.3	9.9
1	2	11.9	15.4
2	1	11.1	13.8
2	2	13.0	17.9

Fit a, b, A, B, C, D by means of the method of least squares.

2. Use the results of Exercise 1 and try to solve the 3D-recovery equations on page 305.
3. What are the values of the invariants in Examples 26.2 and 26.9 in the situation of Exercises 1 and 2? Solve the 3D-recovery equations for these invariants (page 312).
4. Show in Example 26.3 that (a, b) and $\begin{pmatrix} A & B \\ C & D \end{pmatrix}$ are meaningful.
5. Show that h: $\mathrm{SO}(2) \to \{\mathbf{R}_\theta \mid \theta \in [0, 2\pi]\}$ is a faithful representation of SO(2).
6. Show that $\langle ., . \rangle$ in the proof of 26.5(iii) is really an inner product.
7. Prove Theorem 26.7.
8. Suppose h is a faithful representation of a group G such that h is the sum of one-dimensional representations. Show that G must be abelian.
9. Find all representations (via the simple ones) of the cyclic group $\mathbb{Z}_n$.

§27 Symmetry Groups

As we shall see now, group theory is the ideal tool to study symmetries. In this section, any subset of $\mathbb{R}^2$ or $\mathbb{R}^3$ is called an ***object***. We study orthogonal maps from $\mathbb{R}^2$ to $\mathbb{R}^3$ or from $\mathbb{R}^3$ to $\mathbb{R}^3$; from linear algebra, we know that these are rotations or reflections or rotation-reflections. In other terms, orthogonal maps are precisely those linear maps which leave lengths and angles invariant. So we do not consider translations. We now give a preliminary definition, which we shall have to modify.

27.1. Theorem and Definition. *Let M be an object in $\mathbb{R}^2$ or $\mathbb{R}^3$. The set $S(M)$ of all orthogonal maps g with $g(M) = M$ is a group w.r.t. composition, and is called the* ***symmetry group*** *of M. The elements of $S(M)$ are called* ***symmetry operations*** *on M.*

Every object M in $\mathbb{R}^2$ is also an object in $\mathbb{R}^3$. In this case, we write $S_2(M)$ or $S_3(M)$ if we consider orthogonal maps $\mathbb{R}^2 \to \mathbb{R}^2$ or $\mathbb{R}^3 \to \mathbb{R}^3$, respectively.

27.2. Examples.

(i) Let M be the regular n-gon in $\mathbb{R}^2$ with center at $(0, 0)$. By 10.2(vi), $S_2(M) = D_n$, the dihedral group with $2n$ elements. For $S_3(M)$, we also get the reflection about the xy-plane and its compositions with all $g \in S_2(M)$, i.e., $S_3(M) = D_n \times \mathbb{Z}_2$.

(ii) Let M be the regular tetrahedron with center in $(0, 0, 0)$. Then $S_3(M) = S(M) \cong S_4$, the symmetric group on the four corner points.

(iii) Both the symmetry group of the cube and the regular octahedron can be shown to be isomorphic to $S_4 \times \mathbb{Z}_2$; so they each have 48 elements (see Exercise 3).

(iv) Let M be the letter **M**. Then $S_2(M) = \{\text{id}, r\}$, where r is the reflection about the axis a:

So $S_2(M) \cong \mathbb{Z}_2$, and as in (i), $S_3(M) \cong \mathbb{Z}_2 \times \mathbb{Z}_2$.

(v) If M is a circle, then the symmetry group $S_2(M)$ for M is the whole ***orthogonal group*** $O_2(\mathbb{R})$ of all orthogonal maps from $\mathbb{R}^2$ to $\mathbb{R}^2$. In particular, $S_2(M)$ is infinite.

In applications to chemistry and crystallography, M is a molecule. The "larger $S_2(M)$ or $S_3(M)$, the more symmetric is this molecule." In order to find the ***symmetry group*** of M, we have to know the "shape" of this molecule (or conversely, see the lines after 27.19). For example, CH_4 (methane) has the form of a regular tetrahedron with H-atoms at the corners and C in the center (see Figure 27.1).

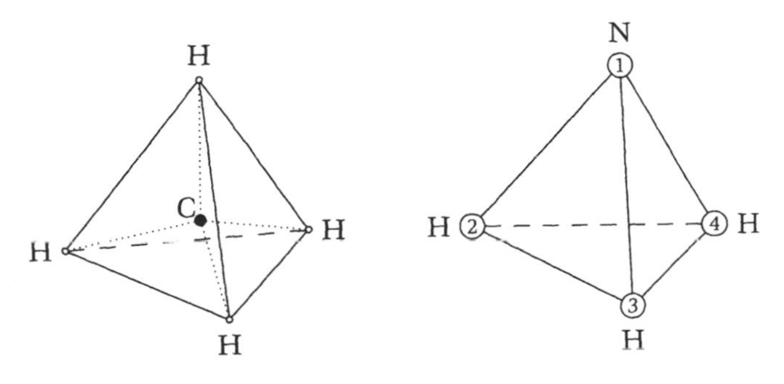

FIGURE 27.1

FIGURE 27.2

If we take C in the place $(0, 0, 0)$, it is clear from 27.2(ii) that the symmetry group of methane is S_4.

Things become a little more difficult if we take a look at ammonia, NH_3 (see Figure 27.2). Here we require a symmetry operation to fix the single N-atom, but the H-atoms can be interchanged. So we have to modify the definition in 27.1:

27.3. Definition. If an object M consists of disjoint parts $M_1, M_2, \dots$ which should be mapped into themselves, then

$$S((M_i)_{i\in I}) := \{g \mid g \text{ is orthogonal and } g(M_i) = M_i \text{ for all } i \in I\}$$

is the symmetry group of $(M_i)_{i\in I}$ or, in another notation, of $M_1 \dot{\cup} M_2 \dot{\cup} \dots$.

If $M_1 = \{①\}$ and $M_2 = \{②, ③, ④\}$ in the case of ammonia, we see that the symmetry operations must consist of the three rotations r_0, r_{120}, r_{240} with angles $0°, 120°, 240°$ through the vertical axis A:

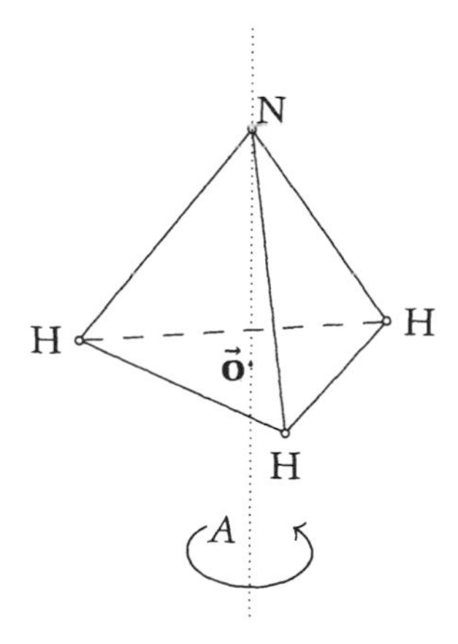

together with the reflections g_1, g_2, g_3 about the three planes through N, $\vec{\mathbf{o}}$, and one of the H-corners. Hence $S(M_1 \cup M_2) \cong D_3 \cong S_3$. Observe that $S(M_1 \cup M_2)$ can be written as $\{r_0 = \mathrm{id}, r_{120}, r_{240}, g_1 \circ r_0, g_1 \circ r_{120}, g_1 \circ r_{240}\}$; so we have, in another point of view, three rotations and three reflection-rotations.

We see that the determination of the symmetry group can be quite involved. For a "search program," we need to look at the fixed points $\mathrm{Fix}(G)$ (cf. 25.17) of the symmetry operations G. We now only work in $\mathbb{R}^3$.

If g is a rotation or a rotation-reflection through angles $\neq 0°$ and $\neq 180°$, $\mathrm{Fix}(g)$ is a line through $(0, 0, 0)$ and called the ***axis*** of g. The number of symmetry operations which have a fixed axis is called the ***order*** of the axis. The order of A in NH_3, for instance was six, because we have three rotations and three rotation-reflections. In the following, we only consider rotations through angles $\neq 0°$ and $\neq 180°$. We write $\perp$ as an abbreviation of "is perpendicular to." The following search program is due to Mathiak & Stingl (1968); see also Sternberg (1994, pp. 27–31).

27.4. Algorithm to find the symmetry group S of a molecule M.

Case 1: There is no axis of rotation.

(1a) There is no reflection possible. Then $S = \{\text{id}\}$.

(1b) There is at least one reflection g. Then $S = \{\text{id}, g\} \cong \mathbb{Z}_2$.

Case 2: There is precisely one axis A of rotation.

(2a) A has infinite order. Then S is infinite.

(2b) A has order $n \in \mathbb{N}$, but A is not an axis of a reflection-rotation of order $2n$. Let r be the rotation through A with angle $360°/n$.

(α) There is no reflection. Then $S = \{\text{id}, r, r^2, \ldots, r^{n-1}\} \cong \mathbb{Z}_n$.

(β) There is a reflection g with $A \subseteq \text{Fix}(g)$. Then $S \cong D_n$.

(γ) There is a reflection g with $A \perp \text{Fix}(g)$. Then $S \cong \mathbb{Z}_n \times \mathbb{Z}_2$.

(2c) A has order n, but A is also the axis of a rotation-reflection g of order $2n$.

(α) There is another reflection h with $A \subseteq \text{Fix}(h)$. Then $S \cong \mathbb{Z}_{2n}$.

(β) Not (α). Then S has order $4n$; for the frequent case $n = 2$ we get $S \cong D_4$. If n is odd, then $S \cong D_n \times \mathbb{Z}_2$.

Case 3: There are at least two axes $A_1, A_2, \ldots$ of rotation.

(3a) A_1 has order $n \in \mathbb{N}$, all others have order 2. Let r_1 be a rotation through A_1 of order n and r_2 a rotation through A_2 of order 2. Then $A_1 \perp A_2$, and

(α) There is no further reflection. Then S is generated by r_1 and r_2, and $S \cong D_n$.

(β) There is a further reflection g. Then S is generated by r_1, r_2, g, and $S \cong D_n \times \mathbb{Z}_2$.

(3b) Not (3a), so we have at least two axes of nontrivial rotations. Then M has either the form of a regular tetrahedron (then $S(M) \cong S_4$), or of a cube or octahedron ($S \cong S_4 \times \mathbb{Z}_2$), or of a regular dodecahedron or icosahedron, in which case S is a nonabelian group of order 120.

27.5. Examples.

(i) Methane, CH_4, clearly belongs to case 3b.

(ii) Ammonia, NH_3, is an example of case 2b(β) (with $n = 3$).

(iii) Water, H_2O, is a "linear" molecule

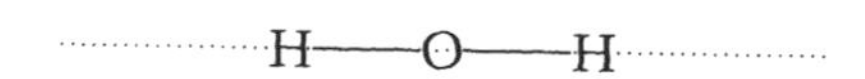

As in case 2a, S has infinite order, consisting of all rotations r_α ($0 \leq \alpha < 360°$) through the axis A, and their products with the reflection on the plane through the oxygen atom which is perpendicular to A.

(iv) Hydrogen peroxide, H_2O_2, has the shape

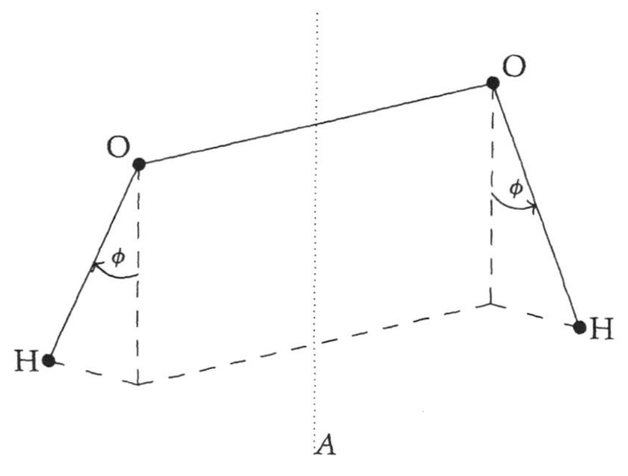

A has order 2. For $0 < \phi < 90°$ ("equilibrium state"), we are in case 2b(α), hence $S \cong \mathbb{Z}_2$. For $\phi = 0°$, we have case 2b(β), hence $S \cong D_4$, while for $\phi = 90°$, case 2b(γ) applies ($S \cong \mathbb{Z}_2 \times \mathbb{Z}_2$).

(v) Naphtalene, $C_2(CH)_8$, is a molecule in a plane P of the form

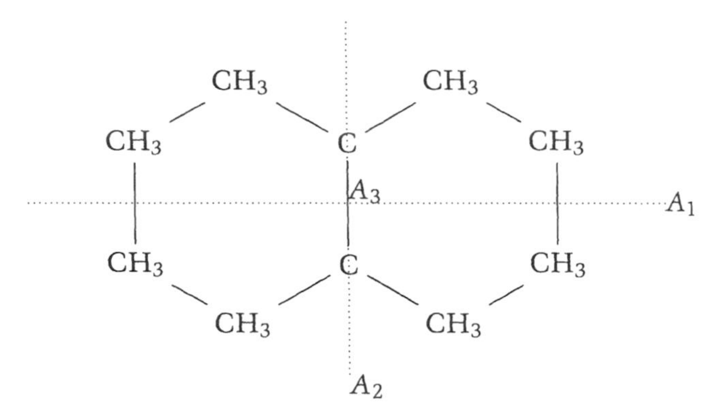

We have three axes of rotation of order 2 (A_3 is perpendicular to P), and a reflection about P. Hence we are in case 3a(β), and $S \cong \mathbb{Z}_2 \times \mathbb{Z}_2 \times \mathbb{Z}_2$.

Sometimes, especially in quantum mechanics, rotations through 360° do *not* act as the identity transformation. This applies in particular to the behavior of neutrons or other fermions with "half integral spin." A nice every-day analogy is an asymmetric wrench which is fastened between two walls with rubber bands (dotted lines):

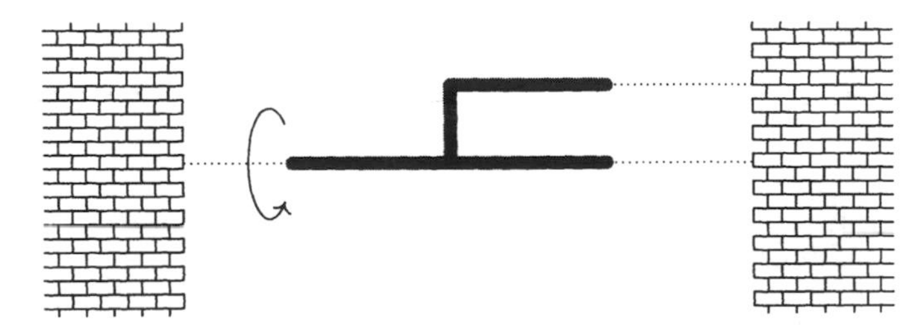

Rotating the wrench in the indicated direction through 360° will entangle the rubber bands, and there is no way to disentangle them without moving the wrench back. But by rotating the wrench twice (i.e., through 720°) the rubber bands can easily be disentangled. Try it yourself, and see the surprise! This experiment was invented by the famous physicist P. A. M. Dirac in 1972—see the amusing article by Bolker (1973).

Molecules or crystals with isomorphic symmetry groups often have similar properties. But the main applications run along the following lines. The next two results are almost obvious.

27.6. Theorem and Definition. *Let $M = \{m_1, \dots, m_n\}$ be finite (maybe partitioned as in 27.3) and let F be a field. If we formally write $(\lambda_1, \dots, \lambda_n) \in$*

F^n as $\lambda_1 m_1 + \cdots + \lambda_n m_n$, then F^n becomes a vector space V over F with basis $\{m_1, \ldots, m_n\} = M$.

Observe that neither λm nor $m_1 + m_2$ have some geometric or physical meaning. Nevertheless, this notation proves to be very useful.

27.7. Theorem. *In the situation of 27.6, every $g \in S(M)$ induces an automorphism α_g of V via*

$$\alpha_g(\lambda_1 m_1 + \cdots + \lambda_n m_n) := \lambda_1 g(m_1) + \cdots + \lambda_n g(m_n).$$

It is customary to identify g and α_g, inasmuch as $\alpha_{gh} = \alpha_g \cdot \alpha_h$ holds for all $g, h \in S(M)$. Suppose now that we have another linear operator h from V to V. Frequently in applications, we have to compute its eigenvalues and eigenvectors. This leads to the problem of solving an equation of degree n, which causes the well-known severe troubles if $n \geq 5$ (see §12).

Recall that a subspace U of V is h-invariant if $h(U) \subseteq U$, and that V is the direct sum of the subspaces $U_1, \ldots, U_k$ if every $v \in V$ is uniquely expressible as $v = u_1 + \cdots + u_k$ with $u_i \in U_i$ for $i \in \{1, \ldots, k\}$.

27.8. Theorem. *Suppose V (as in 27.6) is the direct sum of h-invariant subspaces:*

(i) *If $v = u_1 + \cdots + u_k$ is an eigenvector of h w.r.t. the eigenvalue λ, then each u_i is an eigenvector to λ for the restriction $h/_{U_i}$ of h to U_i.*

(ii) *Conversely, if $w_i \in U_i$ are eigenvectors of $h/_{U_i}$ to some λ ($1 \leq i \leq k$), then $w_1 + \cdots + w_k$ is an eigenvector of h to the eigenvalue λ.*

Proof. (i) If $h(v) = \lambda v$, we get $h(u_1) + \cdots + h(u_k) = \lambda u_1 + \cdots + \lambda u_k$. Since each $h(u_i) \in U_i$ and since the sum is direct, we get $h(u_i) = \lambda u_i$. The converse (ii) is even easier to show. □

This means that we can solve the eigenvalue problem in the smaller components U_i, which is a lot easier. But how can we get V into a direct sum of invariant subspaces? In order to achieve that, this time we make $G := S(M)$ into a vector space

$$F[G] := \{\lambda_1 g_1 + \cdots + \lambda_r g_r \mid \lambda_i \in F\},$$

similar to 27.6, if $G = \{g_1, \ldots, g_r\}$ is finite. We also need a ring structure in V.

27.9. Theorem and Definition. *Let $F[G]$ be as above. Then*

$$\left(\sum_{i=1}^{r} \lambda_i g_i\right) \cdot \left(\sum_{j=1}^{r} \mu_j g_j\right) := \sum_{i=1}^{r}\sum_{j=1}^{r} \lambda_i \mu_j g_i g_j$$

turns $(F[G], +, \cdot)$ into a ring, so $F[G]$ becomes an F-algebra, called the ***group algebra*** *of G over F. As a vector space, $F[G]$ is r-dimensional.*

This is checked by easy computations. Now let k be the class number (see 10.29) of G. We take $F = \mathbb{C}$, and recall the lines after 18.18.

27.10. Theorem and Definition. *Let G and k be as before. Then there are* ***orthogonal idempotents*** *$e_1, \ldots, e_k \in \mathbb{C}[G]$ with*

(i) *$e_i e_j = 0$ for $i \neq j$;*

(ii) *$e_i^2 = e_i$ for all $i \in \{1, \ldots, k\}$;*

(iii) *$e_1 + \cdots + e_k = 1$, the neutral element of G.*

What purpose is served by these orthogonal idempotents? As in the last section on image understanding, representation theory emerges. By 26.5(iv), every representation h of G as a group of matrices is the sum of simple representations $h_1, \ldots, h_k$. We can show that k is again precisely the class number of G. Recall that the trace, $\mathrm{Tr}(A)$, of a (square) matrix $A = (a_{ij})$ is the sum $\sum a_{ii}$ of the entries in the main diagonal.

27.11. Definition. Let G be a group with class number k, and let $h_1, \ldots, h_k$ be its simple representations over $F = \mathbb{C}$. Then the functions $\chi_i \colon G \to \mathbb{C}^*$; $g \mapsto \mathrm{Tr}(h_i(g))$, with $1 \leq i \leq n$, are called the ***irreducible characters*** of G (over $\mathbb{C}$).

If $g, g' \in G$ are conjugated, then obviously $h(g)$ and $h(g')$ are similar matrices. By linear algebra, they must have the same trace. Hence we see that the (irreducible) characters are constant on conjugacy classes. Schur's Lemma 26.10 tells us that precisely for abelian groups G we get $h_1 = \chi_1, \ldots, h_k = \chi_k$. Characters are homomorphisms from G to $(\mathbb{C}^*, \cdot)$.

We illustrate with three examples the calculation of irreducible representations. In general, such calculations can be difficult.

27.12. Example. $G = \mathbb{Z}_n$. Since G is abelian, $r = n$, and the irreducible representations and characters coincide. Let χ be one of them. Since $\chi(1)^n = \chi(n \cdot 1) = \chi(0) = 1$, $\chi(1)$ must be an nth root α of unity, in fact, a primitive one. So we might take $\alpha = e^{2\pi i/n}$. Hence we get the character table

	0	1	2	...	$n-1$
χ_1	1	1	1	...	1
χ_2	1	α	α^2	...	α^{n-1}
$\vdots$	$\vdots$	$\vdots$	$\vdots$	...	$\vdots$
χ_n	1	α^{n-1}	$\alpha^{2(n-1)}$	...	$\alpha^{(n-1)^2}$

27.13. Example. $G = S_3$. This group has three conjugacy classes $\{\mathrm{id}\}$, $\{(1,2),(1,3),(2,3)\}$, $\{(1\,2\,3),(1\,3\,2)\}$. Two irreducible representations are found quickly: $h_1 = \chi_1$, the constant function with value 1, and $h_2 = \chi_2 = \mathrm{sign}$ (see 25.10). The third irreducible representation h_3 cannot be one-dimensional. More thinking (see Exercise 9) gives

π	id	$(1,2)$	$(1,3)$	$(2,3)$	$(1,2,3)$	$(1,3,2)$
$h_3(\pi)$	$\begin{pmatrix}1&0\\0&1\end{pmatrix}$	$\begin{pmatrix}0&1\\1&0\end{pmatrix}$	$\begin{pmatrix}-1&0\\-1&1\end{pmatrix}$	$\begin{pmatrix}1&-1\\0&-1\end{pmatrix}$	$\begin{pmatrix}0&-1\\1&-1\end{pmatrix}$	$\begin{pmatrix}-1&1\\-1&0\end{pmatrix}$
$\chi_3(\pi)$	2	0	0	0	-1	-1

27.14. Example. $G = \mathbb{Z}_2 \times \mathbb{Z}_2$ has the character table

	$(0,0)$	$(0,1)$	$(1,0)$	$(1,1)$
χ_1	1	1	1	1
χ_2	1	1	-1	-1
χ_3	1	-1	1	-1
χ_4	1	-1	-1	1

For extensive lists of character tables, see Leech & Newman (1970) and Sternberg (1994).

27.15. Theorem. *Let G be finite with irreducible characters $\chi_1, \dots, \chi_k$ and $d_i := \chi_i(1)$. Then*

$$e_i := \frac{d_i}{|G|} \sum_{g \in G} \chi_i(g^{-1}) g \qquad (1 \le i \le k)$$

gives orthogonal idempotents $e_1, \dots, e_k$ in $\mathbb{C}[G]$.

We started with a linear operator $h\colon V \to V$, where V is the vector space spanned by $M = \{m_1, \dots, m_n\}$, and wanted to find h-invariant subspaces U_i with $V = U_1 \dotplus \cdots \dotplus U_k$. Then we considered another vector space, in fact an algebra, $\mathbb{C}[G]$. With the help of character tables, we can find orthogonal idempotents in $\mathbb{C}[G]$. But how can we get a decomposition of V? For that, we shall need a further connection between h and the group G.

27.16. Definition. Let V be a vector space, h a linear operator on V, and G a group of automorphisms of V. Then h ***respects*** G if $h \circ g = g \circ h$ holds for all $g \in G$.

27.17. Theorem. *Let V be spanned by M as in 27.6 and $G = S(M)$ be the group of symmetries of M. Let $e_1, \ldots, e_k$ be as in 27.10. If the linear operator $h\colon V \to V$ respects G, then V is the direct sum of the h-invariant subspaces $U_1 := e_1V, \ldots, U_k := e_kV$.*

The proof is done by a direct check of the claimed equations. Observe that U_i is spanned by $\{e_i m_1, \ldots, e_i m_n\}$.

27.18. Example. Let us reconsider naphtalene in 27.5(v). The ten corner points $\{m_1, \ldots, m_{10}\}$ determine a ten-dimensional vector space V over $\mathbb{C}$. For the sake of simplicity, let us consider naphtalene's symmetry group in $\mathbb{R}^2$, which is given by $G := \{g_1, g_2, g_3, g_4\}$, where $g_1 = \mathrm{id}$, and g_2, g_3, g_4 are the indicated reflections:

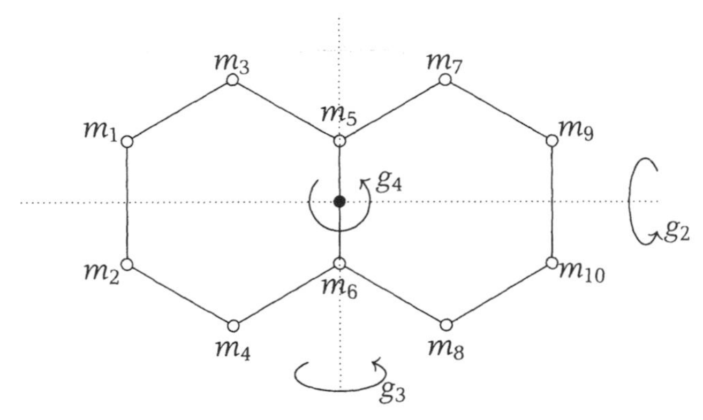

By 27.14 we have the irreducible characters at hand, which give us four idempotents:

$$e_1 = \frac{1}{4}(g_1 + g_2 + g_3 + g_4), \qquad e_2 = \frac{1}{4}(g_1 + g_2 - g_3 - g_4),$$

$$e_3 = \frac{1}{4}(g_1 - g_2 + g_3 - g_4), \qquad e_4 = \frac{1}{4}(g_1 - g_2 - g_3 + g_4).$$

We can easily check that $e_i^2 = e_i$ and $e_i e_j = 0$ for $i \neq j$. Also, $e_1 + e_2 + e_3 + e_4 = g_1 = 1$.

We compute U_1, the linear hull of $\{e_1 m_1, \ldots, e_1 m_{10}\}$:

$$\begin{aligned} e_1 m_1 &= \frac{1}{4}(g_1(m_1) + g_2(m_1) + g_3(m_1) + g_4(m_1)) \\ &= \frac{1}{4}(m_1 + m_2 + m_9 + m_{10}), \end{aligned}$$

$e_1 m_2$ gives the same result, as well as $e_1 m_9$ and $e_1 m_{10}$. Now

$$e_1 m_3 = \frac{1}{4}(m_3 + m_4 + m_7 + m_8) = e_1 m_4 = e_1 m_7 = e_1 m_8,$$

and $e_1 m_5 = e_1 m_6 = \frac{1}{2}(m_5 + m_6)$. So U_1 is the subspace spanned by $\{m_1 + m_2 + m_3 + m_{10},\ m_3 + m_4 + m_7 + m_8,\ m_5 + m_6\}$.

Similarly, U_2 is spanned by $\{m_1 - m_2 - m_9 + m_{10},\ m_3 - m_4 - m_7 + m_8\}$; U_3 is spanned by $\{-m_1 - m_2 + m_9 + m_{10},\ -m_3 - m_4 + m_7 + m_8\}$ and U_4 is the hull of $\{-m_1 + m_2 - m_9 + m_{10},\ -m_3 + m_4 - m_7 + m_8,\ m_5 - m_6\}$.

So V decomposes into $V = U_1 \dotplus U_2 \dotplus U_3 \dotplus U_4$, where U_1 and U_4 are three-dimensional, and U_2, U_3 are two-dimensional. Hence the eigenvalue problem for an operator $h\colon V \to V$ (the solution of an equation of degree 10) which respects G is reduced to solving two cubic and two quadratic equations. Problems of this type occur, for instance, in the study of vibrations (see below).

The decomposition of V into invariant subspaces can again be considered as a "decoupling of overlapping influences" as in §26. And again, it is representation theory which achieves that.

Here is an example of an operator which respects symmetry.

27.19. Example. Let us reconsider water as in 27.5(iii)

but now we consider the "one-dimensional" symmetry group $G = \{\mathrm{id}, r\}$, where r is the reflection about the oxygen atom O, and everything occurs in the line l. The atoms are "tied together" by electromagnetic forces. We assume a constant force k.

$\to a_1$ $\to a_2$ $\to a_3$

H——O——H

If the three atoms are slightly displaced (still on the line l) by amounts a_1, a_2, a_3, then Newton's laws tell us that

$$m\frac{d^2a_1}{dt^2} = k(a_2 - a_1),$$
$$m\frac{d^2a_2}{dt^2} = k(a_3 - a_2) - k(a_2 - a_1),$$
$$m\frac{d^2a_3}{dt^2} = k(a_3 - a_2).$$

If $\mathbf{a} := \begin{pmatrix} a_1 \\ a_2 \\ a_3 \end{pmatrix}$ and $\mathbf{F} = \frac{k}{m}\begin{pmatrix} 1 & -1 & 0 \\ -1 & 2 & -1 \\ 0 & -1 & 1 \end{pmatrix}$, then $\mathbf{F}$ is symmetric and

$$\frac{d^2\mathbf{a}}{dt^2} + \mathbf{F}\mathbf{a} = \mathbf{0}. \tag{27.1}$$

(In general, small oscillations near an equilibrium are described in the form (27.1) with a symmetric $\mathbf{F}$.) Easy calculations show that $\mathbf{F}$ has eigenvalues 0, $\frac{k}{m}$, and $\frac{3k}{m}$ with corresponding eigenvectors $\begin{pmatrix} 1 \\ 1 \\ 1 \end{pmatrix}$, $\begin{pmatrix} 1 \\ 0 \\ -1 \end{pmatrix}$, $\begin{pmatrix} -1 \\ 2 \\ -1 \end{pmatrix}$, respectively. These eigenvalues determine the frequencies of the vibrations, and their eigenvectors yield the corresponding "normal mode configurations." In our case, the eigenvectors describe a uniform translation, symmetric vibration, and antisymmetric vibration, respectively. On the vector space of all displacements $\mathbf{d} = (d_1, d_2, d_3)^{\mathrm{T}} \in \mathbb{R}^3$, the symmetry group G acts as $G = \{\mathbf{I}, \mathbf{R}\}$, where $\mathbf{R} = \begin{pmatrix} 0 & 0 & 1 \\ 0 & 1 & 0 \\ 1 & 0 & 0 \end{pmatrix}$. As symmetries, the two matrices in G should commute with the force operator $\mathbf{F}$. In fact, $\mathbf{IR} = \mathbf{RI}$ and $\mathbf{RF} = \mathbf{FR}$, as we easily see. Observe that the representation of G on $\mathbb{R}^3$ cannot be irreducible, for otherwise $\mathbf{F}$ would be a scalar multiple of $\mathbf{I}$ according to Schur's Lemma 26.10.

If we do not know the "shape" of a molecule M, then "shaking" it and observing the resonant frequencies can yield important information at least on its symmetry group G. Let $\mathbf{E}$ be an operator other than in the preceding example, acting on a vector space V. If G is a possible symmetry group (again acting on V), and if **E** respects G, then G induces a decomposition of V into k **E**-invariant subspaces, where k is the class number of G, according to 27.8 and 27.17. If we observe a number k' of resonant frequencies other than k, then G cannot be the "correct" symmetry group.

We did simplify matters here; for much more on these topics and adaptation to crystallography and quantum mechanics, see Sternberg (1994).

In the algorithm 27.4, we saw that only certain finite groups can be symmetry groups. It can be shown that the only finite rotation groups (i.e., subgroups of SO(3)) are $\mathbb{Z}_n$ (with $n = 1, 2, 3, 4, 6$), D_n (with $n = 2, 3, 4, 6$), A_4, and S_4. Crystals have one of 32 finite symmetry subgroups of SO(3); Sternberg (1994, pp. 42–44) contains a complete list, as well as crystals (if known) for each of these ***geometrical crystal classes***, which were already known around 1830. Up to now, more than 20,000 substances have been classified. If we also allow translations (see below), we altogether get 230 ***crystallographic groups***; they were classified by Fedorov more than 100 years ago.

In the case of translations, the pattern must be infinite, like in:

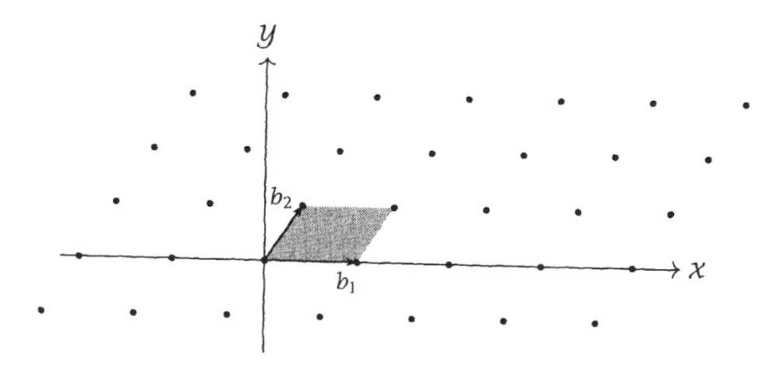

The marked "cell" is called the ***elementary cell*** or ***elementary domain***; it is determined by two basis vectors b_1, b_2 (three basis vectors in $\mathbb{R}^3$). The marked points are precisely the linear combinations of these basis vectors with integral coefficients. Their collection is called a ***net*** or a ***lattice*** (because there are connections to Chapter 1).

Two nets, determined by bases B and B', respectively, are identical iff their bases transformation matrix $M^B_{B'}$ has integral coefficients and determinant ± 1 (see Exercise 12).

What are the possible rotations which map a net into itself? From linear algebra we know that, for a suitable basis, a rotation is done by multiplying with a matrix $\mathbf{R}_\theta = \begin{pmatrix} \cos\theta & -\sin\theta \\ \sin\theta & \cos\theta \end{pmatrix}$ (see page 306) in $\mathbb{R}^2$ or with $\mathbf{R}_\theta = \begin{pmatrix} 1 & 0 & 0 \\ 0 & \cos\theta & -\sin\theta \\ 0 & \sin\theta & \cos\theta \end{pmatrix}$ in $\mathbb{R}^3$. Their trace is $2\cos\theta$ or $1 + 2\cos\theta$, respectively, and this does not change if we take another basis (see the lines after 27.11). This trace must be integral by the preceding paragraph. Now $2\cos\theta$ is an integer precisely for $\theta = \frac{2\pi}{n}$ with $n = 1, 2, 3, 4$, or 6 (see Exercise 13). This was already known to Kepler and partly explains what is so special about the numbers $1, 2, 3, 4, 6$ in the list of finite rotation groups above.

In practice, however, no infinite tilings can exist, of course. In the tiling

which is used on the floor of the cathedral in Pavia (Italy), rotational, reflectional, and translational symmetries are clearly visible. Problems arise on and close to the edges. Some "tile" cannot be rotated or translated like other tiles closer to the center. So, sometimes we allow "partial symmetry operations," which act like symmetry operations wherever they are defined. In this way, we obtain "quasigroups" instead of groups. See the survey article by Weinstein (1996).

There are many more instances where symmetry groups prove to be useful. One instance will be described in §37. Another one is the following. If α is a solution of $ax^6 + bx^4 + cx^2 + d = 0$, then $-\alpha$ is also a solution, since the "symmetry operation" $f\colon x \mapsto -x$ leaves the equation "invariant." Much more generally, let $f(x) = 0$ be an equation where x varies in some set X. If g leaves f "invariant" (i.e., $f(g(x)) = f(x)$ for all x) or if g commutes with f (i.e., $f \circ g = g \circ f$) and fulfills $g(0) = 0$, then for every solution α, each $f(\alpha)$ is also a solution. Similar ideas apply to differential equations (see Fässler & Stiefel (1992)).

Exercises

1. Prove the assertion of Theorem 27.1.
2. Why is the symmetry group of a regular tetrahedron isomorphic to S_4?
3. Determine the symmetry group of a cube.
4. Which of the letters A, B, ..., Z have a nontrivial symmetry group in $\mathbb{R}^2$?
5. Determine the symmetry group of salt: Na—Cl.
6. The same for allene, C_3H_4:

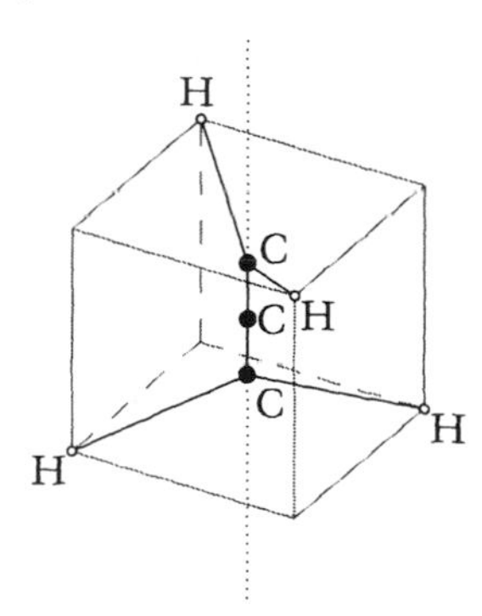

7. The same for PCl_5:

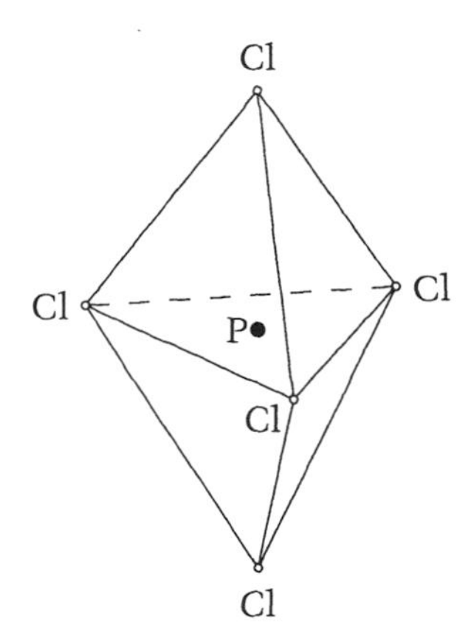

8. Show that α_g in 27.7 is an automorphism of V and that the map $g \mapsto \alpha_g$ establishes a homomorphism i from G into Aut(V). Is i a monomorphism?
9. Let S_3 act on $b_1, b_2, -b_1 - b_2$, where b_1, b_2 are canonical basis vectors of $\mathbb{R}^2$ in the obvious manner. Use this to determine the irreducible representation h_3 in 27.13.
10. Find all orthogonal idempotents in $\mathbb{C}(\mathbb{Z}_n)$ with $n \in \mathbb{N}$.
11. The same for $\mathbb{C}(S_3)$.
12. Convince yourself that two nets, determined by the bases B and B', are identical iff the basis transform matrix $\mathbf{M}_{B'}^{B}$ is invertible in $\mathbb{M}_2(\mathbb{Z})$ or $\mathbb{M}_3(\mathbb{Z})$.
13. Show that the trace of $\mathbf{R}_\theta$ is integral iff $\theta = 0°, 60°, 90°, 120°$, or $180°$.

Notes

The possibility of doing simultaneous computations in $\mathbb{Z}_{p_i^{t_i}}$ instead of $\mathbb{Z}_n$ with $n = p_1^{t_1} p_2^{t_2} \dots p_k^{t_k}$ triggered a whole area of research in Algebraic Computer Science, namely parallel computation. We shall return to this subject in §34, where we shall see that any decomposition of a structure into a direct product of simpler ones gives rise to the possibility of parallel computation.

The central position in the theory of counting is taken by Pólya's Theorem, a very elegant result contained in a paper published in 1937 in *Acta Mathematica* 68. A survey of Pólya's theory of counting is given in De Bruijn (1981), with many further references. Some of the texts on applied algebra also contain a section on applications of the central theory of this paragraph, e.g., Gilbert (1976), Dornhoff & Hohn (1978), Biggs (1985), and van Lint & Wilson (1992). A standard book on Pólya's theory and its applications is Kerber (1991).

Image understanding and image processing are basically old disciplines, which have partly worked (and still do so) with a tremendous numerical apparatus, for instance in Computer Tomography. The algebraic methods in §26 are relatively young and an area of intensive research. The group working under the leadership of Beth in Karlsruhe, Germany and of Kanatani in Gunma, Japan, are certainly "hubs" of these research activities. Samples of good books in this area are Beth (1995),

Clausen & Baum (1993), Kanatani (1990), Lenz (1990), Nagel (1995), and Sung & Zimmermann (1986).

The applications of group theory (in particular, of representation theory) to physics was boosted by H. Weyl's book *Gruppentheorie und Physik* in 1928. Some "classics" are Weyl (1952), the two volumes by Elliot & Dawber (1979), and Sternberg (1994). Molecules, crystals, and elementary particles are clearly different, but very similar theories of symmetry apply to and unify these theories.

It was basically H. Weyl who realized the importance of group theory to study symmetry in Nature. A vast variety of texts are concerned with these connections; some of them are Chen (1989), Elliot & Dawber (1979), Folk, Kartashov, Linsoněk & Paule (1993), Fässler & Stiefel (1992), Kanatani (1990), Lautrup (1989), Leech & Newman (1970), Mathiak & Stingl (1968), Pommaret (1994), Vanasse (1982), Weinstein (1996), and Weyl (1952). Applications of groups to chemistry are described in Cotton (1971). A pleasant new introductory text is Farmer (1996).

It took some time until group theory was really appreciated in physics. In the early 1930s, even the name "Gruppenpest" came up; now group theory is one of the main tools in physics. This history is nicely described in Coleman (1997).

7 CHAPTER Further Applications of Algebra

This chapter contains several topics from various areas in which algebra can be applied. The material is mainly selected to give a brief indication of some further applications of algebraic concepts, mainly groups, semigroups, rings, and fields, without being able to go into much depth. Many results are given without proof. We refer the interested reader to the special literature on these topics in the Bibliography.

We start with semigroups, which are algebraic structures with just one associative operation. They do show many of the features which we met in §10 on groups. Also, interesting new algebraic ideas arise and many applications and connections to other areas of mathematics will show up: we shall describe the connections to the algebra of relations, to input and output sequences for automata (§29), and the use of semigroups to decompose automata into simple parts. §30 contains some of the connections between automata and computer languages ("formal languages"), which also take us back to automata theory. Surprising connections between semigroups, formal languages, and biology are described in §31. In §32 we study semigroups of kinship relations and role structures, thus ending up in sociology. This concludes our journey through the applications of semigroups.

Linear difference equations (the discrete analogue to differential equations) are our next target of investigation (§33). It turns out that the theory of finite fields, which is studied in §13, is the appropriate tool to study these equations, which we treat more from the point of view of recurring sequences. These show a periodic behavior and give rise to numerous applications like measuring distances by radar methods.

A different topic is the technique of Fast Fourier Transforms, which will occupy us in §34. These transforms substantially speed up many computations, especially those with polynomials. The general idea of parallel computing is presented.

Latin squares (§35) and design theory (§36) take us to areas of combinatorics and statistics, where finite fields show their power again. We also use generalizations of finite fields (finite near-rings) to construct efficient designs in an easy way. Latin squares and designs are then used to provide intelligent arrangements for statistical experiments, and we also indicate how to do the statistical analysis. Also, block designs induce (nonlinear) codes of quite a remarkable quality in a very easy way.

In §37, certain matrices over $\{-1, 1\}$ are studied. These "Hadamard matrices" turn out to be surprisingly useful in a variety of areas: codes, weighing problems, and neural networks. A "fast Hadamard transform" (similar to the Fourier transform of §34) enables fast computations and yields an efficient way to decode Reed-Muller codes, so that these codes can be used for compact discs.

Algebraic equations in several unknowns and partial differential equations are our object of study in §38. The idea is to study the involved polynomials, form the ideal they generate, and then find an "easy" basis for this ideal. This often greatly simplifies systems of equations.

In §39, we study linear and nonlinear systems, mostly as "continuous analogues" to automata. Much carries over from automata to systems, and many concepts like controllability and stability, turn out to be of an algebraic nature. Also, the control of (possibly unstable) systems leads to algebraic questions; they start from eigenvalue problems and quickly get us into non-trivial ring theory.

Finally, we study all algebraic structures we have met so far in a unified form as "universal algebras" in §40. We touch model theory briefly and find out that even these "most abstract" theories do have their concrete applications, this time to data types in computer science.

§28 Semigroups

In order to see more connections between algebra and the "real world," we study a generalization of the concept of groups (see §10). We do not require the existence of a neutral element any more, and hence we also do not require the existence of inverses. The resulting "semigroups" are in fact much more general than groups. Consequently, only weaker results can be obtained, but on the other hand, these results apply to a much greater variety of situations.

28.1. Definition. A ***semigroup*** is a set S together with an associative binary operation $\circ$ defined on it. We shall write $(S, \circ)$ or simply S for a semigroup. If $s \circ t = t \circ s$ holds for all $s, t \in S$, we call S a ***commutative semigroup***.

Here are some examples of semigroups.

28.2. Examples.

(i) $(\mathbb{N}, +)$, $(\mathbb{N}_0, +)$, $(\mathbb{N}, \cdot)$, $(\mathbb{Z}, +)$, $(\mathbb{Z}, \cdot)$, $(\mathbb{R}, +)$, and $(\mathbb{R}, \cdot)$ are prominent examples of commutative semigroups. These examples will show up frequently.

(ii) $(\mathbb{R}, \max)$ with $x(\max)y := \max(x, y)$ is another example of a commutative semigroup on the set $\mathbb{R}$.

(iii) Let $\mathbb{M}_n(\mathbb{R})$ denote the set of all $n \times n$ matrices with entries in $\mathbb{R}$. Then $(\mathbb{M}_n(\mathbb{R}), +)$ and $(\mathbb{M}_n(\mathbb{R}), \cdot)$ are semigroups.

(iv) $(\mathbb{R}[x], +)$, $(\mathbb{R}[x], \cdot)$, and $(\mathbb{R}[x], \circ)$ are also examples of semigroups. Again we can see that many different operations can be defined on a set, turning this set into (different) semigroups.

(v) Let S be any nonempty set. Let S^S be the set of all mappings from S to S and let $\circ$ denote the composition of mappings. Then $(S^S, \circ)$ is a semigroup. These types of semigroups turn out to be "universal examples" in the sense that every semigroup "sits inside" such a $(S^S, \circ)$ for a properly chosen S.

(vi) Let V be a vector space and $\mathrm{Hom}(V, V)$ the set of all linear maps from V to V. Then $(\mathrm{Hom}(V, V), +)$ and $(\mathrm{Hom}(V, V), \circ)$ are semigroups.

(vii) Let S again be a set (now S might be empty). Then $(\mathcal{P}(S), \cap)$, $(\mathcal{P}(S), \cup)$, and $(\mathcal{P}(S), \triangle)$ are commutative semigroups whose elements are sets.

(viii) More generally, if $(L, \wedge, \vee)$ is a lattice, then both $(L, \wedge)$ and $(L, \vee)$ are commutative semigroups.

(ix) Every set S can be turned into a semigroup: Define $s_1 * s_2 =: s_1$ for all $s_1, s_2 \in S$. Then $(S, *)$ is a semigroup.

More examples will follow. Sets with operations which together do not form a semigroup are $(\{1, 2, 3, 4, 5\}, +)$, since, for example, $4 + 5 \notin \{1, 2, 3, 4, 5\}$, and $(\mathbb{Z}, -)$, since $-$ is not associative. If the underlying set S is finite, we speak of a ***finite semigroup*** $(S, \circ)$. Similarly to the group case, such a semigroup can be characterized by a ***semigroup table*** (cf. Figure 1.10 and the lines after 10.2 on page 98),

28.3. Definition. Let $(S, \circ)$ be a semigroup and $s \in S$.

(i) s is called a ***zero element*** of $(S, \circ)$ if $x \circ s = s \circ x = s$ holds for all $x \in S$;

(ii) s is called an ***identity element*** of $(S, \circ)$ provided that $x \circ s = s \circ x = x$ holds for all $x \in S$;

(iii) s is called an ***idempotent*** of $(S, \circ)$ if $s \circ s = s$.

As in groups, we write s^2 instead of $s \circ s$, s^3 instead of $s \circ s \circ s$, etc. An element s is an idempotent if and only if $s = s^2 = s^3 = \dots$. Not every semigroup has a zero or an identity. However, similar to the group case (see page 96), no semigroup can have more than one of each of these.

28.4. Definition. If the semigroup $(S, \circ)$ has an identity element, then $(S, \circ)$ is called a ***monoid***. If $(S, \circ)$ is a monoid and $x \in S$, then x is called ***invertible*** if there is some $y \in S$ such that $x \circ y$ and $y \circ x$ are the identity element.

As for groups (page 96) the element y in 28.4 is uniquely determined by x and is called the ***inverse*** of x, denoted by $y = x^{-1}$ or by $y = -x$.

28.5. Examples.

(i) $(\mathbb{N}, +)$ is a semigroup which has neither a zero nor an identity element.

(ii) $(\mathbb{N}_0, +)$ is a monoid with identity 0; only 0 is invertible.

(iii) $(\mathbb{N}, \cdot)$ is a monoid with identity 1; only 1 is invertible.

(iv) $(\mathbb{Z}, \cdot)$ is a monoid with identity 1 and zero 0, in which precisely 1 and -1 are invertible. The only idempotents are given by 0 and 1.

(v) $(\mathbb{Q}, \cdot)$ is also a monoid with 1 as the identity, but this time every element except 0 is invertible. 0 and 1 are still the only idempotents.

(vi) The $n \times n$ matrices over the reals form a monoid $(\mathbb{M}_n(\mathbb{R}), \cdot)$ with the unit matrix as an identity. The invertible elements are precisely the nonsingular matrices. $\mathbf{A} = \begin{pmatrix} 1 & 0 \\ 0 & 0 \end{pmatrix}$ is an example of a nontrivial idempotent.

(vii) $(S^S, \circ)$, as in 28.2(v), is a monoid with the identity function as the identity element. The invertible elements are exactly the invertible (i.e., bijective) functions. The projection $f\colon \mathbb{R} \times \mathbb{R} \to \mathbb{R} \times \mathbb{R};\ (x, y) \mapsto (x, 0)$ is an example of an idempotent element (with $S = \mathbb{R} \times \mathbb{R}$).

(viii) $(\mathcal{P}(S), \cap)$ is a monoid for every set S. The set S, considered as an element of $\mathcal{P}(S)$, serves as the identity; it is the only invertible element. All elements are idempotent.

(ix) Let $(G, \circ)$ be a group. Then $(G, \circ)$ is a monoid in which every element is invertible. No element besides the identity can be idempotent.

From §10 it seems clear how to define subsemigroups, semigroup-homomorphisms, and direct products. Since the intersection of subsemigroups is again a subsemigroup we can talk about subsemigroups $\langle X \rangle$ ***generated*** by a subset X. In this case, $\langle X \rangle$ consists of all finite products of elements of X. For example, $(\mathbb{N}, +)$ is generated by $\{1\}$, while $(\mathbb{N}, \cdot)$ is generated by $\{1\} \cup \mathbb{P}$, where $\mathbb{P}$ is the set of all primes.

28.6. Definition. Let $(S, \circ)$ be a monoid. $G_S := \{x \in S \mid x \text{ is invertible}\}$ is called the ***group kernel*** of $(S, \circ)$.

28.7. Theorem. *G_S is a group within $(S, \circ)$.*

Proof.

(i) If $x, y \in G_S$, then $x \circ y$ has $y^{-1} \circ x^{-1}$ as an inverse (why?). Hence $x \circ y \in G_S$.

(ii) Since the identity element e is invertible ($e = e^{-1}$), e belongs to G_S.

(iii) If $x \in G_S$, then $x^{-1} \circ x = x \circ x^{-1} = e$, so $(x^{-1})^{-1} = x$, and hence $x^{-1} \in G_S$, and so every element in G_S is invertible (in G_S). □

28.8. Examples. The following list gives the group kernels of the monoids in 28.5:

Monoid	Group kernel	Monoid	Group kernel
$(\mathbb{N}_0, +)$	$\{0\}$	$(\mathbb{M}_n(\mathbb{R}), \cdot)$	$GL(n, \mathbb{R})$
$(\mathbb{N}, \cdot)$	$\{1\}$	$(M^M, \circ)$	S_M
$(\mathbb{Z}, \cdot)$	$\{1, -1\}$	$(\mathcal{P}(S), \cap)$	$\{S\}$
$(\mathbb{Q}, \cdot)$	$\mathbb{Q}^* = \mathbb{Q}\setminus\{0\}$	$(G, \circ)$, a group	G

In a certain sense, the larger the group kernel, the closer a monoid resembles a group.

28.9. Example. We study the group kernel $\mathbb{G}_n$ of $(\mathbb{Z}_n, \cdot)$. As in the proof of Theorem 12.1, we see that $x \in \mathbb{Z}_n$ is invertible iff $\gcd(x, n) = 1$. The inverse of x can then be found by representing 1 as $1 = xa + nb$ in $\mathbb{Z}$. This gives $[1] = [x][a]$ in $\mathbb{Z}_n$; so $[a]$ is the inverse of x. Hence $\mathbb{G}_n = \{x \in \mathbb{Z}_n \mid \gcd(x, n) = 1\}$ and $|\mathbb{G}_n| = \varphi(n)$.

28.10. Definition. Let $(S, \circ)$ be a semigroup and $\sim$ an equivalence relation on S. Then $\sim$ is called a ***congruence relation*** on $(S, \circ)$ if $\sim$ is compatible with $\circ$, i.e., if $a \sim a'$ and $b \sim b'$ implies $a \circ a' \sim b \circ b'$.

For congruence relations on $(S, \circ)$, we define (as we have done for other structures) on the factor set $S/\sim$: $[a] \circledcirc [b] := [a \circ b]$. This is well defined: If $[a] = [a']$, $[b] = [b']$, then $a \sim a'$, $b \sim b'$, hence $a \circ b \sim a' \circ b'$, whence $[a \circ b] = [a' \circ b']$.

28.11. Theorem and Definition. *Let $\sim$ be a congruence relation on the semigroup $(S, \circ)$. Then $(S/\sim, \circledcirc)$ is a semigroup, called the* ***factor semigroup*** *(or* ***quotient semigroup****) of $(S, \circ)$ w.r.t. $\sim$. The mapping $\pi\colon S \to S/\sim;\ s \mapsto [s]$ is an epimorphism, called the* ***canonical epimorphism***.

For $(S/\sim, \circledcirc)$, we write $(S/\sim, \circ)$ or simply $S/\sim$. As for groups and rings we get

28.12. Theorem (*Homomorphism Theorem*). *Let $f\colon (S, \circ) \to (S', \circ')$ be an epimorphism. Then $\sim$ defined by $x \sim y :\iff f(x) = f(y)$ is a congruence relation on $(S, \circ)$ and $(S/\sim, \circ) \cong (S', \circ')$.*

This means that all homomorphic images $(S', \circ')$ of $(S, \circ)$ can be found "within" $(S, \circ)$, just by finding all suitable congruence relations.

Let $\sim_1$ and $\sim_2$ be two congruences on a semigroup $(S, \circ)$ such that $\sim_1 \subseteq \sim_2$ (hence for all $x, y \in S$: $x \sim_1 y \Longrightarrow x \sim_2 y$). We say that $\sim_1$ is ***finer*** than $\sim_2$. The equivalence classes with respect to $\sim_1$ and $\sim_2$ are denoted by $[x]_1$ and $[x]_2$, respectively.

28.13. Theorem. *Let $\sim_1$ and $\sim_2$ be two congruence relations on the semigroup $(S, \circ)$. Then $\sim_1 \subseteq \sim_2$ if and only if $[x]_1 \mapsto [x]_2$ defines an epimorphism of $(S/\sim_1, \circ)$ onto $(S/\sim_2, \circ)$.*

Proof. "$\Longrightarrow$" If $\sim_1 \subseteq \sim_2$, then $f\colon [x]_1 \mapsto [x]_2$ is well defined, since $[x]_1 = [y]_1 \Longrightarrow x \sim_1 y$, thus $x \sim_2 y$, and therefore $[x]_2 = [y]_2$. Clearly, f is surjective, and it is an epimorphism, since $f([x]_1 \circledcirc [y]_1) = f([x \circ y]_1) = [x \circ y]_2 = f([x]_1) \circledcirc f([y]_1)$ for all $[x]_1, [y]_1 \in S/\sim_1$.

"$\Longleftarrow$" If $[x]_1 \mapsto [x]_2$ is an epimorphism, then $x \sim_1 y$, i.e., $[x]_1 = [y]_1$, implies $[x]_2 = [y]_2$, thus $x \sim_2 y$. □

In particular, $\sim_1 \subseteq \sim_2$ implies that $(S/\sim_2, \circ)$ is a homomorphic image of $(S/\sim_1, \circ)$ (*"**Dyck's Theorem**"*).

28.14. Example. Let $(S, \circ) = (\mathbb{Z}, \cdot)$ and $\sim_1, \sim_2$ be the congruence relations $\equiv_4, \equiv_2$, respectively. Here we have $\equiv_4 \subseteq \equiv_2$, and also $(S/\sim_1, \circ) = (\mathbb{Z}/\equiv_4, \cdot) \cong (\mathbb{Z}_4, \cdot)$ and $(S/\sim_2, \circ) = (\mathbb{Z}/\equiv_2, \cdot) \cong (\mathbb{Z}_2, \cdot)$. In fact, $h\colon (\mathbb{Z}_4, \cdot) \to (\mathbb{Z}_2, \cdot)$; $[x]_1 \mapsto [x]_2$ is an epimorphism.

28.15. Definition. Let $\sim_1 \subseteq \sim_2$ be congruence relations on $(S, \circ)$. Then the mapping $f\colon [x]_1 \mapsto [x]_2$ of 28.13 is called the ***standard epimorphism*** of $(S/\sim_1, \circ)$ onto $(S/\sim_2, \circ)$.

For later needs, we study two more classes of semigroups: relation semigroups and free semigroups.

If we regard the subsets of $M \times M$ as relations, we write $\mathcal{R}(M)$ instead of $\mathcal{P}(M \times M)$, and for $R, S \in \mathcal{R}(M)$ we define a relation $R \diamond S$ such that $x\,(R \diamond S)\,y$ holds if and only if there is some $z \in M$ satisfying $x R z$ and $z S y$.

The following result is straightforward.

28.16. Theorem. *$(\mathcal{R}(M), \diamond)$ is a monoid, for any set M; the equality $\{(x, x) \mid x \in M\}$ is the identity; the invertible elements are precisely the graphs of bijective functions; $\emptyset$ is the zero element.*

28.17. Definition. $(\mathcal{R}(M), \diamond)$ is called the ***relation semigroup*** on M. The operation $\diamond$ is called the ***relation product***.

It is obvious that $R \in \mathcal{R}(M)$ is transitive if and only if $R \diamond R \subseteq R$.

28.18. Theorem and Definition. *Let $R \in \mathcal{R}(M)$; then the union R^t of the sets in $\langle R \rangle$ is the smallest transitive relation of $\mathcal{R}(M)$ containing R. R^t is called the **transitive hull** of R, and $x R^t y$ iff there are $n \in \mathbb{N}$ and $x_1, \ldots, x_n \in M$ with $x R x_1, x_1 R x_2, \ldots, x_n R y$.*

Proof. We have $R^t \circ R^t \subseteq R^t$, because $\langle R \rangle \leq \mathcal{R}(M)$, and hence R^t is transitive. The characterization of R^t follows from the lines before 28.6 on page 335. Let $U \in \mathcal{R}(M)$ be transitive, with $R \subseteq U$. Then $x R^t y$ implies the existence of $n \in \mathbb{N}$ and $x_1, \dots, x_n \in M$ such that $x R x_1$, $x_1 R x_2, \dots, x_n R y$, i.e., $x U x_1$, $x_1 U x_2, \dots, x_n U y$. So $x U y$. Therefore $R^t \subseteq U$. □

As an application of this result we obtain

28.19. Corollary. *If $R \in \mathcal{R}(M)$, then the smallest equivalence relation $\bar{R}$ containing R is given by $(R \cup R^{-1} \cup \{(x,x) \mid x \in M\})^t$. It is called the* ***equivalence relation generated*** *by R.*

Proof. $\bar{R}$ must contain R, the inverse relation R^{-1}, and the diagonal $D = \{(x,x) \mid x \in M\}$, since $\bar{R}$ is symmetric and reflexive. $\bar{R}$ is also transitive, thus $\bar{R}$ contains $(R \cup R^{-1} \cup D)^t =: U$ because of 28.18. This is equal to $\bar{R}$, since U is an equivalence relation.

Similarly, we can characterize the *congruence* relation $\tilde{R}$, generated by $R \subseteq \mathcal{R}(M)$. For example, if $a R b$ and $c R d$, then $ac \tilde{R} bd$.

28.20. Definition and Theorem. Let A be a nonempty set and let A_* be the set of all finite sequences $(a_1, \dots, a_n)$ with $n \in \mathbb{N}$ and $a_i \in A$, which we—as in coding theory—simply write as $a_1 a_2 \dots a_n$. We define the ***concatenation*** $*$ in A_* by

$$a_1 a_2 \cdots a_n * a_1' a_2' \cdots a_m' := a_1 a_2 \cdots a_n a_1' a_2' \cdots a_m'.$$

Then $(A_*, *)$ is a semigroup, called the ***free semigroup*** on A or the ***word semigroup*** on the ***alphabet*** A. If we add the ***empty word*** Λ to A, we get the ***free*** (or ***word***) ***monoid*** A^* on A.

Recall that, by the definition of sequences, we have

$$a_1 a_2 \cdots a_n = a_1' a_2' \cdots a_m' \iff n = m \text{ and } a_i = a_i' \text{ for all } i \in \{1, \dots, n\}.$$

Every element in A_* then is a unique product of elements of A, and A generates A_*. It is easy to see that every map from A to some semigroup S can be uniquely extended to a homomorphism from A_* to S. In line with linear algebra, A is often called the ***basis*** of A_*.

28.21. Remarks.

(i) If $A = \{a\}$, then $A_* = \{a, a^2, a^3, \dots\}$ is isomorphic to $(\mathbb{N}, +)$ and therefore commutative.

(ii) If $|A| > 1$, then A_* is not commutative, since for $a, b \in A$, $a \neq b$, we get $a * b = ab \neq ba = b * a$.

(iii) A_* is infinite for every A.

So, no finite semigroup can be free; therefore no finite semigroup can have a basis. Thus, free semigroups are rather "rare," but there are still plenty of them, since every semigroup is the homomorphic image of a free semigroup.

28.22. Theorem. *Let $(S, \circ)$ be a semigroup, generated by some $X \subseteq S$. Then S is a homomorphic image of X_*.*

Proof. Let $h\colon X_* \to S$ be given by $h(x_1 x_2 \dots x_n) := x_1 \circ x_2 \circ \cdots \circ x_n$. Then h is a homomorphism and

$$X = h(X) \subseteq h(X_*) \leq S,$$

and thus $S = \langle X \rangle \subseteq h(X_*) \subseteq S$. Hence h is an epimorphism. □

This explains the name "free semigroup": A_* is "free" of all restrictive relations, such as commutativity (see 28.21(ii)), etc. By adding such relations, we can obtain any semigroup. A_* is the most general semigroup which can be obtained from a generating set of cardinality $|A|$. Theorem 28.22 enables us to describe a semigroup in terms of its generating set and relations.

28.23. Definition. Let S be a semigroup generated by X, and let R be a relation on X_*. If $[X, R] := X_*/\tilde{R} \cong S$, the pair $[X, R]$ is called a ***presentation*** of S, and we usually write $S = [X, R]$ in this case.

Elements of R are called ***defining relations***, despite the fact that they are not relations at all. Theorem 28.22 guarantees a presentation for every semigroup; note that a semigroup usually has many different presentations.

28.24. Example. Let $(S, \circ) = (\mathbb{Z}_4, \cdot)$. We choose $X = \{2, 3\}$, and let $x_1 = 2$, $x_2 = 3$. Let R be given by $\{(x_1x_2, x_2x_1), (x_2x_2x_1, x_1), (x_2x_2x_2, x_2), (x_1x_1x_1, x_1x_1), (x_1x_2, x_1)\}$. We consider the equivalence classes $a_2 := [x_1]$, $a_3 := [x_2]$, $a_0 := [x_1x_1]$, $a_1 := [x_2x_2]$ in $X_*/\tilde{R}$. The operations on these classes are as follows: $a_0a_0 = [x_1x_1][x_1x_1] = [x_1x_1x_1x_1] = [x_1x_1x_1] = [x_1x_1] = a_0$ (where the third and fourth equations hold because $(x_1x_1x_1, x_1x_1) \in \tilde{R}$); $a_0a_1 = [x_1x_1][x_2x_2] = [x_1x_1x_2x_2] = [x_1x_1x_2] = [x_1x_1] = a_0$ (where the third and fourth equations hold because $(x_1x_2, x_1) \in \tilde{R}$), etc. We obtain:

	a_0	a_1	a_2	a_3
a_0	a_0	a_0	a_0	a_0
a_1	a_0	a_1	a_2	a_3
a_2	a_0	a_2	a_0	a_2
a_3	a_0	a_3	a_2	a_1

Since the products of a_0, a_1, a_2, a_3 do not yield any new elements, we have $X_*/\tilde{R} = \{a_0, a_1, a_2, a_3\}$. Comparison with the operation table of $(\mathbb{Z}_4, \cdot)$ shows that $\{a_0, a_1, a_2, a_3\} \cong \mathbb{Z}_4$, therefore $X_*/\tilde{R} \cong \mathbb{Z}_4$ and $[\{x_1, x_2\}, R]$ is a presentation of $(\mathbb{Z}_4, \cdot)$. Theoretically, we could have $a_0 = a_1$, etc., but the existence of $(\mathbb{Z}_4, \cdot)$ shows that (x_1, x_2) cannot be in R.

All that can also be done for the classes of monoids, groups, abelian groups, etc., to obtain free monoids (see 28.20), ***free groups***, ***free abelian groups***, respectively. The free monoid with basis X is denoted by X^*, and differs from X_* only by the additional "empty word" Λ, so $X^* = X_* \cup \{\Lambda\}$. While the free semigroup over $A = \{a\}$ is given by $(\mathbb{N}, +)$, the free monoid over $\{a\}$ is (isomorphic to) $(\mathbb{N}_0, +)$. The free group and the free abelian group over $\{a\}$ is $(\mathbb{Z}, +)$.

Presentations are often by far the most concise way to describe semigroups and groups. They are *the* method to investigate (semi)groups on a computer, see Sims (1994).

Exercises

1. (i) Find an isomorphism from $(\mathcal{P}(M), \cap)$ onto $(\mathcal{P}(N), \cap)$ if $|M| = |N|$.
   (ii) For $M \subseteq N$ find an embedding of $(\mathcal{P}(M), \cap)$ into $(\mathcal{P}(N), \cap)$.
2. Determine if $h\colon (\mathbb{Z}, +) \to (\mathbb{Z}_n, +)$; $x \mapsto [x]$ is a homomorphism, an epimorphism, or a monomorphism. What does the Homomorphism Theorem 28.12 mean in this case?
3. Let $(S, \circ)$ be a semigroup and $\emptyset \neq I \subseteq S$ with $IS \subseteq I$ and $SI \subseteq I$. In this case, I is called an ***ideal*** of S. Show that $\sim$ defined by

$$x \sim y :\iff x = y \text{ or } x \in I \text{ and } y \in I$$

is a congruence relation (called the ***Rees congruence***) with respect to I. Describe $(S/\sim, \circledcirc)$.

4. Determine $\mathbb{G}_{20}$ explicitly and find the size of $\mathbb{G}_{100000}$.
5. Prove the Homomorphism Theorem 28.12.
6. How many words of length $\leq n$ are in the free semigroup (the free monoid) over a set B with seven elements?
7. Let $S = (\mathbb{Z}_n, \cdot)$. Find a free semigroup F and an epimorphism from F onto S.
8. Are $(\mathbb{N}, +)$, $(\mathbb{N}_0, \cdot)$, $(\mathbb{Q}, \cdot)$, $(\mathbb{R}, +)$ free semigroups?
9. Let H be presented by $[X, R]$ with $X = \{x_1, x_2\}$. What does it mean for H if (x_1x_2, x_2x_1) is in R?
10. Find the semigroup H presented by $[X, R]$ if $X = \{x_1, x_2, x_3\}$ and $R = \{(x_1x_1, x_1), (x_2x_2, x_1), (x_3x_3, x_1), (x_1x_2, x_1), (x_1x_3, x_1), (x_2x_1, x_2), (x_3x_1, x_1), (x_3x_2, x_2x_3)\}$.

Notes

The concept of a semigroup is relatively young. The first fragmentary studies were carried out early in the twentieth century. Then the necessity of studying general transformations, rather than only invertible transformations (which played a large role in the development of group theory) became clear. During the past few decades, connections in the theory of semigroups and the theory of machines became of increasing importance, both theories enriching each other.

The term "semigroup" first appeared in the literature in L. A. de Séguier's book *Élements de la Théorie des Groupes Abstraites* in 1904. The first paper of fundamental importance for semigroup theory was a publication by K. Suschkewitsch in 1928, in which he showed that every finite semigroup contains a simple ideal. D. Rees, in 1940, introduced the concept of a matrix over a semigroup and studied infinite simple semigroups, and from then on the work on semigroups increased through contributions by Clifford, Dubreil, and many others. In 1948, Hille published a book on the analytic theory of semigroups.

Other books on semigroups which are of interest to research workers in the field as well as being suitable for classroom use are books by Petrich (1973), Howie (1976, 1995), Grillet (1995), and Lallement (1979).

§29 Semigroups and Automata

We meet automata or machines in various forms such as calculating machines, computers, money-changing devices, telephone switchboards, and elevator or lift switchings. All of the above have one aspect in common namely a "box" which can assume various "states." These states can be transformed into other states by outside influences (called "inputs"), for instance by electrical or mechanical impulses. Often the automaton "reacts" and produces "outputs" like results of computations or change.

We shall indicate what is common to all automata and describe an abstract model which will be amenable to mathematical treatment. We shall see that there is a close relationship between automata and semigroups. Essential terms are a set of states and a set of inputs which operate on the states and transform them into other states (and sometimes yield an output). If outputs occur, then we shall speak of automata, otherwise we speak of semiautomata.

29.1. Definition. A semiautomaton is a triple $S = (Z, A, \delta)$, consisting of two nonempty sets Z (the set of ***states***) and A (the ***input alphabet***), and a function $\delta\colon Z \times A \to Z$, called the ***next-state function*** of S.

The above definition is very much an abstraction of automata in the usual sense. Historically, the theory of automata developed from concrete automata in communication techniques, nowadays it is a fundamental science. If we want "outputs," then we have to study automata rather than semiautomata.

29.2. Definition. An ***automaton*** is a quintuple $\mathcal{A} = (Z, A, B, \delta, \lambda)$ where (Z, A, δ) is a semiautomaton, B is a nonempty set called the ***output alphabet***, and $\lambda\colon Z \times A \to B$ is a function, called the ***output function***.

If $z \in Z$ and $a \in A$, then we interpret $\delta(z, a) \in Z$ as the next state into which z is transformed by the input a. We consider $\lambda(z, a) \in B$ as the output of z resulting from the input a. Thus if the automaton is in state z and receives the input a, then it changes to state $\delta(z, a)$, producing an output $\lambda(z, a)$.

29.3. Definition. A (semi-)automaton is ***finite*** if all sets Z, A, and B are finite; finite automata are also called ***Mealy automata***. If a special state $z_0 \in Z$ is fixed, then the (semi-)automaton is called ***initial*** and z_0 is the ***initial state***. We write (Z, A, δ, z_0) in this case. An automaton with λ depending only on z is called a ***Moore automaton***.

In practical examples it often happens that states are realized by collections of switching elements each of which has only two states (e.g., current–no current), denoted by 1 and 0. Thus Z will be the cartesian product of several copies of $\mathbb{Z}_2$. Similarly for A and B. Sometimes δ and λ are given by formulas. Very often, however, in finite automata, δ and λ are given by tables.

29.4. Example (*Cafeteria Automaton*). We consider the following situation in a student's life: The student is angry or bored or happy; the cafeteria is closed, offers junk food or good dishes. If the cafeteria is closed, it does not change the student's mood. Junk food offers "lower" it by one "degree" (if he is already angry, then no change), good food creates general happiness for him. Also, there are two outputs b_1, b_2 with interpretation

b_1 ... "student shouts," $\qquad$ b_2 ... "student is quiet."

We assume that the student only shouts if he is angry and if the cafeteria only offers bad food. Otherwise he is quiet, even in state z_3.

We try to describe this rather limited view of a student's life in terms of an automaton $S = (Z, A, B, \delta, \lambda)$. We define $Z = \{z_1, z_2, z_3\}$ and $A = \{a_1, a_2, a_3\}$, with

z_1 ... "student is angry," $\qquad$ a_1 ... "the cafeteria is closed,"
z_2 ... "student is bored," $\qquad$ a_2 ... "the cafeteria offers junk food,"
z_3 ... "student is happy," $\qquad$ a_3 ... "the cafeteria offers good food."

We then obtain

δ	a_1	a_2	a_3
z_1	z_1	z_1	z_3
z_2	z_2	z_1	z_3
z_3	z_3	z_2	z_3

λ	a_1	a_2	a_3
z_1	b_2	b_1	b_2
z_2	b_2	b_2	b_2
z_3	b_2	b_2	b_2

29.5. Example (*IR Flip-Flop*). $A = \{e, 0, 1\}$, $B = \{0, 1\} = Z$, and

δ	e	0	1
0	0	0	1
1	1	0	1

λ	e	0	1
0	0	0	0
1	1	1	1

The input e does not change anything, the inputs $i \in \{0, 1\}$ change all states into i, the output is equal to the corresponding state. For that reason this automaton is called ***Identity-Reset-Automaton***. We shall need it later on.

For a description of (semi-)automata by state graphs we depict $z_1, \ldots, z_k$ as discs in the plane and draw an arrow labeled a_i from z_r to z_s if $\delta(z_r, a_i) = z_s$. In case of an automaton, we label the arrow by the pair $(a_i, \lambda(z_r, a_i))$. For example, the IR Flip-Flop 29.5 has the state graph:

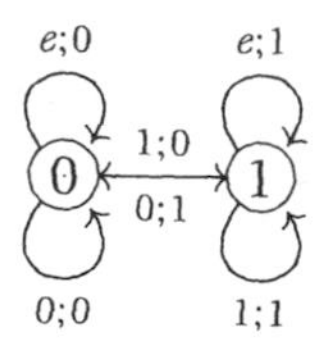

In a computer it would be rather artificial to consider only single input signals. Programs consist of a sequence of elements of an input alphabet. Thus it is reasonable to consider the set of all finite sequences of elements of the set A, including the empty sequence Λ. In other words, in our study of automata we extend the input set A to the free monoid A^* (see 28.20) with Λ as identity.

We also extend δ and λ from $Z \times A$ to $Z \times A^*$ by defining for $z \in Z$ and $a_1, a_2, \ldots, a_r \in A$: $\delta^*(z, \Lambda) := z$, $\delta^*(z, a_1) := \delta(z, a_1)$, $\delta^*(z, a_1a_2) := \delta(\delta^*(z, a_1), a_2)$, etc., and $\lambda^*(z, \Lambda) := \Lambda$, $\lambda^*(z, a_1) := \lambda(z, a_1)$, $\lambda^*(z, a_1a_2) := \lambda(z, a_1)\lambda^*(\delta(z, a_1), a_2)$, and so on. In this way we obtain functions $\delta^*\colon Z \times A^* \to Z$ and $\lambda^*\colon Z \times A^* \to B^*$. The semiautomaton $S = (Z, A, \delta)$ (the automaton $\mathcal{A} = (Z, A, B, \delta, \lambda)$) is thus extended to the new semiautomaton $S^* := (Z, A^*, \delta^*)$ (automaton $\mathcal{A}^* = (Z, A^*, B^*, \delta^*, \lambda^*)$, respectively). We can easily describe the action of S and $\mathcal{A}$ if we let $z \in Z$ and $a_1, a_2, \ldots \in A$:

$$
\begin{aligned}
z_1 &:= z, \\
z_2 &:= \delta(z_1, a_1), \\
z_3 &:= \delta^*(z_1, a_1a_2) = \delta^*(\delta(z_1, a_1), a_2) = \delta(z_2, a_2), \\
z_4 &:= \delta(z_3, a_3), \ldots.
\end{aligned}
$$

If the (semi-)automaton is in state z and an input sequence $a_1a_2 \cdots a_r \in A^*$ operates, then the states are changed from $z = z_1$ to $z_2, z_3, \ldots$ until the final state z_{r+1} is obtained. The resulting output sequence is $\lambda(z_1, a_1)\lambda(z_2, a_2) \cdots \lambda(z_r, a_r)$.

We can imagine that the transition from S to S^* or from $\mathcal{A}$ to $\mathcal{A}^*$ greatly increases the operating power of the semiautomaton or automaton. The case of "continuously changing states" will be covered in §39. Now we shall introduce some more algebraic concepts for automata by showing that any (semi-)automaton determines a certain monoid. Conversely, by Exercise 11, any monoid gives rise to a certain (semi-)automaton. Let $S = (Z, A, \delta)$ be a semiautomaton. We consider $S^* = (Z, A^*, \delta^*)$ as before. For $\alpha \in A^*$ let f_α: $Z \to Z$; $z \mapsto \delta^*(z, \alpha)$.

29.6. Theorem and Definition. $(\{f_\alpha \mid \alpha \in A^*\}, \circ) =: M_S$ *is a monoid (submonoid of* $(Z^Z, \circ)$*), called the* ***syntactic monoid of*** S. *For all* $\alpha, \beta \in A^*$ *we have*

$$f_\alpha \circ f_\beta = f_{\beta\alpha}.$$

The ***monoid of an automaton*** *is defined as the monoid of the "underlying" semiautomaton.*

Proof. Composition of mappings is associative. For $\alpha, \beta \in A^*$ we have

$$(f_\alpha \circ f_\beta)(z) = f_\alpha(f_\beta(z)) = f_\alpha(\delta^*(z, \beta)) = \delta^*(\delta^*(z, \beta), \alpha) = f_{\beta\alpha}(z),$$

thus $f_\alpha \circ f_\beta = f_{\beta\alpha}$. Here $\beta\alpha$ denotes the concatenation of sequences. The collection $\{f_\alpha | \alpha \in A^*\}$ is closed with respect to $\circ$, therefore it is a semigroup and because of $\mathrm{id}_Z = f_\Lambda$ it is a monoid. □

29.7. Definition. Let $(S, \circ)$ and $(S', \circ)$ be semigroups and f: $S \to S'$. f is called an ***antihomomorphism*** if $f(x \circ y) = f(y) \circ f(x)$ is fulfilled for all $x, y \in S$. An ***anti-isomorphism*** is a bijective antihomomorphism.

M_S is an antihomomorphic image of A^* as the following result shows.

29.8. Theorem. *In* A^* *let* $\alpha \equiv \beta :\iff f_\alpha = f_\beta$. *Then* $\equiv$ *is a congruence relation in* A^* *and* $A^*/\equiv$ *is anti-isomorphic to* M_S.

Proof. The mapping f: $A^* \to M_S$; $\alpha \mapsto f_\alpha$ is an antihomomorphism, by 29.6, therefore a homomorphism from A^* into M'_S, where

$$M'_S := (\{f_\alpha \mid \alpha \in A^*\}, \circ') \quad \text{with} \quad f_\alpha \circ' f_\beta := f_\beta \circ f_\alpha.$$

The Homomorphism Theorem 28.12 implies that $A^*/\equiv$ is isomorphic to M'_S, therefore M_S is anti-isomorphic to $A^*/\equiv$. □

What is the meaning of $\equiv$? We have

$$\alpha \equiv \beta \iff f_\alpha = f_\beta \iff f_\alpha(z) = f_\beta(z) \text{ for all } z \in Z$$
$$\iff \delta^*(z, \alpha) = \delta^*(z, \beta) \text{ for all } z \in Z.$$

Thus α and β are equivalent with respect to $\equiv$ if and only if they operate in the same way on each state. If Z is finite, then so is M_S and also $A^*/\equiv$.

The IR Flip-Flop has the monoid $\{f_\Lambda, f_0, f_1\}$. For bigger (semi-)automata it is, however, very cumbersome to compute their syntactic monoid. The use of monoids comes from another area, as we shall see now.

29.9. Definition. $\mathcal{A}_1 = (Z_1, A, B, \delta_1, \lambda_1)$ is called a ***subautomaton*** of $\mathcal{A}_2 = (Z_2, A, B, \delta_2, \lambda_2)$ (in symbols: $\mathcal{A}_1 \leq \mathcal{A}_2$) if $Z_1 \subseteq Z_2$ and δ_1 and λ_1 are the restrictions of δ_2 and λ_2, respectively, on $Z_1 \times A$.

Subautomata of $\mathcal{A}$ thus have the same input and output alphabets as $\mathcal{A}$.

29.10. Definition. Let

$$\mathcal{A}_1 = (Z_1, A_1, B_1, \delta_1, \lambda_1) \quad \text{and} \quad \mathcal{A}_2 = (Z_2, A_2, B_2, \delta_2, \lambda_2)$$

be automata. An ***(automata-) homomorphism*** $\Phi\colon \mathcal{A}_1 \to \mathcal{A}_2$ is a triple $\Phi = (\zeta, \alpha, \beta) \in Z_2^{Z_1} \times A_2^{A_1} \times B_2^{B_1}$ with the property

$$\left.\begin{aligned} \zeta(\delta_1(z, a)) &= \delta_2(\zeta(z), \alpha(a)), \\ \beta(\lambda_1(z, a)) &= \lambda_2(\zeta(z), \alpha(a)), \end{aligned}\right. \qquad (z \in Z_1, a \in A_1).$$

Φ is called a ***monomorphism*** (***epimorphism***, ***isomorphism***) if all functions ζ, α, and β are injective (surjective, bijective).

29.11. Theorem. *If the automaton $\mathcal{A}'$ is a subautomaton or a homomorphic image of $\mathcal{A}$, then $M_{\mathcal{A}'}$ is a homomorphic image of $M_{\mathcal{A}}$.*

Proof. See Lidl & Pilz (1984). □

Two automata $\mathcal{A}_1 = (Z_1, A_1, B_1, \delta_1, \lambda_1)$ and $\mathcal{A}_2 = (Z_2, A_2, B_2, \delta_2, \lambda_2)$ can be connected ***in parallel***, as in Figure 29.1, or ***in series*** (if $A_2 = B_1$), as in Figure 29.2.

The definitions of δ, λ in $\mathcal{A}_1 \times \mathcal{A}_2$ and $\mathcal{A}_1 \vdash\!\dashv \mathcal{A}_2$ are as to be expected. For getting the monoids of the parallel and series connections of automata we need another concept.

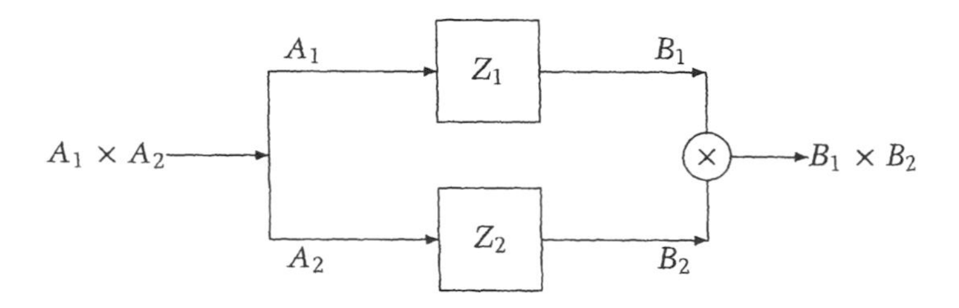

FIGURE 29.1 $\mathcal{A}_1 \times \mathcal{A}_2$.

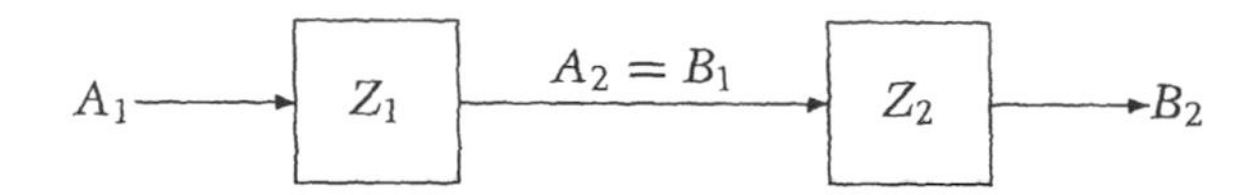

FIGURE 29.2 $\mathcal{A}_1 \vdash\!\dashv \mathcal{A}_2$.

29.12. Definition.

(i) A semigroup S_1 ***divides*** a semigroup S_2, if S_1 is a homomorphic image of a subsemigroup of S_2. In symbols $S_1 \mid S_2$.

(ii) An automaton $\mathcal{A}_1 = (Z_1, A, B, \delta_1, \lambda_1)$ ***divides*** an automaton $\mathcal{A}_2 = (Z_2, A, B, \delta_2, \lambda_2)$ (equal input and output alphabets) if $\mathcal{A}_1$ is a homomorphic image of a subautomaton of $\mathcal{A}_2$. In symbols: $\mathcal{A}_1 \mid \mathcal{A}_2$.

If S_1 divides S_2, then we also say that S_2 ***covers*** S_1 or S_2 ***simulates*** S_1. In this case the multiplication in S_1 is determined by part of the multiplication in S_2 via an epimorphism. Similarly for automata: "$\mathcal{A}_2$ covers $\mathcal{A}_1$" means that $\mathcal{A}_2$ can "do at least as much as" $\mathcal{A}_1$.

29.13. Definition. Let S, T be two semigroups. The ***wreath product*** $S \wr T$ is given as $(S^T \times T, \square)$, where $\square$ is defined by $(f_1, t_1) \square (f_2, t_2) := (f_{f_1,f_2,t_1}, t_1 t_2)$ with $f_{f_1,f_2,t_1}\colon T \to S;\ x \mapsto f_1(x) f_2(x t_1)$.

29.14. Remark. Let $\bar{s}$ be the constant map with value s. Since $S \cong \{\bar{s} \mid s \in S\} =: \bar{S}$, we have $S \times T \cong \{(\bar{s}, t) \mid s \in S,\ t \in T\} \leq S \wr T$. This implies $S \times T \mid S \wr T$ and furthermore $S \mid S \wr T$ and $T \mid S \wr T$. If S contains n elements and T contains m elements ($m, n \in \mathbb{N}$), then $S \wr T$ has $n^m m$ elements. For instance, $(\mathbb{Z}_2, +) \wr (\mathbb{Z}_2, +) \cong D_4$. It can be shown that every finite group can be covered by a wreath product of finite simple groups. We note that $\wr$ is neither commutative nor associative. Much more information about wreath products can be found in Meldrum (1995).

29.15. Theorem. *The monoids of* $\mathcal{A}_1 \times \mathcal{A}_2$ *and of* $\mathcal{A}_1 \mapsto\!\dashv \mathcal{A}_2$ *divide* $M_{\mathcal{A}_2 \wr \mathcal{A}_1}$.

Proof. See, e.g., Arbib (1969). □

Now we have all necessary tools to formulate and sketch the proof of a fundamental theorem for automata. For the proof and more details see Arbib (1968) (several versions of the theorem), Eilenberg (1974) (together with an algorithm for the decomposition), Meldrum (1995), or Wells (1976).

29.16. Theorem (*Decomposition Theorem of Krohn-Rhodes*). *Any finite automaton* $\mathcal{A}$ *can be simulated by series/parallel connections of the following basic types of automata:*

(1) *IR flip-flops.*

(2) *Automata whose monoids are finite simple groups dividing* $M_{\mathcal{A}}$.

So by the classification of finite simple groups (see page 107) all finite automata are "known" in the above sense. The proof of Theorem 29.16 first decomposes $M_{\mathcal{A}}$ into a wreath product of "simple subsemigroups" of $M_{\mathcal{A}}$, using 29.14 and other nontrivial parts of semigroup theory; then this decomposition of $M_{\mathcal{A}}$ is "transferred back" to automata via Theorem 29.15.

Exercises

1. Describe parts of your brain as a semiautomaton such that A is the set of theorems in this book and δ describes "studying."
2. Could you interpret the semiautomaton in 1 as a computer (brain = storage, etc.)?
3. What would be a possible interpretation if the semiautomaton in 1 should be considered as an automaton?
4. Same as Exercise 3, but with the interpretation as an initial automaton.
5. A stamp automaton S has a capacity of ten stamps. Define it as a semiautomaton, according to 29.1, such that the state z means "there are z stamps in S," the inputs are "no coin is inserted," "a correct coin is inserted," and "a wrong coin is inserted." Describe this semiautomaton by a table and a state graph.

6. Generalize the semiautomaton of Exercise 5 to an automaton by adding the outputs "no output," "a stamp is output," and "a coin is output." Complete the table and the state graph.
7. Depict the state graph of the cafeteria automaton 29.4.
8. An automaton $\mathcal{A}$ is given by the state graph:

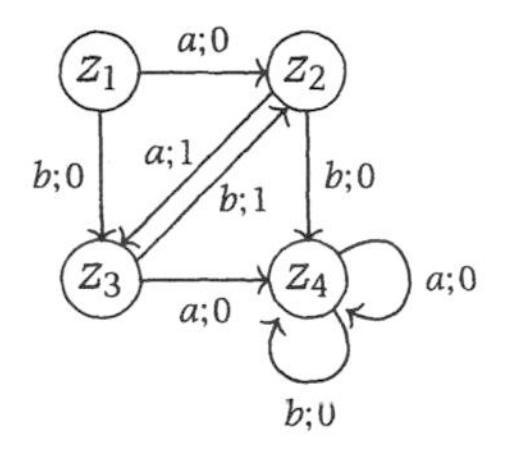

   (i) Describe this automaton by finding $(Z, A, B, \delta, \lambda)$ (in the form of a table).
   (ii) Let $\mathcal{A}$ be in state z_2 when the input sequence *aaabbaa* begins. What is the final state and what is the output sequence?
   (iii) Find all input sequences of shortest length such that x_1 is transformed into z_4 and the corresponding output sequences have exactly two 1's.
   (iv) Why can we describe z_4 as the "dead end" of this automaton?
   (v) Find all input sequences which transform z_2 into z_3.
   (vi) Let 01000 be an output sequence. Find which state has been transformed into which state.
9. In $\mathcal{A} = (Z, A, B, \delta, \lambda)$ let λ_* be given by $\lambda_*\colon Z \times A_* \to B$; $(z, a_1a_2\cdots a_n) \mapsto \lambda(\delta^*(z, a_1a_2\cdots a_{b-1}), a_n)$. Give a good interpretation of λ_* and show that $\lambda_*(z, a_1a_2\cdots a_na_{n+1}) = \lambda_*(\delta^*(z, a_1\cdots a_{n-1}), a_na_{n+1})$ holds for all $z \in Z$, $n \in \mathbb{N}$, and $a_i \in A$.
10. For which δ will $(\{z_0, z_1, z_2, z_3\}, \{a, b, c\}, \delta)$ be a semiautomaton such that exactly those input sequences $a_1a_2\cdots a_n$ change z_1 into z_0 in which $a_i = a$ for every odd i?
11. Let S be a monoid. Demonstrate that S determines a (semi-)automaton in a natural way. Is the monoid of this (semi-)automaton isomorphic to S?
12. Let $\mathcal{A}$ be the automaton of Exercise 8. Determine if $\{z_1\}$ or $\{z_2, z_3, z_4\}$ or $\{z_4\}$ are subautomata.
13. Give the tables of parallel and series compositions of two IR flip-flops.
14. Find the operation table of the wreath product of $(\mathbb{Z}_2, +)$ by itself. Is this wreath product a group? If it is, is it abelian?

15. Show that $(\mathbb{Z}_n, +) \mid (\mathbb{Z}_m, +) \Longrightarrow n \mid m$.
16. If the semigroup H divides a group G, is H a group, too?
17. If $\mathcal{A}_1 \mathcal{A}_2$ and $M_{\mathcal{A}_2}$ is a group, does this hold for $M_{\mathcal{A}_1}$, too?

Notes

The theory of automata has its origins in work by Turing (see papers by Shannon (1948) and Heriken (1994)). Turing developed the theoretical concept of what is now called Turing machines, in order to give computability a more concrete and precise meaning. Shannon investigated the analysis and synthesis of electrical contact circuits using switching algebra. The work of McCulloch and Pitts centers on neuron models to explain brain functions and neural networks by using finite automata. Their work was continued by Kleene. Our description of a finite automaton is due to Moore and Mealy. The development of technology in the areas of electromechanical and electronic machines, and particularly computers, has had a great influence on automata theory, which traces back to the mid-1950s. Many different parts of pure mathematics are used as tools, such as abstract algebra, universal algebra, lattice theory, category theory, graph theory, mathematical logic, and the theory of algorithms. In turn, automata theory can be used, for example, in economics, linguistics, and in learning processes.

Here we are mainly concerned with one subdiscipline of automata theory, namely the algebraic theory of automata, which uses algebraic concepts to formalize and study certain types of finite-state machines. One of the main algebraic tools used to do this is the theory of semigroups.

§30 Semigroups and Formal Languages

Besides the "natural" languages such as English, so-called formal languages are of importance in the formal sciences, especially in computing or information science. The underlying principle is always the same: Given an "alphabet" or "vocabulary" (consisting of letters, words, punctuation symbols, number symbols, programming instructions, etc.), we have a method ("grammar") for constructing "meaningful" words or sentences (i.e., the "language") from this alphabet. This immediately reminds us of the term "word semigroup" and, indeed, these free semigroups will play a major role, the language L constructed will be a subset of the free semigroup A_* or the free monoid A^* on the alphabet A.

There are essentially three ways to construct a language L.

(a) Approach via grammar: given a collection of rules ("grammar"), generate L from A.

(b) Approach via automata: consider an initial semiautomaton which processes the elements of L in a suitable way.

(c) Algebraic approach: L is constructed by the algebraic combination of certain subsets of A^*.

All three approaches use algebraic methods. Let us start with method (a). This approach is largely based on the work of Chomsky (1957).

30.1. Definition. A ***phrase-structure grammar*** (in short: ***grammar***) is a quadruple $\mathcal{G} = (A, G, \twoheadrightarrow, g_0)$, where A and G are nonempty disjoint finite sets, $g_0 \in G$, and $\twoheadrightarrow$ is a finite relation from G into $(A \cup G)^*$.

30.2. Definition. Let $\mathcal{G}$ be a grammar.

(i) A is called the ***alphabet*** of $\mathcal{G}$.

(ii) G is called the set of ***grammar symbols*** (or ***metalinguistic symbols***).

(iii) $V := A \cup G$ is the ***complete vocabulary***.

(iv) The elements (x, y) in $\twoheadrightarrow$ (which is a subset of $G \times V^*$) are also written in the form $x \twoheadrightarrow y$ and are called ***productions*** or ***rewriting rules***.

(v) g_0 is called the ***initial symbol***.

We now show how to obtain a suitable subset of A^*, namely the desired language, by using $\mathcal{G}$. In order to do this, we need another relation, this time on V^*.

30.3. Definition. Let $\mathcal{G} = (A, G, \twoheadrightarrow, g_0)$ be a grammar. For $y, z \in V^*$, we write $y \Rightarrow z$ if there are $v \in G_*$ and $v_1, v_2, w \in V^*$ with

$$y = v_1 v v_2, \quad z = v_1 w v_2, \quad \text{and} \quad v \twoheadrightarrow w.$$

The reason for introducing $\Rightarrow$ is to obtain a new word $v_1 v_2 \cdots v_r' \cdots v_n$ from a given word $v_1 v_2 \cdots v_r \cdots v_n$ and a rule $v_r \twoheadrightarrow v_r'$. Thus we extend $\twoheadrightarrow$

to a compatible relation $\Rightarrow$ on V^*. We recall from 28.18 that the transitive hull $\Rightarrow^{\mathrm{t}}$ of $\Rightarrow$ is given by

$$y \Rightarrow^{\mathrm{t}} z \iff \text{there are } n \in \mathbb{N}_0 \text{ and } x_0, \ldots, x_n \in V^*$$
$$\text{with } y = x_0 \Rightarrow x_1 \Rightarrow \cdots \Rightarrow x_n = z.$$

The sequence $x_0, x_1, \ldots, x_n$ is called a ***derivation*** (or ***generation***) of z from y; n is the ***length*** of the derivation.

30.4. Definition. Let $\mathcal{G} = (A, G, \rightarrow, g_0)$ be a grammar.

$$L(\mathcal{G}) := \{l \in A^* \mid g_0 \Rightarrow^{\mathrm{t}} l\}$$

is called the ***language generated*** by $\mathcal{G}$ (also the ***phrase-structure language***).

We give a few examples.

30.5. Example. Let $A = \{a\}$, $G = \{g_0\}$, $g_0 \neq a$, and $\rightarrow = \{g_0 \rightarrow a\}$, $\mathcal{G} := (A, G, \rightarrow, g_0)$; we find $L(\mathcal{G})$. Let $z \in A^*$ with $g_0 \Rightarrow^{\mathrm{t}} z$. Then there are some $v \in \{g_0\}^*$ and $v_1, v_2, w \in \{a, g_0\}^*$ with $g_0 = v_1 v v_2$, $z = v_1 w v_2$, and $v \rightarrow w$. This implies $v = g_0$ and $w = a$. From $g_0 = v_1 g_0 v_2$, we obtain $v_1 = v_2 = \Lambda$ and also $z = w = a$. From a, we cannot generate anything new, since $a \Rightarrow x$, $g_0 \Rightarrow^{\mathrm{t}} x$ means we had to find $v' \in \{g_0\}^*$, $v_1, v_2 \in \{a, g_0\}^*$ such that $a = v_1' v' v_2$; this is impossible because of $g_0 \neq a$. Hencc $L(\mathcal{G}) = \{a\}$, a very simple language.

The following language can do a little more.

30.6. Example. Let A and G be as in 30.5 and $\rightarrow := \{g_0 \rightarrow a, g_0 \rightarrow ag_0\}$, $\mathcal{G} := (A, G, \rightarrow, g_0)$. Then $g_0 \Rightarrow z$ implies $z = a$ or $z = ag_0$. Again, there is no x with $a \Rightarrow x$. $ag_0 \Rightarrow y$ is only possible for $y = aa$ or $y = aag_0$. Thus $aag_0 \Rightarrow aaa$ and $aag_0 \Rightarrow aaag_0$, etc. In a diagram,

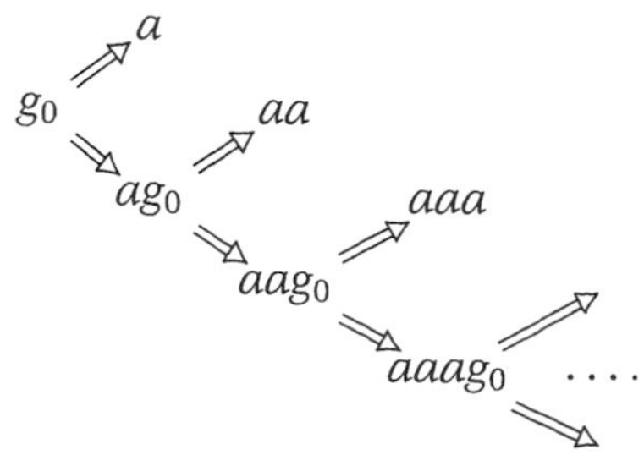

$$\begin{array}{ccccccc}
 & \nearrow & a & & & & \\
g_0 & & & \nearrow & aa & & \\
 & \searrow & ag_0 & & & \nearrow & aaa \\
 & & & \searrow & aag_0 & & \nearrow \\
 & & & & & \searrow & aaag_0 \quad \ldots. \\
 & & & & & & \searrow
\end{array}$$

Cancellation of the elements outside A^* gives us the result $L(\mathcal{G}) = \{a, aa, aaa, \ldots\} = A_*$.

In "natural" languages (more precisely: in excerpts of natural languages), the set A consists of the words of the language (e.g., all words of a dictionary, together with all declinations, etc.), G consists of the names for terms of the grammar (e.g., "sentence," "noun," "adverb," ...), $\rightharpoonup$ consists of rules such as "sentence $\rightharpoonup$ subject, predicate," and substitution rules, e.g., "subject $\rightharpoonup$ wine bottle." In this way in (partial) languages we obtain sets of grammatically correct sentences, which need not be meaningful with respect to content.

30.7. Definition. A grammar $\mathcal{G} = (A, G, \rightharpoonup, g_0)$ and its generated language are called ***right linear*** if all elements in $\rightharpoonup$ are of the forms $g \rightharpoonup a$ or $g \rightharpoonup ag'$, with $g, g' \in G$, $a \in A^*$. $\mathcal{G}$ is called ***context-free*** if for all $x \rightharpoonup y$ in $\rightharpoonup$ we have $l(x) = 1$.

In a right linear language, the length of a word is never shortened by any derivation. The grammars in Examples 30.5 and 30.6 are right linear. All right linear grammars are context-free.

In order to study the approach (b), we now define special semiautomata and introduce the concept of "accepting," so that we obtain "accepted" subsets of free monoids, which turn out to be languages $L(\mathcal{G})$.

30.8. Definition. A finite initial semiautomaton $\mathcal{A} = (Z, A, \delta, z_0)$ is called an ***acceptor*** if, together with $\mathcal{A}$, a set $E \subseteq Z$ is given. We write $\mathcal{A} = (Z, A, \delta, z_0, E)$ and call E the set of ***designated final states***. $W(\mathcal{A}) := \{w \in A^* \mid \delta^*(z_0, w) \in E\}$ is called the ***set of words accepted by*** $\mathcal{A}$.

30.9. Definition. Let A be a set and $L \subseteq A^*$. L is called an ***acceptor language*** if there is an acceptor $\mathcal{A}$ with $W(\mathcal{A}) = L$.

30.10. Example. Let $\mathcal{A}$ be the IR flip-flop of Example 29.5 with $z_0 = 1$ and $E = \{0\}$. Then

$$W(\mathcal{A}) = \{w \in \{0, 1\}^* \mid \delta^*(1, w) = 0\}$$

is the set of words w of the form $w = \alpha 0 \epsilon$ with $\alpha \in A^*$ and $\epsilon \in \{e, 0\}^*$.

Finally, we take a look at the method (c).

30.11. Definition. Let A be a set and $R \subseteq A^*$. R is called ***regular*** or a regular language if R can be obtained from one-element subsets of A^* by a finite number of "admissible operations." Admissible operations are the formations of unions, products (in A^*), and generated submonoids.

30.12. Example. Let $A = \{a_1, a_2, a_3\}$. The following subsets of A^* can be obtained using the operations mentioned in 30.11: $R_1 = \{a_1a_2a_1, a_1a_2a_3, a_1a_1a_2a_2a_3a_3\}$, since R_1 is finite; $R_2 = \{\Lambda, a_1, a_1a_1, a_1a_1a_1, \dots\}$, because R_2 is the submonoid generated by $\{a_1\}$; $R_3 = \{a_1^n a_2^m \mid n, m \in \mathbb{N}_0\}$, since R_3 is the product of submonoids generated by a_1 and a_2, respectively.

A characterization of regular languages over singleton alphabets $A = \{a\}$ is as follows. For a proof of the result, see, e.g., Lidl & Pilz (1984).

30.13. Definition. A subset $P = \{p_1, p_2, \dots\}$ of $\mathbb{N}_0$ is called ***periodic*** if there are $k \in \mathbb{N}$ and $q, n_0 \in \mathbb{N}_0$ such that for all $n \geq n_0$: $p_{n+k} - p_n = q$.

30.14. Theorem. *Let $A = \{a\}$ and $L \subseteq A^*$. L is regular if and only if there is a periodic subset P of $\mathbb{N}_0$ with $L = \{a^p \mid p \in P\}$.*

30.15. Corollary. *Let $A = \{a\}$. Then $\{a^{n^2} \mid n \in \mathbb{N}_0\}$ is not regular.*

Proof. This follows from 30.14 and the fact that $\{1, 4, 9, \dots\}$ is not periodic. □

We also mention that regular sets and related concepts are applicable in mathematical logic (computability, decidability, etc.), where so-called "Turing machines" are studied. See Arbib (1969), or Birkhoff & Bartee (1970).

What are the connections between grammar languages, acceptor languages, and regular languages? Without proofs (see, e.g., Ginsburg (1975) and Arbib (1969)) we mention

30.16. Theorem. *Let A be a set and $L \subseteq A^*$.*

(i) (***Ginsburg's Theorem***) *L is an acceptor language if and only if L is a right linear language.*

(ii) (***Kleene's Theorem***) *L is an acceptor language if and only if L is regular.*

By 30.16 we can draw a sketch of the situation:

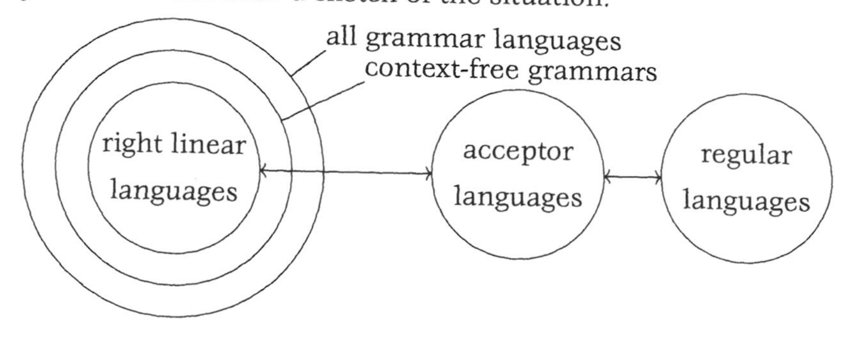

By 30.15, not every subset of A^* is regular, and so not every grammar language is right linear.

Exercises

1. Let $A = \{a\}$, $G = \{g_0\}$, $\rightarrow = \{g_0 \rightarrow aa,\ g_0 \rightarrow g_0a\}$, $\mathcal{G} = (A, G, \rightarrow, g_0)$. Find $L(\mathcal{G})$.
2. Let $A = \{a\}$, $G = \{g_0\}$, $\rightarrow = \{g_0 \rightarrow ag_0,\ g_0 \rightarrow g_0a\}$, $\mathcal{G} = (A, G, \rightarrow, g_0)$. Find $L(\mathcal{G})$.
3. Are the languages in Exercises 1 and 2 above context-free? right-linear?
4. Let $\mathcal{A}$ be the cafeteria automaton (see 29.4), which is an acceptor with respect to $z_0 := z_1$, and $E = \{z_3\}$. What is $W(\mathcal{A})$? Find an interpretation of $W(\mathcal{A})$ in "everyday language".

5.–7. Let $A = \{a, b\}$, $G = \{g_0\}$. For the following definitions of $\rightarrow$, find the generated language.

5. $\rightarrow = \{g_0 \rightarrow ag_0,\ g_0 \rightarrow b\}$.
6. $\rightarrow = \{g_0 \rightarrow g_0g_0,\ g_0 \rightarrow aa\}$.
7. $\rightarrow = \{g_0 \rightarrow aag_0,\ g_0 \rightarrow aa\}$.
8. Compare the results of Exercises 6 and 7. What can you deduce?
9. Find some grammar $\mathcal{G}$ such that $L(\mathcal{G}) = \{b, aba, aabaa, \dots\}$.
10. Find $\mathcal{G}$ such that $L(\mathcal{G}) = \{\Lambda, b, aa, aba, aaaa, aabaa, \dots\}$.
11. Call two grammars equivalent if they generate the same language. Show that this yields an equivalence relation on every set of grammars.
12. Let $A = \{a, b, c\}$, $G = \{g_0, g_1, g_2, g_3\}$, $\rightarrow = \{g_0 \rightarrow g_1,\ g_0 \rightarrow g_2,\ g_1 \rightarrow abc,\ g_2 \rightarrow ab,\ g_2 \rightarrow g_3,\ g_3 \rightarrow c\}$.

    (i) Is this grammar $\mathcal{G}$ context-free? (ii) Is $abc \in L(\mathcal{G})$?
13. Is $M = \{\Lambda, ab, aabb, aaabbb, \dots\}$ a regular set? Does there exist an acceptor $\mathcal{A}$ with $W(\mathcal{A}) = M$?
14. Let $T := \{a_1a_2a_3 \mid a_1, a_2, a_3 \in A\}$. Does there exist a grammar $\mathcal{G}$ such that $T = L(\mathcal{G})$? Does there exist an acceptor $\mathcal{A}$ with $T = W(\mathcal{A})$?
15. Is $\{1, 2, 3, 6, 9, 12, 15, \dots\}$ periodic? If yes, what is its corresponding regular language?
16. Is $\{a^n \mid n \in \mathbb{N} \text{ and } n \equiv 3 \pmod 4\}$ a right linear language?

Notes

The beginning of the study of formal languages can be traced to Chomsky, who in 1957 introduced the concept of a context-free language in order to model natural languages. Since the late 1960s there has been considerable activity in the theoretical development of context-free languages both in connection with natural languages and with the programming languages. Chomsky used Semi-Thue systems to define languages, which can be described as certain subsets of finitely generated free monoids. Chomsky (1957) details a revised approach in the light of experimental evidence and careful consideration of semantic and syntactic structure of sentences.

For a common approach to formal languages and the theory of automata, we refer to Eilenberg (1974). Some of the early books on formal languages from a mathematician's or computer scientist's viewpoint are Arbib (1968) and Ginsburg (1975). See also Ginsburg (1966), Salomaa (1969), Salomaa (1981), and Bobrow & Arbib (1974); the latter is an elementary introduction to this subject.

§31 Semigroups and Biology

Semigroups can be used in biology to describe certain aspects in the crossing of organisms, in genetics, and in considerations of metabolisms.

Let us start with an example.

31.1. Example. In breeding a strain of cattle, which can be black or brown, monochromatic or spotted, it is known that black is dominant and brown recessive and that monochromatic is dominant over spotted. Thus there are four possible types of cattle in this herd:

a...black, monochromatic, c...brown, monochromatic,
b...black, spotted, d...brown, spotted.

Due to dominance, in crossing a black spotted one with a brown monochromatic one, we expect a black monochromatic one. This can be symbolized by "$b * c = a$." The "operation" $*$ can be studied for all possible pairs to obtain the table:

$*$	a	b	c	d
a	a	a	a	a
b	a	b	a	b
c	a	a	c	c
d	a	b	c	d

By checking associativity, we see that $S := (\{a, b, c, d\}, *)$ is a semigroup. Moreover, the table is symmetric with respect to the main diagonal, therefore $*$ is commutative, d is the identity element, and a is the zero element. Clearly S cannot be a group. So S is a commutative monoid with zero element a.

In general, the table for breeding operations is more complicated, see Nahikian (1964). We can ask for connections between hereditary laws and the corresponding semigroups. Of course, we need $S \circ S = S$ for such semigroups S, since $s \in S \backslash S^2$ would vanish after the first generation and would not even be observed (rare cases excepted).

A different biological problem which leads to algebraic problems is as follows: All genetic information in an organism is given in the so-called deoxyribonucleic acid (DNA), which consists of two strands which are combined together to form the famous double helix. Each strand is made up as a polymer of four different basic substances, the nucleotides. We concentrate our attention on one strand only. If the nucleotides are denoted by n_1, n_2, n_3, n_4, then the strand can be regarded as a word over $\{n_1, n_2, n_3, n_3\}$. DNA cannot put the genetic information into effect. By means of a messenger ribonucleic acid, the information contained in the DNA is copied ("transcription") and then transferred onto the protein chains ("translation"). These protein chains are polymers consisting of 21 different basic substances, the amino acids, denoted by $a_1, \ldots, a_{21}$. As with the DNA, each protein chain can be regarded as a word over $\{a_1, \ldots, a_{21}\}$.

In general, it is assumed that the sequence of nucleotides in the DNA is uniquely determined by the sequence of amino acids in a protein chain. In other words, it is assumed that there is a monomorphism from the free semigroup $F_{21} := \{a_1, \ldots, a_{21}\}_*$ into the free semigroup $F_4 := \{n_1, n_2, n_3, n_4\}_*$. The DNA protein-coding problem is the question: How many, if any, monomorphisms are there from F_{21} into F_4?

A first glance would indicate that the answer is a clear "none." How could we have a monomorphism from the "huge" semigroup F_{21} into the "much smaller" F_4? There are plenty of monomorphisms from F_4 into F_{21}—wouldn't this mean that $F_4 \cong F_{21}$? The somewhat surprising answer is given in the following theorem.

31.2. Theorem. *There are infinitely many monomorphisms from F_{21} into F_4. Thus the DNA protein-coding problem has infinitely many solutions.*

Proof. From the lines after 28.20, we only have to define an injective map from $\{a_1, \dots, a_{21}\}$ into F_4, and extend this to a homomorphism h from F_{21} into F_4, which is then easily shown to be injective, too. This can be done via $g(a_1) := n_1n_1n_1$, $g(a_2) := n_1n_1n_2$, $g(a_3) = n_1n_1n_3, \dots$, $g(a_{21}) := n_2n_2n_1$. Then g is injective. Obviously, by different choices of g, e.g., $g(n_1) := n_3n_1n_1$, etc., or by using words of length ≥ 4, we can obtain infinitely many such monomorphisms.

It is well known since 1965 that the DNA protein-coding in Nature is always done by means of words of length 3 ("triplet-code"), maybe with the exception of mitochondria in some "exotic" primitive organisms.

A different aspect of applied algebra appears in problems of metabolisms and processes of cellular growth. Metabolism consists of a sequence of transitions between organic substances (acids), triggered off by enzymes and co-enzymes. Such a process can be described in terms of semiautomata, in which the organic acids form the state set Z, the (co-)enzymes form the inputs A and the chemical reactions "substance + (co-)enzymes $\to$ new substance" define the state transition function δ.

This (finite) semiautomaton can be decomposed into IR flip-flops and simple group-semiautomata, using the theorem of Krohn-Rhodes, see 29.16. This decomposition illustrates also the structure of the metabolic processes (see Krohn, Langer & Rhodes (1976)). As a further illustration, we show by means of an example how cellular growth can be described in terms of formal languages.

31.3. Example. Let $A = \{0, 1, 2, 3, 4, 5, (,)\}$, $G = \{g_0\}$, and $\twoheadrightarrow = \{g_0 \twoheadrightarrow 0g_0, g_0 \twoheadrightarrow 1g_0, \dots, g_0 \twoheadrightarrow 5g_0, g_0 \twoheadrightarrow 3(4)g_0, g_0 \twoheadrightarrow (4)(4)g_0, g_0 \twoheadrightarrow \Lambda\}$. The language $L(\mathcal{G})$ with $\mathcal{G} = (A, G, \twoheadrightarrow, g_0)$ is right linear and contains also

2	$(g_0 = \Lambda g_0 \Lambda \Rightarrow \Lambda 2 g_0 \Lambda \Rightarrow \Lambda 2 \Lambda\Lambda = 2)$,
233	$(g_0 = \Lambda g_0 \Lambda \Rightarrow \Lambda 2 g_0 \Lambda = 2g_0\Lambda \Rightarrow 23g_0\Lambda \Rightarrow 233g_0\Lambda \Rightarrow 233\Lambda\Lambda = 233)$,
2333(4)	$(g_0 = \Lambda g_0 \Lambda \Rightarrow \cdots \Rightarrow 2333(4)g_0\Lambda \Rightarrow 2333(4)\Lambda\Lambda = 2333(4))$.

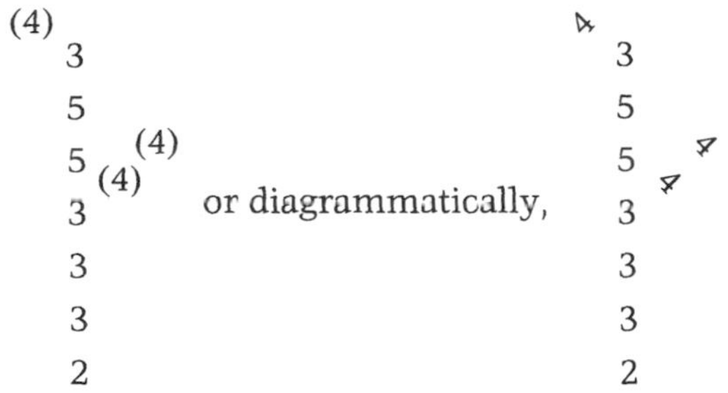

```
(4)
   3                                         3
   5                                         5
   5  (4)                                    5
    (4)                                      3
   3        or diagrammatically,             3
   3                                         3
   3                                         2
   2
```

FIGURE 31.1

We also obtain 2333(4)(4), 2333(4)(4)55, 2333(4)(4)553(4). In Figure 31.1, we interpret 0, 1, . . . , 5 as cells number 0 to 5, e.g., 233 is the sequence of cells No. 2, 3, 3 in the direction of growth (for instance, straight upward). 3(4) is denoted as a branch point, where cell No. 4 branches out from cell No. 3 and forms a "bud" sideways. The form of the last "word" above in this example is shown in Figure 31.1. It looks like the beginning of a plant. In this example, cell No. 3 has the capability to form "buds," the "branches" consist of cells No. 4 only.

Exercises

1. Find a monomorphism from the F_4 into F_{21} (see 31.2).
2. How many different solutions to the DNA protein-coding problem by triplet codes are there?
3. In the sense of Example 31.3, draw the "plant" corresponding to the word 123(4)53(4)3(4)3(4). Are (4)(4)(4)(4), (4)(4)(4), 12345(4) elements of $L(\mathcal{G})$?
4. Is 23(4)3(4)(4) obtainable in Example 31.3? If so, draw the corresponding "plant."

Notes

In the way indicated above, the growth of plants can be described algebraically, see Hermann & Rosenberg (1975). Further material on this subject is contained in Holcombe (1982).

Details on the use of semiautomata in metabolic pathways and the aid of a computer therein, including a theory of scientific experiments, can be found in Krohn, Langer & Rhodes (1976). Rosen (1973) studies ways in which environmental changes can affect the repair capacity of biological systems and considers carcinogenesis and reversibility problems. Language theory is used in cell-development problems, as introduced by Lindenmayer (1968), see also Hermann & Rosenberg (1975). Suppes (1969) and Kieras (1976) develop a theory of learning in which a subject is instructed to behave like a semiautomaton.

§32 Semigroups and Sociology

Are the enemies of my enemies my friends? Is the enemy of my friend a friend of my enemy? We shall see that questions like these can be elegantly formulated in the language of semigroups. We continue our investigation of relation semigroups defined in 28.17. Kinship relationships, such as "being father," "being mother," can be combined to "being the mother of the father," i.e., "grandmother on father's side," etc.

First we define the relation semigroup of all kinship relations. However, we evade the tricky problem of defining the term "kinship relation."

We start with a selected set X of basic relations such that the set of all their relation products yields all remaining kinship relations. In this way, we arrive at the concept of a free (hence infinite) semigroup over X. But there are only finitely many people on Earth, and some kinship relations like "daughter of a mother" and "daughter of a father" might be considered to be "the same."

Hence we might define:

32.1. Definition. A ***kinship system*** is a semigroup $S = [X, R]$, where:

(i) X is a set of "elementary kinship relationships."

(ii) R is a relation on X_*, which expresses equality of kinship relationships.

The product in S is always interpreted as relation product.

32.2. Example. Let $X = \{$"is father of", "is mother of"$\}$ and $R = \emptyset$. Then the kinship system S is the semigroup $\{$"is father of", "is mother of", "is grandfather on father's side of", ...$\}$.

We consider another example.

32.3. Example. Let F := "is father of," M := "is mother of," S := "is son of," D := "is daughter of," B := "is brother of," Si := "is sister of," C := "is child of." Then $X := \{\mathsf{F}, \mathsf{M}, \mathsf{S}, \mathsf{D}, \mathsf{B}, \mathsf{Si}, \mathsf{C}\}$ and

$$R := \{(\mathsf{CM}, \mathsf{CF}), (\mathsf{BS}, \mathsf{S}), (\mathsf{SiD}, \mathsf{D}), (\mathsf{CBM}, \mathsf{CMM}), (\mathsf{MC}, \mathsf{FC}), (\mathsf{SB}, \mathsf{S}), (\mathsf{DSi}, \mathsf{D}), (\mathsf{MBC}, \mathsf{MMC}), \ldots\}.$$

See Boyd, Haehl & Sailer (1972) for the complete list of R. The first pair means that in the semigroup we have $\mathsf{CM} = \mathsf{CF}$, i.e., children of the mother are the same as the children of the father, thus we do not distinguish between brothers and stepbrothers, etc. It can be shown that we obtain a finite semigroup.

Let $\mathcal{G}$ be a "society," i.e., a nonempty set of people and let $S(\mathcal{G})$ be the semigroup of all different kinship relationships of this society.

32.4. Definition. Any set X which generates $S(\mathcal{G})$ is called a ***kinship generating set***. $S(\mathcal{G})$ is called the ***kinship system*** of the society $\mathcal{G}$.

32.5. Example. It is difficult to imagine that the semigroup in Example 32.2 is the kinship system of an interesting society. But the semigroup S of Example 32.3 is "very close to" the kinship system of the society of the Fox Indians in North America, described by S. Tax in 1937.

The semigroups $S(\mathcal{G})$ often have special properties, e.g., "is a son of" and "is father of" are nearly inverse relations. The framework for the investigation of special $S(\mathcal{G})$ would be the theory of ***inverse semigroups***, i.e., semigroups S such that for all $s \in S$ there is a unique $s' \in S$ with $ss's = s$ and $s'ss' = s'$. For instance, $(\mathbb{M}_n(\mathbb{R}), \cdot)$ is an inverse semigroup for all $n \in \mathbb{N}$. If s denotes "is son of" and s' denotes "is father of," then these two equations hold in most (monogamous) societies.

We now study similar questions, but with different interpretations.

Sociology includes the study of human interactive behavior in group situations, in particular, in underlying structures of societies. Such structures can be revealed by mathematical analysis. This indicates how algebraic techniques may be introduced into studies of this kind.

Imagine, for a moment, an arbitrary society, such as your family, your circle of friends, your university colleagues, etc. In such societies, provided they are large enough (at least ≥ 3 members), coalitions can be formed consisting of groups of people who like each other, who have similar interests, or who behave similarly. These coalitions are either known or can be found by questionnaires using suitable computer programs (see White, Boorman & Breiger (1976) or Lidl & Pilz (1984)).

32.6. Example. In 1969, the sociologist S. F. Sampson investigated an American monastery. He found three coalitions: (1) = "conservative monks," (2) = "progressive monks," (3) = "outsiders." He asked each monk x to evaluate each other monk y according to (E) "Do you have high esteem for y?" and (D) "Do you dislike y?". As it should be with coalitions, members of the same coalition showed a very similar voting behavior and got similar votes, so we could treat them really as one block. If coalition i answered question (E) to coalition j overwhelmingly by "yes," we denote this by $e_{ij} = 1$, and take $e_{ij} = 0$ otherwise $(i, j \in \{1, 2, 3\})$. From this we get a 3×3 matrix $\mathbf{E} = (e_{ij})$, and similarly we get a matrix $\mathbf{D}$ for question (D). The observed values were

$$\mathbf{E} = \begin{pmatrix} 1 & 0 & 0 \\ 0 & 1 & 0 \\ 0 & 1 & 1 \end{pmatrix} \text{ and } \mathbf{D} = \begin{pmatrix} 0 & 1 & 1 \\ 1 & 0 & 1 \\ 1 & 1 & 0 \end{pmatrix}.$$

So every coalition liked itself and disliked the others. Only the outsiders also liked the progressives. Not a good sign for a monastery!

The relation products "x has an esteem for z who dislikes y" can be easily seen to correspond to the "Boolean matrix product" $\mathbf{E} \circ \mathbf{D}$, which is defined as follows.

32.7. Definition. Let $\mathbf{A}$ and $\mathbf{B}$ be $k \times k$ matrices over $\{0, 1\}$. Then $\mathbf{A} \circ \mathbf{B}$ is the usual product of matrices if we operate with 0 and 1 as in $\mathbb{B}$ (see 3.3(ii)); so $1 + 1 = 1$.

+	0	1
0	0	1
1	1	1

·	0	1
0	0	0
1	0	1

32.8. Example ***(32.6 continued).***

$$\mathbf{E} \circ \mathbf{D} = \begin{pmatrix} 1 & 0 & 0 \\ 0 & 1 & 0 \\ 0 & 1 & 1 \end{pmatrix} \circ \begin{pmatrix} 0 & 1 & 1 \\ 1 & 0 & 1 \\ 1 & 1 & 0 \end{pmatrix} = \begin{pmatrix} 0 & 1 & 1 \\ 1 & 0 & 1 \\ 1 & 1 & 1 \end{pmatrix},$$

while the matrix corresponding to "x dislikes z who esteem y" gives

$$\mathbf{D} \circ \mathbf{E} = \begin{pmatrix} 0 & 1 & 1 \\ 1 & 0 & 1 \\ 1 & 1 & 0 \end{pmatrix} \circ \begin{pmatrix} 1 & 0 & 0 \\ 0 & 1 & 0 \\ 0 & 1 & 1 \end{pmatrix} = \begin{pmatrix} 0 & 1 & 1 \\ 1 & 1 & 1 \\ 1 & 1 & 0 \end{pmatrix} \neq \mathbf{E} \circ \mathbf{D}.$$

So, in brief, in this society "friend of an enemy" is different from "enemy of a friend." Furthermore,

$$\mathbf{D} \circ \mathbf{D} = \begin{pmatrix} 0 & 1 & 1 \\ 1 & 0 & 1 \\ 1 & 1 & 0 \end{pmatrix} \circ \begin{pmatrix} 0 & 1 & 1 \\ 1 & 0 & 1 \\ 1 & 1 & 0 \end{pmatrix} = \begin{pmatrix} 1 & 1 & 1 \\ 1 & 1 & 1 \\ 1 & 1 & 1 \end{pmatrix} \neq \mathbf{D}.$$

Thus the "enemy of an enemy" is not necessarily a friend in this society.

Boorman & White (1976) introduced the term "role structure," which is an abstraction of concrete situations, like the one in a monastery or in a commune.

32.9. Definition. Let $\mathcal{G}$ be a community partitioned into coalitions $C_1, \ldots, C_k$, and let $R_1, \ldots, R_s$ be relations on $\mathcal{G}$ with corresponding $k \times k$ matrices $\mathbf{M}_1, \ldots, \mathbf{M}_s$ over $\{0, 1\}$. The ***role structure*** corresponding to $\mathcal{G}, C_1, \ldots, C_k, R_1, \ldots, R_s, \mathbf{M}_1, \ldots, \mathbf{M}_s$ is the subsemigroup generated by $\mathbf{M}_1, \ldots, \mathbf{M}_s$ of the finite semigroup of all $k \times k$ matrices over $\mathbb{B}$ with respect to $\circ$.

32.10. Example ***(32.8 and 32.6 continued).*** We take all products of $\mathbf{E}$ and $\mathbf{D}$ until we get a multiplication table with no other entries as in the margins. It is advisable to use a computer. We get the role structure of this monastic society as the subsemigroup S generated by $\{\mathbf{E}, \mathbf{D}\}$:

$$S = \{\mathbf{E}, \mathbf{D}, \mathbf{E} \circ \mathbf{D}, \mathbf{D} \circ \mathbf{E}, \mathbf{D} \circ \mathbf{D}, \mathbf{E} \circ \mathbf{D} \circ \mathbf{E}\}$$

with six elements.

The introduction of finite semigroups for role structures enables us to compare different societies. Boorman & White (1976) defined the concept of a "distance" between semigroups, and hence a distance between different societies. See also Lidl & Pilz (1984). This was applied to the society of monks in Example 32.6 at different times. We could see that the distances between their semigroups at different times became bigger and bigger, and in fact, about a year later, the monastery was dissolved. This is perhaps a rather strange application of semigroups!

Exercises

1. Describe the kinship system S of 32.2 in terms of semigroup theory.
2. Are the sisters of daughters in 32.3 considered to be "the same" as the daughters of sisters?
3. What does it mean for a kinship system to be a group?
4. Let $X = \{\mathsf{F}, \mathsf{M}\}$ and $R = \{(\mathsf{FFF}, \mathsf{F}), (\mathsf{MM}, \mathsf{M}), (\mathsf{FM}, \mathsf{MF})\}$. Which kinship system S do we get?
5. Let X be as above and $R' = \{(\mathsf{FF}, \mathsf{F}), (\mathsf{MM}, \mathsf{M}), (\mathsf{FM}, \mathsf{MF})\}$. Describe the kinship system S' as a semigroup.

6–7. Are the semigroups S, S' in Exercises 4 and 5 commutative? monoids? groups? abelian groups?

8. Let the kinship system $S = [X, R]$ be defined by $X = \{\mathsf{P}(=$ "is parent of"), $\mathsf{C}(=$ "is child of")$\}$, $R = (\mathsf{PP}, \mathsf{P}), (\mathsf{CC}, \mathsf{C}), (\mathsf{PC}, \mathsf{CP})$. How many elements are there in S? Find its operation table. Determine whether S is an (abelian) group.
9. The members of a mathematics department belong to two different groups A, B. A hates B and loves itself, while members of B hate everybody (including themselves), but love nobody. What is their semigroup with respect to "love" and "hate"?
10. The members of a department of psychology fall into three coalitions A, B, C. All of them love members of their own group, but nobody else. Members of A hate members of the other groups, and conversely. There is no hate between B and C. Again, work out their role structure.

11–12. Find presentations of the role structures S_1 and S_2 in the departments of mathematics and psychology (Exercises 9 and 10, respectively).

Notes

The study of kinship goes back to a study by A. Weil in response to an inquiry by the anthropologist C. Levi-Strauss (see the appendix to Part I of *Elementary Structures of Kinship* by Levi-Strauss, 1949; see also White (1963)). For examples similar to the one described in this section, we refer to White, Boorman & Breiger (1976), and Breiger, Boorman & Arabie (1975), see also Kim & Roush (1980) for an elementary account of such examples. Ballonoff (1974) presents several of the fundamental papers on kinship; Carlson (1980) gives elementary examples of applications of groups in anthropology and sociology.

§33 Linear Recurring Sequences

Linear recurring sequences and properties of finite fields are closely related. Examples of applications of linear recurring sequences occur in radar and communications systems. Radar is an abbreviation for "radio detection and ranging." It operates by the emission of electromagnetic energy with radio frequency, which is reflected by an object "hit" by radar waves. One possible way of measuring the distance of an object (airplane, satellite, etc.) from a radar station consists of emitting a long periodic sequence of signals of electromagnetic energy, which is reflected back by the object and received by the station. The time t which elapses between transmission of the signals from the radar station and their return after reflection from the object can be used to calculate the distance of the object from the station. For a discrete system with "period" r, we must have $rc > t$, where c is the time unit used. If the transmitter decreases the time interval between sending signals below this elapsed-time between transmission and reception, a condition called "range ambiguity" results, making it impossible to accurately determine the actual range of the target without further information being available. It is also desirable that the transmitted pulses be of short duration. We therefore try to use encoding techniques for transmitting signals such that these signals may be decoded without range ambiguity and with an increase in the total on-time for the transmitter. So-called maximal-length linear recurring sequences (or M-sequences) present a method for encoding the transmitter for a tracking radar. These sequences have very long "periods." They may be generated by shift registers.

Let $f \in \mathbb{F}_q[x]$ be an irreducible polynomial with $(q^n - 1)$th roots of unity as its roots. In calculating the formal power series

$$\frac{1}{f} = \sum_{i=0}^{\infty} c_i x^i,$$

the c_i are in $\mathbb{F}_q$, and they form a "pseudo-random sequence" of elements of $\mathbb{F}_q$. We can show that $\mathbf{c} = (c_0, c_1, \ldots)$ is periodic with period $q^n - 1$ (see 33.19). Within a period, every subsequence of length n with elements in $\mathbb{F}_q$ occurs exactly once, except the null-sequence.

Consider the primitive polynomial $f = x^3 + x + 1$ over $\mathbb{F}_2$. Using the techniques of §14, we find $1 + x^7 = (x^3 + x + 1)(x^3 + x^2 + 1)(x + 1)$. Since $(1 + x^7)(1 + x^7 + x^{14} + x^{21} + \cdots) = 1 + x^7 + x^7 + x^{14} + x^{14} + \cdots = 1$, we get

$$\frac{1}{f} = (1 + x + x^2 + x^4)(1 + x^7 + x^{14} + x^{21} + \cdots).$$

Hence, the sequence of coefficients c_i in this case is periodic with period 7, the first cocfficients are 1110100.

We construct such sequences which are of interest in various branches of information theory and electrical engineering. As before, $\mathbb{F}_q$ denotes the finite field with $q = p^t$ elements.

33.1. Definition. Let $k \in \mathbb{N}$. A sequence $s_0, s_1, \ldots$ of elements in $\mathbb{F}_q$ is called a ***linear recurring sequence of order*** k in $\mathbb{F}_q$, if there are elements $a, a_0, a_1, \ldots, a_{k-1} \in \mathbb{F}_q$,

$$s_{n+k} = a_{k-1}s_{n+k-1} + a_{k-2}s_{n+k-2} + \cdots + a_0 s_n + a. \tag{33.1}$$

The terms $s_0, s_1, \ldots, s_{k-1}$ determining the sequence are called ***initial values***. The equation (33.1) is called a ***linear recurrence relation of order*** k.

Sometimes, also the expressions ***difference equation*** or ***linear recursive relation*** are used for (33.1). In the case $a = 0$, the linear recurrence relation (33.1) is called ***homogeneous***, otherwise it is called ***inhomogeneous***. The vector $\mathbf{s}_n = (s_n, s_{n+1}, \ldots, s_{n+k-1})$ is called an nth ***state vector***, and $\mathbf{s}_0 = (s_0, s_1, \ldots, s_{k-1})$ is called an ***initial state vector***. It is a characteristic of linear recurring sequences in $\mathbb{F}_q$ that they are periodic.

As was indicated in §18, the generation of linear recurring sequences can be implemented on a feedback shift register. The building blocks

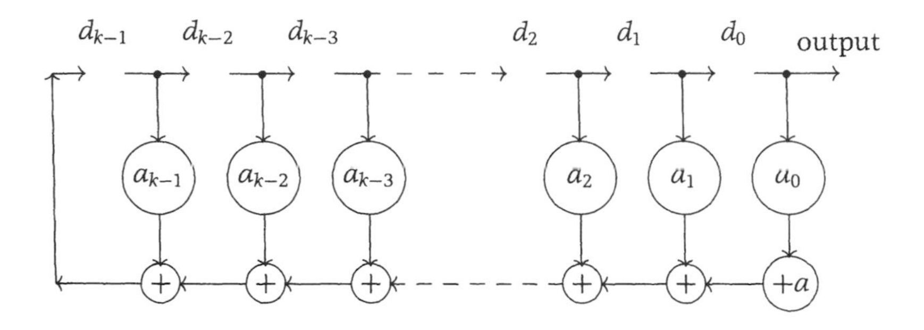

FIGURE 33.1

of a feedback shift register are delay elements, adders (see §18 for a description), and constant multipliers and adders, which multiply and add constant elements of $\mathbb{F}_q$ to the input. In the case of binary homogeneous linear recurring sequences, we only need delay elements, adders, and wire connections for their implementation. A feedback shift register that generates a linear recurring sequence (33.1) is given in Figure 33.1. At the start, each delay element $d_j, j = 0, 1, \dots, k-1$, contains the initial value s_j. After one time unit, each d_j will contain s_{j-1}. The output of the feedback shift register is the string of elements $s_0, s_1, s_2, \dots$ received in intervals of one time unit.

33.2. Example. In order to generate a linear recurring sequence in $\mathbb{F}_5$ which satisfies the homogeneous linear recurrence relation

$$s_{n+5} = s_{n+4} + 2s_{n+3} + s_{n+1} + 3s_n \quad \text{for } n = 0, 1, \dots,$$

we may use the feedback shift register of Figure 33.2.

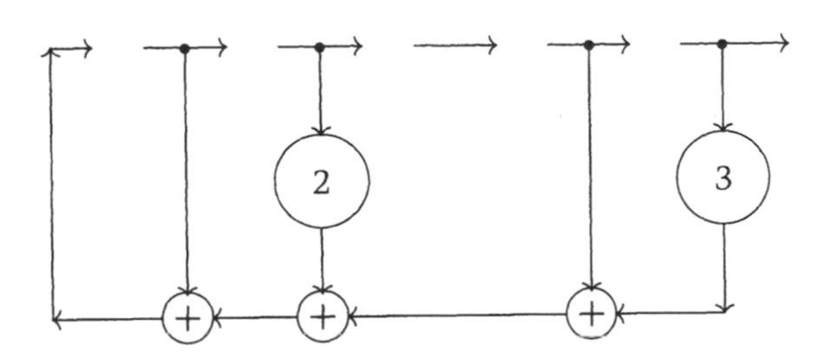

FIGURE 33.2

33.3. Definition. Let S be an arbitrary set, and let $s_0, s_1, \dots$ be a sequence in S. The sequence is called ***ultimately periodic*** if there are numbers $r \in \mathbb{N}$

and $n_0 \in \mathbb{N}_0$ such that $s_{n+r} = s_n$ for all $n \geq n_0$. The number n_0 is called the ***preperiod***. The number r is called the ***period*** of the sequence, and the smallest of all possible periods of an ultimately periodic sequence is called the ***least period*** of the sequence. An ultimately periodic sequence with period r is called ***periodic*** if $s_{n+r} = s_n$ holds for all $n = 0, 1, \ldots$.

33.4. Theorem. *Every linear recurring sequence $s_0, s_1, \ldots$ of order k in $\mathbb{F}_q$ is ultimately periodic with least period $r \leq q^k$; and if the sequence is homogeneous, then $r \leq q^k - 1$.*

Proof. There are q^k distinct k-tuples of elements of $\mathbb{F}_q$. So, in the list $\mathbf{s}_0, \mathbf{s}_1, \ldots, \mathbf{s}_{q^k}$ of state vectors of the sequence $s_0, s_1, \ldots$ there must be some i, j in $\{0, \ldots, q^k\}$ with $i < j$ and $\mathbf{s}_i = \mathbf{s}_j$. By induction, we get $\mathbf{s}_{n+j-i} = \mathbf{s}_n$ for all $n \geq i$, so $s_0, s_1, \ldots$ is ultimately periodic with least period $r \leq j - i \leq q^k$. If the sequence $s_0, s_1, \ldots$ is homogenous and if no state vector $\mathbf{s}_t$ is the zero vector (otherwise $s_{t+1} = s_{t+2} = \cdots = 0$), then we can use the same argument with $q^k - 1$ instead of q^k.

33.5. Examples.

(i) $s_0, s_1, \ldots \in \mathbb{F}_q$, where $s_{n+1} = s_n + 1$ for $n = 0, 1, \ldots$, has least period q.

(ii) It can be shown that for homogeneous linear recurring sequences of first order in $\mathbb{F}_q$ the least period is a divisor of $q - 1$. For $k \geq 2$, the least period of a kth order homogeneous linear recurring sequence need not divide $q^k - 1$. Let $s_0, s_1, \ldots \in \mathbb{F}_5$ with $s_0 = 0, s_1 = 1$, and $s_{n+2} = s_{n+1} + s_n$ for $n = 0, 1, \ldots$. The least period for this sequence is 20.

33.6. Theorem. *If $s_0, s_1, \ldots$ is a linear recurring sequence in $\mathbb{F}_q$ defined by (33.1) with $a_0 \neq 0$, then the sequence is periodic.*

Proof. See the proof of 33.7. □

Let $\mathbf{s} = (s_0, s_1, \ldots)$ be a homogeneous recurring sequence of order k in $\mathbb{F}_q$ satisfying the relation

$$s_{n+k} = a_{k-1}s_{n+k-1} + a_{k-2}s_{n+k-2} + \cdots + a_0 s_n \tag{33.2}$$

for $n = 0, 1, \ldots, a_j \in \mathbb{F}_q$, $0 \leq j \leq k - 1$. With this sequence, we associate a $k \times k$ matrix

$$\mathbf{A} = \begin{pmatrix} 0 & 0 & \dots & 0 & a_0 \\ 1 & 0 & \dots & 0 & a_1 \\ 0 & 1 & \dots & 0 & a_2 \\ \vdots & \vdots & \ddots & \vdots & \vdots \\ 0 & 0 & \dots & 1 & a_{k-1} \end{pmatrix}. \tag{33.3}$$

For $k = 1$, let $\mathbf{A} = (a_0)$. This matrix $\mathbf{A}$ is called the ***companion matrix*** of (33.2).

The proof of the next theorem is straightforward.

33.7. Theorem. *Let $s_0, s_1, \dots$ be a homogeneous linear recurring sequence in $\mathbb{F}_q$ satisfying* (33.2), *and let $\mathbf{A}$ be its companion matrix. Then the state vector of the sequence satisfies*

$$\mathbf{s}_n = \mathbf{s}_0 \mathbf{A}^n \quad \textit{for } n = 0, 1, \dots.$$

33.8. Theorem. *Let $s_0, s_1, \dots$ be a homogeneous linear recurring sequence in $\mathbb{F}_q$ satisfying* (33.2) *with $a_0 \neq 0$. Then the least period of the sequence is a divisor of the order of the companion matrix $\mathbf{A}$ in the group* $\mathrm{GL}(k, \mathbb{F}_q)$ *of all invertible $k \times k$ matrices over $\mathbb{F}_q$.*

Proof. We have $\det \mathbf{A} = (-1)^{k-1} a_0 \neq 0$, thus $\mathbf{A}$ is nonsingular. Let m be the order of $\mathbf{A}$. Then Theorem 33.7 implies $\mathbf{s}_{n+m} = \mathbf{s}_0 \mathbf{A}^{n+m} = \mathbf{s}_0 \mathbf{A}^n = \mathbf{s}_n$ for all $n \geq 0$. Therefore m is a period of the given sequence. The proof that m is divisible by the least period is left as an exercise. □

33.9. Definition. Let $s_0, s_1, \dots$ be a homogeneous linear recurring sequence of order k in $\mathbb{F}_q$ satisfying (33.2). Then the polynomial

$$f = x^k - a_{k-1}x^{k-1} - a_{k-2}x^{k-2} - \dots - a_0 \in \mathbb{F}_q[x]$$

is called the ***characteristic polynomial*** of the sequence.

The polynomial

$$f^{\perp} = 1 - a_{k-1}x - a_{k-2}x^2 - \dots - a_0 x^k \in \mathbb{F}_q[x]$$

is the reciprocal characteristic polynomial (see 18.15).

It can be shown that the characteristic polynomial f is also the characteristic polynomial of the companion matrix, so that, in the sense of linear algebra,

$$f = \det(x\mathbf{I} - \mathbf{A}).$$

We can evaluate the terms of a linear recurring sequence by using the characteristic polynomial, so we get a "solution formula" for s_n.

33.10. Theorem. *Let $s_0, s_1, \dots$ be a homogeneous linear recurring sequence of order k over $\mathbb{F}_q$ with characteristic polynomial f. If all zeros $\alpha_1, \dots, \alpha_k$ of f are distinct, then*

$$s_n = \sum_{j=1}^{k} \beta_j \alpha_j^n \quad \text{for } n = 0, 1, \dots,$$

where the elements β_j are uniquely determined by the initial values of the sequence. Moreover, they are elements of the splitting field of f over $\mathbb{F}_q$.

Proof. The elements $\beta_1, \dots, \beta_k$ can be determined from the system of linear equations

$$\sum_{j=1}^{k} \alpha_j^n \beta_j = s_n \quad \text{for } n = 0, 1, \dots, k-1. \tag{33.4}$$

Cramer's rule shows that $\beta_1, \dots, \beta_k$ are elements in the splitting field $\mathbb{F}_q(\alpha_1, \dots, \alpha_k)$ of f over $\mathbb{F}_q$. Since the determinant of (33.4) is the Vandermonde determinant, which is nonzero, the elements β_j are uniquely determined. Direct substitution shows that the elements s_n in (33.4) satisfy the recurrence relation (33.2).

33.11. Example. Consider the linear recurring sequence $s_0, s_1, \dots$ in $\mathbb{F}_2$ defined by $s_0 = s_1 = 1$ and $s_{n+2} = s_{n+1} + s_n$ for $n = 0, 1, \dots$. The characteristic polynomial is $f = x^2 - x - 1 = x^2 + x + 1 \in \mathbb{F}_2[x]$. If $\mathbb{F}_{2^2} = \mathbb{F}_2(\alpha)$, then $\alpha_1 = \alpha$ and $\alpha_2 = 1 + \alpha$ are zeros of f. Then we obtain $\beta_1 + \beta_2 = 1$ and $\beta_1\alpha + \beta_2(1 + \alpha) = 1$, thus $\beta_1 = \alpha$ and $\beta_2 = 1 + \alpha$. Theorem 33.10 implies $s_n = \alpha^{n+1} + (1 + \alpha)^{n+1}$ for all $n \geq 0$. We have $\beta^3 = 1$ for all $0 \neq \beta \in \mathbb{F}_{2^2}$, and therefore $s_{n+3} = s_n$ for all $n \geq 0$, which agrees with the fact that by 33.4, the least period of the sequence is 3.

Some rather exhausting comparisons of coefficients prove

33.12. Theorem. *Let $s_0, s_1, \dots$ be a homogeneous linear recurring sequence of order k and least period r over $\mathbb{F}_q$ satisfying (33.2). Let f be its characteristic polynomial. Then*

$$fs = (1 - x^r)h, \tag{33.5}$$

with $s = s_0x^{r-1} + s_1x^{r-2} + \cdots + s_{r-1} \in \mathbb{F}_q[x]$ and

$$h = \sum_{j=0}^{k-1} \sum_{i=0}^{k-1-j} a_{i+j+1} s_i x^j \in \mathbb{F}_q[x], \quad a_k = -1.$$

We recall from 14.17 that the order of $f \in \mathbb{F}_q[x]$ is the least natural number e such that $f \mid x^e - 1$, if f is not divisible by x. The case $x \mid f$ can be dealt with separately.

33.13. Theorem. *Let $s_0, s_1, \ldots$ be a homogeneous linear recurring sequence in $\mathbb{F}_q$ with characteristic polynomial $f \in \mathbb{F}_q[x]$. Then the least period of the sequence is a divisor of the order of f.*

Proof. It is easily verified that the order of f equals the order of its companion matrix. The result then follows from 33.8. □

33.14. Theorem. *Let $s_0, s_1, \ldots$ be a homogeneous linear recurring sequence in $\mathbb{F}_q$ with nonzero initial state vector and irreducible characteristic polynomial f over $\mathbb{F}_q$. Then the sequence is periodic with least period equal to the order of f.*

Proof. Theorem 33.13 shows that the sequence is periodic with least period r being a divisor of the order of f. On the other hand, equation (33.5) implies $f \mid (x^r - 1)h$. Since $s, h \neq 0$ and $\deg h < \deg f$, the irreducibility of f ensures that $f \mid x^r - 1$, and thus $r \geq$ order of f. □

In §14, we saw that the order of an irreducible polynomial f of degree k over $\mathbb{F}_q$ is a divisor of $q^k - 1$. Hence 33.14 implies that the least period of a homogeneous linear recurring sequence in $\mathbb{F}_q$ can be at most $q^k - 1$. In order to construct sequences with least period equal to $q^k - 1$, we use primitive polynomials (see 14.21). Such sequences with "long" periods are of particular importance in applications.

33.15. Definition. A homogeneous linear recurring sequence in $\mathbb{F}_q$ is called a ***sequence with maximal period*** if its characteristic polynomial is primitive over $\mathbb{F}_q$ and its initial state vector is nonzero.

33.16. Theorem. *A sequence of order k with maximal period in $\mathbb{F}_q$ is periodic with least period $q^k - 1$.*

Proof. We need to know (see 33.13) that a monic polynomial f of degree k over $\mathbb{F}_q$ is primitive if and only if $f(0) \neq 0$ and the order of f is $q^k - 1$. The result then follows from 33.14 and 33.4. □

33.17. Example. The sequence $s_{n+7} = s_{n+4} + s_{n+3} + s_{n+2} + s_n$, $n = 0, 1, \ldots$, in $\mathbb{F}_2$ has the characteristic polynomial $f = x^7 - x^4 - x^3 - x^2 - 1 \in \mathbb{F}_2[x]$. Since f is primitive over $\mathbb{F}_2$, any sequence defined by the linear recurrence equation and having nonzero initial state vector is a sequence of maximal period. The least period of such sequences is $2^7 - 1 = 127$. All possible nonzero vectors of $\mathbb{F}_2^7$ are state vectors of this sequence. Any other sequence of maximal period which can be obtained from the given linear recurrence is a "shifted" copy of the sequence $s_0, s_1, \ldots$.

Next we formalize the approach indicated at the beginning of this section by using formal power series.

33.18. Definition. Let $s_0, s_1, \ldots$ be an arbitrary sequence of elements in $\mathbb{F}_q$, then the formal power series

$$G = s_0 + s_1 x + s_2 x^2 + \cdots + s_n x^n + \cdots = \sum_{n=0}^{\infty} s_n x^n$$

is called the ***generating function associated*** with $s_0, s_1, \ldots$.

We know that the set of all power series forms an integral domain with respect to the operations addition and multiplication. It can be shown (see Exercise 11.36) that the power series $\sum_{n=0}^{\infty} b_n x^n \in \mathbb{F}_q[[x]]$ has a multiplicative inverse if and only if $b_0 \neq 0$.

33.19. Theorem. *Let $s_0, s_1, \ldots$ be a homogeneous linear recurring sequence of order k in $\mathbb{F}_q$ satisfying* (33.2), *let $f^\perp$ be its reciprocal characteristic polynomial in $\mathbb{F}_q[x]$, and let $G \in \mathbb{F}_q[[x]]$ be its generating function. Then*

$$G = \frac{g}{f^\perp}, \tag{33.6}$$

where

$$g = -\sum_{j=0}^{k-1} \sum_{i=0}^{j} a_{i+k-j} s_i x^j \in \mathbb{F}_q[x], \quad a_k := -1.$$

Conversely, if g is an arbitrary polynomial over $\mathbb{F}_q$ of degree less than k and if $f^\perp \in \mathbb{F}_q[x]$ is as in Definition 33.9, *then the formal power series G defined by* (33.6) *is the generating function of a homogeneous linear recurring sequence of order k in $\mathbb{F}_q$, which satisfies* (33.2).

Proof. We have

$$f^\perp G = g - \sum_{j=k}^{\infty} \left(\sum_{i=0}^{k} a_i s_{j-k+i} \right) x^j. \tag{33.7}$$

If the sequence $s_0, s_1, \ldots$ satisfies (33.2), then $f^{\perp}G = g$ holds because $\sum_{i=0}^{k} a_i s_{n+i} = 0$. Since $f^{\perp}$ has a multiplicative inverse, equation (33.6) follows.

Conversely, (33.7) implies that $f^{\perp}G$ can only be a polynomial of degree $< k$ if

$$\sum_{i=0}^{k} a_i s_{j-k+i} = 0 \quad \text{for all } j \geq k.$$

This means that the sequence of coefficients $s_0, s_1, \ldots$ satisfies (33.2). □

Theorem 33.19 implies that there is a one-to-one correspondence between homogeneous linear recurring sequences of order k with reciprocal characteristic polynomial $f^{\perp}$ and the rational functions $g/f^{\perp}$ with $\deg g < k$.

33.20. Example. Consider $s_{n+4} = s_{n+3} + s_{n+1} + s_n$, $n = 0, 1, \ldots$, in $\mathbb{F}_2$. The reciprocal characteristic polynomial is $f^{\perp} = 1 + x + x^3 + x^4 \in \mathbb{F}_2[x]$. For the initial state vector $(1, 1, 0, 1)$ we obtain $g = 1 + x^2$ from Theorem 33.19. Thus the generating function G is of the form

$$G = \frac{1 + x^2}{1 + x + x^3 + x^4} = 1 + x + x^3 + x^4 + x^6 + \cdots,$$

which corresponds to the binary sequence 1101101... with least period 3.

Theorem 33.16 has practical applications in systems of information transmission, in locating satellites, and the like. For instance, a primitive polynomial of degree 20 over $\mathbb{F}_2$, such as $x^{20} + x^3 + 1$, generates a periodic sequence of period $2^{20} - 1$, which is approximately of length 10^6, just suitable for radar observations of the Moon. For satellite communication systems, primitive polynomials of degree 50 or higher must be used.

A linear recurring sequence is defined by a specific linear recurrence relation, but satisfies many other linear recurrence relations as well.

The next theorem establishes the connection between the various linear recurrence relations valid for a given homogeneous linear recurring sequence.

33.21. Theorem. *Let $s_0, s_1, \ldots$ be a homogeneous linear recurring sequence in $\mathbb{F}_q$. Then there exists a uniquely determined monic polynomial $m \in \mathbb{F}_q[x]$ with the following property: a monic polynomial $f \in \mathbb{F}_q[x]$ of positive degree is a characteristic polynomial of $s_0, s_1, \ldots$ if and only if m divides f.*

Proof. Let I be the ideal of $\mathbb{F}_q[x]$ generated by all characteristic polynomials of $s_0, s_1, \ldots$; since $\mathbb{F}_q[x]$ is a PID, $I = (m)$ for some monic m, which is unique with the required property. □

33.22. Definition. The uniquely determined polynomial m over $\mathbb{F}_q$ associated with the sequence $s_0, s_1, \ldots$ as described in Theorem 33.21 is called the ***minimal polynomial*** of the sequence.

If $s_n = 0$ for all $n \geq 0$, the minimal polynomial is equal to the constant polynomial 1. For all other homogeneous linear recurring sequences, m is that monic polynomial with $\deg m > 0$ which is the characteristic polynomial of the linear recurrence relation of least possible order satisfied by the sequence.

33.23. Example. Let $s_0, s_1, \ldots$ be the linear recurring sequence in $\mathbb{F}_2$ with $s_{n+4} = s_{n+3} + s_{n+1} + s_n$, $n = 0, 1, \ldots$, and initial state vector $(1, 1, 0, 1)$. To find the minimal polynomial, take $f = x^4 - x^3 - x - 1 = x^4 + x^3 + x + 1 \in \mathbb{F}_2[x]$. Now $f = (x^2 + x + 1)(x + 1)^2$ is the factorization of f. Certainly, $x + 1$ is not the characteristic polynomial, but $x^2 + x + 1$ is, since it generates the same sequence $s_0, s_1, \ldots$. Hence the sequence satisfies the linear recurrence relation $s_{n+2} = s_{n+1} + s_n$, $n = 0, 1, \ldots$, as it should according to the general theory. We note that $\operatorname{ord}(m) = 3$, which is identical with the least period of the sequence. We shall see in Theorem 33.24 below that this is true in general.

The minimal polynomial plays an important role in the determination of the least period of a linear recurring sequence. This is shown in the following result.

33.24. Theorem. *Let $s_0, s_1, \ldots$ be a homogeneous linear recurring sequence in $\mathbb{F}_q$ with minimal polynomial $m \in \mathbb{F}_q[x]$. Then the least period of the sequence is equal to* $\operatorname{ord}(m)$.

Proof. If r is the least period of the sequence and n_0 is its preperiod, then we have $s_{n+r} = s_n$ for all $n \geq n_0$. Therefore, the sequence satisfies the homogeneous linear recurrence relation $s_{n+n_0+r} = s_{n+n_0}$ for $n = 0, 1, \ldots$. Then, according to Theorem 33.21, m divides $x^{n_0+r} - x^{n_0} = x^{n_0}(x^r - 1)$, so that m is of the form $m = x^h g$ with $h \leq n_0$ and $g \in \mathbb{F}_q[x]$, where $g(0) \neq 0$ and g divides $x^r - 1$. It follows from the definition of the order of a polynomial that $\operatorname{ord}(m) = \operatorname{ord}(g) \leq r$. On the other hand, r divides $\operatorname{ord}(m)$, by Theorem 33.13, and so $r = \operatorname{ord}(m)$. □

33.25. Example. Let $s_0, s_1, \ldots$ be the linear recurring sequence in $\mathbb{F}_2$ with $s_{n+5} = s_{n+1} + s_n$, $n = 0, 1, \ldots$, and initial state vector $(1, 1, 1, 0, 1)$. We take $f = x^5 - x - 1 = x^5 + x + 1 \in \mathbb{F}_2[x]$ and get $h = x^4 + x^3 + x^2$ from Theorem 33.12. Then $d = x^2 + x + 1$, and so the minimal polynomial m of the sequence is given by $m = f/d = x^3 + x^2 + 1$. We have $\operatorname{ord}(m) = 7$, and so Theorem 33.24 implies that the least period of the sequence is 7.

The details in the example above show how to find the least period of a linear recurring sequence without evaluating its terms. The method is particularly effective if a table of the order of polynomials is available.

There is an interesting connection between linear recurring sequences and cryptography, namely in ***stream ciphers***. A stream cipher is a conventional (single-key) cryptosystem in which messages and ciphertexts are strings of elements of a finite field $\mathbb{F}_q$, and enciphering and deciphering proceed by termwise addition and subtraction, respectively, of the same secret string of elements of $\mathbb{F}_q$. This secret string, called the ***key stream***, is generated by a (possibly known) deterministic algorithm from certain secret seed data and should possess good statistical randomness properties and high complexity, so that the key stream cannot be inferred from a small portion of its terms. Many key-stream generators use linear recurring sequences in $\mathbb{F}_q$ as building blocks. A relevant measure of complexity is the ***linear complexity*** $L(s)$, which is simply the degree of the minimal polynomial of a periodic, and thus linear recurring, sequence s of elements of $\mathbb{F}_q$. In the periodic case, only sequences with a very large linear complexity are acceptable as key streams. A fundamental problem in the analysis of key streams is that of bounding the linear complexity of sequences obtained by algebraic operations on periodic sequences (see Rueppel (1986)).

Linear recurring sequences are not suitable for constructing secure stream ciphers for the following reason. If a homogeneous linear recurring sequence is known to have a minimal polynomial of degree $k \geq 1$, then the minimal polynomial is determined by the first $2k$ terms of the sequence. To see this, we write the equations

$$s_{n+k} = a_{k-1}s_{n+k-1} + a_{k-2}s_{n+k-2} + \cdots + a_0 s_n$$

for $n = 0, 1, \ldots, k-1$, thereby obtaining a system of k linear equations for the unknown coefficients $a_0, a_1, \ldots, a_{k-1}$ of the minimal polynomial. The determinant of this system is $\neq 0$ and therefore the system can be solved uniquely.

Exercises

1. Give an example of a linear recurring sequence in $\mathbb{F}_q$, which is ultimately periodic but not periodic.
2. Determine the linear recurring sequence which generates

$$100010011010111000$$

and has period 15.

3. A homogeneous linear recurring sequence $d_0, d_1, \ldots$ of order k in $\mathbb{F}_q$ is called the ***impulse response sequence*** corresponding to the sequence $s_0, s_1, \ldots$ with recursion (33.2), if

$$d_0 = d_1 = \cdots = d_{k-2} = 0, \qquad d_{k-1} = 1 \qquad (d_0 = 1 \text{ for } k = 1),$$

and

$$d_{n+k} = a_{k-1}d_{n+k-1} + a_{k-2}d_{n+k-2} + \cdots + a_0 d_n$$

for $n = 0, 1, \ldots$.

   (i) Find the impulse response sequence $d_0, d_1, \ldots$ for the linear recurring sequence $s_{n+5} = s_{n+1} + s_n$ $(n = 0, 1, \ldots)$ in $\mathbb{F}_2$, and determine the least period.

   (ii) Let $d_0, d_1, \ldots$ be an impulse response sequence and let $\mathbf{A}$ be the matrix defined in (33.3). Show that two state vectors $\mathbf{d}_m$ and $\mathbf{d}_n$ are identical if and only if $\mathbf{A}_m = \mathbf{A}_n$. Determine $\mathbf{A}$ for the sequence in (i) and discuss the sequences resulting from all possible initial state vectors.

4. Design a feedback shift register corresponding to the linear binary recurrence relation $s_{n+7} = s_{n+4} + s_{n+3} + s_{n+2} + s_n$.
5. Determine the impulse response sequence $d_0, d_1, \ldots$ corresponding to the linear recurring sequence of Example 33.20. Find the generating polynomials of this sequence $d_0, d_1, \ldots$. (See Exercise 3 for the definition.)
6. Design a feedback shift register which multiplies polynomials with a fixed polynomial in $\mathbb{F}_q[x]$.
7. Construct a shift register for dividing by $q = 1 + x + x^3 \in \mathbb{F}_2[x]$, and list the contents of the shift register when $1 + x^3 + x^4$ is divided by q.
8. Consider the impulse response sequence (see Exercise 3) $d_0, d_1, \ldots$ corresponding to the linear recurring sequence $s_{n+5} = s_{n+1} + s_n$ $(n = 0, 1, \ldots)$ in $\mathbb{F}_2$. Design a feedback shift register generating the sequence.

9. Design a feedback shift register which yields the elements of $\mathbb{F}_{2^4}$, if ζ is a primitive element, where $\zeta^4 = \zeta + 1$.

10. Let $\alpha_1, \ldots, \alpha_m$ be the distinct roots of the characteristic polynomial f of a kth order homogeneous linear recurring sequence in $\mathbb{F}_q$, and suppose that each α_i has multiplicity $e_i \leq \operatorname{char} \mathbb{F}_q$. Let P_i be a polynomial of degree less than e_i whose coefficients are uniquely determined by the initial values of the sequence and belong to the splitting field of f over $\mathbb{F}_q$. Prove that $s_n = \sum_{i=0}^{m} P_i(n)\alpha_i^n$ for $n = 0, 1, \ldots$ satisfy the homogeneous linear recurring relation with characteristic polynomial f. (The integer n is identified with the element $n \cdot 1$ of $\mathbb{F}_q$.)

11. Find the characteristic polynomial f of the recurring sequence $s_{n+6} = s_{n+4} + s_{n+2} + s_{n+1} + s_n$, $n = 0, 1, \ldots$, in $\mathbb{F}_2$. Determine the order of f and find the impulse response sequence corresponding to the given linear recurrence relation. Take 000011 as the initial state vector and find the corresponding output sequence.

12. Find a linear recurring sequence of least order in $\mathbb{F}_2$ whose least period is 31.

13. Find the order of $f = x^4 + x^3 + 1$ over $\mathbb{F}_2$, determine the homogeneous linear recurrence and the corresponding sequence with characteristic polynomial f and initial values $s_0 = 1$, $s_1 = s_2 = s_3 = 0$.

14. Given the linear recurring sequence in $\mathbb{F}_3$ with $s_0 = s_1 = s_2 = 1$, $s_3 = s_4 = -1$, and $s_{n+5} = s_{n+4} + s_{n+2} - s_{n+1} + s_n$ for $n = 0, 1, \ldots$, represent the generating function of this sequence in the form (33.6) of Theorem 33.19.

15. Determine the multiplicative inverse of $3 + x + x^2$ in $\mathbb{F}_5[[x]]$.

16. Compute a/b in $\mathbb{F}_2[x]$, where $a = \sum_{n=0}^{\infty} x^n$, $b = 1 + x + x^3$.

17. A nonlinear binary shift register can be obtained from two linear ones by multiplication, that is, if one linear shift register generates a sequence s_i and the other generates a sequence t_i, then the output of their product is $s_i t_i$. Prove: If $\gcd(m, n) = 1$ and the sequences generated by two linear shift registers have periods $2^m - 1$ and $2^n - 1$, respectively, then the output sequence of their product has period $(2^m - 1)(2^n - 1)$. Verify this in case $m = 4$, $n = 3$, and characteristic polynomials $f = x^4 + x^3 + 1$ and $g = x^3 + x^2 + 1$. Why is multiplication of two linear shift registers not good for cryptanalytic purposes? (Another operation of combining linear shift registers is multiplexing, see Beker & Piper (1982).)

18. A ***k-stage circulating shift register*** over $\mathbb{F}_q$ is a feedback shift register over $\mathbb{F}_q$ with characteristic polynomial $x^k - 1$. Design its circuit, find its period and its companion matrix.

19. Let $s_0, s_1, \ldots$ be an inhomogeneous linear recurring sequence in $\mathbb{F}_2$ with

$$s_{n+4} = s_{n+3} + s_{n+1} + s_n + 1 \quad \text{for } n = 0, 1, \ldots$$

and initial state vector $(1, 1, 0, 1)$. Find its least period.

20. Is the linear recurring sequence $s_0, s_1, \ldots$ in $\mathbb{F}_2$, defined by $s_{n+4} = s_{n+2}+s_{n+1}$ for $n = 0, 1, \ldots$ and initial state vector $(1, 0, 1, 0)$, a periodic sequence?

Notes

The history of linear recurring sequences from 1202 to 1918 has been traced by Dickson (1901). The celebrated ***Fibonacci sequence*** $F_0, F_1, F_2, \ldots$ defined by $F_0 = 0$, $F_1 = 1$, $F_{n+2} = F_{n+1} + F_n$ for $n = 0, 1, \ldots$ attracted a great deal of attention; there also exists a *Fibonacci Quarterly* journal. Then, linear recurring sequences of real and complex numbers were considered, and, more recently, linear sequences in $\mathbb{Z}$ have been considered modulo a prime p. Such sequences became important in electrical engineering because of their connection with switching circuits and coding theory. Selmer (1966) gives a brief survey of the history of the subject with particular emphasis on the developments after 1918. The basic paper for the modern theory of linear recurring sequences in finite fields is Zierler (1959), and expository accounts of this theory can be found in Birkhoff & Bartee (1970), Dornhoff & Hohn (1978), Golomb (1967), Gill (1976), Lidl & Niederreiter (1994), Lüneburg (1979), and Peterson & Weldon (1972). Some of the pioneering work on the interplay between linear recurring sequences, feedback shift registers, and coding theory is presented in Lin (1970) and Peterson & Weldon (1972). Linear recurring sequences in cryptography are discussed in Beker & Piper (1982), Simmons (1992), McEliece (1977), and Rueppel (1986). Other applications of such sequences are presented in Golomb (1967). Linear recurring sequences are also used in the construction of Hadamard matrices, in the theory of difference sets, in pseudorandom number generators, and in simplifying computations in $\mathbb{F}_q$ and $\mathbb{F}_q[x]$.

§34 Fast Fourier Transforms

We are now going to generalize the idea of parallel computation of §24, and at the end of §37 we shall introduce the Hadamard transform. Before doing so, we take a closer look at evaluating polynomials economically. Instead of $\bar{p}(r)$, we again write $p(r)$.

34.1. Idea 1. Let $p = a_0 + a_1x + \cdots + a_nx^n \in R[x]$, R a commutative ring with identity. If $r \in R$, then

$$p(r) = a_0 + a_1r + \cdots + a_nr^n = a_0 + r(a_1 + \cdots + r(a_{n-1} + ra_n))\ldots).$$

This method avoids that in evaluating $p(r)$ we have to raise r to successive powers over and over again.

So, for instance, $p = 3 + 4x + 5x^2 + 6x^3 + 7x^4$ gives

$$p(2) = 3 + 2(4 + 2(5 + 2(6 + 7 \cdot 2))) = 191.$$

For many purposes (like evaluations at many points), however, another method is much better. Let us first start with the field $\mathbb{C}$.

34.2. Idea 2. Let $p = a_0 + a_1x + \cdots + a_{n-1}x^{n-1} \in \mathbb{C}[x]$ and suppose we want (for later purposes) to evaluate p at n points, where we have the freedom to choose these points. The idea is best described by an example. Let $p = a_0+a_1x+a_2x^2+a_3x^3$, so $n = 4$. Choose one point $c_1 \in \mathbb{C}$ arbitrarily, and choose c_2 as $c_2 := -c_1$. Now $p = p_{\text{even}} + p_{\text{odd}}$, where $p_{\text{even}} = a_0 + a_2x^2$, $p_{\text{odd}} = a_1x + a_3x^3$. Clearly

$$p(c_1) = p_{\text{even}}(c_1) + p_{\text{odd}}(c_1),$$
$$p(c_2) = p_{\text{even}}(c_2) + p_{\text{odd}}(c_2).$$

Now, $p_{\text{even}} = (a_0+a_2x) \circ x^2 =: q_{\text{even}}(x^2)$ with $q_{\text{even}} = a_0+a_2x$, and similarly $p_{\text{odd}} = x((a_1+a_3x) \circ x^2) =: x \cdot q_{\text{odd}}(x^2)$. Then q_{even} and q_{odd} are polynomials of degree ≤ 2 (in general, of degree $\leq \frac{n}{2}$) and

$$p(c_1) = q_{\text{even}}(c_1^2) + c_1q_{\text{odd}}(c_1^2),$$
$$p(c_2) = q_{\text{even}}(c_1^2) - c_1q_{\text{odd}}(c_1^2).$$

So, evaluating two polynomials of degree $\leq \frac{n}{2}$ gives the two values of p at c_1 and $c_2 = -c_1$. We apply the same idea to q_{even} and q_{odd}, reducing the degree to $\approx \frac{n}{4}$, and so on. Hence instead of evaluating p at c_1 and $-c_1$ we do it at $c_1, ic_1, -c_1, -ic_1$. We use our freedom and choose $c_1 = 1$, select a primitive root ω of unity (for instance, $\omega = e^{\frac{2\pi i}{n}}$) and evaluate at $1, \omega, \omega^2, \dots, \omega^{n-1}$ (these points are distinct since ω is primitive).

This is the strategy and procedure of the famous ***Cooley-Tuckey Algorithm***, and it works best if n is a power of 2. Let us briefly calculate what it costs. Computing values of a polynomial p of degree n needs n additions and n multiplications if we use Idea 1. So a single evaluation is of ***complexity of order*** n, briefly written as $O(n)$, which means that there is a constant k, independent of n, such that the evaluation needs $\leq kn$ operations. Computing n values of p is then of complexity $O(n^2)$, if we stick to Idea 1. Now let $\mathrm{CT}(n)$ be the number of operations needed to evaluate a polynomial of degree $< n$ at n places according to the Cooley-Tuckey algorithm. We get the recursion formula $\mathrm{CT}(n) \leq 2\,\mathrm{CT}(\frac{n}{2})$ (for evaluating the even and odd parts) $+\frac{3n}{2}$ (to combine these points to get the values for p). Clearly, $\mathrm{CT}(1) = 0$, and we get $\mathrm{CT}(n) \leq 1.5\, n \log n$ (where $\log n$ is the logarithm with base 2), as we can see by induction:

$$\begin{aligned}\mathrm{CT}(2n) &\leq 2\,\mathrm{CT}(n) + 3n \leq 3n \log n + 3n = 1.5(2n)\log n + 1.5(2n)\\ &\leq 1.5(2n)\log(2n).\end{aligned}$$

Hence, Cooley-Tuckey has a complexity of $O(n \log n)$, which really is an improvement over $O(n^2)$: for $n = 1{,}000{,}000$, $n^2 = 10^{12}$ is more than 50,000 times larger than $n \log n$. If n is not a power of 2, adjustments to Cooley-Tuckey can be made, see Clausen & Baum (1993).

34.3. Idea 3. Do the same as in Idea 2 for a field F (instead of $\mathbb{C}$) which has a primitive nth root of unity.

No problems arise in this generalization. For the rest of this section, F stands for such a field and ω is a primitive nth root of unity. For practical purposes it is, however, inconvenient to keep track of the ongoing decompositions $p = p_{\mathrm{even}} + p_{\mathrm{odd}}$, etc., in Ideas 2 and 3. So we improve the situation.

34.4. Theorem. *Using the notation above, let $p \in F[x]$. If r is the remainder of p after division by $x - \omega^i$, then*

$$p(\omega^i) = r(\omega^i).$$

Proof. If $p = (x - \omega^i)q + r$, we get $p(\omega^i) = (\omega^i - \omega^i)q(\omega^i) + r(\omega^i) = r(\omega^i)$. □

Now recall that $x^n - 1 = \prod_{i=0}^{n-1}(x - \omega^i)$. So in $\mathbb{C}[x]$, for example,

$$x^4 - 1 = (x^2 - 1)(x^2 - i^2) = (x - 1)(x - (-1))(x - i)(x - (-i)).$$

If k and m are even, then

$$x^k - \omega^m = (x^{k/2} - \omega^{m/2})(x^{k/2} + \omega^{m/2}) = (x^{k/2} - \omega^{m/2})(x^{k/2} - \omega^{m/2+n/2}).$$

34.5. Theorem. *If we successively divide p by $x^k - \omega^m$ and its remainder by $x^{k/2} - x^{m/2}$ (k, m even), we get the same remainder as by a single division by $x^{k/2} - x^{m/2}$.*

Proof. $p = (x^{k/2} - \omega^{m/2})q + r$ and $p = (x^k - \omega^m)q' + r'$, $r' = (x^{k/2} - \omega^{m/2})q'' + r''$, gives

$$\begin{aligned} p &= (x^{k/2} - \omega^{m/2})(x^{k/2} + \omega^{m/2})q' + (x^{k/2} - \omega^{m/2})q'' + r'' \\ &= (x^{k/2} - \omega^{m/2})q''' + r,'' \end{aligned}$$

for some q''' and with $\deg r'' < k/2$ and $\deg r < k/2$. By the uniqueness of the remainder, $r = r''$. □

34.6. Theorem. *Dividing $p = a_{n-1}x^{n-1} + \cdots + a_1x + a_0 \in F[x]$ by $x^k - c$ with $k = \frac{n}{2}$ gives the remainder $r = c(a_{n-1}x^{k-1} + \cdots + a_{k+1}x + a_k) + (a_{k-1}x^{k-1} + \cdots + a_1x + a_0)$.*

Proof. The proof is left as Exercise 4. □

We demonstrate the use of these results by two examples.

34.7. Example. Let $p = (1 + i)x^3 + 5ix^2 + 4x + 3 \in \mathbb{C}[x]$. So $n = 4$.

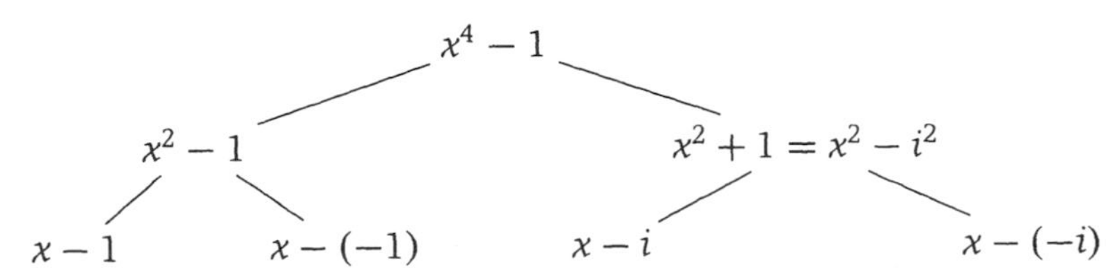

Dividing p by $x^4 - 1$, $x^2 - 1$, $x^2 - i^2$, $x - 1$, etc., gives the remainders (again in tree-form), using 34.6:

$$p$$

$$((1 + i)x + 5i) + (4x + 3) = (5 + i)x + (3 + 5i) \qquad -((1 + i)x + 5i) + (4x + 3) = (3 - i)x + (3 - 5i)$$

$$8 + 6i \qquad -2 + 4i \qquad 4 - 2i \qquad 2 - 8i$$

So, on the bottom level, we have the desired values $p(1), p(-1), p(i), p(-i)$.

34.8. Example. Take $p = 4x^2 + 3x + 2 \in \mathbb{Z}_5[x]$. To get p up to powers of the form x^{2^m-1}, we write $p = 0x^3 + 4x^2 + 3x + 2$. So $n = 4$, and we find $\omega = 2$ as a primitive fourth root of unity, i.e., as a generator of the cyclic group $(\mathbb{Z}_5^*, \cdot)$ (see 13.2(i) and the table after 13.2). We get $2^1 = 2$, $2^2 = 4$, $2^3 = 3$, $2^4 = 1$, so

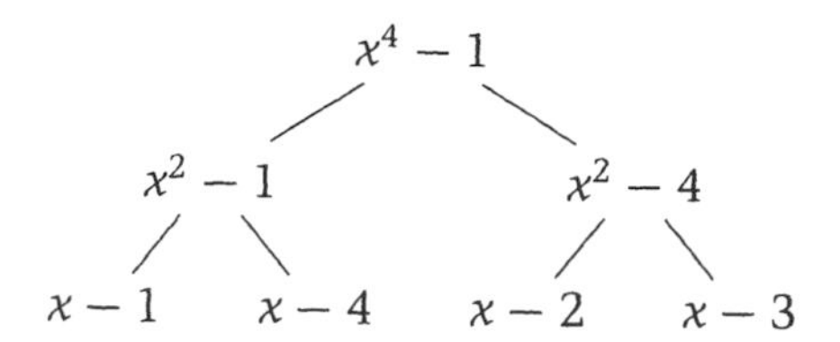

Then, using 34.6 again,

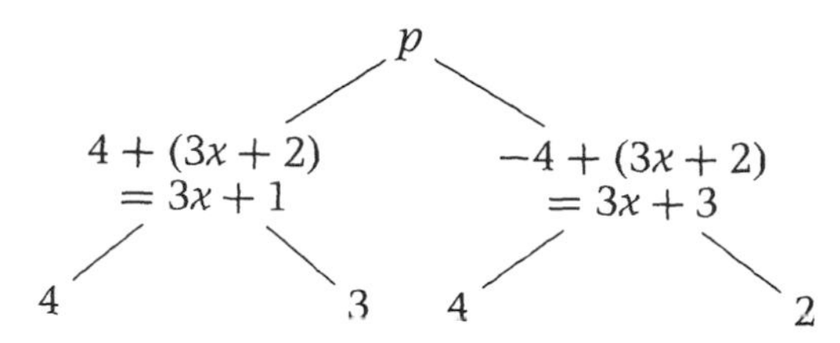

So $p(1) = 4$, $p(4) = 3$, $p(2) = 4$, $p(3) = 2$.

Observe that, due to 11.30, p is uniquely determined by the values $p(1) = p(\omega^0), p(\omega^1), \dots, p(\omega^{n-1})$. So we might replace p by the sequence $(p(1), \dots, p(\omega^{n-1}))$.

34.9. Definition. Let $p = a_0 + a_1x + \dots + a_{n-1}x^{n-1} \in F[x]$ and $\omega \in F$ a primitive nth root of unity. Then

$$\hat{p} := p(\omega^0) + p(\omega^1)x + \dots + p(\omega^{n-1})x^{n-1} =: \hat{a}_0 + \hat{a}_1x + \dots + \hat{a}_{n-1}x^{n-1}$$

is called the ***discrete Fourier transform*** **(DFT)** of p, and the $\hat{a}_i$ are the ***Fourier coefficients*** of p.

Recall again that by interpolation we can reconstruct p from $\hat{p}$. We can write any vector $(a_0, \dots, a_{n-1}) \in F^n$ as a polynomial $a_0 + a_1x + \dots + a_{n-1}x^{n-1} \in F[x]$ (see 11.17). This (or the writing as $a_0 + a_1z^{-1} + \dots + a_{n-1}z^{-(n-1)}$) is sometimes called the ***z-transformation*** (see, e.g., Szidarovzky & Bahill (1992)), which is a little odd, since $(a_0, \dots, a_{n-1})$ and $a_0 + \dots + a_{n-1}x^{n-1}$ *are* the same. So it is clear what the ***Fourier transform*** of a vector or a sequence means. Since for $\mathbf{a} = (a_0, \dots, a_{n-1}) =$

$a_0 + \cdots + a_{n-1}x^{n-1} =: p$ we get

$$p(\omega^k) = a_0 + a_1\omega^k + a_2\omega^{2k} + \cdots + a_{n-1}\omega^{(n-1)k},$$

we can write

$$\widehat{\mathbf{a}} = \begin{pmatrix} \hat{a}_0 \\ \hat{a}_1 \\ \hat{a}_2 \\ \vdots \\ \hat{a}_{n-1} \end{pmatrix} = \begin{pmatrix} 1 & 1 & 1 & \dots & 1 \\ 1 & \omega & \omega^2 & \dots & \omega^{n-1} \\ 1 & \omega^2 & \omega^4 & \dots & \omega^{2(n-1)} \\ \dots & \dots & \dots & \dots & \dots \\ 1 & \omega^{n-1} & \omega^{2(n-1)} & \dots & \omega^{(n-1)(n-1)} \end{pmatrix} \begin{pmatrix} a_0 \\ a_1 \\ a_2 \\ \vdots \\ a_{n-1} \end{pmatrix} =: \mathbf{D}_n\mathbf{a}. \tag{34.1}$$

34.10. Definition. $\mathbf{D}_n$ is called the ***DFT matrix*** of order n, $\widehat{\mathbf{a}}$ is the ***spectral vector*** to the ***signal vector*** $\mathbf{a}$.

Compare $\mathbf{D}_n$ with the character table of the cyclic group $\mathbb{Z}_n$, see 27.12. Observe that $\mathbf{D}_n$ is symmetric and also depends on the choice of ω; we denote this by $\mathbf{D}_{n,\omega}$ if necessary. If we work with polynomials, we also write $\hat{p} = \mathbf{D}_n p$. Computing $\widehat{\mathbf{a}}$ (or $\hat{p}$) via (34.1) is of complexity $O(n^2)$. If we use the faster algorithm as in 34.7 and 34.8, we speak of the ***fast Fourier transform*** **(FFT)**. Since the DFT is bijective, $\mathbf{D}_n$ is nonsingular. What about the inverse?

34.11. Theorem. *We have*

$$\mathbf{D}_n^2 = n \begin{pmatrix} 1 & 0 & 0 & \dots & 0 \\ 0 & 0 & 0 & \dots & 1 \\ \dots & \dots & \dots & \dots & \dots \\ 0 & 0 & 1 & \dots & 0 \\ 0 & 1 & 0 & \dots & 0 \end{pmatrix} \quad \text{and} \quad \mathbf{D}_{n,\omega}^{-1} = \frac{1}{n}\mathbf{D}_{n,\omega^{-1}},$$

provided that $\frac{1}{n}$ *exists in* F.

Proof. $\mathbf{D}_n \cdot \mathbf{D}_n = (d_{ij})$, where

$$\begin{aligned} d_{ij} &= 1 \cdot 1 + \omega^i\omega^j + \omega^{2i}\omega^{2j} + \cdots + \omega^{(n-1)i}\omega^{(n-1)j} \\ &= \sum_{k=0}^{n-1} \omega^{(i+j)k} \\ &= \frac{\omega^{(i+j)n} - 1}{\omega^{i+j} - 1} \\ &= \frac{(\omega^n)^{i+j} - 1}{\omega^{i+j} - 1} = 0 \text{ if } i + j \neq n, \end{aligned}$$

and $d_{ij} = 1+0+\cdots+0 = 1$ otherwise. Similarly, we see that $\mathbf{D}_{n,\omega} \cdot \mathbf{D}_{n,\omega^{-1}} = n\mathbf{I}$, from which we get the other result. □

Hence we can compute the inverse Fourier transform $\mathbf{a} \to \mathbf{D}_{n,\omega}^{-1}\mathbf{a}$ either:

- by applying $\mathbf{D}_{n,\omega}$ again, dividing through n and permuting the indices; or
- by dividing through n and applying $\mathbf{D}_{n,\omega^{-1}}$ (observe that ω^{-1} is also a primitive root of unity), which can be done again as a fast Fourier transform.

So $\mathbf{D}_n$ is "essentially self-inverse," which means that we can transform back and forth with "essentially the same device." It also shows that the inverse Fourier transform can be done in complexity $O(n \log n)$.

As a first application, we develop a fast algorithm to multiply polynomials p, q. The somewhat "exotic" idea is to evaluate p, q at sufficiently many points $c_1, c_2, \ldots$ and to find a polynomial f by interpolation at the points $(c_i, p(c_i)q(c_i))$. Then $f = pq$. At a first glance, it looks like a cumbersome idea to do the job, but the work we have done so far provides a lot of sense to support this idea.

If $p = a_0 + a_1x + \cdots + a_{n-1}x^{n-1}$, $q = b_0 + b_1x + \cdots + b_{m-1}x^{m-1} \in F[x]$, then $pq = c_0 + c_1x + \cdots + c_{n+m-2}x^{n+m-2}$, where

$$c_k = \sum_{i+j=k} a_i b_j = \sum_{i=0}^{k} a_i b_{k-i}. \tag{34.2}$$

So pq is sometimes also called the ***convolution*** of p and q, analoguous to the convolution $f * g(t) = \int_{-\infty}^{\infty} f(x)g(t-x)dx$ of functions f and g. Similarly, the ***convolution*** of sequences (vectors) $(a_0, \ldots, a_{n-1}) * (b_0, \ldots, b_{m-1})$ is defined as $(c_0, \ldots, c_{n+m-2})$, with c_k as in (34.2). With $(a_0, \ldots, a_{n-1}) \cdot (b_0, \ldots, b_{n-1}) := (a_0b_0, \ldots, a_{n-1}b_{n-1})$, we get from our discussion:

34.12. Theorem (*Fast Multiplication, Fast Convolution*). *Let $p, q \in F[x]$. Then*

$$pq = \mathbf{D}_n^{-1}(\mathbf{D}_n p \cdot \mathbf{D}_n q),$$

where n is sufficiently large. This multiplication can be done with complexity $O(n \log n)$. Ordinary multiplication via (34.2) has complexity $O(n^2)$.

34.13. Remark. We can use the same ideas as above in any commutative ring R with identity with some restrictions (n must be invertible, there must be a primitive nth root ω of unity, and for $0 < k < n$, $\omega^k - 1$ must not be a zero divisor in R). Sufficiently many $\mathbb{Z}_m$ have these properties. We shall not pursue that direction, however.

34.14. *The Schönhage-Strassen method* *for fast multiplication of (large) integers a, b:*

1. *Represent a, b in binary form as $a = a_0 + a_1 2 + a_2 2^2 + \cdots$, $b = b_0 + b_1 2 + b_2 2^2 + \cdots$.*
2. *Multiply $(\sum a_i x^i)(\sum b_j x^j)$ "quickly" using* 34.12.
3. *Evaluate this product at $x = 2$.*

This algorithm needs several restrictions and shortcuts. For instance, it is not a good idea to do it over $F = \mathbb{C}$, because ω has usually infinitely many nonzero digits, but in some $\mathbb{Z}_m$, where m is sufficiently large (cf. §24 and 34.13). It then can be done with complexity $O(m \log m \log \log m)$. See Clausen & Baum (1993) for more details.

Without proof (see Beth (1995)), we mention three important equations involving Fourier transformations (the first one is just a reformulation of 34.12):

34.15. Theorem. *Let $\mathbf{a} = (a_0, a_1, \ldots, a_{n-1})$, $\mathbf{b} = (b_0, b_1, \ldots, b_{n-1})$ be sequences, vectors, or polynomials, and $\hat{\mathbf{a}} = (\hat{a}_0, \ldots)$, $\hat{\mathbf{b}} = (\hat{b}_0, \ldots)$ their Fourier transforms.*

(i) ***(Convolution property)*** $\widehat{\mathbf{ab}} = (\hat{a}_0\hat{b}_0, \hat{a}_1\hat{b}_1, \ldots)$;

(ii) ***(Parseval identity)*** $\sum \hat{a}_i\hat{b}_i = n \sum a_{n-1-i} b_i$;

(iii) ***(Plancherel identity*** *or* ***Conservation of energy)*** $\sum |\hat{a}_i|^2 = n \sum |a_i|^2$ *(if $F = \mathbb{R}$ or $\mathbb{C}$);*

(iv) ***Phase property:*** *Let $\mathbf{T}$ be the matrix*

$$\mathbf{T} = \begin{pmatrix} 0 & 1 & 0 & \ldots & 0 \\ 0 & 0 & 1 & \ldots & 0 \\ \multicolumn{5}{c}{\dotfill} \\ 0 & 0 & 0 & \ldots & 1 \\ 1 & 0 & 0 & \ldots & 0 \end{pmatrix}.$$

Then $\mathbf{T}$ is called the ***translation*** *or* ***shift operation*** *since*

$$(a_0, \ldots, a_{n-1})\mathbf{T} = (a_{n-1}, a_0, a_1, \ldots, a_{n-2}),$$

and

$$\mathbf{T}^k\mathbf{D}_n = \mathbf{D}_n \begin{pmatrix} 1 & 0 & 0 & \dots & 0 \\ 0 & \omega^k & 0 & \dots & 0 \\ 0 & 0 & \omega^{2k} & \dots & 0 \\ \dots & \dots & \dots & \dots & \dots \\ 0 & 0 & 0 & \dots & \omega^{(n-1)k} \end{pmatrix}.$$

For $F = \mathbb{C}$, the last result (iv) means that shifting the phase does not effect the norm of the spectral vectors.

In §24 we replaced computations in $\mathbb{Z}_n$ (n large) by "parallel computation" in each $\mathbb{Z}_{p_i^{k_i}}$, where $n = p_1^{k_1} \dots p_t^{k_t}$. The striking similarity between $\mathbb{Z}$ and $F[x]$ (see the table after 11.24) indicates that we can do similar things with polynomials, hence we might go from $F[x]$ to $F[x]/(p)$, where p has a sufficiently large degree. We shall be especially well off if $p = x^n - 1$, since we have:

34.16. Theorem. *Let F contain a primitive nth root of unity ω, so that $\prod_{i=0}^{n-1}(x - \omega^i) = x^n - 1$. Then $\mathbf{D}_n$ gives an isomorphism*

$$\mathbf{D}_n\colon F[x]/(x^n - 1) \to \bigoplus_{k=0}^{n-1} F[x]/(x - \omega^k)$$

by $\mathbf{D}_n(p) = ([p]_{x-\omega^0}, \dots, [p]_{x-\omega^{n-1}}) = (p(\omega^0), \dots, p(\omega^{n-1})) = \hat{p}$.

So we can do parallel computations in the fields $F[x]/(x - \omega^k)$ (observe that $x - \omega^k$ is irreducible). The inverse transformation $\mathbf{D}_n^{-1}$ serves the same purpose as the Chinese Remainder Theorem in §24. To make things better, we might do the calculation with the coefficients in parallel as well (§24, again), if these coefficients are integers.

The ring $F[x]/(x^n - 1)$ occurs in several other places as well. We know from §18 that this is the "correct" domain to study cyclic codes. The shift operator in 34.15(iv) gives yet another approach: $\mathbf{T}$ transfers $(a_0, \dots, a_{n-1}) = a_0 + a_1x + \dots + a_{n-1}x^{n-1} =: p$ into $(a_0, \dots, a_{n-1})\mathbf{T} = (a_{n-1}, a_0, \dots, a_{n-2}) = x \cdot p$, reduced modulo $x^n - 1$. And $\mathbf{T}^2$: $p \to x^2p$ (computed in $F[x]/(x^n - 1)$) yields cyclic shifts by two "time units," and so on. Hence it makes sense to consider $F[x]$ as an $F[x]/(x^n - 1)$-algebra. We presented this idea already on page 206.

Still another point of view is that the cyclic group $\mathbb{Z}_n$ operates on F^n by permuting the components cyclically (cf. 25.15(iii)). Let $\mathbb{Z}_n$ be represented by $\{\text{id}, t, t^2, \dots, t^{n-1}\}$, where t is the "action of T," the cyclic shift by one place. This leads to the group algebra $F\mathbb{Z}_n := \{\sum_{i=0}^{n-1} f_i t^i \mid f_i \in F\}$ (which we have met in 27.9), where the multiplication is convolution mod $(t^n - 1)$. So $F\mathbb{Z}_n$ is a ring (even an algebra) which is isomorphic to $F[x]/(x^n - 1)$, and the Fourier transform changes the "unpleasant" convolution into componentwise multiplications of the components (see 34.16).

In a more general situation, we study again image understanding (see §26). A screen might consist of nm pixels

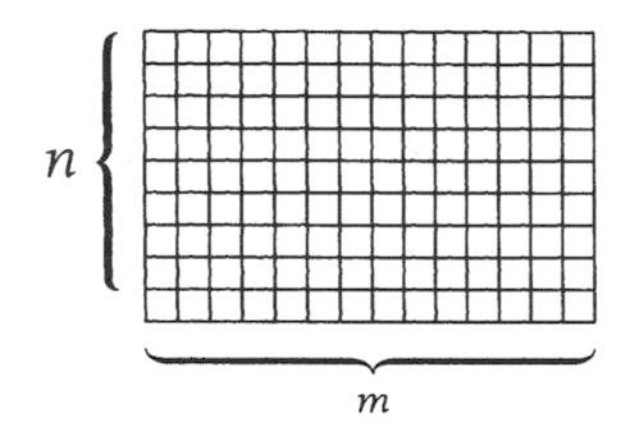

If an object moves along this screen, we have a natural action of the elements of the group $\mathbb{Z}_n \times \mathbb{Z}_m$ as translation operators. If rotations are also possible (like in computer tomography), we get more complicated groups G like groups of orthogonal or affine transformations. As above, this leads to group algebras FG, and the places the group G operates upon (the nm pixel places in our example) become a $K[G]$-module. Here, once again, we have to use the deep theory of group representations in order to get decompositions of FG into direct sums of "simple" rings R_i, which will turn the multiplication in FG (the "simple" convolution) into componentwise multiplications in R_i. For instance, $\mathbb{F}_{p^n}[x]/(x^n - 1)$ is the direct sum of fields $\mathbb{F}_{p^n}[x]/(p_i)$, where p_i are the irreducible factors of $x^n - 1$, iff $\gcd(p, n) = 1$ (***Maschke's Theorem***). In general, every isomorphism between a ring and a direct sum of simple rings might be called a ***general Fourier transform***. Any decomposition of a ring into a finite direct sum gives the chance to do parallel computations.

Exercises

1. Let $n = 4$ and $\omega = i \in \mathbb{C}$. Compute the Fourier transform of $(1, 1, 1, 1)$, $(1, 0, 0, 0)$, and $(1, i, i, -i)$.
2. Compute $\mathbf{D}_{4,2}$ over $\mathbb{F}_5$ according to Example 34.8.
3. Do the same for $\mathbf{D}_7$ over $\mathbb{Z}_2$ with α as in Example 18.33.

4. Prove Theorem 34.6.
5. Let $p = 3 + 2x$, $q = 1 + 4x \in \mathbb{Z}_5[x]$. Use Exercise 2 to compute their Fourier transforms f, g, compute $f * g$, and then $p \cdot q$ using the inverse Fourier transform.
6. Verify that 3 is a primitive nth root of unity in $\mathbb{Z}_{17}$, and do the same as in Exercise 5 with $p = 1 + x + x^2 + x^3$, $q = 1 + 2x + 4x^2 + 5x^3$.
7. Use the Schönhage-Strassen method for multiplying $123 \cdot 234$ in $\mathbb{Z}_{257}$.
8. For $F = \mathbb{Z}_2$, write down the elements and the addition and multiplication tables of $F\mathbb{Z}_3$. [Hint: Use different symbols to distinguish the elements of F and of $\mathbb{Z}_3$.]

Notes

Fourier transforms have many more applications. The largest Fourier coefficients contain the most important information of a signal, throwing away the others leads to efficient data compression. Signals $(a_0, a_1, \dots)$ which are distorted by noise might be Fourier transformed into $(\hat{a}_0, \hat{a}_1, \dots)$ and then multiplied point-by-point by a "filter function" $(f_0, f_1, \dots)$ which brings the high-frequency components, which are responsible for the noise, close to zero (as $f_i \hat{a}_i$); then we apply $\mathbf{D}_n^{-1}$ and get a filtered signal, which is much less distorted. The shortest length $A(s)$ which generates a recurring periodic sequence s (see §33) turns out to be the Hamming weight (16.5) of $\hat{s}$. For that and many more applications, see the excellent book by Clausen & Baum (1993). For the computational problems and shortcuts connected with the fast Fourier transform, see Geddes, Czapor & Labahn (1993), Winkler (1996), and Zippel (1993).

§35 Latin Squares

35.1. Definitions. An $n \times n$ matrix $\mathbf{L} = (a_{ij})$ over $A = \{a_1, \dots, a_n\}$ is called a ***Latin square*** of order n if each row and column contains each element of A exactly once. Two Latin squares (a_{ij}) and (b_{ij}) over A of order n are called ***orthogonal*** if all n^2 ordered pairs (a_{ij}, b_{ij}) of entries of $\mathbf{L}$ are distinct.

35.2. Example. The two Latin squares

$$\mathbf{L}_1 = \begin{pmatrix} a & b & c \\ c & a & b \\ b & c & a \end{pmatrix} \quad \text{and} \quad \mathbf{L}_2 = \begin{pmatrix} a & b & c \\ b & c & a \\ c & a & b \end{pmatrix}$$

superimposed give the square

$$\begin{pmatrix} aa & bb & cc \\ cb & ac & ba \\ bc & ca & ab \end{pmatrix},$$

where aa stands for the pair (a, a), etc. Thus $\mathbf{L}_1$ and $\mathbf{L}_2$ are orthogonal.

Latin squares first appeared in parlor games, e.g., in the problem of arranging the Jacks, Queens, Kings, and Aces of a card deck into a 4×4 matrix, such that in each row and column each suit and each value of a card occurs exactly once.

The famous ***Euler's officers problem*** goes back to Leonard Euler, who posed the following question (in 1779, and more generally in 1782): Is it possible to arrange 36 officers of six regiments with six ranks into a 6×6 square in a parade, such that each "row" and each "column" of the square has exactly one officer of each regiment and each rank? In the formulation of Definition 35.1, do there exist two orthogonal Latin squares of order 6? Euler conjectured that there are no two orthogonal Latin squares of order 6. More generally, he conjectured: There are no two orthogonal Latin squares of order n, where $n \equiv 2 \pmod 4$.

In 1899, G. Tarry proved that Euler's officers problem cannot be done. But Euler's more general conjecture was totally disproved by Bose, Shrikhande, and Parker in 1959. They showed that for any $n > 6$ there are at least two orthogonal Latin squares of order n.

35.3. Theorem. *For any positive integer n, there is a Latin square of order n.*

This follows immediately, since Latin squares are essentially the operation tables of (not necessarily associative) finite "groups" (called finite ***loops***). We might take $(\mathbb{Z}_n, +)$, for instance.

Next we show that for $n = p$, a prime, there are $n - 1$ mutually orthogonal Latin squares (**MOLS**) of order n. We form

$$\mathbf{L}_j := \begin{pmatrix} 0 & 1 & \dots & p-1 \\ j & 1+j & \dots & p-1+j \\ 2j & 1+2j & \dots & p-1+2j \\ \dots & \dots & \dots & \dots \\ (p-1)j & 1+(p-1)j & \dots & p-1+(p-1)j \end{pmatrix}$$

for $j = 1, 2, \ldots, p-1$, where all elements in $\mathbf{L}_j$ are reduced (mod p). The $\mathbf{L}_j$'s are Latin squares and $\mathbf{L}_i$ and $\mathbf{L}_j$ are orthogonal if $i \neq j$. For suppose (a, b) were a pair which appeared twice, say in row α and column β as well as in row γ and column δ, then

$$\beta + \alpha j \equiv \delta + \gamma j \equiv a \pmod{p},$$
$$\beta + \alpha i \equiv \delta + \gamma i \equiv b \pmod{p}.$$

Thus $\alpha(i-j) \equiv \gamma(i-j) \pmod{p}$, i.e., $\alpha \equiv \gamma$, $\beta \equiv \delta \pmod{p}$.

This construction can be generalized to Latin squares over $\mathbb{F}_q$, $q = p^e$.

35.4. Theorem. *Let $0, 1, a_2, \ldots, a_{q-1}$ be the elements of $\mathbb{F}_q$. Then the Latin squares $\mathbf{L}_i$, $1 \le i \le q-1$, form a set of $q-1$ MOLS, where*

$$\mathbf{L}_i := \begin{pmatrix} 0 & 1 & \ldots & a_{q-1} \\ a_i & a_i + 1 & \ldots & a_i + a_{q-1} \\ a_i a_2 & a_i a_2 + 1 & \ldots & a_i a_2 + a_{q-1} \\ \cdots & \cdots & \cdots & \cdots \\ a_i a_{q-1} & a_i a_{q-1} + 1 & \ldots & a_i a_{q-1} + a_{q-1} \end{pmatrix}.$$

If ζ is a primitive element of $\mathbb{F}_q$, then the following $q-1$ Latin squares $\mathbf{L}'_i$ with $i = 0, 1, \ldots, q-2$ are also MOLS:

$$\mathbf{L}'_i := \begin{pmatrix} 0 & 1 & \ldots & a_{q-1} \\ \zeta^{0+i} & 1 + \zeta^{0+i} & \ldots & a_{q-1} + \zeta^{0+i} \\ \zeta^{1+i} & 1 + \zeta^{1+i} & \ldots & a_{q-1} + \zeta^{1+i} \\ \cdots & \cdots & \cdots & \cdots \\ \zeta^{q-2+i} & 1 + \zeta^{q-2+i} & \ldots & a_{q-1} + \zeta^{q-2+i} \end{pmatrix}.$$

$\mathbf{L}'_{i+1}$ can be obtained from $\mathbf{L}'_i$ by a cyclic exchange of the last $q-1$ rows.

Proof. We prove the first part of the theorem. Each $\mathbf{L}_i$ is a Latin square. Let $a^{(i)}_{mn} = a_i a_{m-1} + a_{n-1}$ be the (m, n) entry of $\mathbf{L}_i$. For $i \neq j$, suppose $(a^{(i)}_{mn}, a^{(j)}_{mn}) = (a^{(i)}_{rs}, a^{(j)}_{rs})$ for some $1 \le m, n, r, s \le q$. Then

$$(a_i a_{m-1} + a_{n-1},\ a_j a_{m-1} + a_{n-1}) = (a_i a_{r-1} + a_{s-1},\ a_j a_{r-1} + a_{s-1}),$$

so

$$a_i(a_{m-1} - a_{r-1}) = a_{s-1} - a_{n-1}$$

and

$$a_j(a_{m-1} - a_{r-1}) = a_{s-1} - a_{n-1}.$$

Since $a_i \neq a_j$, it follows that $a_{m-1} = a_{r-1}$, $a_{s-1} = a_{n-1}$ hence $m = r$, $n = s$. Thus the ordered pairs of corresponding entries from $\mathbf{L}_i$ and $\mathbf{L}_j$ are all different and so $\mathbf{L}_i$ and $\mathbf{L}_j$ are orthogonal. □

35.5. Example. Three MOLS of order 4 are given by

$$\mathbf{L}_1 = \begin{pmatrix} 0 & 1 & \zeta & \zeta^2 \\ 1 & 0 & \zeta^2 & \zeta \\ \zeta & \zeta^2 & 0 & 1 \\ \zeta^2 & \zeta & 1 & 0 \end{pmatrix}, \quad \mathbf{L}_2 = \begin{pmatrix} 0 & 1 & \zeta & \zeta^2 \\ \zeta & \zeta^2 & 0 & 1 \\ \zeta^2 & \zeta & 1 & 0 \\ 1 & 0 & \zeta^2 & \zeta \end{pmatrix}, \quad \mathbf{L}_3 = \begin{pmatrix} 0 & 1 & \zeta & \zeta^2 \\ \zeta^2 & \zeta & 1 & 0 \\ 1 & 0 & \zeta^2 & \zeta \\ \zeta & \zeta^2 & 0 & 1 \end{pmatrix}.$$

Here ζ is a primitive element of $\mathbb{F}_4$ and a zero of $f = x^2 + x + 1$. $\mathbf{L}_1$ is the addition table of $\mathbb{F}_4$.

35.6. Example. As an example of the use of Latin squares in experimental design, we study gasoline additives and their effects on the nitrogen oxide emission of cars. Suppose we have three additives A_1, A_2, A_3 which should be tested in three cars C_1, C_2, C_3 by three different drivers D_1, D_2, D_3. No matter how carefully the drivers and the cars are chosen, some effects of the individual drivers and the different cars will be present in the results of the experiment. By the statistical analysis of variance, it is possible to eliminate these effects to a large extent.

We could test the effect of every driver with every car and every additive. However, using Latin squares and statistics, it suffices that each additive is tested once by each driver and in each car. This reduces the number of tests required from 27 to 9.

Let us take the first Latin square in Example 35.2, where we replace a, b, c by A_1, A_2, A_3.

	C_1	C_2	C_3
D_1	A_1	A_2	A_3
D_2	A_3	A_1	A_2
D_3	A_2	A_3	A_1

	C_1	C_2	C_3
D_1	17	10	12
D_2	11	18	7
D_3	9	14	19

The second table gives the emission results of the nine experiments. For instance, the first entry in the tables tells us that the first driver used the first additive in the first car and produced an emission value of 17. A careful look at the tables indicates that the second additive produced the lowest results, but usually the picture is not that clear.

Now suppose that the performance of the additives also varied with the weather situations, in this case. So three weather situations W_1, W_2, and W_3 should also be taken into account. This can be achieved by selecting a second Latin square which is orthogonal to the first one. We take the second square in Example 35.2 and superimpose them as we did there. In statistics it is customary to use Latin letters a, b, c for the additives and Greek letters α, β, γ for the weather situations. The entire experiment can then be described by the table

	C_1	C_2	C_3
D_1	$a\alpha$	$b\beta$	$c\gamma$
D_2	$c\beta$	$a\gamma$	$b\alpha$
D_3	$b\gamma$	$c\alpha$	$a\beta$

	C_1	C_2	C_3
D_1	16	10	13
D_2	11	19	8
D_3	7	15	18

which now tells that the first driver produced an emission value of 16 using the first additive in the first car during weather W_1. This explains the names "Latin square" (for one square) and "Latin-Graeco square" (which is also used for a pair of orthogonal Latin squares).

It is always a good idea to randomize these squares (i.e., to rearrange randomly the order of the "ingredients") and it is often a good idea to repeat the experiments, especially when so few tests are performed as in our example.

Latin squares should only be used if the effects of the "ingredients" are approximately additive and if there is no essential interaction between the ingredients. For tests if these assumptions are satisfied and how the statistical analysis of variance is performed, we refer the reader to books on statistics, for instance, to Box, Hunter & Hunter (1993).

It may happen that only two cars are available or, if machines are involved, that only a limited number of these machines can be switched on at a time to avoid a breakdown of the power supply. Then we have to work with "Latin rectangles," or we use balanced incomplete block designs (see §36).

We let $N(n)$ denote the maximum number of MOLS of order n, with the convention that $N(1) = \infty$. Since a Latin square exists for every order, we have $N(n) \geq 1$. Exercise 6 demonstrates that $N(n) \leq n - 1$ for $n > 1$. The reader can verify $N(2) = 1$, $N(3) = 2$, and with the aid of a computer, $N(6) = 1$. It is also known that $N(10) \geq 2$ and that $N(n) \geq 3$ for all $n \notin \{2, 3, 6, 10\}$.

Theorem 35.4 provides us with a technique for producing MOLS of prime-power order. For composite orders, we can show, see Street & Street (1987), that

35.7. Theorem. *$N(nm) \geq \min(N(m), N(n))$ for $m, n > 1$.*

The Kronecker product of matrices can be used to construct MOLS for composite order n. Let $\mathbf{A}$ and $\mathbf{B}$ be Latin squares of orders m and n, respectively. Let a_{ij} in $\mathbf{A}$ denote the entry at row i and column j. Then the ***Kronecker product*** $\mathbf{A} \otimes \mathbf{B}$ is the $mn \times mn$ square given by

$$\mathbf{A} \otimes \mathbf{B} = \begin{pmatrix} a_{11}\mathbf{B} & \dots & a_{1m}\mathbf{B} \\ \vdots & & \vdots \\ a_{m1}\mathbf{B} & \dots & a_{mm}\mathbf{B} \end{pmatrix}.$$

35.8. Theorem. *If there is a pair of MOLS of order n and a pair of MOLS of order m, then there is a pair of MOLS of order mn.*

Proof. Let $\mathbf{A}_1$ and $\mathbf{A}_2$ be MOLS of order m, and $\mathbf{B}_1$ and $\mathbf{B}_2$ MOLS of order n. The $mn \times mn$ squares $\mathbf{A}_1 \otimes \mathbf{B}_1$ and $\mathbf{A}_2 \otimes \mathbf{B}_2$ are easily seen to be Latin squares of order mn. The orthogonality of $\mathbf{A}_1 \otimes \mathbf{B}_1$ and $\mathbf{A}_2 \otimes \mathbf{B}_2$ follows from the orthogonality of $\mathbf{A}_1$ and $\mathbf{A}_2$ and of $\mathbf{B}_1$ and $\mathbf{B}_2$. □

See Exercise 10 for a numerical example. In conclusion of this section we mention further examples of applications of Latin squares. We follow the forthcoming book by C. F. Laywine and G. L. Mullen entitled *Discrete Mathematics Using Latin Squares*.

35.9. Examples.

(i) ***Tennis tournament.***
Suppose h_i and w_i are husband and wife, and are one of $2k$ couples entered in a mixed-doubles tennis tournament. A round in the tournament consists of k matches so that every player plays once in every round. Then a tournament consists of $2k - 1$ rounds in which each player has every other player but his or her spouse as an opponent once; and each player has every player of the opposite sex other than his or her spouse as a partner exactly once.

Suppose such a schedule is feasible. Then consider two $2k \times 2k$ arrays, $\mathbf{L} = (l_{im})$ and $\mathbf{M} = (m_{ij})$, such that l_{ij}, $i \neq j$, specifies h_i's partner in the match in which he opposes h_j, and m_{ij} gives the round in which this match takes place. Additionally, let $m_{ii} = 2k$ and $l_{ii} = i$ for $i = 1, \dots, 2k$.

It follows that $\mathbf{L}$ and $\mathbf{M}$ are Latin squares and that $\mathbf{M}$ is symmetric since m_{ij} and m_{ji} both specify the round in which h_i and h_j are opponents and both $\mathbf{L}$ and $\mathbf{L}^{\mathrm{T}}$ are orthogonal to $\mathbf{M}$.

For $i \neq j$, l_{ij} and l_{ji} are female opponents who meet exactly once in round m_{ij}. For $i = j$, $l_{ij} = l_{ji} = i$, so that $\mathbf{L}$ and $\mathbf{L}^T$ are othogonal. Then a solution to the scheduling of a spouse-avoiding mixed-doubles tournament for $2k$ couples would be given by $\mathbf{L}$, $\mathbf{L}^T$, and $\mathbf{M}$, three mutually orthogonal Latin squares of order $2k$ where the second is the transpose of the first, and the third is symmetric.

In the case of $2k+1$ couples the ith couple is designated to sit out round i and $\mathbf{M}$ is modified so that $m_{ii} = i$. Then a similar solution is constructed using orthogonal Latin squares whereby $2k$ couples play in each round.

(ii) ***Golf tournament for trios.***
Suppose $3n$ players wish to play golf for n weeks. There are n starting times available in the morning before the Sun gets too hot and so the golfers play in threesomes (groups of three) so that:

- Each golfer plays exactly once a week;
- No golfer has the same tee time twice;
- No two golfers play in the same threesome more than once.

Using MOLS on order n, such an arrangement is easily described. Let $\mathbf{L}_1, \mathbf{L}_2, \mathbf{L}_3$ be three MOLS of order n where the set of symbols for $\mathbf{L}_i$ is taken from the symbol set P_i, $i = 1, 2, 3$. If we superimpose the three squares, the triples that lie in each of the n^2 cells are the threesomes, the columns correspond to the weeks of the tournament, and the rows correspond to the n starting times. The reader should verify that such a tournament cannot be constructed for $n = 2$ or $n = 3$. Whether it can be done for $n = 10$ is unknown, but it can be done for $n = 6$.

A surprising connection between Latin squares and error-correcting codes can be established as follows. Let $\mathbf{L}_1, \ldots, \mathbf{L}_m$ be a collection of mutually orthogonal $q \times q$ Latin squares over $\mathbb{F}_q$, and let $\mathbf{L}_k = (l_{ij}^{(k)})$ for $1 \leq k \leq m$. By Exercise 6, we have $m \leq q-1$. For simplicity of notation, we write $\mathbb{F}_q = \{0, 1, \ldots, q-1\}$. Now we form the code $C(\mathbf{L}_1, \ldots, \mathbf{L}_m)$ consisting of all words $(i, j, l_{ij}^{(1)}, \ldots, l_{ij}^{(m)})$ with $1 \leq i, j \leq q$.

35.10. Example. We take $q = 4$ and the $m = 3$ mutually orthogonal Latin squares of Example 35.5, in which we replace ζ by 2 and ζ^2 by 3. So

$$\mathbf{L}_1 = \begin{pmatrix} 0 & 1 & 2 & 3 \\ 1 & 0 & 3 & 2 \\ 2 & 3 & 0 & 1 \\ 3 & 2 & 1 & 0 \end{pmatrix}, \quad \mathbf{L}_2 = \begin{pmatrix} 0 & 1 & 2 & 3 \\ 2 & 3 & 0 & 1 \\ 3 & 2 & 1 & 0 \\ 1 & 0 & 3 & 2 \end{pmatrix}, \quad \mathbf{L}_3 = \begin{pmatrix} 0 & 1 & 2 & 3 \\ 3 & 2 & 1 & 0 \\ 1 & 0 & 3 & 2 \\ 2 & 3 & 0 & 1 \end{pmatrix}.$$

Then $C(\mathbf{L}_1, \mathbf{L}_2, \mathbf{L}_3)$ is given by the 16 words $(i, j, l_{ij}^{(1)}, l_{ij}^{(2)}, l_{ij}^{(3)})$ with $1 \le i, j \le 4$, which we briefly write without parentheses and commas as in Chapter 4:

00000	10123	20231	30312
01111	11032	21320	31203
02222	12301	22013	32130
03333	13210	23102	33021

If two words start with the same two entries i, j, they clearly must be the same. If the first one starts with $(i, j, \ldots)$ and the second one with $(i, j', \ldots)$, where $j \neq j'$, then they cannot coincide in one of the remaining entries $a_{ij}^{(k)}$ and $a_{ij'}^{(k)}$, since $\mathbf{L}_k$ is a Latin square. The same holds if they start with $(i, j, \ldots)$ and $(i', j, \ldots)$ with $i \neq i'$. Finally, if two words start with (i, j) and (i', j') with $i \neq i'$, $j \neq j'$, they cannot coincide in two (or more) places k, k', since $\mathbf{L}_k$ and $\mathbf{L}_{k'}$ are orthogonal. This shows

35.11. Theorem. *If $\mathbf{L}_1, \ldots, \mathbf{L}_m$ are mutually orthogonal Latin squares of size $q \times q$ over $\mathbb{F}_q$, then $C(\mathbf{L}_1, \ldots, \mathbf{L}_m)$ is a code over $\mathbb{F}_q$ which consists of q^2 codewords of length $m + 2$ and with minimum distance $m + 1$.*

So the code in Example 35.10 has 16 codewords of length 5 and minimum distance 4. Some other good codes can be obtained from $C(\mathbf{L}_1, \ldots, \mathbf{L}_m)$ by taking only those codewords which have the same kth entry (and deleting this kth entry). See Denes and Keedwell (1974, Chap. 10).

In particular, $m = 1$ gives us a q-ary code of q^2 codewords of length 3 with minimum distance 2; this code is single error detecting and cannot correct any errors, but has the following nice interpretation.

We still write $\mathbb{F}_q = \{0, 1, \ldots, q-1\}$ and consider $\mathbb{F}_q \times \mathbb{F}_q$ as "general chess board." Suppose a rook is placed on one of the squares $(i, j) \in \mathbb{F}_q^2$. Then its "rook domain" is given by all $(k, l) \in \mathbb{F}_q^2$ which differ from (i, j) in at most one place:

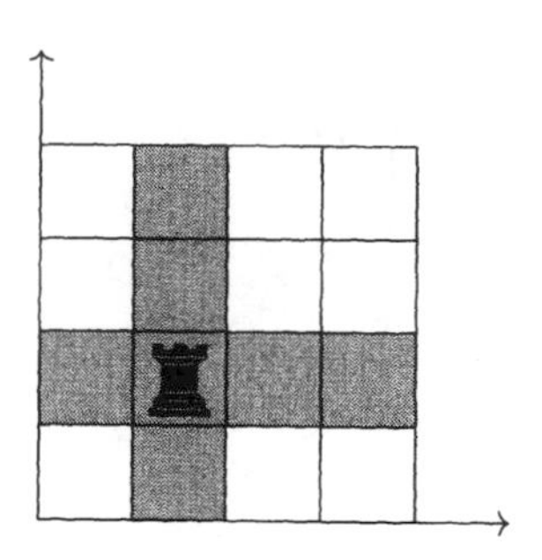

In general, a ***rook domain*** in $\mathbb{F}_q^n$ determined by $(f_1, \dots, f_n)$ is the set of all $(g_1, \dots, g_n) \in \mathbb{F}_q^n$ which differ from $(f_1, \dots, f_n)$ in at most one place.

A Latin square L determines a single-error-detecting code $C(L)$, as we have seen above. Since $d_{\min}C(\mathbf{L}) = 2$, we can place q^2 mutually non-attacking rooks on a three-dimensional chess board $\mathbb{F}_q^3$. So we see

35.12. Theorem. *The following entities determine each other:*

(i) *a Latin square of order q;*

(ii) *a set of q^2 mutually nonattacking rooks on the chess board $\mathbb{F}_q^3$;*

(iii) *a single-error-detecting code of length* 3 *over $\mathbb{F}_q$, consisting of q^2 codewords.*

This connection can be generalized to $C(\mathbf{L}_1, \dots, \mathbf{L}_m)$; see Denes & Keedwell (1974), which also offers other connections between Latin squares and codes.

Exercises

1. Give a Latin square of order 6.
2. Find two orthogonal Latin squares of order 7.
3. An $r \times n$ Latin rectangle is an $r \times n$ array made out of the integers $\{1, 2, \dots, n\}$ such that no integer is repeated in any row or in any column. Give an example of a 3×5 Latin rectangle with $r \leq n$, and extend it to a Latin square of order 5. Can this Latin rectangle be extended to a Latin square?
4. Let $\mathbf{L}^{(k)} = (a_{ij}^{(k)})$, where $a_{ij}^{(k)} = i + jk \pmod 9$. Which of the $\mathbf{L}^{(k)}$, $1 \leq k \leq 8$, are Latin squares? Are $\mathbf{L}^{(2)}$ and $\mathbf{L}^{(5)}$ orthogonal?

5. A Latin square over $\{1, \ldots, n\}$ is called ***normalized*** if in the first row and column the integers are in the usual order. How many normalized Latin squares of order $n \leq 5$ are there?

6. Show that for $n \geq 2$ there can be at most $n - 1$ mutually orthogonal Latin squares of order n.

7. A ***magic square*** of order n consists of the integers 1 to n^2 arranged in an $n \times n$ array such that the sums of entries in rows, columns, and diagonals are all the same. Let $\mathbf{A} = (a_{ij})$ and $\mathbf{B} = (b_{ij})$ be two orthogonal Latin squares of order n with entries in $\{0, 1, \ldots, n-1\}$ such that the sum of entries in each of the diagonals of $\mathbf{A}$ and $\mathbf{B}$ is $n(n-1)/2$. Show that $\mathbf{M} = (na_{ij} + b_{ij} + 1)$ is a magic square of order n. Construct a magic square of order 4 from two orthogonal Latin squares.

8. De la Loubère's method for constructing magic squares of odd order is as follows: Place a 1 in the middle square of the top row. The successive integers are then placed in their natural order in a diagonal line which slopes upward to the right with the following modifications.

   (i) When the top row is reached, the next integer is placed in the square in the bottom row as if it came immediately above the top row.

   (ii) When the right-hand column is reached, the next integer is placed in the left-hand column as if it immediately succeeded the right-hand column.

   (iii) When a square already contains a number or when the top right-hand square is reached, the next integer is placed vertically below it in the next row.

   Illustrate this construction for a 5×5 array.

9. (i) Solve the following special case of the "Kirkman Schoolgirl Problem": A schoolmistress took her nine girls for a daily walk, the girls arranged in rows of three girls. Plan the walk for four consecutive days so that no girl walks with any of her classmates in any triplet more than once.

   (ii) Try to solve the same problem for 15 girls and five rows of three girls for seven consecutive days.

10. Take a pair of MOLS of order 3 and a pair of MOLS of order 4 and use the Kronecker product to construct a pair of MOLS of order 12.

11. A ***spouse avoiding mixed doubles round robin tournament for*** n ***couples*** is a tournament played in tennis so that:

    - A match consists of two players of different genders playing against two other players of different genders;
    - Every pair of players of the same gender plays against each other exactly once;
    - Every player plays with each player of the opposite gender (except their spouse) exactly once as a partner and exactly once as an opponent.

    Use two self-orthogonal Latin squares to construct such a tournament. (A Latin square $\mathbf{L}$ is ***self-orthogonal*** if $\mathbf{L}$ and its transpose are orthogonal.)

12. Determine $C(\mathbf{L}_1)$, $C(\mathbf{L}_2)$, and $C(\mathbf{L}_1, \mathbf{L}_3)$ in Example 35.10.

Notes

Good books on Latin squares are, e.g., Biggs (1985), Denes & Keedwell (1974), Denes & Keedwell (1991), and van Lint & Wilson (1992). The state of the art, along with many useful tables, can be found in Colbourn & Dinitz (1996). The applications of Latin squares go deep into Combinatorics. Denes & Keedwell (1991) describe further applications of Latin squares to areas such as Satellite-Switch Time Division Multiple Access Systems (where a satellite should observe different geographical zones in contact with receiving stations on the ground on a time-sharing basis), synchronized "alarm codes", pseudorandom permutations, systems to share secrets (see 11.34), authentication schemes in public key cryptosystems (see §§22–23), and to ***local permutation polynomials*** $f \in \mathbb{F}_q[x, y]$; they are defined by the property that for all $a, b \in \mathbb{F}_q$, $f(a, x)$ and $f(x, b)$ are permutation polynomials (see 21.12). The number of local permutation polynomials can be shown to be equal to the number of Latin squares of order q. An amazing picture on the impossibility of a pair of orthogonal Latin squares of order 6 was the "logo" of a conference on combinatorics in Gaeta, Italy (1990):

14 20 41 05 32 53
30 15 04 51 23 42
21 03 12 40 54 35
02 31 50 13 45 24
43 52 25 34 10 01
55 44 33 22

1 1 0 0

Up to the last two pairs, everything works fine; just the last entries won't fit!

§36 Block Designs

R. A. Fisher in his book *The Design of Experiments* was the first to indicate how experiments can be organized systematically so that statistical analysis can be applied. In the planning of experiments, it often occurs that results are influenced by phenomena outside the control of the experimenter. We have met this situation already in §35 on Latin squares. As we have seen in Example 35.6, sometimes an "incomplete" testing is desirable or economically necessary. For instance, in wine tasting, the number of wines a tester can test has its natural bounds. The first applications of "incomplete" experimental design methods were in the design of agricultural and biological experiments. Suppose v types of fertilizers have to be tested on b different plants. In ideal circumstances, we would test each plant with respect to each fertilizer, i.e., we would have b plots (blocks) each of size v. However, this often is not economical, and so we try to design experiments such that each plant is tested with k fertilizers and any two fertilizers together are applied on the same plant λ times, so that each two fertilizers are tested λ times in a "direct competition." Such an experiment is called "balanced." It is a "$2-(v, k, \lambda)$ design" with b blocks, and is called "incomplete" if $k < v$.

36.1. Definition. A pair $(P, \mathcal{B})$ is called a ***balanced block design*** with parameters (v, b, r, k, λ) if it has the following properties:

(i) $P = \{p_1, \dots, p_v\}$ is a set with v elements (called ***points***);

(ii) $\mathcal{B} = \{B_1, \dots, B_b\}$ is a subset of $\mathcal{P}(P)$ consisting of b ***blocks***;

(iii) Each B_i has exactly k elements, where $k < v$;

(iv) Each point belongs to precisely r blocks;

(v) Each combination of two different elements of P occurs in exactly λ blocks.

Such a design is also called a (v, b, r, k, λ) ***configuration*** or $2-(v, k, \lambda)$ ***design***. The term "balanced" indicates that each pair of elements occurs in exactly the same number of blocks. A design is ***symmetric*** if $v = b$.

We follow a convention in combinatorics and call a set with k elements a k-***set***.

36.2. Example. Let P be a v-set, and take the collection $\mathcal{P}_k P$ of all k-subsets of P as blocks, $2 \leq k \leq v$. Then $b = \binom{v}{k}$ and $\lambda = \binom{v-2}{k-2}$ (fill up every 2-set by $k-2$ elements to get a k-set). In general, this design is not symmetric.

It is much harder to get designs which are less "trivial" than the ones in 36.2. If $b < \binom{v}{k}$, not all k-subsets are blocks and we then call the design ***incomplete*** and speak of a ***balanced incomplete block (BIB-)design***. We now present an cxample of a $(7, 7, 3, 3, 1)$ configuration "out of the blue" and give a systematic construction method later on.

36.3. Example. Let $\{1, 2, 3, 4, 5, 6, 7\}$ be the set of points (varieties) and $\{\{1, 2, 3\}, \{3, 4, 5\}, \{1, 5, 6\}, \{1, 4, 7\}, \{3, 6, 7\}, \{2, 5, 7\}, \{2, 4, 6\}\}$ be the set of blocks. Then we obtain a $(7, 7, 3, 3, 1)$ block design. This is the so-called ***Fano geometry*** (Fano plane), the simplest example of a finite "projective plane" over $\mathbb{F}_2$. The elements are the seven points of the plane, the blocks are the sets of points on lines (including the "inner circle").

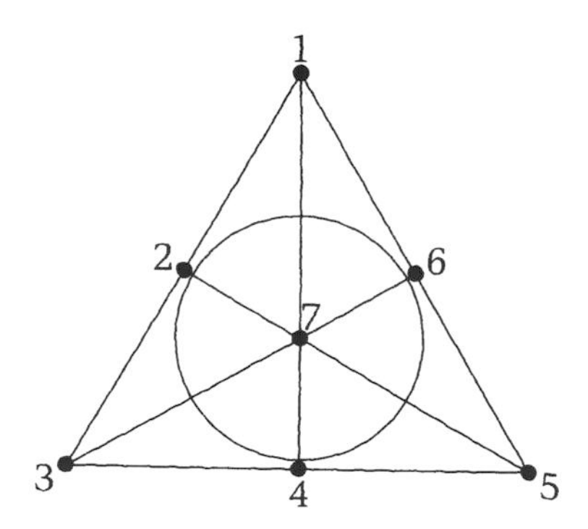

36.4. Definition. The ***incidence matrix*** of a (v, b, r, k, λ) configuration is the $v \times b$ matrix $\mathbf{A} = (a_{ij})$, where

$$a_{ij} = \begin{cases} 1 & \text{if } p_i \in B_j, \\ 0 & \text{otherwise.} \end{cases}$$

Note that designs and incidence matrices determine each other uniquely.

36.5. Example. Take $v = 5$ and $k = 3$ in Example 36.2, so the design consists of all 3-subsets of a 5-set. There are $b = \binom{5}{3} = 10$ blocks, and $\lambda = \binom{3}{1} = 3$. We take the points $p_1 = 1, \ldots, p_5 = 5$ and choose the following numbering of the blocks:

$$\begin{aligned} B_1 &= \{1, 2, 3\}, & B_6 &= \{1, 4, 5\}, \\ B_2 &= \{1, 2, 4\}, & B_7 &= \{2, 3, 4\}, \\ B_3 &= \{1, 2, 5\}, & B_8 &= \{2, 3, 5\}, \\ B_4 &= \{1, 3, 4\}, & B_9 &= \{2, 4, 5\}, \\ B_5 &= \{1, 3, 5\}, & B_{10} &= \{3, 4, 5\}. \end{aligned}$$

This yields the 5×10 incidence matrix

$$\mathbf{A} = \begin{pmatrix} 1 & 1 & 1 & 1 & 1 & 1 & 0 & 0 & 0 & 0 \\ 1 & 1 & 1 & 0 & 0 & 0 & 1 & 1 & 1 & 0 \\ 1 & 0 & 0 & 1 & 1 & 0 & 1 & 1 & 0 & 1 \\ 0 & 1 & 0 & 1 & 0 & 1 & 1 & 0 & 1 & 1 \\ 0 & 0 & 1 & 0 & 1 & 1 & 0 & 1 & 1 & 1 \end{pmatrix}.$$

Observe that each row has $r = 6$ ones and each column has $k = 3$ ones. Each pair of points is in precisely $\lambda = 3$ blocks.

We summarize some properties of incidence matrices.

36.6. Theorem. *Let* $\mathbf{A}$ *be the incidence matrix of a* (v, b, r, k, λ) *configuration. Then:*

(i) $\mathbf{A}\mathbf{A}^{\mathrm{T}} = (r - \lambda)\mathbf{I} + \lambda\mathbf{J}$;

(ii) $\det(\mathbf{A}\mathbf{A}^{\mathrm{T}}) = \big(r + (v-1)\lambda\big)(r - \lambda)^{v-1}$;

(iii) $(\mathbf{A}\mathbf{A}^{\mathrm{T}})^{-1} = \frac{1}{r-\lambda}\mathbf{I} - \frac{\lambda}{(r-\lambda)(r-\lambda+\lambda v)}\mathbf{J}$;

where $\mathbf{J}$ *is the* $v \times v$ *matrix with all entries equal to* 1 *and* $\mathbf{I}$ *is the* $v \times v$ *identity matrix.*

Proof. We prove (ii) and leave the other parts as Exercise 1 to the reader. If we subtract the first column from all other columns and then add the second, third,..., vth row to the first, then all elements above the main diagonal are zero. On the main diagonal, the first entry is $r + (v-1)\lambda$, and the others are $r - \lambda$. □

36.7. Theorem. *The following conditions are necessary for the existence of a BIB-design with parameters* v, b, r, k, λ*:*

(i) $bk = vr$;

(ii) $r(k-1) = \lambda(v-1)$;

(iii) $b \geq v$ (***Fisher's inequality***).

Proof. (i) $bk = vr$ is the total number of ones in the incidence matrix (counted by rows and by columns). See Example 36.5.

(ii) This follows by a similar argument.

(iii) Since $r > \lambda$, using (ii) and 36.6(ii) shows that $\mathbf{AA}^{\mathrm{T}}$ has rank v. On the other hand, since the rank of $\mathbf{A}$ can be at most b, and since $\mathrm{Rk}\,\mathbf{A} \geq \mathrm{Rk}\,\mathbf{AA}^{\mathrm{T}} = v$, we have $b \geq v$. □

There are many ways to construct designs. We use a special type of algebraic structure, near-rings, which yield very "efficient" designs that are at the same time easy to construct.

36.8. Definition. A set N together with two operations $+$ and $\circ$ is a ***near-ring*** provided that:

(i) $(N, +)$ is a group (not necessarily abelian);

(ii) $(N, \circ)$ is a semigroup;

(iii) $(n + n') \circ n'' = n \circ n'' + n' \circ n''$ holds for all $n, n', n'' \in N$.

Of course, every ring is a near-ring. Hence near-rings are generalized rings. Two ring axioms are missing: the commutativity of addition and (much more important) the other distributive law.

36.9. Example. The set $\mathbb{R}^{\mathbb{R}}$ of all real functions, together with pointwise addition and composition of functions, forms a near-ring. Observe that $(f+g) \circ h = f \circ h + g \circ h$ holds trivially due to the definition of addition, while $f \circ (g+h) = f \circ g + f \circ h$ holds for all $g, h \in \mathbb{R}^{\mathbb{R}}$ if and only if f is linear, i.e., $f(x+y) = f(x) + f(y)$ for all $x, y \in \mathbb{R}$. This and other instances show that near-rings form the "nonlinear counterpart to ring theory." This will be considered again in §39. More generally, take a group $(G, +)$; then $(G^G, +, \circ)$ is a near-ring, and every near-ring can be shown to be "embeddable" in some $(G^G, +, \circ)$. More examples of near-rings are given by $(R[x], +, \circ)$, where R is a commutative ring with identity, by the continuous real functions, the differentiable ones, by polynomial functions, etc.(see Exercises 2 and 3).

However, we need other types of near-rings for our purposes.

36.10. Definition. Let N be a near-ring and $n_1, n_2 \in N$.

$$n_1 \equiv n_2 :\Longleftrightarrow n \circ n_1 = n \circ n_2 \text{ for all } n \in N.$$

The relation $\equiv$ is an equivalence relation on N.

36.11. Definition. A near-ring N is called ***planar*** if there are at least three equivalence classes in N w.r.t $\equiv$ and if all equations $x \circ a = x \circ b + c$ $(a, b, c \in N,\ a \not\equiv b)$ have exactly one solution $x \in N$.

This condition is motivated by geometry. It implies that two "non-parallel lines $y = x \circ a + c_1$ and $y = x \circ b + c_2$ have exactly one point of intersection." One example (see Exercise 4) is $(\mathbb{C}, +, *)$, where $a*b := a|b|$. We need finite examples, however. There are various methods to construct planar near-rings (see Clay (1992), Pilz (1983), or Pilz (1996)). We present the easiest one.

36.12. Construction method for finite planar near-rings. Take a finite field $\mathbb{F}_q$ and choose a generator g for its multiplicative group $(\mathbb{F}_q^*, \cdot)$. Choose a nontrivial factorization $q-1 = st$. Define a new multiplication $*_t$ in $\mathbb{F}_q$ as

$$g^a *_t g^b := g^{a+b-[b]_s},$$

where $[a]_s$ denotes the remainder of a on division by s; if one factor is 0, the product should be 0. Then $(\mathbb{F}_q, +, *_t)$ is a planar near-ring.

36.13. Example. We choose $\mathbb{F}_7$, and $t = 3$ as divisor of $7 - 1$, so $g = 3$ serves well as a generator. From this, we get the multiplication table

$*_3$	0	1	2	3	4	5	6
0	0	0	0	0	0	0	0
1	0	1	2	1	4	4	2
2	0	2	4	2	1	1	4
3	0	3	6	3	5	5	6
4	0	4	1	4	2	2	1
5	0	5	3	5	6	6	3
6	0	6	5	6	3	3	5

The equivalence classes w.r.t. $\equiv$ can be seen immediately from the table, since $n_1 \equiv n_2$ iff the columns of n_1, n_2 in the multiplication table coincide. They are

$$\{0\},\ \{1,3\},\ \{2,6\},\ \{4,5\}.$$

We are going to see now that planar near-rings produce BIB-designs with excellent "efficiency" $E := \lambda v/rk$. This E is a number between 0 and 1, and it estimates, for the statistical analysis of experiments, the "quality" of the design; designs with efficiency $E \geq 0.75$ are usually considered to be "good" ones.

We get BIB-designs from finite planar near-rings $(N, +, \circ)$ in the following way. We take N as the set of points and the blocks as $a \circ N^* + b$ (with $a, b \in N$, $a \not\equiv 0$).

36.14. Theorem. *$(N, \{a \circ N^* + b \mid a, b \in N,\ a \not\equiv 0\})$ is a BIB-design with parameters*

$$v = |N|, \qquad b = \frac{v(v-1)}{k}, \qquad r = v - 1,\ k,\ \lambda = k - 1,$$

where k is the cardinality of each $a \circ N^*$ with $a \not\equiv 0$. As $k \to \infty$, the efficiency $E \to 1$.

36.15. Corollary. *If N is constructed via 36.12 (with the parameters mentioned there), then $k = t$, so we get a design with parameters $(p^n, p^n s, p^n - 1, t, t-1)$. This is a BIB-design iff $t \neq 2$, and it is never symmetric.*

36.16. Example. We consider the problem of testing combinations of 3 out of seven fertilizers such that each fertilizer is applied on six fields. Definition 36.1 tells us that we have to look for a BIB-design with parameters $(7, b, 6, 3, \lambda)$. By 36.7, we get $b \cdot 3 = 7 \cdot 6$, so $b = 14$, and $6 \cdot 2 = \lambda \cdot 6$, so $\lambda = 2$ (all under the assumption that there actually exists a design with these parameters). Hence we are searching for a planar near-ring of order 7, and the near-ring N of 36.13 comes to mind.

We construct the blocks $B_i = a *_3 N^* + b$, $a \not\equiv 0$, and abbreviate $a *_3 b$ by ab:

$$\begin{aligned}
1N^* + 0 &= \{1,2,4\} =: B_1, & 3N^* + 0 &= \{3,5,6\} =: B_8,\\
1N^* + 1 &= \{2,3,5\} =: B_2, & 3N^* + 1 &= \{4,6,0\} =: B_9,\\
1N^* + 2 &= \{3,4,6\} =: B_3, & 3N^* + 2 &= \{5,0,1\} =: B_{10},\\
1N^* + 3 &= \{4,5,0\} =: B_4, & 3N^* + 3 &= \{6,1,2\} =: B_{11},
\end{aligned}$$

$$1N^* + 4 = \{5, 6, 1\} =: B_5, \qquad 3N^* + 4 = \{0, 2, 3\} =: B_{12},$$
$$1N^* + 5 = \{6, 0, 2\} =: B_6, \qquad 3N^* + 5 = \{1, 3, 4\} =: B_{13},$$
$$1N^* + 6 = \{0, 1, 3\} =: B_7, \qquad 3N^* + 6 = \{2, 4, 5\} =: B_{14}.$$

By Theorem 36.14, we get a BIB-design with:

(i) $v = 7$ points (namely $0, 1, 2, 3, 4, 5, 6$);

(ii) $b = 14$ blocks;

(iii) Each point lies in exactly $r = 6$ blocks;

(iv) Each block contains precisely $k = 3$ elements;

(v) Every pair of different points appears in $\lambda = 2$ blocks.

In order to solve our fertilizer problem, we divide the whole experimental area into 14 experimental fields, which we number by $1, 2, \ldots, 14$. Then apply precisely the fertilizers F_i to a field k if $i \in B_k$, $i = 1, 2, \ldots, 6$. The last row indicates the yields on the experimental fields after performing the experiment:

Field / Fertilizer	1	2	3	4	5	6	7	8	9	10	11	12	13	14
F_0				x		x	x		x	x		x		
F_1	x				x		x			x	x		x	
F_2	x	x				x					x	x		x
F_3		x	x				x	x				x	x	
F_4	x		x	x					x				x	x
F_5		x		x	x			x		x				x
F_6			x		x	x		x	x		x			
Yields:	2.8	6.1	1.2	8.5	8.0	3.9	7.7	5.5	4.0	10.7	3.2	4.9	3.8	5.5

Then we get: every field contains exactly three fertilizers, and every fertilizer is applied to six fields. Finally, every pair of different fertilizers is applied precisely twice in "direct competition." The efficiency E of this BIB-design is $E = \lambda v / rk = 0.78$.

The construction method 36.12 was discovered (in a much more general form) by J. R. Clay and G. Ferrero in the 1970s. See Clay (1992) for an excellent account on this subject. Shortly, we shall discuss another construction method via "difference sets," but before that we take a look at the statistical evaluation of those experiments which use designs like the one above.

36.17. Statistical Analysis of experiments using designs:

(i) Form the incidence matrix $\mathbf{A}$ as in 36.4. Let $\mathbf{y}$ be the ***yield vector*** $(y_1, \dots, y_b)$, where y_i is the yield on the ith experimental field.

(ii) Compute $(\mathbf{A}\mathbf{A}^{\mathrm{T}})^{-1}\mathbf{A}\mathbf{y}^{\mathrm{T}} =: (\beta_0, \dots, \beta_{v-1})^{\mathrm{T}}$. Then linear algebra tells us that $\beta_0, \dots, \beta_{v-1}$ are the best estimates for the effects of the ingredients $I_0, \dots, I_{v-1}$ in the sense that they minimize the sums of the squares of the errors. $\beta_0, \dots, \beta_{v-1}$ are also the maximum likelihood estimators. Observe that by 36.6(iii), we don't really have to invert $\mathbf{A}\mathbf{A}^{\mathrm{T}}$.

(iii) More precisely, statistics tells us how to compute approximate confidence intervals $[\underline{\beta}_i, \overline{\beta}_i]$ for β_i to a significance level of $1-\alpha$ (usually, $1-\alpha = 0.9$ or $= 0.95$ or $= 0.99$, corresponding to a significance level of 90% (95%, 99%, respectively)): $\underline{\beta}_i = \beta_i - \gamma_i$, $\overline{\beta}_i = \beta_i + \gamma_i$, where

$$\gamma_i = z_{1-\alpha/2}\sigma\sqrt{d_{ii}}.$$

In here, $z_{1-\alpha}$ is the $(1-\alpha)$-quantile of the normal distribution, $\sigma^2 = \frac{\|\mathbf{y}-\mathbf{A}^{\mathrm{T}}\beta\|^2}{b-v}$, and d_{ii} = the ith diagonal element in $(\mathbf{A}\mathbf{A}^{\mathrm{T}})^{-1} = \frac{r-\lambda v-2\lambda}{(r-\lambda)(r-\lambda+\lambda v)}$, by 36.6(iii). Here are some values of $z_{1-\alpha/2}$:

$1-\alpha$	90%	95%	99%
$z_{1-\alpha/2}$	1.65	1.96	2.58

For small values of b the normal distribution should be replaced by the Student distribution.

36.18. Example (*36.16 continued*). The incidence matrix is basically the design scheme in 36.16 if we replace x by 1:

$$\begin{pmatrix}
0 & 0 & 0 & 1 & 0 & 1 & 1 & 0 & 1 & 1 & 0 & 1 & 0 & 0\\
1 & 0 & 0 & 0 & 1 & 0 & 1 & 0 & 0 & 1 & 1 & 0 & 1 & 0\\
1 & 1 & 0 & 0 & 0 & 1 & 0 & 0 & 0 & 0 & 1 & 1 & 0 & 1\\
0 & 1 & 1 & 0 & 0 & 0 & 1 & 1 & 0 & 0 & 0 & 1 & 1 & 0\\
1 & 0 & 1 & 1 & 0 & 0 & 0 & 0 & 1 & 0 & 0 & 0 & 1 & 1\\
0 & 1 & 0 & 1 & 1 & 0 & 0 & 1 & 0 & 1 & 0 & 0 & 0 & 1\\
0 & 0 & 1 & 0 & 1 & 1 & 0 & 1 & 1 & 0 & 1 & 0 & 0 & 0
\end{pmatrix}.$$

For $\mathbf{y}$, we get $\mathbf{y} = (2.8, 6.1, 1.2, 8.5, 8.0, 3.9, 7.7, 5.5, 4.0, 10.7, 3.2, 4.9, 3.8, 5.5)$. Then $(\mathbf{A}\mathbf{A}^{\mathrm{T}})^{-1}\mathbf{A}\mathbf{y}^{\mathrm{T}} = (3.6, 2.7, 0.3, 1, 0.1, 4.8, 0.1)^{\mathrm{T}}$, and we get confidence intervals for $\alpha = 95\%$ around these values for the contributions of the ingredients $I_0, \ldots, I_6$ (i.e., fertilizers $F_0, \ldots, F_6$) as contribution of

$$\begin{array}{ll} F_0 \in [3.4, 3.8], & F_4 \in [-0.1, 0.3], \\ F_1 \in [2.5, 2.9], & F_5 \in [4.6, 5.0], \\ F_2 \in [0.1, 0.5], & F_6 \in [-0.1, 0.3]. \\ F_3 \in [0.8, 1.2], & \end{array}$$

Hence, the contributions of F_4 and F_6 are doubtful, they might even be counterproductive. We also might exclude F_2, since it costs money, but does not achieve a lot. If we take F_0, F_1, F_2, F_3, F_5, we can expect a yield of $3.6 + 2.7 + 0.3 + 1.0 + 4.8 = 12.4$, which is considerably better than the yields y_i which we got in our experiment.

In this experiment, each fertilizer was applied on six fields. If we had applied single fertilizers, each six times, we would have had to use $7 \cdot 6 = 42$ experiments, three times more than in our case.

As mentioned in 35.6, statistical experiments using BIB-designs can also handle the case when there are dependencies among the "ingredients" by simply regarding these dependencies as "new ingredients". Again, it is a good idea to randomize the design before starting the experiment in order to avoid systematic errors. Much more on the statistical analysis of experiments based on Latin squares and BIB-designs can be found in Box, Hunter & Hunter (1993). Let us now generalize the definition of a design and look for another construction method.

36.19. Definition. Let A be a v-set, $\mathcal{D}$ a collection of k-subsets of A, and $t, \lambda \in \mathbb{N}$. The pair $(A, \mathcal{D})$ is called a ***t-(v, k, λ)-design***, or briefly a ***t-design***, if any t-subset of A is part of precisely λ sets of $\mathcal{D}$.

The Fano geometry in 36.3 is a $2-(7, 3, 1)$ design. Any $2-(v, k, \lambda)$ design is a block design.

36.20. Definition. A ***Steiner system*** is a t-design with $\lambda = 1$, denoted by $S(t, k, v)$. Note the new order of the parameters.

A ***projective geometry of order*** n is a Steiner system $S(2, n+1, n^2+n+1)$ with $n \geq 2$.

An ***affine geometry of order*** n is a Steiner system $S(2, n, n^2)$ with $n \geq 2$.

The Fano geometry is a Steiner system $S(2,3,7)$, hence a projective geometry of order 2. Example 36.16 is a $2-(7,3,2)$ design, but not a Steiner system, because $\lambda \neq 1$.

36.21. Example. We consider the Fano geometry as $S(2,3,7)$. For $v = 1+q+q^2$ and $k = 1+q$, we obtain a projective geometry, denoted by $\mathrm{PG}(2,q)$. This is a finite projective geometry of order q, q a prime power.

The word "geometry" in 36.20 comes from the following connection. Consider the points of a Steiner system $S(t,k,v)$ as points in a "plane" and the "blocks" of $\mathcal{D}$ as "lines." Then all these lines consist of precisely k of the v points. If $t = 2$, each two different points determine a unique line through them, and since each $S(2,k,v)$ is a block design, the relations $vr = bk$ and $r(k-1) = \lambda(v-1) = v-1$ of 36.7 are still valid. If $k = n+1$ and $v = n^2+n+1$ this forces $b = n^2+n+1 = v$, so we have a symmetric design, and we have the same number of lines as we have points. This and other results tell that, in a projective geometry, points and lines "have the same rights and equal opportunities." This does not hold in affine geometries. For more on these fascinating topics see e.g., Beth, Jungnickel & Lenz (1985) or Batten (1993). In the example of the Fano geometry we see that we have

1. seven points and seven lines;
2. each point is on three lines, and each line consists of three points;
3. two different lines always intersect in precisely one point, and two different points determine precisely one line.

We could have used the Fano plane for the design of statistical experiments as well. Let us denote the lines (blocks) by $B_1, \dots, B_7$ and replace each point by the blocks it belongs to. Then we see that we have to give the ingredients I_4, I_5, I_6 to the center point (i.e., field), and so on, and a line connects precisely those points with a common ingredient.

For a long time, it was an open problem if there exists a projective geometry of order 10 (see page 260 in the first edition of this book). However (see Lam, Thiel & Swiercz (1989)), this problem was solved in the negative by the use of a very large computer search and the connection to codes (see 36.25): we cannot find a geometry with 111 points and 111 lines where each line contains 11 points and each point lies on 11 lines.

We denote the points of the Fano geometry by the elements $0, 1, \ldots, 6$ of $\mathbb{Z}_7$; then the lines can be obtained by successively adding the elements of $\mathbb{Z}_7$. This approach can be generalized.

36.22. Definition. A subset $\{d_1, \ldots d_k\}$ of $\mathbb{Z}_v$ is called a ***difference set modulo*** v, or a (v, k, λ) ***difference set***, if the list of all differences $d_i - d_j$ in $(\mathbb{Z}_v, +)$, $i \neq j$, represents all the nonzero elements of $\mathbb{Z}_v$ with each difference occurring exactly λ times.

We have $k(k-1) = \lambda(v-1)$. There is a simple connection between designs and difference sets.

36.23. Theorem. *Let $\{d_1, \ldots, d_k\}$ be a (v, k, λ) difference set. Then $(\mathbb{Z}_v, \{B_0, \ldots, B_{v-1}\})$ is a symmetric (v, k, λ) design, where*

$$B_j := \{j + d_i \mid i = 1, \ldots, k\}, \quad j = 0, \ldots, v-1,$$

(computations are done in $(\mathbb{Z}_v, +)$).

Proof. An element $a \in \mathbb{Z}_v$ occurs precisely in those blocks with subscripts $a - d_1, \ldots, a - d_k$ modulo v, and thus each of the v points lies in the same number k of blocks. For a pair of distinct elements a, c, we have $a, c \in B_j$ if and only if $a = d_r + j$ and $c = d_s + j$ for some d_r, d_s. Therefore, $a - c = d_r - d_s$, and conversely, for every solution (d_r, d_s) of this, both a and c occur in the block with subscript $a - d_r$ modulo v. By hypothesis, there are exactly λ solutions (d_r, d_s) of this congruence and so all the conditions for a symmetric (v, k, λ) block design are satisfied. □

36.24. Example. $\{0, 1, 2, 4, 5, 8, 10\} \subseteq \mathbb{Z}_{15}$ forms a $(15, 7, 3)$ difference set. The blocks

$$B_j = \{j, j+1, j+2, j+4, j+5, j+8, j+10\}, \quad 0 \leq j \leq 14,$$

form a $(15, 7, 3)$ design, which can be interpreted as a projective geometry PG(3, 2). It consists of 15 lines and 15 points. Each of these lines can again be interpreted as a Fano geometry.

It is quite hard to construct t-designs, in particular Steiner systems, with $t \geq 3$. Recently a highly nontrivial 7-design was constructed, see Betten, Kerber, Kohnert, Laue & Wassermann (1996).

Let us go back to block designs and study their incidence matrices $\mathbf{A}$, like the one in 36.18. The rows in the matrix $\mathbf{A}$ are 0–1-sequences, so they can be taken as a binary code $C_{\mathbf{A}}$. This is a nonlinear code (the zero vector does not appear among the rows) which has amazing properties:

36.25. Theorem. *Let $(P, \mathcal{B})$ be a block design with parameters (v, b, r, k, λ) and let $C_{\mathbf{A}} =: C$ be the corresponding code. Then:*

(i) *C consists of v codewords, each having length b;*

(ii) *Each $c \in C$ has the constant weight r;*

(iii) *For each $c_1, c_2 \in C$, $c_1 \neq c_2$, we have $d(c_1, c_2) = 2(r - \lambda) = d_{\min}(C)$.*

Proof. (i) and (ii) are clear. If $c_1 \neq c_2$ are in C, they correspond to points $p_1, p_2 \in P$. Now $d(p_1, p_2)$ is the number of blocks which contain p_1, but not p_2, plus the number of blocks which contain p_2, but not p_1. This is the number of blocks containing $p_1 - 2 *$ (number of blocks containing $\{p_1, p_2\}) = r + r - 2\lambda = 2(r - \lambda)$. □

So all codewords have the same weight, and every two codewords differ in the same number of entries. $C_{\mathbf{A}}$ is hence called a ***constant weight*** and ***equal distance code***.

The complete design in 36.5 produces five codewords of length 10, with minimal distance $2(6 - 3) = 6$. The BIB-design in 36.16 gives seven codewords of length 14, weight 6, and minimal distance 8. Of course, for practical purposes, we need much bigger designs. Observe the large minimal distance of these codes; they are especially well suited for "high security purposes."

In fact, the minimal distances are very large compared with most other codes. Similar to the lines after 16.9, define $A(n, d, w)$ as the maximal number of codewords in a binary equal $(= w)$ weight code of length n and minimal distance d. An equal weight code C with these parameters is ***maximal*** if $|C| = A(n, d, w)$. It seems hard to find good formulas for $A(n, d, w)$, and to construct the corresponding codes. For a recent list of known values and upper and lower bounds on $A(n, d, w)$, see Brouwer, Shearer, Sloane & Smith (1990).

36.26. Theorem. *If $(P, \mathcal{B})$ is a block design with parameters (v, b, r, k, λ) and incidence matrix $\mathbf{A}$, then the code $C_{\mathbf{A}}$ is maximal and we get*

$$A(b, 2(r - \lambda), r) = \left\lfloor \frac{(r - \lambda)b}{r^2 - \lambda b} \right\rfloor.$$

Proof. Since $r(k - 1) = \lambda(v - 1)$, we get $r - \lambda = rk - \lambda v$. Hence $v(r^2 - rb + (r - \lambda)b) = vr^2 - \lambda vb = b(\frac{vr^2}{b} - \lambda v) = b(rk - \lambda v) = b(r - \lambda)$. The "Johnson bound" (see MacWilliams & Sloane (1977, p. 525, vol. 2) implies $A(n, d, w) \leq \left\lfloor \frac{dn}{2w^2 - 2wn + dn} \right\rfloor$. In our case we get $\left\lfloor \frac{b(r-\lambda)}{r^2 - rb + (r-\lambda)b} \right\rfloor$ as the upper bound. Since $v(r^2 - rb + (r - \lambda)b) = b(r - \lambda)$, we get

$$v \leq A(n, d, w) \leq \left\lfloor \frac{dn}{2w^2 - 2wn + dn} \right\rfloor = v,$$

and hence the result. □

Examples 36.5 and 36.16 give $A(10, 6, 6) = 5$ and $A(14, 8, 6) = 7$.

36.27. Corollary. *The construction method in* 36.12 *yields, for a prime* p *and* $p^n - 1 = st$,

$$A(p^n s, 2(p^n - t), p^n - 1) = p^n,$$

together with an easy method to construct these maximal codes.

Some of the considerations above can be done for general $t-(v, k, \lambda)$ designs. We can also take the columns of the incidence matrix of a block design, instead of the rows, to get another binary equal weight code. This code is not maximal, however, unless $\lambda = 1$, i.e., for $S(2, k, v)$ Steiner systems. The corresponding formula for $A(n, d, w)$, in the case when the design is constructed via 36.12, is then given by

$$A(p^n, 2(t - 1), t) = p^n s.$$

These formulas for $A(n, d, w)$ can be interpreted in yet another way. If we look back to Figure 16.3 on page 189, we see that a formula $A(n, d)$ gives the maximal number of nonintersecting spheres in $\mathbb{F}_2^n$ with radius $\lfloor \frac{d-1}{2} \rfloor$, thereby solving the "discrete sphere packing problem" in $\mathbb{F}_2^n$. The numbers $A(n, d, w)$ are an important tool for getting information about $A(n, d)$ (see MacWilliams & Sloane (1977, vol. 2)).

Exercises

1. Prove (i) and (iii) in Theorem 36.6.
2. Show that, for any group $(G, +)$, $(G^G, +, \circ)$ is a near-ring. When is it a ring?
3. Do the same for $(R[x], +, \circ)$, R a commutative ring with identity.
4. Show that $(\mathbb{C}, +, *)$ with $a * b := a|b|$ is a planar near-ring that is not a ring. What do the blocks of Theorem 36.14 look like?
5. Which multiplication table do you get in Example 36.13 with $*_2$ instead of $*_3$?

6. Construct the block design out of $(\mathbb{F}_7, +, *_2)$ (see the previous exercise).
7. Show that the following system of blocks from elements $1, 2, \ldots, 9$ is a design and determine the parameters v, b, r, k, λ:

123	147	159	168
456	258	267	249
789	369	348	357

8. Let $P_1, \ldots, P_t$ be distinct points in a t-(v, k, λ) design, let λ_i be the number of blocks containing $P_1, \ldots, P_i$, $1 \leq i \leq t$, and let $\lambda_0 = b$ be the total number of blocks. Prove that λ_i is independent of the choice of $P_1, \ldots, P_i$ and that
$$\lambda_i = \frac{\lambda\binom{v-i}{t-i}}{\binom{k-i}{t-i}} = \lambda\frac{(v-i)(v-i-1)\ldots(v-t+1)}{(k-i)(k-i-1)\ldots(k-t+1)}$$
for $0 \leq i \leq t$. Show that this implies that a t-(v, k, t) design is an i-(v, k, λ) design for $1 \leq i \leq t$. What is the meaning of λ_1?
9. Set up an "incidence table" for PG(2, 2) assuming that the point P_r and the line L_s are incident if $r + s \equiv 0$, 1, or 3 (mod 7). Does this geometry satisfy the following axioms?

   Axiom 1: Any two distinct points are incident with just one line.

   Axiom 2: Any two lines meet in at least one point.

   Axiom 3: There exist four points no three of which are collinear.

   Axiom 4: The three diagonal points of a quadrangle (intersect nonadjacent sides and the diagonals) are never collinear.
10. Show that $\{0, 4, 5, 7\}$ is a difference set of residues modulo 13 which yields a projective geometry PG(2, 3).
11. Verify that $\{0, 1, 2, 3, 5, 7, 12, 13, 16\}$ is a difference set of residues modulo 19. Determine the parameters v, k, and λ.
12. Decide whether $D = \{0, 1, 2, 32, 33, 12, 24, 29, 5, 26, 27, 22, 18\}$ is a difference set (mod 40). Determine the parameters v, k, and λ.
13. Compute $A(n, 2, w)$. Which formulas for $A(n, d, w)$ do we get from this and 36.27 for $n \leq 20$?

Notes

The methods presented here are by far not the only ways to construct BIB-designs, Steiner systems, etc.. A comprehensive work is Beth, Jungnickel & Lenz (1985). This book also contains extensive lists of designs and

Steiner systems. Hirschfeld (1979) and Batten (1993) are also good sources of information concerning the geometric aspects of block designs. The connections between codes and designs is nicely descibed in Cameron & van Lint (1991). The best reference to what can be done with near-rings is Clay (1992). Brouwer, Shearer, Sloane & Smith (1990) contains an updated list of values of $A(n, d)$ and $A(n, d, w)$, together with methods on how to construct these maximal codes. A very detailed account of projective geometries over finite fields is given in Hirschfeld (1979). We also recommend Cameron (1994), Colbourn & Dinitz (1996), and van Lint & Wilson (1992).

§37 Hadamard Matrices, Transforms, and Networks

Hadamard matrices are certain matrices with entries from $\{1, -1\}$. Because of the use of Hadamard transforms, these matrices are important in algebraic coding theory, information science, physics, and in pattern recognition. They are also useful in various problems for determining weights, voltage or resistance, concentration of chemicals, or frequencies of spectra.

37.1. Definition. A ***Hadamard matrix* H *of order*** n is an $n \times n$ matrix with entries in $\{1, -1\}$, such that

$$\mathbf{H}^{\mathrm{T}}\mathbf{H} = \mathbf{H}\mathbf{H}^{\mathrm{T}} = n\mathbf{I}.$$

This implies that the scalar product of any two distinct rows (or columns) of **H** is equal to 0, i.e., any two distinct rows (or columns) are orthogonal.

37.2. Examples.

$$(1), \quad \begin{pmatrix} 1 & 1 \\ 1 & -1 \end{pmatrix}, \quad \begin{pmatrix} 1 & 1 & 1 & 1 \\ 1 & -1 & 1 & -1 \\ 1 & 1 & -1 & -1 \\ 1 & -1 & -1 & 1 \end{pmatrix},$$

$$\begin{pmatrix} 1 & 1 & 1 & 1 & 1 & 1 & 1 & 1 \\ 1 & -1 & 1 & -1 & 1 & -1 & 1 & -1 \\ 1 & 1 & -1 & -1 & 1 & 1 & -1 & -1 \\ 1 & -1 & -1 & 1 & 1 & -1 & -1 & 1 \\ 1 & 1 & 1 & 1 & -1 & -1 & -1 & -1 \\ 1 & -1 & 1 & -1 & -1 & 1 & -1 & 1 \\ 1 & 1 & -1 & -1 & -1 & -1 & 1 & 1 \\ 1 & -1 & -1 & 1 & -1 & 1 & 1 & -1 \end{pmatrix},$$

are Hadamard matrices of orders 1, 2, 4, 8, respectively.

In Exercise 2, we shall see that a Hadamard matrix of order n has $|\det \mathbf{H}| = n^{n/2}$. This value is an upper bound for the magnitude of the determinant of any real $n \times n$ matrix with entries no more than 1 in absolute value, a result originally given by J. Hadamard. Since transition from $\mathbf{H}$ to $-\mathbf{H}$ and the interchanging of signs of rows or columns of $\mathbf{H}$ does not alter the defining property of $\mathbf{H}$, we may assume that the first row and first column consist of 1's only. In this case, $\mathbf{H}$ is called ***normalized***.

37.3. Theorem. *If a Hadamard matrix* $\mathbf{H}$ *of order* n *exists, then* n *is equal to* 1 *or* 2, *or* n *is a multiple of* 4.

Proof. Let $n \geq 3$ and suppose the first three rows of $\mathbf{H}$ are of the form

$$\underbrace{\begin{matrix} 1 & 1 & \dots & 1 \\ 1 & 1 & \dots & 1 \\ 1 & 1 & \dots & 1 \end{matrix}}_{i \text{ times}} \quad \underbrace{\begin{matrix} 1 & 1 & \dots & 1 \\ 1 & 1 & \dots & 1 \\ -1 & -1 & \dots & -1 \end{matrix}}_{j \text{ times}} \quad \underbrace{\begin{matrix} 1 & 1 & \dots & 1 \\ -1 & -1 & \dots & -1 \\ 1 & 1 & \dots & 1 \end{matrix}}_{k \text{ times}} \quad \underbrace{\begin{matrix} 1 & 1 & \dots & 1 \\ -1 & -1 & \dots & -1 \\ -1 & -1 & \dots & -1 \end{matrix}}_{l \text{ times}}$$

Since the rows are orthogonal, we have

$$\begin{aligned} i+j-k-l&=0, \\ i-j+k-l&=0, \\ i-j-k+l&=0. \end{aligned}$$

This implies that $i = j = k = l$, and thus $n = 4i$. □

It is an unsolved problem whether or not Hadamard matrices of orders $n = 4k$ exist for all $k \in \mathbb{N}$. Hadamard matrices can be constructed in the following way. If $\mathbf{H}$ is a Hadamard matrix of order n, then

$$\overline{\mathbf{H}} := \begin{pmatrix} \mathbf{H} & \mathbf{H} \\ \mathbf{H} & -\mathbf{H} \end{pmatrix}$$

is a Hadamard matrix of order $2n$. So if we start with $\mathbf{H}_0 := (1)$, we get $\mathbf{H}_1 = \begin{pmatrix} 1 & 1 \\ 1 & -1 \end{pmatrix}$, $\mathbf{H}_2 = \begin{pmatrix} \mathbf{H}_1 & \mathbf{H}_1 \\ \mathbf{H}_1 & -\mathbf{H}_1 \end{pmatrix}$, and so on (see 37.2).

37.4. Definition. These Hadamard matrices $\mathbf{H}_0, \mathbf{H}_1, \dots$ are called the ***standard Hadamard matrices***.

Observe that $\mathbf{H}_n$ is a $2^n \times 2^n$ matrix.

In developing further examples of Hadamard matrices, we make use of the following notation and terminology. The ***Legendre symbol*** χ is defined by

$$\chi(a) = \begin{cases} 1 & \text{if } a \neq 0 \text{ is a square,} \\ -1 & \text{if } a \text{ is not a square,} \\ 0 & \text{if } a = 0, \end{cases}$$

for any $a \in \mathbb{F}_q$.

37.5. Theorem (*Payley Construction*). *If*

$$\mathbb{F}_q = \{a_0 = 0, a_1, \dots, a_{q-1}\}, \quad q = p^e = 4t - 1,$$

then the following $(q+1) \times (q+1)$ *matrix is a Hadamard matrix:*

$$\begin{pmatrix} 1 & 1 & 1 & \dots & 1 & 1 \\ 1 & -1 & \chi(a_1) & \dots & \chi(a_{q-2}) & \chi(a_{q-1}) \\ 1 & \chi(a_{q-1}) & -1 & \dots & \chi(a_{q-3}) & \chi(a_{q-2}) \\ \dots & \dots & \dots & \dots & \dots & \dots \\ 1 & \chi(a_1) & \chi(a_2) & \dots & \chi(a_{q-1}) & -1 \end{pmatrix}.$$

The proof is left to the reader. Hadamard matrices of orders 2^k with $k \in \mathbb{N}_0$ are obtainable by the standard examples in 37.4. Examples of orders $2^k(q+1)$ with $k \geq 0$ and prime powers q of the form $q = 4t - 1$ can be obtained by using 37.5 and the construction mentioned before this theorem. At present, the smallest order for which the existence of a Hadamard matrix is not known is 428.

Hadamard matrices obtained by the Payley construction above give rise to an important class of nonlinear codes. Let $\mathbf{H}$ be a normalized Hadamard matrix of order n. We replace all $+1$'s in $\mathbf{H}$ by 0's and all -1's by 1's, and then $\mathbf{H}$ is changed into a ***binary Hadamard matrix*** $\mathbf{A}$. The rows of $\mathbf{H}$ are orthogonal, therefore any two rows of $\mathbf{A}$ agree in $\frac{n}{2}$ places and differ in $\frac{n}{2}$ places and so are Hamming distance $\frac{n}{2}$ apart. A nonlinear code of size M, length n, and minimal distance d is called an ***(n, M, d) code***. The following nonlinear codes can be obtained from the matrix $\mathbf{A}$:

1. An $(n-1, n, \frac{1}{2}n)$ code $\mathcal{A}_n$ consisting of the rows of $\mathbf{A}$ with the first column deleted;
2. An $(n-1, 2n, \frac{1}{2}n-1)$ code $\mathcal{B}_n$ consisting of $\mathcal{A}_n$ together with all the complements of all its codewords;
3. An $(n, 2n, \frac{1}{2}n)$ code C_n consisting of the rows of $\mathbf{A}$ and their complements.

37.6. Example. The codewords of a $(7, 8, 4)$ code are given by $\mathcal{A}_8$ as the rows of

$$\mathbf{A} = \begin{pmatrix} 0 & 0 & 0 & 0 & 0 & 0 & 0 \\ 1 & 0 & 0 & 1 & 0 & 1 & 1 \\ 1 & 1 & 0 & 0 & 1 & 0 & 1 \\ 1 & 1 & 1 & 0 & 0 & 1 & 0 \\ 0 & 1 & 1 & 1 & 0 & 0 & 1 \\ 1 & 0 & 1 & 1 & 1 & 0 & 0 \\ 0 & 1 & 0 & 1 & 1 & 1 & 0 \\ 0 & 0 & 1 & 0 & 1 & 1 & 1 \end{pmatrix}.$$

In general, from the table on page 190 and Exercise 16.6 we know that $A(7, 4) = A(7, 3) = 16$. For $\mathcal{B}_8$, we get the codewords as the rows of

$$\begin{pmatrix} \mathbf{A} \\ \mathbf{A}' \end{pmatrix}$$

with $\mathbf{A}$ as above, where $\mathbf{A}'$ is the complement of $\mathbf{A}$. This is a $(7, 16, 3)$ code. Since $A(7, 3) = 16$, it is an "optimal" code, which also solves the sphere packing problem. Theorem 16.10 shows that this code is perfect, so by 20.9 it must be linear and be the $(7, 4)$ Hamming code. See Exercises 4–6.

Usually, these codes turn out to be "very good." For further properties and examples see MacWilliams & Sloane (1977). We now describe another area of applications.

37.7. Example. We study the problem of determining the weight of four objects by using a chemical balance with two pans. Suppose we know that the scales have error ε at each weighing. This ε is assumed to be a random variable with mean 0 and variance σ^2, independent of the weight of the objects. Let x_i denote the exact weights of the objects, let y_i be the weights found by reading off the scales, and let ε_i be the (unknown) errors, i.e.,

$$x_i = y_i + \varepsilon_i \quad \text{for } i = 1, 2, 3, 4.$$

Estimates for the unknown weights x_i are

$$\hat{x}_i = y_i = x_i - \varepsilon_i, \quad i = 1, 2, 3, 4,$$

where each has variance σ^2.

Alternatively, we can use the following weighing design. We make the measurements z_i:

$$\begin{aligned} z_1 &= x_1 + x_2 + x_3 + x_4 + \varepsilon_1, \\ z_2 &= x_1 - x_2 + x_3 - x_4 + \varepsilon_2, \\ z_3 &= x_1 + x_2 - x_3 - x_4 + \varepsilon_3, \\ z_4 &= x_1 - x_2 - x_3 + x_4 + \varepsilon_4, \end{aligned}$$

where again the x_i denote the true weights, and ε_i denote errors. These equations indicate that in the first weighing all four objects are placed in the left pan, in the second weighing objects 1 and 3 are in the left pan and 2 and 4 in the right, and so on. The coefficient matrix of the x_i is the 4×4 Hadamard matrix $\mathbf{H}_2$ of Example 37.2. The estimates for the weights are

$$\begin{aligned} \hat{x}_1 &= \tfrac{1}{4}(z_1 + z_2 + z_3 + z_4), \\ \hat{x}_2 &= \tfrac{1}{4}(z_1 - z_2 + z_3 - z_4), \\ \hat{x}_3 &= \tfrac{1}{4}(z_1 + z_2 - z_3 - z_4), \\ \hat{x}_4 &= \tfrac{1}{4}(z_1 - z_2 - z_3 + z_4), \end{aligned}$$

which implies that

$$\hat{x}_1 = x_1 + \tfrac{1}{4}(\varepsilon_1 + \varepsilon_2 + \varepsilon_3 + \varepsilon_4), \quad \text{etc.}$$

The variances in the estimates are, according to statistics, $4(\frac{\sigma}{4})^2 = \frac{\sigma^2}{4}$, which is an improvement by a factor of 4 over weighing the objects one at a time. This holds under the assumption that the errors ε_i in the ith measurement are independent of the quantity being measured, i.e., that the weight of the objects should be light in comparison with the mass of the balance. We also assumed that $\text{mean}(\varepsilon_i \varepsilon_j) = 0$.

Observe that we used an idea similar to the above (making simultaneous estimates for the values of interesting entities) in 36.17, using BIB-designs.

MacWilliams & Sloane (1977) describes an application of so-called S-matrices in measuring the spectrum of a beam of light. Instead of n objects whose weights are to be determined, a beam of light has to be divided into n components of different wavelengths and their intensities have to be found. This is done by using a multiplexing spectrometer with several exit slits allowing several components to pass freely through a mask and be measured. The ***S-matrices*** $\mathbf{S}_n$ used for such a device are defined by beginning with a normalized Hadamard matrix $\mathbf{H}$ of order n. Then an S-matrix of order $n-1$ is the $(n-1)\times(n-1)$ matrix of 0's and 1's obtained by omitting the first row and column of $\mathbf{H}$ and then (as for getting binary Hadamard matrices) changing 1's to 0's and -1's to 1's. The S-matrices $\mathbf{S}_1$, $\mathbf{S}_3$, and $\mathbf{S}_7$ are obtained from Example 37.2.

$$\mathbf{S}_1 = (1), \qquad \mathbf{S}_3 = \begin{pmatrix} 1 & 0 & 1 \\ 0 & 1 & 1 \\ 1 & 1 & 0 \end{pmatrix}, \qquad \mathbf{S}_7 = \begin{pmatrix} 1 & 0 & 1 & 0 & 1 & 0 & 1 \\ 0 & 1 & 1 & 0 & 0 & 1 & 1 \\ 1 & 1 & 0 & 0 & 1 & 1 & 0 \\ 0 & 0 & 0 & 1 & 1 & 1 & 1 \\ 1 & 0 & 1 & 1 & 0 & 1 & 0 \\ 0 & 1 & 1 & 1 & 1 & 0 & 0 \\ 1 & 1 & 0 & 1 & 0 & 0 & 1 \end{pmatrix}.$$

We leave it to the reader to show:

37.8. Lemma. *Let $\mathbf{S}_n$ be an S-matrix of order n. Then $\mathbf{S}_n$ satisfies:*

(i) $\mathbf{S}_n\mathbf{S}_n^{\mathrm{T}} = \frac{1}{4}(n+1)(\mathbf{I}_n + \mathbf{J}_n)$;

(ii) $\mathbf{S}_n\mathbf{J}_n = \mathbf{J}_n\mathbf{S}_n = \frac{1}{2}(n+1)\mathbf{J}_n$;

(iii) $\mathbf{S}_n^{-1} = \frac{2}{n+1}(2\mathbf{S}_n^{\mathrm{T}} - \mathbf{J}_n)$;

where $\mathbf{J}_n$ is the $n \times n$ matrix with all entries $= 1$. (Properties (i) *and* (ii) *for an arbitrary $n \times n$ matrix of 0's and 1's imply that $\mathbf{S}_n$ is an S-matrix.)*

The most important examples for practical purposes are cyclic S-matrices, which have the property that each row is a cyclic shift to the left or to the right of the previous row,

$$\begin{pmatrix} 1 & 1 & 1 & 0 & 1 & 0 & 0 \\ 0 & 1 & 1 & 1 & 0 & 1 & 0 \\ 0 & 0 & 1 & 1 & 1 & 0 & 1 \\ 1 & 0 & 0 & 1 & 1 & 1 & 0 \\ 0 & 1 & 0 & 0 & 1 & 1 & 1 \\ 1 & 0 & 1 & 0 & 0 & 1 & 1 \\ 1 & 1 & 0 & 1 & 0 & 0 & 1 \end{pmatrix}.$$

Cyclic S-matrices can be constructed by using maximal length shift-register sequences (see §33). In this construction, the first row of the

$n \times n$ matrix is taken to be one period's worth of the output from a shift-register whose feedback polynomial is a primitive polynomial (see 14.21) of degree m. In §33, we saw that the period of such a shift-register is $n = 2^m - 1$.

37.9. Example. The polynomial $x^4 + x + 1$ over $\mathbb{F}_2$ determines the following shift-register:

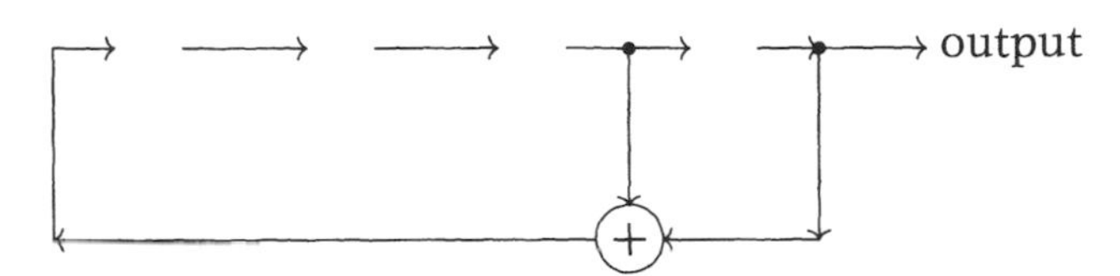

The output from this shift-register has period $n = 2^m - 1 = 15$. If it initially contains 1000 in the delay elements, one period's worth of the output is

$$0 \quad 0 \quad 0 \quad 1 \quad 0 \quad 0 \quad 1 \quad 1 \quad 0 \quad 1 \quad 0 \quad 1 \quad 1 \quad 1 \quad 1.$$

An S-matrix of order 15 having this sequence as its first row is

$$\mathbf{S}_{15} = \begin{pmatrix} 0&0&0&1&0&0&1&1&0&1&0&1&1&1&1 \\ 0&0&1&0&0&1&1&0&1&0&1&1&1&1&0 \\ 0&1&0&0&1&1&0&1&0&1&1&1&1&0&0 \\ 1&0&0&1&1&0&1&0&1&1&1&1&0&0&0 \\ 0&0&1&1&0&1&0&1&1&1&1&0&0&0&1 \\ 0&1&1&0&1&0&1&1&1&1&0&0&0&1&0 \\ 1&1&0&1&0&1&1&1&1&0&0&0&1&0&0 \\ 1&0&1&0&1&1&1&1&0&0&0&1&0&0&1 \\ 0&1&0&1&1&1&1&0&0&0&1&0&0&1&1 \\ 1&0&1&1&1&1&0&0&0&1&0&0&1&1&0 \\ 0&1&1&1&1&0&0&0&1&0&0&1&1&0&1 \\ 1&1&1&1&0&0&0&1&0&0&1&1&0&1&0 \\ 1&1&1&0&0&0&1&0&0&1&1&0&1&0&1 \\ 1&1&0&0&0&1&0&0&1&1&0&1&0&1&1 \\ 1&0&0&0&1&0&0&1&1&0&1&0&1&1&1 \end{pmatrix}.$$

In 17.24(vii) we met first-order Reed-Muller codes RM(m). A message $\mathbf{a} = a_0a_1\dots a_m$ is here encoded into $F(\mathbf{a}) := \big(f_\mathbf{a}(0,\dots,0), f_\mathbf{a}(0,\dots,0,1), f_\mathbf{a}(0,\dots,1,0),\dots,f_\mathbf{a}(1,\dots,1)\big)$, where $f_\mathbf{a}$ is the function $\mathbb{F}_2^m \to \mathbb{F}_2$; $(x_1,\dots,x_m) \mapsto a_0 + a_1x_1 + \dots + a_mx_m$, and $\mathbf{v}_0 = (0,\dots,0)$, $\mathbf{v}_1 = (0,\dots,1)$, $\dots, \mathbf{v}_{2^m-1} = (1,\dots,1)$ are the elements of $\mathbb{F}_2^m$ in lexicographic order. For instance, $\mathbf{a} = 101$ means $m = 2$, hence $f_{\mathbf{a}}\colon (x_1, x_2) \mapsto a_0 + a_1x_1 + a_2x_2 = 1 + x_2$, and $\mathbf{a}$ is encoded into $F(\mathbf{a}) = 10101010$.

This clearly gives a linear code $\mathrm{RM}(m) = \{F(\mathbf{a}) \mid \mathbf{a} \in \mathbb{F}_2^{m+1}\} \subseteq \mathbb{F}_2^{2^m}$ of length 2^m and dimension $m+1$. In order to find its minimal distance, we compute the minimal weight w (see 17.10). If $F(\mathbf{a})$ is not constant, some a_i $(1 \leq i \leq m)$ must be $= 1$; w.l.o.g let $a_1 = 1$. For all $x_2, \dots, x_m \in \mathbb{F}_2$, we then get $f_{\mathbf{a}}(1, x_2, \dots, x_m) = a_0 + 1 + a_2x_2 + \cdots + a_mx_m = 1 + a_0 + a_2x_2 + \cdots + a_mx_m = 1 + f_{\mathbf{a}}(0, x_2, \dots, x_m)$. So $f_{\mathbf{a}}$ assumes the values 0 and 1 equally often (as does every nonconstant affine map on a group), which shows $d_{\min} = w = 2^{m-1}$.

In order to decode an incoming message $\mathbf{b} = b_0b_1 \cdots b_{2^m-1}$, take any $\mathbf{c} = c_1c_2 \cdots c_m \in \mathbb{F}_2^m$, and define $\widehat{\mathbf{c}} \in \mathbb{Z}$ as the number of 0's minus the number of 1's in $\mathbf{b} + F(0c_1 \cdots c_m)$. Note that if $\mathbf{c} = c_1 \cdots c_m$ is considered as the vector $(c_1, \dots, c_m)$ and if $\cdot$ denotes the dot product, then $F(0c_1 \dots c_m) = (c \cdot v_0, c \cdot v_1, \dots, c \cdot v_{c^m-1}) \in \mathbb{F}_2^{2^m}$. The vector $(0, c_1, \dots, c_m)$ and the sequence $0c_1 \cdots c_m$ will both be denoted by $0\mathbf{c}$. As the next step, we get $\widehat{\mathbf{c}} = 2^m - 2\,d(\mathbf{b}, F(0c_1 \dots c_m))$, since for every component in which $\mathbf{b}$ and $F(0c_1 \dots c_m)$ differ, we get one 0 less and one more 1. So

$$d(\mathbf{b}, F(0c_1 \dots c_m)) = \frac{1}{2}(2^m - \widehat{\mathbf{c}})$$

and

$$d(\mathbf{b}, F(1c_1 \dots c_m)) = \frac{1}{2}(2^m + \widehat{\mathbf{c}}).$$

From this we see that the closest codeword to $\mathbf{b}$ is that $F(a_0a_1 \dots a_m) =: F(a_0\mathbf{a})$ for which $|\widehat{\mathbf{a}}|$ is maximal. More precisely, if $\widehat{\mathbf{a}} \geq 0$, we decode to $0\mathbf{a}$, otherwise to $1\mathbf{a}$.

The problem arises how to compute all $\widehat{\mathbf{c}}$ (for $\mathbf{c} \in \mathbb{F}_2^m$). Surprisingly enough, we can get all $\widehat{\mathbf{c}}$ simultaneously. Note that $f_{0\mathbf{c}}$ denotes the function $\mathbb{F}_2^m \to \mathbb{F}_2$; $\mathbf{v} = (v_1, \dots, v_m) \mapsto c_1v_1 + \cdots + c_mv_m = \mathbf{c} \cdot \mathbf{v}$. So we have, by definition,

$$\widehat{\mathbf{c}} = \sum_{i=0}^{2^m-1} (-1)^{\mathbf{c}\cdot\mathbf{v}_i + b_i} = \sum_{i=0}^{2^m-1} (-1)^{\mathbf{c}\cdot\mathbf{v}_i}(-1)^{b_i}.$$

Let $\beta_i := (-1)^{b_i}$. Then, if $\mathbf{c}$ runs through $\{\mathbf{v}_0, \dots, \mathbf{v}_{2^m-1}\}$,

$$(\widehat{\mathbf{c}})_{\mathbf{c}\in\mathbb{F}_2^m} = \left((-1)^{\mathbf{c}\cdot\mathbf{v}}\right)_{\mathbf{c},\mathbf{v}\in\mathbb{F}_2^m} \begin{pmatrix} \beta_0 \\ \vdots \\ \beta_{2^m-1} \end{pmatrix}.$$

But the $2^m \times 2^m$ matrix $\left((-1)^{\mathbf{c}\cdot\mathbf{v}}\right)_{\mathbf{c},\mathbf{v}\in\mathbb{F}_2^m}$ is just the standard Hadamard matrix $\mathbf{H}_m$. So we get all $\widehat{\mathbf{c}}$ in the following way:

(1) In $\mathbf{b} = b_0b_1\dots b_{2^m-1}$, replace 0's by 1's and 1's by -1's, to get $\boldsymbol{\beta} = (\beta_0, \beta_1, \dots, \beta_{2^m-1})^{\mathrm{T}}$.

(2) Compute $\mathbf{H}_m \cdot \boldsymbol{\beta}$ to get all $\widehat{\mathbf{c}}$ for $\mathbf{c} \in \mathbb{F}_2^m$. Then we look (as indicated above) for the largest among all $|\widehat{\mathbf{c}}|$; if it is in place i, we decode it to $0\mathbf{v}_i$ or to $1\mathbf{v}_i$, as mentioned above.

The transformation

$$\begin{pmatrix}\beta_0\\ \vdots\\ \beta_{2^m-1}\end{pmatrix} \mapsto \mathbf{H}_m \begin{pmatrix}\beta_0\\ \vdots\\ \beta_{2^m-1}\end{pmatrix}$$

is called the ***Hadamard transform***. Since $\mathbf{H}_m = \begin{pmatrix}\mathbf{H}_{m-1} & \mathbf{H}_{m-1}\\ \mathbf{H}_{m-1} & -\mathbf{H}_{m-1}\end{pmatrix}$, evaluation of $\mathbf{H}_m \begin{pmatrix}\beta_0\\ \vdots\\ \beta_{2^m-1}\end{pmatrix}$ reduces to the two evaluations

$$\mathbf{H}_{m-1}\begin{pmatrix}\beta_0\\ \vdots\\ \beta_{2^{m-1}-1}\end{pmatrix} \quad\text{and}\quad \mathbf{H}_{m-1}\begin{pmatrix}\beta_{2^{m-1}}\\ \vdots\\ \beta_{2^m-1}\end{pmatrix}.$$

If we do this recursively, we get the ***fast Hadamard transform*** with complexity $O(m2^m)$ (see §34, in particular, the lines between "Idea 2" and "Idea 3"). In fact, the (fast) Hadamard transform can be viewed as a special (fast) Fourier transform (see Clausen & Baum (1993)).

Finally, we take a look at another interpretation of Hadamard matrices. It concerns the modeling of the brain. Information in the brain is transmitted by a system (network) of neurons, nerve cells which receive numerous inputs from dendrites and produce outputs via axons:

dendrites $\rightrightarrows$ neuron $\rightrightarrows$ axons

Neurons are not physically connected; between the axon of one neuron and the dendrite of the next one, there is a small joint, called a synapse:

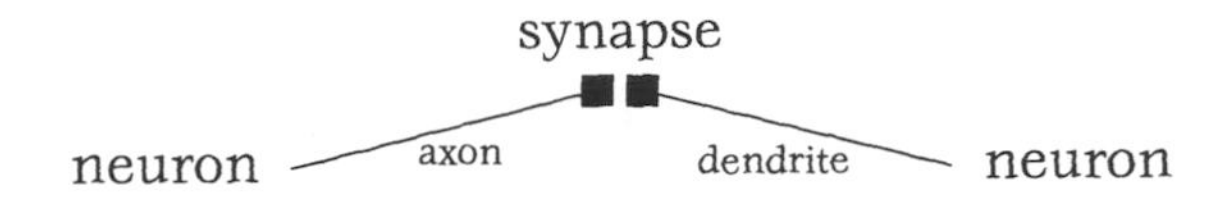

If an electrical impulse reaches a neuron at one of its dendrites, it is passed to the axons; if the "transmitting potential" there is high enough, the axon "fires" by emitting neurotransmitters, chemical substances which allow the signal to pass through the synapse and reach the next dendrite. In Figure 37.1, an impulse coming to neuron 1 is passed to neurons 2, 3, and 4.

The brain of human beings contains about 10^{17} such synapses, and each synapse is a highly complex entity. So the brain is an extremely complicated network, and relatively very little is actually known about it. How, for instance, can intelligence in a brain arise "by itself"? So, obtaining good models of brain activity represents a very active area of research; the main area being the one of "neural networks."

A neuron is not simply an automaton in the sense of §29; signals processing the neurons may influence the synapses. This allows the storage of information (memory) and the impression of neural activity patterns to make these patterns recallable (learning). So neural networks are trained, not "programmed." Information is distributed throughout the whole network; hence even the destruction of parts of thc brain still allows the brain to work fairly well. The systematic structure of the network in the brain and the possibility of extreme "parallel computation" largely contributes to the sometimes incredible "thinking power" of the brain. For example, recognizing that L and $\mathcal{L}$ are the same letters is still a hard task for computers, but a very easy one for our brains. A more thorough understanding of the brain is vital not only for brain surgeons, neurologists, and psychologists, but also for developing much more powerful computers.

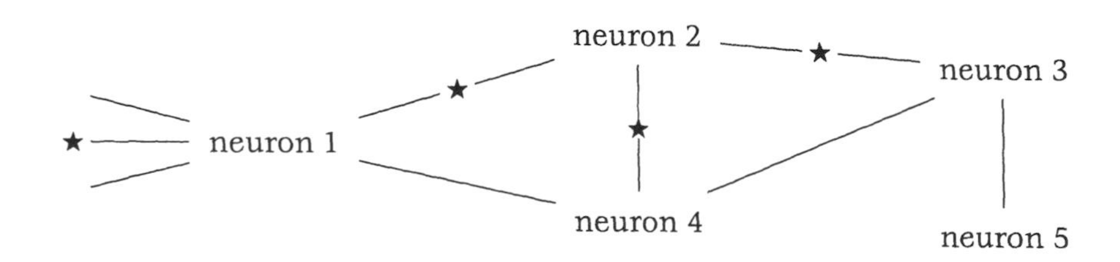

FIGURE 37.1

Since the neural activity at an axon is usually either very high or very low, we might simplify the activity at a synapse by -1 (inactive) or by 1 (active). Also, we simplify neurons so that they have single inputs and single outputs. The numbers $1, -1$ indicate the state of the neuron. A ***neural network*** is a collection $N_1, \dots, N_k$ of neurons which are in the states $s_1, \dots, s_k$. The k-tuple $(s_1, \dots, s_k) \in \{-1, 1\}^k$ is called the ***state of the network***. For an excellent account see Lautrup (1989). Often we omit consideration of the neurons if they are given in a fixed order and are simply interested in the state $(s_1, \dots, s_k)$. The rows of a Hadamard matrix are such k-tuples consisting of -1's and 1's.

37.10. Definition. If $\mathbf{H}$ is a Hadamard matrix, each row of it is called a ***Hadamard pattern***. The set $P(\mathbf{H})$ of the rows of $\mathbf{H}$ is called a ***Hadamard network***.

Since we have the freedom to renumber the neurons, we may consider the symmetric group S_k, which acts on the positions of the neurons. Each $\pi \in S_k$ gives rise to a permutation $\bar{\pi}$ of the corresponding states by $\bar{\pi}(s_1, \dots, s_k) = (s_{\pi(1)}, \dots, s_{\pi(k)})$. For example, $\pi = (2, 3) \in S_4$ induces $\bar{\pi}\colon (s_1, s_2, s_3, s_4) \mapsto (s_1, s_3, s_2, s_4)$.

37.11. Definition. Let $\mathbf{H}$ be a $k \times k$ Hadamard matrix with Hadamard pattern set $P = P(\mathbf{H})$. Then

$$S(\mathbf{H}) := \{\pi \in S_k \mid \bar{\pi}(P) = P\}$$

is called the ***symmetry group*** of $\mathbf{H}$.

So, we relate to §25 and §27. If $\mathbf{H}$ is the standard Hadamard matrix $\mathbf{H}_n$ (see 37.4), we have $k = 2^n$ neurons, and each Hadamard pattern starts with 1 (hence each $\bar{\pi} \in S(\mathbf{H})$ has $\pi(1) = 1$). Obviously $S(\mathbf{H}_0) = \{\mathrm{id}\}$. For $\mathbf{H}_1$, the pattern set is $P(\mathbf{H}_1) = \{(1, 1), (1, -1)\}$. Each $\bar{\pi} \in S(\mathbf{H}_1)$ must have $\pi(1) = 1$, hence $\pi(2) = 2$, so $S(\mathbf{H}_1) = \{\mathrm{id}\}$ as well. If we consider $P(\mathbf{H}_2)$, take, e.g., $\pi = (2, 4)$. Then $\bar{\pi}$ acts on $P(\mathbf{H}_2) = \{(1, 1, 1, 1), (1, -1, 1, -1), (1, 1, -1, -1), (1, -1, -1, 1)\} =: \{h_1, h_2, h_3, h_4\}$ by $\bar{\pi}(h_1) = h_1$, $\bar{\pi}(h_2) = h_2$, $\bar{\pi}(h_3) = (1, -1, -1, 1) = h_4$ and $\bar{\pi}(h_4) = h_3$. So $\bar{\pi} \in S(\mathbf{H}_2)$, and similarly we get $S(\mathbf{H}_2) = \{\pi \in S_4 \mid \pi(1) = 1\} \cong S_3$. Observe that, for instance, $\pi = (1, 2)$ induces $\bar{\pi}$, which has the property that $\pi(h_2) = (-1, 1, 1, -1) \notin P(\mathbf{H}_2)$. Folk, Kartashov, Linoseněk & Paule (1993) have shown:

37.12. Theorem. *For the standard Hadamard matrices $\mathbf{H}_n$, we have $S(\mathbf{H}_n) \cong GL(n, \mathbb{Z}_2)$.*

So, for $\mathbf{H}_3$, the 8×8 matrix in 37.2, we get $S(\mathbf{H}_3) \cong GL(3, \mathbb{Z}_2)$, a famous group of order 168, namely the second smallest nonabelian simple group (see 10.31). This is much smaller than the group $\{\pi \in S_8 \mid \pi(1) = 1\} \cong S_7$, which is of order $7! = 5040$.

In fact, more can be shown in 37.12: there is an isomorphism φ: $S(\mathbf{H}_n) \to GL(n, \mathbb{Z}_2)$ together with a natural bijection ψ: $P(\mathbf{H}_n) \to \mathbb{F}_2^n$ such that $\bar{\pi}(h)$ is given by the usual matrix multiplication $\varphi(\pi)\psi(h)$.

It turns out that neural networks with isomorphic symmetry groups have a similar behavior, similar "learning rules," etc. Hadamard networks have the property that each pattern (except the pattern $(1, 1, \ldots, 1)$) has the same number of -1's and 1's, i.e., the same number of quiet and firing neurons. The same applies to the expected values for random networks. Hence Hadamard networks turn out to be an excellent tool to study random networks. Also of interest are the orbits (see 25.14(iii)) of $P_k(\mathbf{H})$, the set of k-subsets of $P(\mathbf{H})$, under the action of $S(\mathbf{H})$. They are also described in Folk, Kartashov, Linoseněk & Paule (1993), using Burnside's Lemma 25.17. Almost all features of patterns in the same orbit coincide, because they just differ by a rearrangement of the neurons. The number of these orbits increases very quickly if n gets larger. Interestingly enough, for any fixed k there emerges (as n gets larger) a certain orbit which contains almost all of the Hadamard patterns and is therefore called the "winner."

Exercises

1. Prove that if Hadamard matrices of orders m and n exist, then there is also a Hadamard matrix of order mn.
2. Let $\mathbf{H}$ be a Hadamard matrix of order n. Prove:

   (i) $|\det \mathbf{H}| = n^{1/2}n$;

   (ii) Hadamard matrices may be changed into other Hadamard matrices by permuting rows and columns and by multiplying rows and columns by -1;

(iii) If $\mathbf{H}$ is a normalized Hadamard matrix of order $4n$, then every row (column) except the first contains $2n - 1$'s and $2n + 1$'s, and moreover $n - 1$'s in any row (column) overlap with $n - 1$'s in each other row (column).

3. Construct a 12×12 Hadamard matrix $\mathbf{H}$.
4. Use $\mathbf{H}$ of Exercise 3 to construct $(11, 12, 6)$, $(11, 24, 5)$, and $(12, 24, 6)$ codes.
5. Show that the codes of Exercise 4 are not linear. Are they perfect?
6. We can show that $A(11, 6) \leq 12$. Compute $A(11, 6)$, $A(12, 6)$, $A(11, 5)$ and $A(10, 5)$.
7. Design the weighing of eight objects similar to Example 37.7.
8. Construct two more **S**-matrices other than those given in the text.
9. Use the fast Hadamard transform to compute $\mathbf{H}_4 \cdot (1, -1, 1, 1, 1, -1, 1, -1, 1, 1, 1, 1, 1, -1, 1, -1)^{\mathrm{T}}$.
10. Consider the Reed-Muller code RM(4). Encode 01010 and replace the first three digits in the resulting codeword by their complements. Decode the distorted message by using the fast Hadamard transform and convince yourself that the original message is obtained.
11. Show that the symmetry group of $\mathbf{H}_2$ is isomorphic to S_3.
12. Why has $\mathrm{GL}(3, \mathbb{Z}_2)$ precisely 168 elements?

Notes

Hadamard matrices make the best balance weighing designs, i.e., they are the best matrices to be used to design the weighing of objects as described above, see Sloane & Harwit (1976). If there are n objects and a Hadamard matrix of order n is used, then the mean squared error in each unknown x_i is reduced by a factor of n.

MacWilliams & Sloane (1977) contains a lot of information on how Hadamard matrices can be used to get excellent codes. This can be achieved by taking the rows of a Hadamard matrix (or slight variations thereof) as a (nonlinear) code. A taste of these ideas can be found in Exercises 4–6. Also, we can take the linear hull of the rows of a Hadamard matrix, and often we get interesting linear codes in this way.

Hadamard patterns are also useful in optics and pattern recognition for recovering blurred images, see Vanasse (1982). A nice book on Hadamard matrices and their uses is van Lint & Wilson (1992).

§38 Gröbner Bases for Algebraic and Differential Equations

In many areas of applications of mathematics, we arrive at systems of algebraic equations such as

$$\begin{aligned} x^2 + y^2 &= 25, \\ xy &= 1. \end{aligned} \tag{38.1}$$

In some cases, we can solve such a system by some special methods. Since we know that even for real equations in one unknown there cannot be any "formula" for solving it if the degree of the equation is ≥ 5 (see page 178 and 25.13(iii)), it is hopeless to search for a general method to solve systems of algebraic equations in more than one unknown. But we can at least try to simplify the equations (in a somewhat similar way to the Gaussian elimination process for systems of linear equations).

First let us consider the system (38.1) over the reals and bring it into the form

$$\begin{aligned} x^2 + y^2 - 25 &= 0, \\ xy - 1 &= 0. \end{aligned} \tag{38.2}$$

Starting from a commutative ring R with identity, we can form the polynomial ring $R[x]$, then in turn $(R[x])[y] =: R[x, y]$, and so on, to get ***multivariate polynomials*** (or ***polynomials in several indeterminates***). $R[x, y]$ and, more generally, $R[x_1, \ldots, x_n]$ are again commutative rings with identity. Observe, however, that composition poses a problem now. As in §11, to every $p \in R[x_1, \ldots, x_n]$ there corresponds a polynomial function $\bar{p}$: $R^n \to R$, which we often simply write as p again.

The system (38.2) can then be written in the form

$$\begin{aligned} \bar{p}_1(r, s) &= 0, \\ \bar{p}_2(r, s) &= 0, \end{aligned} \tag{38.3}$$

with $r, s \in \mathbb{R}$, where $p_1 = x^2 + y^2 - 25$ and $p_2 = xy - 1$ are in $\mathbb{R}[x, y]$. This is still a bit clumsy, and it is tempting to simply write (38.3) as $p_1 = p_2 = 0$, which, however, contradicts our definition of equality of polynomials. A clever idea is to simply write (38.2) as a *pair* of polynomials:

$$\{x^2 + y^2 - 25, xy - 1\},$$

which seems to be some cumbersome notation at a first glance, but it will be useful in a moment.

38.1. Definition. A ***system of algebraic equations in*** n ***"variables"*** over a commutative ring R with identity is a subset $S = \{p_i \mid i \in I\}$ of $R[x_1, \dots, x_n]$. A ***solution*** of S in R (or in some super-ring $\bar{R} \geq R$) is an n-tuple $(r_1, \dots, r_n) \in R^n$ (or $\bar{R}^n$) with $p_i(r_1, \dots, r_n) = 0$ for all $i \in I$. The system S is ***solvable*** if it has a solution in some super-ring of R.

So $\{x^2+1\} \subseteq \mathbb{R}[x]$ is solvable, since there exists a solution in $\mathbb{C} > \mathbb{R}$. The next idea is to observe that if $\mathbf{r} = (r_1, \dots, r_n)$ is a solution of $S = \{p_1, p_2, \dots\}$, then $\mathbf{r}$ is also a solution of $\{f_1 p_1\}$ for any $f_1 \in R[x_1, \dots, x_n]$, and, more generally, for $\{f_1 p_1 + f_2 p_2 + \cdots \mid f_i \in R[x_1, \dots, x_n]\}$, which is just the ideal $(p_1, p_2, \dots)$ generated by $\{p_1, p_2, \dots\}$. Hence we see:

38.2. Observation. $(r_1, \dots, r_n) \in R^n$ is a solution of $S \subseteq R[x]$ iff it is a solution of (S), the ideal generated by S.

Hence, the study of systems of equations is equivalent to looking for "zeros common to an ideal." This might sound impractical since (S) is usually very large, but here we use another idea: let S, T be subsets of $R[x_1, \dots, x_n]$ which generate the same ideal $(S) = (T)$, then S and T have the same solutions. Hence, to solve S it is natural to find a small and "well-behaved" T which generates the same ideal as S. Before turning to an algorithm which does this, we can use 38.2 to get an easy criterion when S has a solution at all in R or in some extension ring $\bar{R}$: the ideal (S) must not contain any nonzero constant.

38.3. Theorem. *If* $S \subseteq R[x_1, \dots, x_n]$, *then* S *is solvable in some extension* $\bar{R}$ *of* R *iff* $(S) \cap R = \{0\}$.

Proof. (i) Suppose S has a solution $(r_1, \dots r_n) \in \bar{R}^n$ for some $\bar{R} \geq R$. If $c \in (S) \cap R$ is not zero, we get the contradiction $c = \bar{c}(r_1, \dots, r_n) = 0$.

(ii) Suppose (S) does not contain a nonzero constant element. Then $r \mapsto [r]$ is an embedding from R into $R[x_1, \dots, x_n]/(S) =: \bar{R}$, because

$[r] = [r']$ implies $r - r' \in (S)$, whence $r = r'$. Now, for every $p \in (S)$, $p([x_1], \dots, [x_n]) = [p] = [0]$, so $([x_1], \dots, [x_n])$ is a solution of S in $\bar{R}$. □

Observe that the proof of (ii) is similar to that of Kronecker's Theorem 12.13.

38.4. Example. $T = \{x^2 + y^2 - 25, xy - 1, x^2 - y^2 - 1\}$ cannot have any solution, neither in $\mathbb{R}$ nor in any extension of $\mathbb{R}$, since $50(x^2+y^2-25)+(x^2+y^2-25)^2-8(xy-1)-4(xy-1)^2-2(x^2-y^2-1)-(x^2-y^2-1)^2 = -620 \in (T)$.

Of course, it can be hard to find such a suitable combination as in 38.4. Our next target is to overcome this difficulty, and to achieve our goal which was stated before 38.3. First we state two very famous theorems, both due to D. Hilbert.

38.5. Theorem (*Hilbert's Basis Theorem*). *Let F be a field. Then every ideal of $F[x_1, \dots, x_n]$ is finitely generated.*

38.6. Theorem (*Hilbert's Nullstellensatz*). *Let F be an algebraically closed field and $S \subseteq F[x_1, \dots, x_n]$ a system of equations with the set Z of all elements in F^n which are solutions of every $s \in S$. If $p \in F[x_1, \dots, x_n]$ satisfies $p(z) = 0$ for all $z \in Z$, then some power of p is in S.*

Proofs can be found in any standard text on commutative ring theory, for instance in Cox, Little & O'Shea (1992).

Many results can be generalized from fields to commutative rings with identity in which every ideal is finitely generated (these are called ***Noetherian rings***); $\mathbb{Z}$ is such a Noetherian ring, but not a field. Theorem 38.5 tells us that any infinite system of equations can be replaced by a finite one with the same solutions.

Let us try to simplify a system S of equations to an "equivalent" system T, i.e., such that $(S) = (T)$. Let us start with a simple case: $S = \{p, q\} \subseteq F[x]$. By 18.18(ii), $(S) = (p, q) = (p) + (q) = (d)$, where $d = \gcd(p, q)$. Recall how the last equation can be seen: If $p = qg + r$, $q = rs + t, \dots$ (see 11.22), then $(p, q) = (q, r) = (r, t)$, and so on, until $(p, q) = (q, r) = \dots = (d, 0) = (d)$. So $\{p, q\}$ reduces to the single equation $\{d\}$. This procedure of continued Euclidean divisions is the basic ingredient for the following simplification procedure; we follow the beautiful book by Sturmfels (1993).

First we explain the principles of the algorithm by examples in $F[x, y]$, where F is a field. If $p, q \in F[x, y]$ are given by $p = x^2y + 1$, $q = xy + x$, then $p - qx = x^2y + 1 - x^2y - x = -x + 1$ is in (p, q) again, and $I :=$ $(p, q) = (p, q, p - qx) = (q, p - qx)$ since $p = qx + (p - qx)$ is in $(q, p - qx)$. So $I = (xy + x, 1 - x)$, and we may apply this idea again to get rid of high powers x^iy^j: $x+y = xy+x+(1-x)y \in I$, so $I = (1 - x, x + y) = (1 - x, 1 + y)$; hence $x = 1$, $y = -1$.

If, however, we start with our introductory example (38.2), namely $p = x^2 + y^2 - 25$, $q = xy - 1$, we have a "critical pair," where our idea does not work. But we can argue that

$$r = px - qy = x^3 + xy^2 - 25x - xy^2 + y = x^3 - 25x + y$$

is in $I = (p, q)$, too. Then also

$$s := rx - q = x^4 - 25x^2 + 1 \in I,$$

so $I = (p, q, r, s)$. But with r and s, we also have $q = rx - s \in (r, s)$. It is not so clear (see Exercise 6) that also $p \in (r, s)$, and we get

$$I = (x^2 - y^2 - 25, xy - 1) = (x^3 - 25x + y, x^4 - 25x^2 + 1),$$

from which we have reached the result 38.8(ii) below.

These steps of "eliminating the largest monomials" are the basis of an algorithm to turn every finite generating set of an ideal into one of a special type, called a "Gröbner basis" (see 38.7). We define these Gröbner bases now.

First we establish a linear order relation for the set M of monomials $x_1^{k_1}x_2^{k_2}\dots x_n^{k_n}$ in $R[x_1, \dots, x_n]$ by setting $x_1 < x_2 < x_3 < \cdots$ and extending this lexicographically to M. So $1 < x_1 < x_1^2 < x_1^3 < \cdots < x_1x_2 < x_1x_2x_1 < \cdots < x_2 < \cdots$. (Sometimes, another order $\preceq$ in M is more suitable and works as well, as long as 1 is the smallest element and $\preceq$ is "compatible," i.e., $m_1 \prec m_2 \Longrightarrow m_1m_3 \prec m_2m_3$.) If $p \in R[x]$, $p \neq 0$, let init(p), the ***initial monomial*** of p, be the largest monomial appearing in p. If $S \subseteq R[x]$, let init(S) be the ideal generated by $\{\text{init}(s) \mid s \in S\}$.

38.7. Definition. Let $I \trianglelefteq R[x_1, \dots, x_n]$, then a finite subset $G = \{g_1, \dots, g_k\}$ of I is called a ***Gröbner basis*** of I if

(i) $\{\text{init}(g_1), \dots, \text{init}(g_k)\}$ generates init(I).

If, moreover,

(ii) init(g_i) does not divide any monomial in g_j, for any distinct $i, j \in \{1, \dots, s\}, i \neq j$,

holds, we call the Gröbner basis ***reduced***.

38.8. Examples.

(i) If $d = \gcd(p, q)$ with $p, q \in F[x]$, F a field, then $\operatorname{init}(r) = x^n$ if $r \in F[x]$, $\deg r = n$, and $\{d\}$ clearly is a Gröbner basis of (p, q).

(ii) A Gröbner basis for $I = (x^2 + y^2 - 25, xy - 1)$ over $\mathbb{R}$ is $\{y + x^3 - 25x, x^4 - 25x^2 + 1\}$ (cf. the comments above and see Exercise 6).

(iii) A Gröbner basis for $J = (x^2 + y^2 - 25, xy - 1, x^2 - y^2 - 1)$ over $\mathbb{R}$ is $\{1\}$.

By Hilbert's Basis Theorem 38.5, every ideal in $F[x_1, \dots, x_n]$, F a field, is finitely generated by some $\{p_1, \dots, p_k\}$. Then $\{p_1, \dots, p_k\}$ can be turned into a Gröbner basis (as we have indicated above) by substituting and canceling elements and a careful treatment of "critical pairs." This motivates the first part of

38.9. Theorem. *Let F be a field and $I \trianglelefteq F[x_1, \dots, x_n]$. Then:*

(i) *I has a Gröbner basis;*

(ii) *every Gröbner basis G of I generates I.*

Proof of (ii). Suppose $I \setminus (G) \neq \emptyset$. By the construction of the linear order $\leq$, the initial monomials of polynomials in $I \setminus (G)$ contain a minimal element $\operatorname{init}(p)$. But $\operatorname{init}(p)$ is in the ideal generated by $\operatorname{init}(G)$. So $\operatorname{init}(g) \mid \operatorname{init}(p)$ for some $g \in G$. But $p_1 := p - \frac{\operatorname{init}(p)}{\operatorname{init}(g)} g$ is in $I \setminus (G)$ and $\operatorname{init}(p_1) < \operatorname{init}(p)$, a contradiction. □

We now give three applications of Gröbner bases and Buchberger's algorithm. Many more can be found, e.g., in Buchberger (1985) and Sturmfels (1993). The first application, of course, deals with equations.

38.10. Application 1: Algebraic Equations.

(i) We want to solve the system (38.1): $x^2 + y^2 = 25$, $xy = 1$. By 38.8(ii), $\{y + x^3 - 25x,\ x^4 - 25x^2 + 1\}$ is a Gröbner basis for $(x^2 + y^2 - 25, xy - 1)$. So $y = -x^3 + 25x$, $x^4 = 25x^2 - 1$ has the same solutions as (38.1) and is "triangularized" since the second equation is an equation only in x (observe the similarity to the Gaussian method for systems of linear equations). The four solutions are $(\pm\sqrt{a}, \pm\sqrt{a}(a - 25))$ and $(\pm\sqrt{b}, \pm\sqrt{b}(b - 25))$ with $a = \frac{1}{2}(25 + \sqrt{621})$, $b = \frac{1}{2}(25 - \sqrt{621})$.

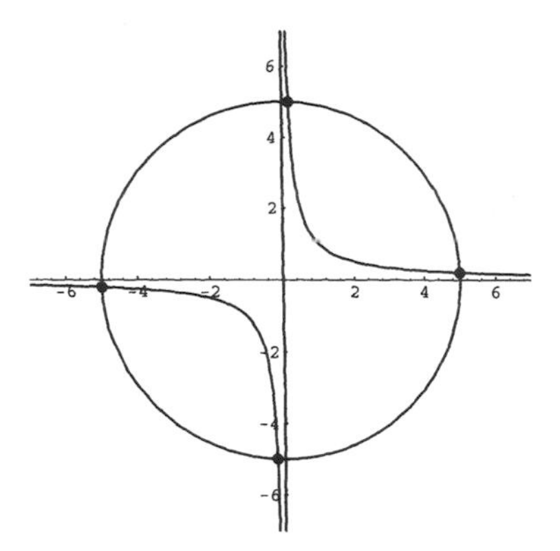

FIGURE 38.1 Four solutions.

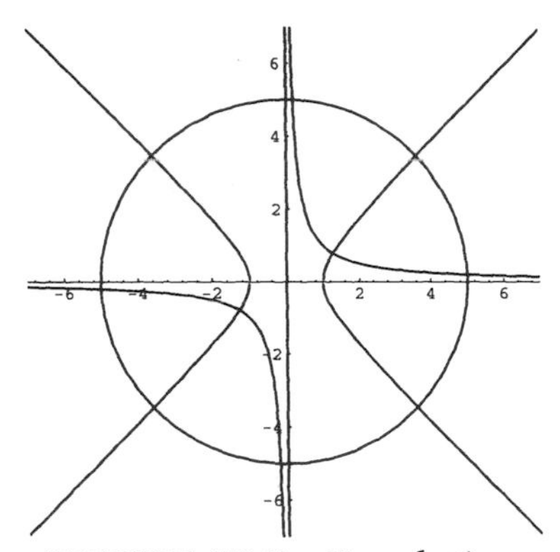

FIGURE 38.2 No solution.

(ii) The system

$$\begin{aligned} x^2 + y^2 &= 25, \\ xy &= 1, \\ x^2 - y^2 &= 1, \end{aligned} \tag{38.4}$$

leads to the Gröbner basis $\{1\}$, by 38.8(iii). By Theorem 38.3, (38.4) has no solution in any $R \geq \mathbb{R}$.

In these easy examples, we can also find the solutions graphically. $x^2+y^2 = 25$ "is" a circle with radius 5, $xy = 1$ a hyperbola. Thus Figure 38.1 shows that (38.1) yields the four points of intersections, as computed in 11.10(i). Adding the equation $x^2 - y^2 = 1$, Figure 38.2 shows that the resulting system has no solution any more.

38.11. Application 2: Partial Differential Equations. Consider the system of linear homogeneous partial differential equations with constant coefficients:

$$\begin{aligned}\frac{\partial^2 f(x,y)}{\partial x^2}+\frac{\partial^2 f(x,y)}{\partial y^2}-25f(x,y)&=0,\\ \frac{\partial^2 f(x,y)}{\partial x\partial y}-f(x,y)&=0.\end{aligned}\tag{38.5}$$

Replace $f(x,y)$ by 1, $\frac{\partial f(x,y)}{\partial x}$ by x, $\frac{\partial^2 f(x,y)}{\partial x \partial y}$ by xy, and so on. Then (38.5) becomes our algebraic system (38.2):

$$\begin{aligned}x^2+y^2-25&=0,\\ xy-1&=0.\end{aligned}$$

This system was simplified by multiplying the left sides, adding them, and so on. But if we take the derivative $\frac{\partial}{\partial x}$ of the second equation in (38.5), for instance, we get

$$\frac{\partial^3 f(x,y)}{\partial x^2\partial y}-\frac{\partial f(x,y)}{\partial x}=0,$$

which corresponds to

$$x^2y-x=0.$$

So, taking the derivative $\frac{\partial}{\partial x}$ induces the multiplication of the corresponding polynomial by x, and so on. Hence, the same Buchberger method which simplifies algebraic equations also simplifies partial differential equations. So (38.5) is equivalent to

$$\begin{aligned}\frac{\partial f(x,y)}{\partial y}-\frac{\partial^3 f(x,y)}{\partial x^3}-25\frac{\partial f(x,y)}{\partial x}&=0,\\ \frac{\partial^4 f(x,y)}{\partial x^4}-25\frac{\partial^2 f(x,y)}{\partial x^2}+1&=0.\end{aligned}\tag{38.6}$$

The second equation can be substantially simplified via $g(x,y):=\frac{\partial^2 f(x,y)}{\partial y^2}$, since this gives

$$\frac{\partial^2 g(x,y)}{\partial x^2}=25g(x,y)-1,$$

and then the system can be solved. The "extended" system

$$\frac{\partial^2 f(x,y)}{\partial x^2} + \frac{\partial^2 f(x,y)}{\partial y^2} - 25f(x,y) = 0,$$
$$\frac{\partial^2 f(x,y)}{\partial x \partial y} - f(x,y) = 0, \qquad (38.7)$$
$$\frac{\partial^2 f(x,y)}{\partial x^2} - \frac{\partial^2 f(x,y)}{\partial y^2} - f(x,y) = 0,$$

has only the trivial solution $f = 0$. Even better: (38.7) can instantly be "drawn" as in Figure 38.2 from which we see at once that (38.7) has no nontrivial solution.

38.12. Application 3: Syzygies. Let us go back to the system (38.4). Since it is unsolvable, the circle $x^2 + y^2 = 25$ and the hyperbolas $xy = 1$, $x^2 - y^2 = 1$ might be "algebraically dependent." To find out more about that let us introduce new "slack variables" a, b, c (with $c < b < a < x < y$) and apply Buchberger's algorithm to the ideal generated by

$$\{x^2 + y^2 - 25 - a, xy - 1 - b, x^2 - y^2 - 1 - c\}.$$

We get the Gröbner basis

$$\begin{aligned}
\{&2y^2 - a + c - 24,\\
&xy - b - 1,\\
&ay + cy + 26y - 2bx - 2x,\\
&2by + 2y - ax + cx - 24x,\\
&2x^2 - a - c - 26,\\
&a^2 + 50a - 4b^2 - 8b - c^2 - 2c + 620\}.
\end{aligned}$$

The last basis element contains neither x nor y; so we see immediately that $(x^2 + y^2 - 25)^2 + 50(x^2 + y^2 - 25) - 4(xy - 1)^2 - 8(xy - 1) - (x^2 - y^2 - 1)^2 - 2(x^2 - y^2 - 1) + 620 = 0$, which gives us the "relation between the circle and the hyperbolas," as well as 38.4. Such a "relation between polynomials" is called a ***syzygy***.

Exercises

1. Is $S = \{x^2 + 1,\ y^2 + 1\}$ solvable in $\mathbb{R}$? In some extension of $\mathbb{R}$?
2. Reduce the system S in Exercise 1 to a single equation.

3. Let $I := \{\sum a_{ij}x^i y^j \mid a_{ij} \in \mathbb{R},\ i \geq 3,\ j \geq 2\}$. Is I an ideal of $\mathbb{R}[x, y]$? Is it finitely generated? A principal ideal?
4. Compute a Gröbner basis for the ideal generated by S in Exercise 1.
5. Compute a Gröbner basis for the ideal generated by $\{x^3y^3 + x^2y^2 + xy,\ (xy)^4\}$.
6. Check 38.8(ii), by hand and by using a computer algebra package.
7. Show 38.8(iii).
8. Check 38.12, using a computer.

Notes

The Gröbner basis algorithm was developed by B. Buchberger in 1965 in his dissertation under the guidance of W. Gröbner. The formalization and implementation of this algorithm and the proof that it always terminates are not easy; we refer the reader to Buchberger (1985) or Winkler (1996) for details. Buchberger's algorithm is now implemented in all major computer algebra packages. The method we presented to solve systems of partial differential equations was already known at the beginning of this century (see Janet (1929) and Pommaret (1994)).

Standard textbooks on Gröbner bases and their applications are Becker & Weisspfenning (1993), Cox, Little & O'Shea (1992), Sturmfels (1993), and Winkler (1996).

For more information on syzygies see, e.g., Cox, Little & O'Shea (1992) or Sturmfels (1993). See Hong (1997) for conditions when a Gröbner basis for $I \circ f = \{i \circ f \mid i \in I\}$ can be easily computed from a basis of the ideal I ($\circ$ denotes the composition of polynomials).

§39 Systems Theory

Systems theory is closely related to automata theory; the main difference is that for systems we consider "time" as a much more essential feature. If time is discrete, we are basically back to automata, where time is relevant by considering input sequences.

Let us consider automata (i.e., "discrete systems") $\mathcal{A} = (Z, A, B, \delta, \lambda)$ in which Z, A, B are vector spaces over the same field F, and for which δ and λ are linear maps (acting on the corresponding product spaces). We then call $\mathcal{A}$ a ***linear automaton*** (or a ***linear discrete system***). Now, due to

$$\delta(\mathbf{z}, \mathbf{a}) = \delta(\mathbf{z}, \mathbf{0}) + \delta(\mathbf{0}, \mathbf{a}) =: \delta_1(\mathbf{z}) + \delta_2(\mathbf{a}),$$

we get linear maps $\delta_1 \colon Z \to Z$ and $\delta_2 \colon A \to Z$, which may be represented by matrices $\mathbf{F}$, $\mathbf{G}$ in the finite-dimensional case. Hence

$$\delta(\mathbf{z}, \mathbf{a}) = \mathbf{Fz} + \mathbf{Ga}$$

and similarly

$$\lambda(\mathbf{z}, \mathbf{a}) = \mathbf{Hz} + \mathbf{Ka}.$$

So we simply refer to the automaton (discrete system) $[\mathbf{F}, \mathbf{G}, \mathbf{H}, \mathbf{K}]$. Often $\mathbf{K}$ can be taken to be $\mathbf{0}$; then we have the automaton $[\mathbf{F}, \mathbf{G}, \mathbf{H}]$.

From page 344, $\delta^*(\mathbf{z}, \mathbf{a}_1\mathbf{a}_2) := \delta(\delta(\mathbf{z}, \mathbf{a}_1), \mathbf{a}_2)$, so, in the linear case, we get $\delta(\mathbf{z}, \mathbf{a}_1\mathbf{a}_2) = \mathbf{F}(\mathbf{Fz} + \mathbf{Ga}_1) + \mathbf{Ga}_2 = \mathbf{F}^2\mathbf{z} + \mathbf{FGa}_1 + \mathbf{Ga}_2$, and, in general,

$$\delta(\mathbf{z}, \mathbf{a}_1\mathbf{a}_2 \cdots \mathbf{a}_n) = \mathbf{F}^n\mathbf{z} + \mathbf{F}^{n-1}\mathbf{Ga}_1 + \cdots + \mathbf{FGa}_{n-1} + \mathbf{Ga}_n. \tag{39.1}$$

Now turning to "nondiscrete" (i.e., continuous) systems, let us take a look at the pendulum.

39.1. Example. Let a point of mass $m = 1$ be the end of a pendulum of length r:

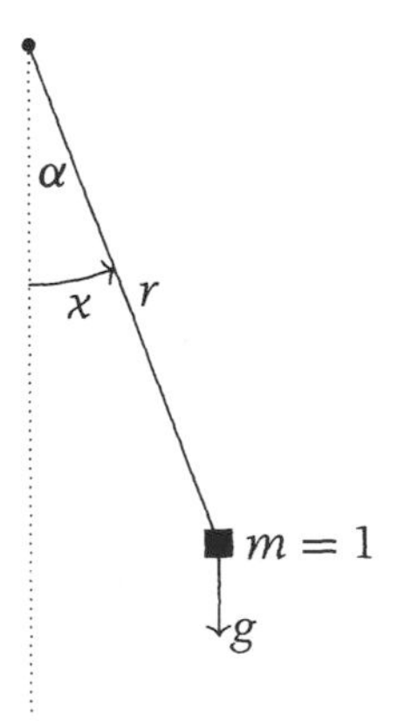

Here, g is gravity and x the radian measure of α. If b denotes air resistance, Newton's second law tells us that x, depending on the time t (so we write

$x(t)$) is governed by the differential equation

$$r\ddot{x}(t) + br\dot{x}(t) + g \sin x(t) = 0, \tag{39.2}$$

where $\dot{x}(t)$, $\ddot{x}(t)$ denote the derivatives w.r.t time. For small x, (39.2) can be replaced by the linear equation

$$r\ddot{x}(t) + br\dot{x}(t) + gx(t) = 0. \tag{39.3}$$

For $\mathbf{x}(t) := \begin{pmatrix} x(t) \\ \dot{x}(t) \end{pmatrix}$, we then get

$$\dot{\mathbf{x}}(t) = \begin{pmatrix} 0 & 1 \\ -g/r & -b \end{pmatrix} \mathbf{x}(t) =: \mathbf{F}\mathbf{x}(t).$$

Suppose now that the pendulum is hanging on a bar which can be turned at an angular speed of $u(t)$ and has the friction coefficient k (between the bar and the "rope"):

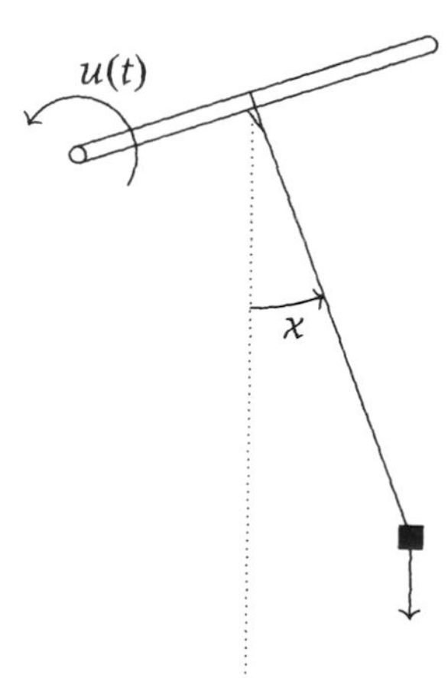

For small x, (39.3) now turns into

$$r\ddot{x}(t) + br\dot{x}(t) + gx(t) = k(u - \dot{x}). \tag{39.4}$$

With $\bar{\mathbf{F}} := \begin{pmatrix} 0 & 1 \\ -g/r & -b - k/r \end{pmatrix}$ and $\mathbf{G} := \begin{pmatrix} 0 \\ k/r \end{pmatrix}$, the equation becomes

$$\dot{\mathbf{x}}(t) = \bar{\mathbf{F}}\mathbf{x}(t) + \mathbf{G}u(t). \tag{39.5}$$

This is a typical example of a ***linear continuous system***. We shall not introduce a formal definition of a system as a 7-tuple here (similar to 29.2, with several axioms, like "consistency"); see, e.g., Kalman, Falb & Arbib (1969) for precise definitions. The main terms are ***states*** $\mathbf{x}(t)$ depending on a ***time*** t via a difference equation (discrete systems) or a differential equation (continuous systems), involving input functions $\mathbf{u}(t)$. In the linear continuous finite-dimensional case, we have

$$\dot{\mathbf{x}}(t) - \mathbf{F}\mathbf{x}(t) + \mathbf{G}\mathbf{u}(t),$$

with suitable matrices $\mathbf{F}$ and $\mathbf{G}$. Often we also study an ***output map*** $\mathbf{y}(t)$, which in the linear case is given by

$$\mathbf{y}(t) = \mathbf{H}\mathbf{x}(t).$$

Here $\mathbf{x}, \mathbf{y}, \mathbf{u}$ are elements of properly chosen vector spaces. In Example 39.1, we might take $\mathbf{y}(t) = \mathbf{x}(t)$, i.e., $\mathbf{H} = \mathbf{I}$. If $\mathbf{F}$, $\mathbf{G}$, and $\mathbf{H}$ are matrices with entries not depending on t, we call the system a ***constant*** or ***time-invariant linear system***. Note that replacing $\dot{\mathbf{x}}(t)$ by $\mathbf{x}(t+1)$ brings us to the discrete case, in which the time set is a subset of $\mathbb{Z}$ rather than of $\mathbb{R}$. We shall also speak of "the system $[\mathbf{F}, \mathbf{G}, \mathbf{H}]$." Often, the word "***plant***" is used instead of "system." Here, we do not consider systems that involve both discrete and continuous variables, like birth-death processes.

In all cases (discrete or continuous, linear or nonlinear), it is intuitively clear that for times t_1, t_2 with $t_1 \leq t_2$, an input function acting in the interval $[t_1, t_2]$ will turn the state $\mathbf{x}(t_1)$ into a state $\mathbf{x}(t_2)$. This yields a map Φ_{t_1,t_2} from $X \times U$ into X, where X and U are the sets of states (state functions) and inputs (input functions), respectively. In the discrete linear constant case, Φ is computed by summation due to equation 39.1:

$$\Phi_{t_1,t_n}(\mathbf{x}, \mathbf{u}_{t_1} \dots \mathbf{u}_{t_n}) = \mathbf{F}^n\mathbf{x} + \mathbf{F}^{n-1}\mathbf{G}\mathbf{u}_{t_1} + \cdots + \mathbf{G}\mathbf{u}_{t_n}, \qquad (39.6)$$

if $\mathbf{u}_{t_i}$ are the inputs at times t_i with $t_1 \leq t_i \leq t_n$. For continuous linear constant systems, we get

$$\Phi_{t_1,t_2}(\mathbf{x}, \mathbf{u}) = e^{(t_2-t_1)\mathbf{F}}\mathbf{x} + \int_{t_1}^{t_2} e^{(t_2-\tau)\mathbf{F}\mathbf{x}}\mathbf{G}\mathbf{u}(\tau)\, d\tau; \qquad (39.7)$$

observe the similarity in these expressions, and note that $e^{\mathbf{A}} := \sum_{k=0}^{\infty} \frac{\mathbf{A}^k}{k!}$, for a square matrix $\mathbf{A}$ over $\mathbb{R}$ or $\mathbb{C}$.

If we fix the input sequence (input function), then Φ_{t_1,t_2} can be considered as a map $X \to X$. It is intuitively clear that the ***semigroup property*** $\Phi_{t_2,t_3} \circ \Phi_{t_1,t_2} = \Phi_{t_1,t_3}$ holds; this generalizes the concept of syntactic semigroups, as defined in 29.6.

We now define an important property for general systems.

39.2. Definition. A system is ***controllable*** if for every t in the time set T ($\subseteq \mathbb{R}$) and for every pair $\mathbf{x}_1, \mathbf{x}_2$ of states there is some input (function) $\mathbf{u}$ in the collection U of all "allowed" input functions and some time $s \geq t$ such that $\mathbf{u}$ "moves" $\mathbf{x}_1$ into $\mathbf{x}_2$ between the times t and s, i.e., such that

$$\Phi_{t,s}(\mathbf{x}_1, \mathbf{u}) = \mathbf{x}_2.$$

There are several slight variations of the definition of controllability in the literature, careful checking is required before applying the results. From (39.1) and (39.6), the following criterion does not come as a surprise in the discrete case (recall that $\mathbf{F}^n$ and higher powers of $\mathbf{F}$ can be reduced to combinations of $\mathbf{I}, \mathbf{F}, \mathbf{F}^2, \ldots, \mathbf{F}^{n-1}$, by using the Cayley-Hamilton Theorem):

39.3. Theorem (*Kalman's Theorem*). *Let S be the constant linear finite-dimensional system determined by the matrices $\mathbf{F}, \mathbf{G}, \mathbf{H}$ of sizes $n \times n$, $n \times m$, and $p \times n$, respectively. Then S is controllable iff the $n \times nm$ matrix $(\mathbf{G}, \mathbf{FG}, \mathbf{F}^2\mathbf{G}, \ldots, \mathbf{F}^{n-1}\mathbf{G})$ has rank n.*

For a proof see, e.g., Kalman, Falb & Arbib (1969) or Szidarovzky & Bahill (1992); these books also contain characterizations in the nonconstant case. It is most remarkable that the same criterion 39.3 holds in the discrete and in the continuous case, and that it is purely algebraic. This triggered the far-reaching algebraic approach to systems theory (see Kalman, Falb & Arbib (1969), especially Chapters 6 and 10). Observe that controllability only depends on $\mathbf{F}$ and $\mathbf{G}$; so we simply say that "$(\mathbf{F}, \mathbf{G})$ is controllable" in this case.

39.4. Example. The system in Example 39.1 has

$$(\mathbf{G}, \bar{\mathbf{F}}\mathbf{G}) = \begin{pmatrix} 0 & \frac{k}{r} \\ \frac{k}{r} & \frac{-bkr - k^2}{r^2} \end{pmatrix}.$$

The rank is $n = 2$ iff $k \neq 0$ (clearly, if $k = 0$, rotating the bar has no influence on the pendulum.). So the system is controllable iff $k \neq 0$.

Similar to the procedure for image understanding (§26), we prepare the inputs by representing them in a suitable basis in order to get "nicer" forms for $\mathbf{F}$ and $\mathbf{G}$; usually we want $\mathbf{F}$ to be a companion matrix (see equation 33.3 on page 369) or at least a block diagonal form of companion matrices (rational canonical form); for this and the rest of this section, see Szidarovzky & Bahill (1992) or Knobloch & Kwakernaak (1985).

For many practical purposes, however, controllability is a bit too strong. In many cases (as in Example 39.4 with $k = 0$) it suffices to know that the system does not "explode," or, even better, that it "converges" to some stable "equilibrium state." For the sake of simplicity, we only study linear constant continuous systems (the discrete ones are again treated almost identically).

39.5. Definition. Consider the system described by $\dot{\mathbf{x}}(t) = \mathbf{F}\mathbf{x}(t)+\mathbf{G}\mathbf{u}(t)$, with appropriate time set T. In this context, outputs are of no importance. This system is ***stable*** if for all $t_0 \in T$ and states $\mathbf{x}$ the function $t \mapsto \Phi_{t_0,t}(\mathbf{x}, \mathbf{0})$ is bounded for $t \geq t_0$. The system is ***asymptotically stable*** if moreover $\Phi_{t_0,t}(\mathbf{x}, \mathbf{0}) \to \mathbf{0}$ as $t \to \infty$.

Observe that (asymptotical) stability only depends on $\mathbf{F}$ and does not depend on $\mathbf{G}$ at all. Due to equation (39.7), $\Phi_{t_0,t}(\mathbf{x}, \mathbf{0}) = e^{(t-t_0\mathbf{F})}\mathbf{x} = \mathbf{x}(t)$, with $\mathbf{x} = \mathbf{x}(t_0)$. So stability means that the states do not "move away too far from $\mathbf{0}$," while under asymptotic stability they even converge to $\mathbf{0}$.

As for controllability, the reader should be warned that different authors may use different concepts of stability; they include "BIBO-stability" (bounded-input–bounded-output), concepts which require the output $\mathbf{y}(t)$ to "behave properly," "Lyapunov stability," etc.

Let us take $\mathbb{C}$ as the base field and let us have a look at the simplest case $n = 1$. We write $\mathbf{F}$ just as a complex number f. Then $\Phi_{t_0,t}(x, 0) = e^{(t-t_0)f}$, arising from the differential equation $\dot{x}(t) = fx(t)$. If we write f as $f_1 + if_2 \in \mathbb{C}$, then $e^{(t-t_0)f} = (e^{f_1})^{t-t_0}(\cos f_1 + i \sin f_2)^{t-t_0}$. Now $\cos f_1 + \sin f_2$, and hence $(\cos f_1 + i \sin f_2)^{t-t_0}$ always has norm $= 1$, but $(e^{f_1})^{(t-t_0)}$ "explodes" if $f_1 > 0$. Hence in the one-dimensional case, the system is stable if the real part $\Re(f)$ is ≤ 0, and asymptotically stable if $\Re(f) < 0$. For higher-dimensional systems, we have to consider the eigenvalues of $\mathbf{F}$.

39.6. Theorem. *Consider a system as in* 39.5, *and let* $\lambda_1, \lambda_2, \ldots$ *be the eigenvalues of* $\mathbf{F}$. *Then the system is:*

(i) *asymptotically stable iff each* $\Re(\lambda_i) < 0$;

(ii) *stable iff each* $\Re(\lambda_i) \leq 0$ *and, in the case* $\Re(\lambda_i) = 0$, *the algebraic and geometric multiplicities of* λ_i *coincide.*

For a proof, see, e.g., Szidarovzky & Bahill (1992). Observe that again we have a purely algebraic characterization, and it works both in the continuous and discrete case. In case (i), the system is also BIBO-stable, as mentioned above.

39.7. Example. In Examples 39.1 and 39.4, for $\mathbf{F}$ we get the real eigenvalues $-\frac{b}{2} \pm \sqrt{(\frac{b}{2})^2 - \frac{g}{r}}$. Since $\sqrt{(\frac{b}{2})^2 - \frac{g}{r}} < \frac{b}{2}$, both eigenvalues have negative real parts. Hence the system is asymptotically stable, as everybody knows: after a while, the pendulum will stop.

39.8. Example. Let $\mathbf{F} = \begin{pmatrix} 0 & \omega \\ -\omega & 0 \end{pmatrix} \in \mathbb{M}_2(\mathbb{R})$. This matrix is important for studying harmonic motions. The eigenvalues of $\mathbf{F}$ are $\pm i\omega$; hence they have 0 as real parts. So this system is stable, but not asymptotically stable (it oscillates forever).

In this context, it might be useful to note:

39.9. Theorem. *Suppose that a matrix $\mathbf{F}$ over $\mathbb{R}$ has negative real parts in all its eigenvalues. Then all coefficients in its characteristic polynomial $c_{\mathbf{F}}$ are > 0.*

The converse is not true, but a refinement of this result (the "Hurwitz criterion"), which uses determinants involving the coefficients of $c_{\mathbf{F}}$, gives an "if and only if condition" for all eigenvalues of $\mathbf{F}$ having negative real parts. For this and the next example (which requires some knowledge of physics), see Szidarovzky & Bahill (1992), where also applications to social sciences (a predator–prey model) are given.

39.10. Example. A (very simple) nuclear reactor has the states $\mathbf{x} = (x_1, x_2)$, where x_1 is the neutron density, x_2 is the precursor density, and matrix $\mathbf{F} = \begin{pmatrix} (\rho - \beta)/l & \lambda \\ \beta/l & -\lambda \end{pmatrix}$, where ρ is the fractional change in neutron reproduction, l is the neutron generation time, β is the delayed-neutron factor, and λ is the decay constant. Then

$$c_{\mathbf{F}} = x^2 - \left(\frac{\rho - \beta}{l} - \lambda\right) x - \frac{\lambda \rho}{l}.$$

If $\rho > 0$, the reactor is unstable ("supercritical") for $\rho = 0$, we have eigenvalues 0 and $-\frac{\beta + l\lambda}{l} < 0$, hence stability, but not an asymptotical one ("critical reactor"), otherwise we have asymptotical stability ("subcritical" case) (see Exercise 6).

One of the main topics in systems theory is control theory: Given an unstable system S, can we modify it so that it will become stable? In the

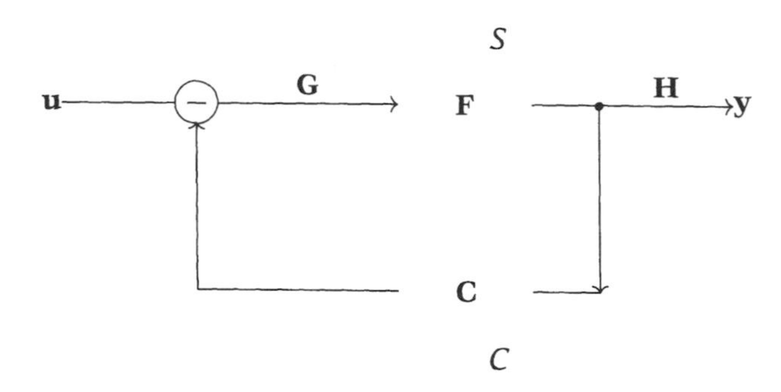

FIGURE 39.1

linear case, we have seen that stability only depends on **F**. So we might try a "closed-loop configuration," as shown in Figure 39.1.

C is another system, called the ***controller***, ***compensator***, ***regulator***, or the ***feedback***. It takes **x**, transforms it properly, and subtracts it from the input **u**. In the linear case, we might represent C by a matrix **C**, so that the new equation reads

$$\dot{\mathbf{x}}(t) = (\mathbf{F} - \mathbf{GC})\mathbf{x}(t) + \mathbf{Gu}(t).$$

So, given **F** and **G**, we look for a matrix **C** such that the real parts of all eigenvalues of $\mathbf{F} - \mathbf{GC}$ are (strictly) negative. In fact, "the more negative," the better, since "very negative" real parts will quickly bring the system to the desired stable state. Or we can start with a stable system and improve its behavior by switching from **F** to $\mathbf{F} - \mathbf{GC}$. An unskillful choice of C, however, can turn a stable system into an unstable one. Efficient methods are available to actually construct **C**. Recall that we had a control situation already in Example 39.1 when we used the turnable bar to influence the states. Among the many excellent texts on control theory, we mention Wonham (1974), Knobloch, Isidori & Flockerzi (1993), and, for the nonlinear case, de Figueiredo & Chen (1993) and Isidori (1989).

So we want to find out if, given two matrices **F** and **G**, there is a matrix **C** such that all eigenvalues of $\mathbf{F} - \mathbf{GC}$ have negative real parts. This can be formulated if we work over $\mathbb{C}$. Often, however, we work with fields other than $\mathbb{C}$, or even with rings. If, for example, control starts with some time delay, we know from the previous section that a multiplication with x will represent the shift operator (cf. the first lines in §18). So very soon we need **F** and **G** to be matrices over rings like $\mathbb{R}[x]$, $\mathbb{C}[x]$, or $\mathbb{Z}[x]$.

Then "negative real parts" have no meaning any more. We now give some conditions which assure the existence of "efficient" controllers C.

39.11. Definition. Let R be a commutative ring with identity, $\mathbf{F}$ an $n \times n$ matrix, and $\mathbf{G}$ an $n \times m$ matrix over R.

(i) $(\mathbf{F}, \mathbf{G})$ has the ***pole assignment property* (PA)** if for all $\lambda_1, \ldots, \lambda_n \in R$ there is some $m \times n$ matrix $\mathbf{C}$ such that $c_{\mathbf{F}-\mathbf{GC}} = (x - \lambda_1) \ldots (x - \lambda_n)$. The reason for this name will be clear shortly.

(ii) $(\mathbf{F}, \mathbf{G})$ is ***coefficient assignable (CA)*** if for all $c_0, c_1, \ldots, c_{n-1} \in R$ there is some $\mathbf{C}$ with $c_{\mathbf{F}-\mathbf{GC}} = c_0 + c_1 x + \cdots + c_{n-1} x^{n-1} + x^n$.

(iii) If $\mathbf{v} \in R^n$ has the property that $\{\mathbf{v}, \mathbf{Fv}, \ldots, \mathbf{F}^{n-1}\mathbf{v}\}$ is a basis for R^n, $\mathbf{v}$ is said to be ***cyclic for* F**. The pair $(\mathbf{F}, \mathbf{G})$ has a ***feedback to a cyclic vector* (FC)** if there are $\mathbf{v} \in R^n$ and $\mathbf{C}$ (as above) such that $\mathbf{Gv}$ is cyclic for $\mathbf{F} - \mathbf{GC}$.

In case (iii), we can replace $\mathbf{G}$ by a column vector $\mathbf{Gv}$, so we can "reduce our system to one with one-dimensional inputs." For a proof of the following results, see Brewer, Katz & Ullery (1987) and Kautschitsch (1991).

39.12. Theorem. *Let R, $\mathbf{F}$, $\mathbf{G}$ be as in* 39.11. *Then, in an obvious notation,*

$$\text{(FC)} \Longrightarrow \text{(CA)} \Longrightarrow \text{(PA)} \Longrightarrow \text{controllability.}$$

Controllability and (PA) can be shown to be equivalent for $R = \mathbb{R}$ and $R = \mathbb{R}[x]$. They are not equivalent for $\mathbb{R}[x, y]$ and $\mathbb{Z}[x]$. More generally:

39.13. Theorem. *If R is a principal ideal domain, then controllability and (PA) are equivalent.*

39.14. Definition. A ring R has (PA), (CA), (FC) if for every pair of matrices $(\mathbf{F}, \mathbf{G})$ over R, controllability and (PA) ((CA), (FC), respectively) are equivalent.

Hence (FC) $\Longrightarrow$ (CA) $\Longrightarrow$ (PA). Every field has (FC); from above, every principal ideal domain has (PA). $\mathbb{Z}$ and $\mathbb{R}[x]$ have (PA), but not (CA). Up to now, it is an open question if $\mathbb{C}[x]$ has (CA). More generally, it is unknown if (FC) and (CA) are really different. Concerning (FC), we have

39.15. Theorem (*Kautschitsch's Theorem*). *Let R and R' be commutative rings with identity and let $h\colon R \to R'$ be an epimorphism such that every*

element in $1 + \ker h$ *is invertible in* R. *Then*

$$R \text{ has (FC)} \iff R' \text{ has (FC)}.$$

For a proof, see Kautschitsch (1991). If we take the ring $R[[x]]$ of formal power series over R (see 11.17) and h: $R[[x]] \to R$; $\sum_{i\geq 0} u_i x^i \mapsto u_0$, we get

39.16. Corollary. *R has (FC) $\iff$ $R[[x]]$ has (FC). In particular, $F[[x]]$ has (FC), for every field F.*

Because it is often impossible to measure all the relevant components of the states, it is often desirable to compute the output sequence (function) directly from the input sequence (function) without referring to the states. For instance, transmitting to the cockpit all relevant information on closed parts inside the jet engines in an airplane would result in such heavy radio noise that no communication between the aircraft and the ground stations would be possible any more. For linear systems, this can be done in the following way:

39.17. Theorem. *For a linear constant discrete system* $(\mathbf{F}, \mathbf{G}, \mathbf{H})$ *with* $T \subseteq \mathbb{N}_0$ *and initial state* $\mathbf{0}$, *we write the sequence* $\mathbf{u}$ *as* $\mathbf{u} = \sum u_t x^t$ *and the output sequence as* $\mathbf{y} = \sum y_t x^t$. *Then*

$$\mathbf{y} = \mathbf{H}(x\mathbf{I} - \mathbf{F})^{-1}\mathbf{G}\mathbf{u}.$$

In the continuous case, we first use the Laplace transform $f \mapsto L(f)$, where $L(f)(s) = \int_0^\infty f(t)e^{-st}\,dt$ for $s \in \mathbb{C}$. Similar to the Fourier transform, the Laplace transform transfers a linear differential equation (and some nonlinear ones as well) into an algebraic one. The new variable is usually denoted by s. Then a result similar to 39.17 applies.

39.18. Theorem. *Let* $(\mathbf{F}, \mathbf{G}, \mathbf{H})$ *be a continuous constant linear system with* $x(0) = 0$. *Then*

$$\mathbf{y}(s) = \mathbf{H}(s\mathbf{I} - \mathbf{F})^{-1}\mathbf{G}\mathbf{u}(s).$$

39.19. Definition. In the cases 39.17 and 39.18, $\mathbf{T} = \mathbf{H}(x\mathbf{I} - \mathbf{F})^{-1}\mathbf{G}$ or $\mathbf{T}$: $s \mapsto \mathbf{H}(s\mathbf{I} - \mathbf{F})^{-1}\mathbf{G}$ is called the ***transfer function*** of the system.

We remark that transfer functions can also be computed from a system by performing "enough" stimulus-response experiments ("frequency responses").

If $(\mathbf{F}, \mathbf{G})$ is controllable and if the output is one-dimensional, the poles of the transfer function are precisely the zeros of $\det(x\mathbf{I} - \mathbf{F}) = c_{\mathbf{F}}$, the characteristic polynomial of $\mathbf{F}$; so the poles of T are precisely the eigenvalues of $\mathbf{F}$. This explains the name "pole assignment property" in 39.11. Typical transfer functions are $T(s) = \frac{s}{s^2-1}$, which describes an unstable system (since 1 is a pole with a positive real part), while $\frac{s+4}{s^2+2s+2}$ is even asymptotically stable, since the zeros of $s^2 + 2s + 2$, namely $-1 \pm i$, have negative real parts -1.

39.20. Example. If

$$\mathbf{x}(t) = \begin{pmatrix} 0 & \omega \\ -\omega & 0 \end{pmatrix} \mathbf{x}(t) + \begin{pmatrix} 1 \\ 0 \end{pmatrix} \mathbf{u}(t), \quad \mathbf{y}(t) = (1 \quad 1)\, \mathbf{x}(t),$$

(cf. Example 39.8), then

$$T(s) = (1 \quad 1) \begin{pmatrix} s & -\omega \\ \omega & s \end{pmatrix}^{-1} \begin{pmatrix} 1 \\ 0 \end{pmatrix} = \frac{s - \omega}{s^2 + \omega^2}.$$

As in 39.8, we see that this system is stable, but not asymptotically stable.

This leads to the study of functions which are transfer functions of stable systems. We also don't want them to "explode."

39.21. Definition. Let S be the set of all rational functions $T = \frac{p(x)}{q(x)}$, where p and q are polynomial functions from $\mathbb{R}$ to $\mathbb{R}$ whose poles have negative real parts and which stay bounded as $x \to \infty$. The functions in S are called ***stable functions***.

So $x \to \frac{x+4}{x^2+2x+2}$ is stable. Stable functions turn out to be precisely the transfer functions of BIBO-stable linear time-invariant systems. The nontrivial parts of the next theorems can be found in Vidyasagar (1987) and Vidyasagar (1989).

39.22. Theorem.

(i) *S is a subring of the field $\mathbb{R}(x)$ of all rational functions from $\mathbb{R}$ to $\mathbb{R}$ (see 12.5(ii)).*

(ii) *S is even an integral domain. Its quotient field is precisely $\mathbb{R}(x)$, which consists of all transfer functions (of possibly unstable systems). Also, S is "Euclidean," since it has an analogue of the Euclidean division algorithm.*

For feedback constructions as in Figure 39.1, we can easily see that if T_1 is the transfer function of the "uncompensated system" $(\mathbf{F}, \mathbf{G}, \mathbf{H})$ and T_2 the one of the compensator C, then the "total" transfer function of the compensated system is given by $\frac{T_1}{1+T_1T_2}$.

39.23. Theorem. *Let $T = \frac{p}{q} \in S$, where $\gcd(p, q) = 1$. As in Theorem 11.21, there are $a, b \in S$ with $ap + bq = 1$ (**Bezout identity**). If a system with one-dimensional output determined by T can be stabilized, it can be done by the compensator with transfer function $\frac{a}{b}$. More generally, all compensators which make the system stable are given by*

$$r = \frac{a + qc}{b - pc} \qquad (c \in S,\ b - pc \neq 0).$$

So we can choose c to get compensation with "nice" properties ("optimal control"). In the case of more-dimensional output, we need matrices $\mathbf{A}, \mathbf{B}$, all of whose entries are stable functions (see Exercise 5), and we look for matrices $\mathbf{X}, \mathbf{Y}$ such that $\mathbf{XA} + \mathbf{YB} = \mathbf{I}$ ("Right Bezout Identity," see ch. 4 of Vidyasagar (1987)).

This allows easy algebraic computations of combinations of systems. What about series and parallel connections (which are defined as expected, similar to page 346)?

39.24. Theorem. *If two systems have transfer functions T_1, T_2, respectively, then their parallel connection has the transfer function $T_1 + T_2$, while their series connection has the transfer functions $T_1 T_2$. Linear time-invariant systems themselves form a ring w.r.t. parallel and series connection.*

For a proof, see, e.g., Szidarovzky & Bahill (1992). Observe that, from 39.23 and the lines before, this ring is "closed w.r.t. feedback." So it seems that ring theory can contribute a lot to the theory of linear systems.

What about generalizations? The matrices $\mathbf{F}, \mathbf{G}, \mathbf{H}$ might be time-dependent, the changes of states might be given by partial differential equations, and/or the system might be nonlinear. Sometimes, we can approximate a nonlinear system by a linear one; we did that in Example 39.1, when we replaced $\sin x$ by x.

Studying these more general (but more realistic) systems leads to numerous open questions, and makes up a vast area of active research. A class of "tame" nonlinear systems is the one of ***affine-input systems*** (***AI systems***), where the corresponding equations are

$$\begin{aligned}\dot{\mathbf{x}}(t) &= \mathbf{f}(\mathbf{x}(t)) + \mathbf{g}(\mathbf{x}(t))\mathbf{u}(t),\\ \mathbf{y}(t) &= \mathbf{h}(\mathbf{x}(t)),\end{aligned}$$

with sufficiently smooth functions **f**, **g**, **h** in several variables. Many realistic models of industrial plants can be described in this way (see Kugi & Schlacher (1996), Schlacher, Kugi & Irschik (1996), and Haas & Weinhofer (1996)). AI systems are also closed w.r.t. parallel and series connections. But if T_1, T_2 are the transfer functions of two AI systems then their series connection has the transfer map $T_1 \circ T_2$, where $\circ$ denotes the composition of mappings. Why do we get composition now, where we had multiplication before in the linear case?

Consider the composition of the maps $x \to ax$ and $x \to bx$, which is given by $x \to abx$. More generally, we know from linear algebra that the matrix representation of the composition of linear maps is just the product of the matrices corresponding to these maps. Hence, in our context, *multiplication of maps is just the specialization of composition to the linear case*. This fact seems to be widely unknown to system theorists, and it might contribute to several unsuccessful attempts in nonlinear systems theory. Arbitrary maps together with addition and composition form near-rings instead of rings (see 36.8 and 36.9). So, the generalization of 39.22 and 39.24 comes as no surprise.

39.25. Theorem. *AI systems form a near-ring w.r.t. parallel and series connections, if we identify systems with the same input–output behavior. This near-ring is closed w.r.t. AI feedbacks.*

Exercises

1. Prove equation 39.1.
2. Compute $e^{\mathbf{A}}$ for $\mathbf{A} = \begin{pmatrix} 0 & 1 \\ 1 & 0 \end{pmatrix}$, $\mathbf{A} = \begin{pmatrix} 0 & -1 \\ 1 & 0 \end{pmatrix}$, $\mathbf{A} = \begin{pmatrix} 0 & \omega \\ -\omega & 0 \end{pmatrix}$.
3. Let $\mathbf{F} = \begin{pmatrix} 1 & 1 & 0 \\ 0 & 1 & 1 \\ 1 & 0 & 1 \end{pmatrix}$, $\mathbf{G} = \begin{pmatrix} 1 & 1 \\ 0 & 0 \\ 1 & 0 \end{pmatrix}$, and $\mathbf{H}$ be any $p \times 3$ matrix, all with $F = \mathbb{Z}_2$. Is $(\mathbf{F}, \mathbf{G}, \mathbf{H})$ controllable? Can we transform the state $\begin{pmatrix} 0 \\ 1 \\ 0 \end{pmatrix}$ to the state $\begin{pmatrix} 1 \\ 0 \\ 0 \end{pmatrix}$ in ≤ 2 time units?

4. Let $\mathbf{F} = \begin{pmatrix} 1 & 1 & 0 \\ 0 & 1 & -1 \\ 1 & 0 & 1 \end{pmatrix}$, and $\mathbf{G}$ be as in Exercise 3, but now over $F = \mathbb{R}$. Is $(\mathbf{F}, \mathbf{G})$ stable? Asymptotically stable?
5. Construct a compensator which stabilizes the system in Exercise 4, and compute the transfer functions of $(\mathbf{F}, \mathbf{G}, \mathbf{I})$ and $(\mathbf{F} - \mathbf{GC}, \mathbf{G}, \mathbf{I})$. Compare the result with 39.21 and 39.22.
6. Show the assertions in Example 39.10.
7. With $(\mathbf{F}, \mathbf{G})$ as in Exercise 4, does it have a feedback to a cyclic vector?
8. Work out the details in 39.16. Why does the method not work for $R[x]$?
9. Let a system have the transfer function $\mathbf{T}\colon x \mapsto \frac{x+1}{x^2-2}$. Is the system stable? Can it be stabilized?
10. Prove Theorem 39.25.

Notes

Among the many fine books on mathematical (in particular, algebraic) systems theory, we mention Isidori (1989), Kalman, Falb & Arbib (1969), Knobloch, Isidori & Flockerzi (1993), Knobloch & Kwakernaak (1985), Szidarovzky & Bahill (1992), Vidyasagar (1987), Vidyasagar (1989), and Wonham (1974). The pole assignment problem is treated, for instance, in Brewer, Katz & Ullery (1987). Applications of AI systems were mentioned in the text above. Pilz (1996) contains more on the role of near-rings in systems theory. The problem of linearization of systems is treated in Szidarovzky & Bahill (1992), for example.

§40 Abstract Data Types

In our journey through algebra, we have seen quite a number of algebraic structures, such as lattices, groups, and rings. In all of these cases, we have studied substructures, homomorphisms, direct products, congruence relations, and the like. It should be possible to unify these different treatments through a general approach. So let us climb a mountain from which we can see all these structures below, and let us learn from these structures how to build up a general theory. Surprisingly enough, this general theory will have some quite practical applications.

Our first guess might be to define a "general algebra" as a set A equipped with some (binary) operations $\circ_1, \circ_2, \ldots, \circ_n$. Lattices and rings would fit nicely into this scheme with $n = 2$. But we run into difficulties with vector spaces $(V, +)$ over a field F. The group part $(V, +)$ causes no difficulty (with $n = 1$ above), but what should we do with the multiplication of vectors $\mathbf{v}$ with scalars λ? One way out (later we will meet another way to solve the problem) is to consider, for each $\lambda \in F$, the map $\omega_\lambda\colon V \to V$; $\mathbf{v} \mapsto \lambda\mathbf{v}$. The operations on a set A that we have met so far were maps from A^2 into A. So why not allow maps from A into A as "unary operations" and maps from A^3 into A as "ternary operations":

40.1. Definition. Let n be in $\mathbb{N}_0$. An ***n-ary operation*** on A is a map from A^n to A.

Here, $A^0 := \{\emptyset\}$, so a "nullary" operation is a map $f\colon A^0 \to A$, which can be identified with the element $f(\emptyset) \in A$. Hence nullary operations simply "single out" a certain element of A.

40.2. Definition. A ***(universal) algebra*** is a set A together with a family $\Omega = (\omega_i)_{i \in I}$ of n_i-ary operations ω_i. We call $\tau := (n_i)_{i \in I}$ the ***type*** of Ω. Let $\mathcal{K}(\tau)$ be the class of all algebras of type τ.

If $\tau = (2)$, then $\mathcal{K}(\tau)$ is the class of all algebras with one binary operation; it includes all groups and all semigroups. More precisely, $\mathcal{K}((2))$, briefly written as $\mathcal{K}(2)$, is the class of all "***groupoids***" (i.e., sets with a binary operation which is not necessarily associative). Observe that we might have infinitely many operations.

Let (A, Ω) be an algebra as in 40.2. If $B \subseteq A$, then B is called a ***subalgebra*** of A if $\omega(b_1, \ldots, b_n) \in B$ for $b_i \in B$ and all n-ary operations $\omega \in \Omega$. Then (B, Ω) is again in $\mathcal{K}(\tau)$, and we write $A \leq B$, as we did so often before. An equivalence relation $\sim$ in A is called ***compatible*** or a ***congruence relation*** in (A, Ω) if $a_1 \sim b_1, \ldots, a_n \sim b_n$ implies $\omega(a_1, \ldots, a_n) \sim \omega(b_1, \ldots, b_n)$ for all n-ary $\omega \in \Omega$. In this case, the equivalence classes $[a]$ again form an algebra in $\mathcal{K}(\tau)$, denoted by $A/\sim$, and called the ***factor algebra*** of A w.r.t. $\sim$. The operations in $A/\sim$ are, of course, defined by $\omega([a_1], \ldots, [a_n]) := [\omega(a_1, \ldots, a_n)]$. A map h from $(A, \Omega) \in \mathcal{K}(\tau)$ to $(A', \Omega') \in \mathcal{K}(\tau)$ is called a ***homomorphism*** if $h(\omega_1(a_1, \ldots, a_n)) = \omega(h(a_1), \ldots, h(a_n))$ always holds when the expressions make sense. We leave it to the reader to define direct products of algebras (of the same type). So, all these constructions go through without any difficulty. The notions "monomorphism," $A \cong A'$, etc., seem to be clear as well.

However, problems await us just around the corner. $(\mathbb{Z}, +)$ is, as a group, an algebra in $\mathcal{K}(2)$. Observe that $(\mathbb{N}, +)$ is a subalgebra, but, of course, not a subgroup of $(\mathbb{Z}, +)$. More generally, the subalgebras of groups G are precisely the sub*semi*groups of G. That can be resolved as follows. We can consider a group $(G, \circ)$ also as being equipped with a binary operation $\circ$, a unary operation $^{-1}$: $G \to G$ sending each $g \in G$ into its inverse g^{-1}, and a nullary operation $\overline{1}$ selecting the neutral element. Then a group has $\Omega = (\circ, ^{-1}, \overline{1})$ and the type $(2, 1, 0)$. Now, the subalgebras of a group must contain the identity and must be closed w.r.t. inversion; hence the subalgebras are precisely the subgroups. So we shall adopt this view, and rings will have type $(2, 1, 0, 2)$, rings with identity are of type $(2, 1, 0, 2, 0)$, lattices are of type $(2, 2)$, and Boolean algebras have $\Omega = (\cap, \cup, \overline{0}, \overline{1}, ')$, where $'$ is complementation, and their type is $(2, 2, 0, 0, 1)$. In particular, this means that homomorphisms of Boolean algebras must preserve zero, one, and the complements, i.e., they must be just Boolean homomorphisms.

Many theorems we have met in "individual structures" carry over. The reader is invited to prove

40.3. Theorem (*Homomorphism Theorem*). *If $(A, \Omega) \in \mathcal{K}(\tau)$, and h is an epimorphism from A to $(A', \Omega') \in \mathcal{K}(\tau)$, then $\sim$ with $a \sim b :\iff h(a) = h(b)$ is a congruence in A and*

$$A/{\sim} \cong A'.$$

Conversely, if $\sim$ is a congruence in A, Ω, then π: $A \to A/{\sim}$; $a \mapsto [a]$ is an epimorphism.

How are groups specified within the class $\mathcal{K}(\tau)$ with $\tau = (2, 1, 0)$? The answer is, by "equations": associativity, $x \circ 1 = 1 \circ x$, and $x \circ x^{-1} = x^{-1} \circ x = 1$ for all x. The definition of "equation" is surprisingly difficult, but we have some experience from §5, where we briefly studied "Boolean equations." First we build up "terms." We fix a countable set $X = \{x_1, x_2, \ldots\}$.

40.4. Definition. Let Ω be a family of operations. A ***term*** over Ω and X is defined by:

(i) Every nullary operation in Ω and every x_i is a term.

(ii) If ω is an n-ary operation in Ω and if $t_1, \ldots, t_n$ are terms, so is $\omega(t_1, \ldots, t_n)$.

More precisely, a term is what can be constructed by (i) and (ii) in a finite number of steps. Equality is formal equality. The set $T_X(\Omega)$ of all terms is again an algebra of $\mathcal{K}(\tau)$ in a natural way, if τ is the type of Ω, and is called the ***term algebra*** over τ and X.

So, some terms over $(2, 1, 0)$ are

$$x_1 \circ (x_2 \circ x_3),\ (x_1 \circ x_2) \circ x_3,\ (x_1 \circ x_1^{-1}) \circ \overline{1} \circ \overline{1}^{-1},\ \text{etc.}$$

Recall that $x_1 \circ (x_2 \circ x_3) \neq (x_1 \circ x_2) \circ x_3$, but we might use that to define associativity.

40.5. Definition. An ***algebraic equation*** over τ is a pair (t_1, t_2) of terms over τ. The ***equation*** (t_1, t_2) ***holds*** in an algebra $A \in \mathcal{K}(\tau)$ if every replacement of $x_1, x_2, \ldots \in X$ by elements $a_1, a_2, \ldots \in A$ gives the same element in A.

So, some equations which hold in groups are: $(x_1 \circ (x_2 \circ x_3), (x_1 \circ x_2) \circ x_3)$, $(\overline{1} \circ x_1, x_1)$, $(x_1 \circ x_1^{-1}, \overline{1})$, while $(x_1 \circ x_2, x_2 \circ x_1)$ is an additional equation, which holds precisely for abelian groups.

Many classes of algebraic structures, like semigroups, groups, abelian groups, lattices, distributive lattices, rings, or vector spaces, can be characterized as subclasses of precisely those algebras in $\mathcal{K}(\tau)$ which fulfill a certain set of equations:

40.6. Definition. A class $\mathcal{K} \subseteq \mathcal{K}(\tau)$ is ***equationally definable*** or a ***variety*** if there is a set E of equations over τ such that

$$\mathcal{K} = \{A \in \mathcal{K}(\tau) \mid \text{every equation of } E \text{ holds in } A\}.$$

The following characterization is most famous.

40.7. Theorem (*Birkhoff's Theorem*). *$\mathcal{K} \subseteq \mathcal{K}(\tau)$ is a variety iff $\mathcal{K}$ is closed w.r.t. forming subalgebras, homomorphic images, and (arbitrary) direct products.*

So the classes mentioned before 40.6 are varieties. For distributive lattices, see Exercise 2.7. Exercise 2.8 shows that the class of complemented lattices is not a variety. $\mathcal{K}(\tau)$ itself is a variety with $E = \emptyset$ in 40.6. The class of finite groups is not a variety, since, e.g., $\mathbb{Z}_1 \times \mathbb{Z}_2 \times \mathbb{Z}_2 \times \cdots$ is infinite. Hence "finiteness" for groups cannot be defined by equations.

Commutative rings with identity are also a variety. A subclass of it is the class $\mathcal{F}$ of fields—one of our favorite structures in this book. Here, however, they cause trouble. The existence of x^{-1} is only to be required if $x \neq 0$, and $x_1 \neq 0 \Longrightarrow x_1x_1^{-1} = \bar{1}$ is not an equation in our sense. But maybe we have chosen the wrong type and a "wrong axiom system." No; since a direct product of fields is not a field any more, $\mathcal{F}$ is not a variety, and we cannot define the existence of inverses for nonzero elements by equations. In varieties, we have an analogue of free semigroups (28.20).

40.8. Definition. A subset B of an algebra $A \in \mathcal{K} \subseteq \mathcal{K}(\tau)$ is called a ***base*** w.r.t. $\mathcal{K}$ if every map from B to some $A' \in \mathcal{K}$ can uniquely be extended to a homomorphism from A to A':

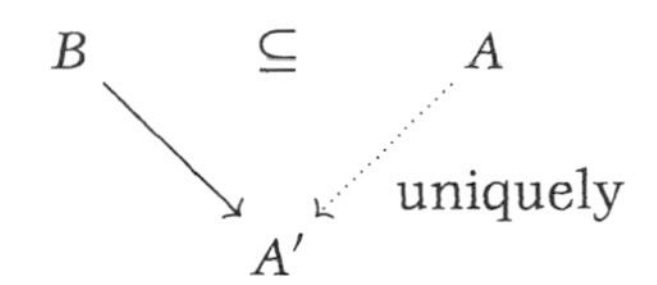

A is ***free*** in $\mathcal{K}$ if it has a base B; then A is also called ***free on*** B.

For the class S of all semigroups (as subclass of $\mathcal{K}(2)$), we get the concept of a free semigroup. For the class $\mathcal{M}$ of monoids with $\mathcal{M} \subseteq \mathcal{K}(2,0)$, we get precisely free monoids, etc. If ${}_F\mathcal{V}$ is the class of all vector spaces over the (same) field F, then every algebra in ${}_F\mathcal{V}$ is free—a famous fact in linear algebra, which gives rise to matrix representations of linear maps.

40.9. Theorem. *Let $\mathcal{V}$ be a variety in $\mathcal{K}(\tau)$, and define $\sim_{\mathcal{V}}$ in the term algebra $T_X(\tau)$ via*

$$t_1 \sim_{\mathcal{V}} t_2 :\Longleftrightarrow (t_1, t_2) \text{ holds in every algebra of } \mathcal{V}.$$

Then $\sim_{\mathcal{V}}$ is a congruence, $F_X(\mathcal{V}) := T_X(\tau)/\sim_{\mathcal{V}}$ is in $\mathcal{V}$, and it is the free algebra in $\mathcal{V}$ with basis $\{[x] \mid x \in X\}$. Every free algebra in $\mathcal{V}$ is isomorphic to some $F_X(\mathcal{V})$, and every algebra A in $\mathcal{V}$ is the homomorphic image of a free algebra $F_X(\mathcal{V})$ if A can be generated by a subset of the same cardinality as X.

So, we have free distributive lattices, but not "free fields" (cf. Exercise 7).

If E is a set of equations over τ,

$$\mathcal{K}_E := \{A \in \mathcal{K}(\tau) \mid \text{all equations of } E \text{ hold in } A\}$$

is then a variety, and every algebra in $\mathcal{K}_E$ is a ***model*** of E. Conversely, we might start with some $\mathcal{K} \subseteq \mathcal{K}(\tau)$ and look for the ***theory*** $\mathrm{Th}(\mathcal{K})$, i.e., the set of all equations which hold in each $A \in \mathcal{K}$. Clearly, $\mathcal{K} \subseteq \mathcal{K}_{\mathrm{Th}(\mathcal{K})}$, with equality iff $\mathcal{K}$ is a variety. $\mathcal{K}_{\mathrm{Th}(\mathcal{K})}$ is also called the ***variety generated*** by $\mathcal{K}$.

We can form ***relational algebras*** (i.e., algebras with some relations) and we can do similar things: study free relational algebras, consider models for a theory, etc. Model theory has developed into a deep and difficult branch of mathematics.

Algebras with relations are most important in the design of computer programs. Extremely important, of course, is computation in $\mathbb{N}_0$. Here we might need the special role of 0 (nullary operation $\overline{0}$), the ***successor function*** succ: $n \mapsto n+1$ (unary), addition (binary), and $\leq$ in order to compare numbers. So we have the type $(0, 1, 2)$ and one relation, together with some equations ($0 + x = x$, etc.) and "other laws," like "$\leq$ is a linear order relation." In computer science, it is useful to extract the essential features of the types of data we want to calculate with (like natural numbers above). Representing them (as strings, graphs, or so) in a special way usually points to a specific computer language, often at a too early stage. It is advisable just to describe the features of the data and to select a representation of the data much later when we can see better what might be the most economical way of doing it. It is easier to work with the "structure" before it is written down in a very implementation-oriented way. With our knowledge now, we can define concepts of Computer Science very concisely.

40.10. Definition. A ***data type*** is a relational algebra.

nat $:= (\mathbb{N}_0, 0, \mathrm{succ}, +, \leq)$ is an example of a data type. Another example is, for an alphabet A, queue $:= (A^*, \Lambda, \min, \mathrm{remove}, *, \sim)$, where A^* is the free monoid on A (whose elements we interpret as a queue of elements of A), Λ is the empty word, $*$ the concatenation, $\min(a_1 a_2 \dots a_n)$ is the queue $\{a_1\}$ with $\min(\Lambda) = \Lambda$, $\mathrm{remove}(a_1 \dots a_n) = a_1 \dots a_{n-1}$ with $\mathrm{remove}(\Lambda) := \Lambda$, and $a_1 \dots a_n \sim a'_1 \dots a'_m :\iff n = m$. So we get a data type of type $(0, 1, 1, 2)$ and an equivalence relation. We might add some equations like $\min(x * y) = \min(\min(x) * \min(y))$ and "laws" like $x \sim y \Longrightarrow \mathrm{remove}(x) \sim \mathrm{remove}(y)$, etc.

Very often we come to situations like "min(Λ)," which we might want to be undefined, like 0^{-1} in a field. The role of relations is still not quite clear. Also, the view of the multiplication of vectors with scalars as a (possibly infinite) family of unary operations, as mentioned earlier in this section, might look a bit strange. All the problem areas can be resolved simultaneously by generalizing the concept of an operation.

40.11. Definition. Let S be a collection of nonempty sets, which are called ***sorts***. A ***(general) operation*** on S is a partial map from $S_1 \times \cdots \times S_n \to S$, where $S_1, \ldots, S_n, S \in S$. A ***(general) data type*** is a set S with a family Ω of (general) operations. If $S = \{S\}$, the data type is called ***homogeneous***, otherwise ***inhomogeneous***. The pair (S, Ω) is called the ***signature*** of Ω.

Here, a partial map $f\colon A \to B$ is a map which is defined on a subset of A. A ring R is a homogeneous data type with $S = \{R\}$, while a vector space ${}_FV$ over a field F is inhomogeneous with $S = \{F, V\}$. For a field F, the inversion $i\colon x \mapsto x^{-1}$ is now also an operation, since it is a partial map from $F \to F$. And what about relations? If R is a relation on A, we can characterize it via the operation $\omega_R\colon A \times A \to \{0, 1\} = \mathbb{B}$, where $\omega_R(a, b) = 1 :\iff a\,R\,b$. If we write "general programs" and if we, for instance, want to leave open which order relation should be specified, we are in the same situation as before for universal algebras. It seems clear how to define when two data types are of the same type; it is technically quite involved to write it down, so we omit it.

40.12. Definition. An ***abstract data type*** is a class $\mathcal{A}$ of algebras of the same data type. $\mathcal{A}$ is called ***monomorphic*** if all algebras in $\mathcal{A}$ are isomorphic, otherwise $\mathcal{A}$ is ***polymorphic***.

The classes of groups or ordered fields are polymorphic. The same applies to all algebras of the same type as $(\mathbb{N}_0, 0, \mathrm{succ}, +, \leq)$, since $(\mathbb{R}, 0, \mathrm{succ}, +, \leq)$ is a nonisomorphic model. The class of all Boolean algebras with 16 elements (as algebras of type $(2, 2, 0, 0, 1)$, as on page 449) is monomorphic due to Theorem 3.12.

Now we can do everything again for our more general concepts of operations and (abstract) data types: we can define terms, equations, varieties, models, and so on. Then $(\mathbb{N}_0, 0, \mathrm{succ}, +, \leq)$ would, for instance, be a model in the variety of all $(A, \overline{0}, \mathrm{succ}, +, \leq)$, where $\overline{0}$, succ, $+$ are operations as before of type $0, 1, 2$, respectively, and $\leq$ is a map from $A \times A \to \mathbb{B} = \{0, 1\}$, and equations like $((x_1 + x_2) + x_3, x_1 + (x_2 + x_3))$,

$(x_1 + 0, x_1)$, $(x_1 + x_2, x_2 + x_1)$, $(\leq (0, x_1), \text{true})$, $(\leq (\text{succ}(x), 0), \text{false})$. These equations are usually written as $(x_1 + x_2) + x_3 = x_1 + (x_2 + x_3)$, $x_1 + 0 = x_1$, $x_1 + x_2 = x_2 + x_1$, $(0 \leq x_1) = \text{true}$, $(\text{succ}(x) \leq 0) = \text{false}$, etc.

Finding enough equations (or other "laws") to describe an abstract data type properly is the problem of ***specification***. This clearly can be a highly nontrivial problem.

As Ehrich, Gogolla & Lipeck (1989) point out, every signature determines a family of context-free grammars in the sense of Definition 30.7, where the productions are basically the operations, and a kind of converse is also true. The generated language is just the term algebra over the signature.

Again, if $\mathcal{A}$ is an abstract data type, every data type $A \in \mathcal{A}$ is the homomorphic image of a term algebra T, so $A \cong T/\sim$ for a suitable congruence $\sim$. This is of enormous practical importance, since we can handle huge data types (even some infinite ones) on a computer. We have had this situation already in the idea of presentation of a semigroup (see 28.23).

Computers can handle terms and the computation with them usually very efficiently. But a new and serious problem arises. If $A \cong T/\sim$ as before, and if t_1, t_2 are terms with $t_1 \sim t_2$, then $[t_1] = [t_2]$ represents the same element in A. But how can we decide if $t_1 \sim t_2$? This is known as the ***word problem*** and it is hard, because it can be shown to be "undecidable" in its general form.

The terms in Boolean algebras (with type $(2, 2, 0, 0, 1)$) are precisely the Boolean polynomials. Here we had a normal form (the disjunctive normal form), and, since $\mathbb{B}^n \cong P_n/\sim$, two terms (polynomials) in P_n are equivalent iff they have the same DNF. So it would be great if we had normal forms for terms in other situations as well, plus an efficient algorithm to reduce a term into its normal form. It will not always be possible.

The basic idea is to establish a "rewriting system" which reduces each term to a "shortest" one, and to hope that these shortest terms form a normal system. The first problem is how to define "shorter." Is, in the case of groups, $(x_1 \circ x_2) \circ x_3$ shorter than $x_1 \circ (x_2 \circ x_3)$? Suppose we have overcome this difficulty. Then we will try to use the equations (or, more generally, the "laws" which hold in A) to reduce terms into equivalent ones which are shorter. This "greedy" strategy is too simple, as we shall see now. Let us work again in the variety of groups. Let T be the term algebra over $X = \{x_1, x_2\}$ and A the symmetric group S_4. Since S_4 is generated by

$a_1 = \begin{pmatrix} 1 & 2 & 3 & 4 \\ 2 & 3 & 4 & 1 \end{pmatrix} = (1,2,3,4)$ and $a_2 = \begin{pmatrix} 1 & 2 & 3 & 4 \\ 2 & 1 & 3 & 4 \end{pmatrix} = (1,2)$, we might extend the map $x_1 \mapsto a_1, x_2 \mapsto a_2$ to an epimorphism h from T to A. The generators a_1, a_2 fulfill the equations $a_1^4 = \text{id}$, $a_2^2 = \text{id}$, $(a_1a_2)^3 = \text{id}$. Suppose we have the terms $t_1 = x_2x_1x_2x_1$, $t_2 = x_1^3x_2$. Does $t_1 \sim t_2$ hold, i.e., is $h(t_1) = h(t_2)$? No immediate simplifications seem to be possible. But if we first make t_1 longer, we get:

$$\begin{aligned} t_1 &= x_2x_1x_2x_1 \sim x_1^4x_2x_1x_2x_1x_2^2 \sim x_1^3(x_1x_2x_1x_2x_1x_2)x_2 \\ &\sim x_1^3 1 x_2 \sim x_1^3x_2 = t_2. \end{aligned}$$

Despite this we need a better method than some undirected trial-and-error search. Presently, a very active area of research is the development of good "rewriting systems," because if we cannot decide the equivalence of terms (i.e., the equality in A), we cannot work in A with a computer in a satisfactory manner. Recall that for Boolean polynomials, and also for the elements of a finite field $\mathbb{F}_q[x]/(f)$ (see page 127), we do have a good rewriting system.

Let us close with some elementary information on possible rewriting systems. As before, we do not give mathematically precise definitions in order to avoid a huge technical apparatus. Let us again consider the situation $A \cong T/\!\sim$, and let us write $t_1 \to t_2$ if $t_1 \sim t_2$, t_2 is shorter than t_1 (or $t_1 = t_2$), and t_2 is obtained from t_1 by applying the "laws" in A a finite number of times (as we did above with t_1, t_2 in $A = S_4$). Let N be the set of all shortest terms. The whole setup might be called a "rewriting system." It is called ***confluent*** if for all $t, u, v \in T$ with $t \to u$ and $t \to v$ there is some $w \in T$ with $u \to w$ and $v \to w$:

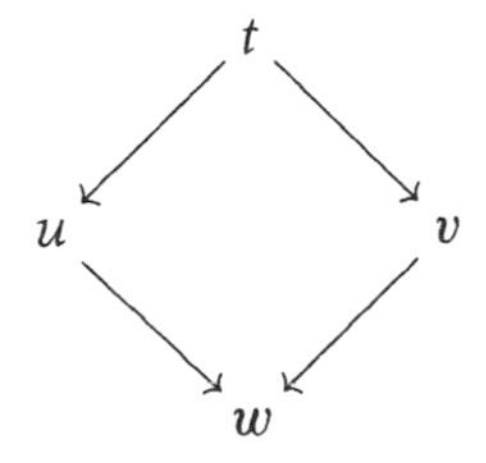

The system is called ***terminating*** if for each $t \in T$ there is some $n \in N$ with $t \to n$, and it has the ***Church-Rosser property*** if $t_1 \sim t_2$ always implies the existence of some $t \in T$ with $t_1 \to t$ and $t_2 \to t$:

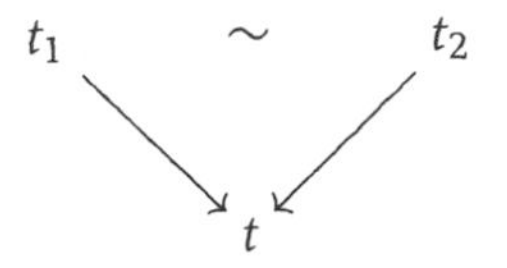

40.13. Proposition. A rewriting system is confluent iff it has the Church-Rosser property.

We indicate the proof by an example. Suppose we have confluence and $t_1 \sim t_2$. We use the fact that we can get a "chain" from t_1 to t_2 (and back), like

$$t_1 \longrightarrow t_3 \longrightarrow t_4 \longleftarrow t_5 \longrightarrow t_6 \longleftarrow t_7 \longrightarrow t_2$$

By confluence, t_4 and t_6 "reduce" to some t_8, etc.,

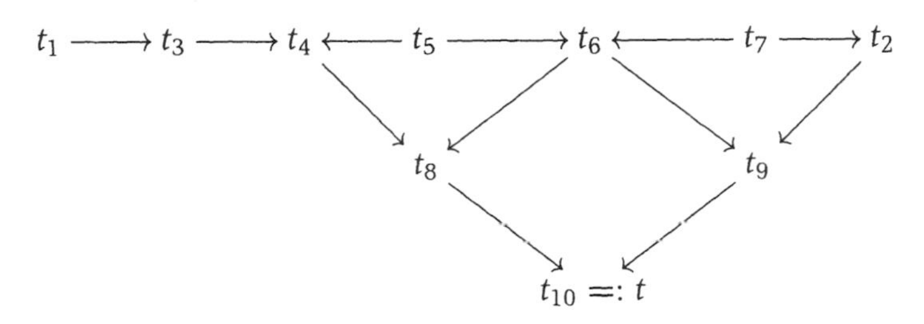

so

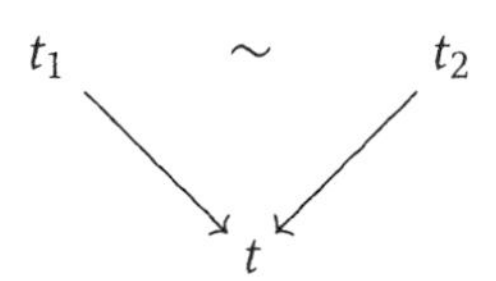

The converse is immediate since $u \leftarrow t \rightarrow v$ implies $u \sim v$.

40.14. Theorem. *Let the rewriting system be confluent and terminating. Then for each $t \in T$ there is precisely one $n \in N$ (its **normal form**) with $t \rightarrow n$. Two terms are equivalent iff they have the same normal form.*

Proof. If $t \in T$, we reach some $n \in N$ with $t \rightarrow n$ in finitely many steps, since the system is terminating. If also $t \rightarrow n' \in N$, then $n \rightarrow n'' \leftarrow n'$ for some n''. By minimality, $n = n'$. If $t_1 \sim t_2$ we will find some $n \in N$ after finitely many steps (due to the Church-Rosser property) with $t_1 \rightarrow n \leftarrow t_2$. So t_1 and t_2 have the same normal form. Of course, if two terms have the same normal form, they must be equivalent. □

Abstract data types are also helpful in converting "object-oriented" data structures into "machine-oriented" ones. Also, algebra tells us how to identify objects with the "same properties" properly: switch to a quotient structure in the same abstract data type. So, even the most abstract algebraic concepts have practical uses!

Exercises

1. Let groups be first considered as algebras of type (2) and then as ones of type (2, 1, 0), according to the lines before 40.3 on page 449. Write down the definitions for group homomorphism in both cases. Do these definitions mean the same?
2. Give a term over some $\Omega = (\omega_1, \omega_2, \omega_3)$ of type (2, 1, 0) which is not a term over the type $\Omega' = (\omega_1)$.
3. What are terms in the case of Boolean algebras? Are the equations in the sense of 5.9 the same as in the sense of 40.5?
4. What does the free ring over $\{a\}$ look like?
5. Is the class of all integral rings a variety?
6. Let $\mathcal{V}$ be the variety of groups with the additional equation $x^2 = 1$. Is $\mathcal{V}$ a subvariety of the variety of all abelian groups? What do the free algebras over $\{a\}$ and over $\{a, b\}$ look like?
7. Do there exist free fields?
8. Give examples of algebras in the variety generated by all fields.
9. What does 40.9 mean for the variety of Boolean algebras?
10. Determine the free Boolean algebras over a finite basis.
11. Do near-rings form a variety?
12. Find an equation in the theory of groups which is not an axiom.
13. Let $\mathcal{K}$ be the class of all semigroups with the equations $x_1x_2 = x_2x_1$, $x_2^2x_1 = x_1$, $x_2^3 = x_2$, $x_1^3 = x_1^2$, $x_1x_2 = x_1$. Find a model for $\mathcal{K}$ with four elements.
14. Define an abstract data type for ***ordered fields*** (i.e., fields F with a linear order relation $\leq$ such that $a \leq b$ and $c \in F$ imply $a+c \leq b+c$ and, if $c \geq 0$, $ac \leq bc$). Give two nonisomorphic models for this data type.
15. Indicate what the word problem, confluences, the Church-Rosser property, and 40.14 mean for Boolean algebras.

Notes

After Van der Waerden's book on modern algebra the concept of universal algebra was not far away. Birkhoff developed the concepts of an "algebra" from the approach of Van der Waerden and took the name "universal algebra" from Whitehead's book. In 1934, MacLane also stated some ideas on universal algebra, influenced by his stay in Göttingen, but did not publish them. Kurosh (1965), Grätzer (1979), Burris & Sankappanavar (1981), and Ihringer (1988) are standard texts on universal algebras. This theory, especially concerning varieties and equations, can also be found, for instance, in Lausch & Nöbauer (1973). Abstract data types are nicely described in Ehrich, Gogolla & Lipeck (1989). A good source for rewriting rules is Winkler (1996) and the literature cited there. Perhaps the most important result for rewriting systems is the Knuth-Bendix algorithm (see Knuth (1970)). For the interested reader, we also recommend Le Chenadec (1986) and Leech (1970).

Bibliography

Abbott, J. C. (Ed.) (1970). *Trends in Lattice Theory*. New York: Van Nostrand.

Arbib, M. A. (1968). *Algebraic Theory of Machines, Languages, and Semigroups*. New York: Academic Press.

Arbib, M. A. (1969). *Theories of Abstract Automata*. Englewood Cliffs, N.J.: Prentice Hall.

Ballonoff, P. (1974). *Genetics and Social Structure*. Stroudsburg, Penn.: Dowden, Hutchinson & Ross.

Barnes, D. W. & J. M. Mack (1975). *An Algebraic Introduction to Mathematical Logic*. New York: Springer-Verlag.

Batten, L. M. (1993). *The Theory of Finite Linear Spaces*. Cambridge: Cambridge University Press.

Becker, T. & V. Weisspfenning (1993). *Gröbner Bases: A Computational Approach to Commutative Algebra*. New York: Springer-Verlag.

Beker, H. & F. Piper (1982). *Cipher Systems, The Protection of Communications*. London: Northwood Books.

Berlekamp, E. R. (1984). *Algebraic Coding Theory* (revised ed.). Laguna Hills: Algean Park Press.

Beth, T. (1995). Arbeitsbericht der Forschungsgruppen. Technical report, University of Karlsruhe (Germany).

Beth, T., M. Frisch & G. J. Simmons (1992). *Public-Key Cryptography: State of the Art and Future Directions*. Berlin: Springer-Verlag.

Beth, T., P. Hess & K. Wirl (1983). *Kryptographie*. Stuttgart: Teubner.

Beth, T., D. Jungnickel & H. Lenz (1985). *Design Theory*. Mannheim: Bibliographisches Institut.

Betten, A., A. Kerber, H. Kohnert, R. Laue & A. Wassermann (1996). The discovery of a simple 7-design with automorphism group $P\Gamma L(2, 32)$. See Cohen et al. (1996), pp. 131–145.

Bhargava, V. K., D. Haccoun, R. Matyas & P. P. Nuspl (1981). *Digital Communications by Satellite*. New York: Wiley.

Biggs, N. L. (1985). *Discrete Mathematics*. Oxford: Clarendon Press.

Binder, F. (1995). Polynomial decomposition: Theoretical results and algortihms. Master's thesis, Johannes Kepler Universität Linz, Austria, electronically available from ftp://bruckner.stoch.uni-linz.ac.at/pub/decomposition.

Binder, F. (1996). Fast computations in the lattice of polynomial rational function fields. In *Proceedings of the ISSAC'96, Zürich, Switzerland*.

Birkhoff, G. (1967). *Lattice Theory*. Providence, R.I.: Amer. Math. Soc.

Birkhoff, G. (1976). The rise of modern algebra. In J. D. Tarwater, J. T. White, and J. D. Miller (Eds.), *Man and Institutions in American Mathematics*, Number 13 in Graduate Studies, pp. 41–85. Lubbock, Texas: Texas Technical University.

Birkhoff, G. & T. C. Bartee (1970). *Modern Applied Algebra*. New York: McGraw-Hill.

Birkhoff, G. & S. MacLane (1977). *A Survey of Modern Algebra* (4th ed.). New York: Macmillan.

Blahut, R. E. (1983). *Theory and Practice of Error Control Codes*. Reading, Mass.: Addison-Wesley.

Blake, I. F. (1973). *Algebraic Coding Theory: History and Development*. Stroudsburg, Penn.: Dowden, Hutchinson & Ross.

Blake, I. F. & R. C. Mullin (1975). *The Mathematical Theory of Coding*. New York: Academic Press.

Blakley, G. R. & I. Borosh (1996). Codes, *Pragocrypt '96*, J. Pribyl (ed.), Prague: CTU Publishing House.

Bobrow, L. S. & M. A. Arbib (1974). *Discrete Mathematics*. Philadelphia: Saunders.

Bolker, E. (1973). The spinor spanner. *Amer. Math. Monthly 80*, 977–984.

Boorman, S. A. & H. C. White (1976). Social structure from multiple networks, II: Role structures. *Amer. J. Sociol. 81*, 1384–1466.

Box, G. E. P., W. G. Hunter & J. Hunter (1993). *Statistics for Experimenters*. New York: Wiley.

Boyd, J. P., J. H. Haehl & L. D. Sailer (1972). Kinship systems and inverse semigroups. *J. Math. Sociol. 76*, 37–61.

Breiger, R. L., S. A. Boorman & P. Arabie (1975). An algorithm for clustering relational data with applications to social network analysis and comparison with multi-dimensional scaling. *J. Math. Psychol. 12*, 328–383.

Brewer, J., D. Katz & W. Ullery (1987). Pole assignability in polynomial rings, power series rings, and Prüfer domains. *J. Algebra 106*, 265–286.

Brouwer, A. E., J. B. Shearer, N. J. A. Sloane & W. D. Smith (1990). A new table of constant weight codes. *IEEE Trans. Inform. Theory 36*, 1334–1380.

Buchberger, B. (1985). Gröbner bases: An algorithmic method in polynomial ideal theory. In N. K. Bose (Ed.), *Multidimensional Systems Theory*, pp. 184–232. Dordrecht: Reidel.

Burris, S. & H. P. Sankappanavar (1981). *A Course in Universal Algebra*. New York: Springer-Verlag.

Cameron, P. J. (1994). *Combinatorics*. Cambridge: Cambridge University Press.

Cameron, P. J. & J. H. van Lint (1991). *Designs, Graphs, Codes, and their Links*. London Math. Soc. Student Texts. Cambridge: Cambridge University Press.

Carlson, R. (1980). *An Application of Mathematical Groups to Structures of Human Groups*, Volume 1(3) of *UMAP Module 476*. Boston: Birkhäuser.

Chen, J. Q. (1989). *Group Representation Theory for Physicists*. Singapore: World Scientific.

Childs, L. A. (1995). *A Concrete Introduction to Higher Algebra* (2nd ed.). New York: Springer-Verlag.

Chomsky, N. (1957). *Syntactic Structures*. Den Haag: Mouton.

Clausen, M. & U. Baum (1993). *Fast Fourier Transforms*. Mannheim: Bibliographisches Institut.

Clay, J. R. (1992). *Nearrings: Geneses and Applications*. Oxford: Oxford University Press.

Cohen, G., M. Giusti & T. Mora (Eds.) (1996). *Applied Algebra, Algebraic Algorithms, and Error-Correcting Codes*, Volume 948 of Lecture Notes in Computer Science. New York: Springer-Verlag.

Cohen, S. & H. Niederreiter (Eds.) (1996). *Finite Fields and Applications*. Cambridge: Cambridge University Press.

Colbourn, C. J. & J. H. Dinitz (1996). *The CRC Handbook of Combinatorial Designs*. Boca Raton, FL: CRC Press.

Coleman, A. J. (1997). Groups and physics—dogmatic opinions of a senior citizen. *Notices Amer. Math. Soc. 44*, 8–17.

Conway, J. H. & N. J. A. Sloane (1988). *Sphere Packings, Lattices, and Groups*. New York: Springer-Verlag.

Cotton, F. A. (1971). *Chemical Applications of Group Theory*. New York: Wiley.

Cox, D., J. Little & D. O'Shea (1992). *Ideals, Varieties, and Algorithms*. Undergraduate Texts in Mathematics. New York: Springer-Verlag.

Davio, M., J. P. Deschaps & A. Thayse (1978). *Discrete and Switching Functions*. New York: McGraw-Hill.

De Bruijn, N. G. (1981). Pólya's theory of counting. In E. F. Beckenbach (Ed.), *Applied Combinatorial Mathematics*. Florida: Malabar.

de Figueiredo, R. J. P. & G. Chen (1993). *Nonlinear Feedback Control Systems*. Boston: Academic Press.

Denes, J. & A. D. Keedwell (1974). *Latin Squares and Their Applications*. London: English Universities Press.

Denes, J. & A. D. Keedwell (1991). *Latin Squares: New Developments in the Theory and Applications*. Amsterdam: North-Holland.

Denning, D. E. R. (1983). *Cryptography and Data Security*. Reading, Mass.: Addison-Wesley.

Dickson, L. E. (1901). *Linear Groups with an Exposition of the Galois Field Theory*. Leipzig: Teubner. Reprint: Dover, 1958.

Diffie, W. & M. E. Hellman (1976). New directions in cryptography. *IEEE Trans. Infom. Theory 22*, 644–684.

Dokter, F. & J. Steinhauer (1972). *Digitale Elektronik*. Hamburg: Philips GmbH.

Dornhoff, L. & F. E. Hohn (1978). *Applied Modern Algebra*. New York: Macmillan.

Dorninger, D. W. & W. B. Müller (1984). *Allgemeine Algebra und Anwendungen*. Stuttgart: Teubner.

Dworatschek, S. (1970). *Schaltalgebra und digitale Grundschaltungen*. Berlin: De Gruyter.

Ehrich, H. D., M. Gogolla & U. W. Lipeck (1989). *Algebraische Spezifikation abstrakter Datentypen*. Stuttgart: Teubner.

Eilenberg, S. (1974). *Automata, Languages, and Machines*, Vols. I, II. New York: Academic Press.

Elliot, J. P. & P. G. Dawber (1979). *Symmetry in Physics*. New York: Macmillan.

Farmer, D. W. (1996). *Groups and Symmetry*. Number 5 in Mathematical World Series. Providence, R.I.: Amer. Math. Soc.

Fässler, A. & E. Stiefel (1992). *Group Theoretical Methods and Their Applications*. Basel: Birkhäuser.

Fisher, J. L. (1977). *Application-Oriented Algebra*. New York: Dun-Donnelley.

Folk, R., A. Kartashov, P. Linsoněk & P. Paule (1993). Symmetries in neural networks: a linear group action approach. *J. Phys. A: Math. Gen. 26*, 3159–3164.

Franser, D. A. S. (1976). *Probability and Statistics*. Wadsworth, Mass.: Duxbury Press.

Friedell, M. F. (1967). Organizations as semilattices. *Amer. Math. Soc. Rev. 31*, 46–54.

Gallager, R. G. (1968). *Information Theory and Reliable Communication*. New York: Wiley.

Galland, J. E. (1970). *Historical and Analytical Bibliography of the Literature of Cryptology*. Providence, R.I.: Amer. Math. Soc.

Galois, E. (1897). *Sur la théorie des nombres*, pp. 15–23. Paris: Gauthier-Villars.

Garfinkel, S. (1996). *PGP*. Cambridge, Mass.: O'Reilly.

Geddes, R. O., S. R. Czapor & G. Labahn (1993). *Algorithms for Computer Algebra*. Dordrecht: Kluwer.

Gilbert, W. J. (1976). *Modern Algebra with Applications*. New York: Wiley.

Gill, A. (1976). *Applied Algebra for the Computer Sciences*. Englewood Cliffs, N.J.: Prentice Hall.

Ginsburg, S. (1966). *The Mathematical Theory of Context-Free Languages*. New York: McGraw-Hill.

Ginsburg, S. (1975). *Algebraic and Automata-Theoretic Properties of Formal Languages*. Amsterdam: North-Holland.

Golomb, S. W. (1967). *Shift Register Sequences*. San Francisco: Holden-Day.

Goult, R. Z. (1978). *Applied Linear Algebra*. New York: Horwood.

Graham, R. L., M. Grötschel & L. Lovász (Eds.) (1994). *Handbook of Combinatorics*, Vols. I, II. Amsterdam: North-Holland.

Grätzer, G. (1971). *Lattice Theory*. San Francisco: Freeman.

Grätzer, G. (1978). *General Lattice Theory*. Basel: Birkhäuser.

Grätzer, G. (1979). *Universal Algebra*. New York: Springer-Verlag.

Grillet, P. A. (1995). *Semigroups*. New York: Dekker.

Guiasu, S. (1977). *Information Theory with Applications*. New York: McGraw-Hill.

Gumm, H. P. & W. Poguntge (1981). *Boolesche Algebra*. Mannheim: Bibliographisches Institut.

Haas, W. & J. Weinhofer (1996). Multivariable compensator designs using polynomial matrices and interval arithmetic. In *Proceedings of the IEEE-International Symposion on Computer-Aided-Control-System-Design*, Dearborn, pp. 194–199.

Halmos, P. (1967). *Boolean Algebras*. Princeton, N.J.: Van Nostrand.

Harrison, M. A. (1970). *Introduction to Switching and Automata Theory*. New York: McGraw-Hill.

Hazewinkel, M. (Ed.) (1996). *Handbook of Algebra*, Volume I. Amsterdam: North-Holland.

Heise, W. & P. Quattrocchi (1995). *Informations- und Codierungstheorie* (3rd ed.). Berlin: Springer-Verlag.

Hellman, M. E. (1979). The mathematics of public-key cryptography. *Sci. Amer. 241*, 130–139.

Heriken, R. (1994). *The Universal Turing Machine: A Half-Century Survey*. Wien: Springer-Verlag.

Hermann, G. T. & G. Rosenberg (1975). *Developmental Systems and Languages*. Amsterdam: North-Holland.

Herstein, I. N. (1975). *Topics in Algebra* (2nd ed.). New York: Wiley.

Hill, L. S. (1929). Cryptography in an algebraic alphabet. *Amer. Math. Monthly 36*, 306–312.

Hirschfeld, J. W. P. (1979). *Projective Geometries over Finite Fields*. Oxford: Clarendon Press.

Hohn, F. E. (1970). *Applied Boolean Algebra* (2nd ed.). New York: Macmillan.

Holcombe, W. M. L. (1982). *Algebraic Automata Theory*. Cambridge: Cambridge University Press.

Hong, H. (1997). Gröbner bases under composition I. *J. Symb. Comp. 23*.

Howie, J. M. (1976). *An Introduction to Semigroup Theory*. London: Academic Press.

Howie, J. M. (1995). *Fundamentals of Semigroup Theory*. Oxford: Clarendon Press.

Ihringer, T. (1988). *Allgemeine Algebra*. Stuttgart: Teubner.

Ireland, K. & M. I. A. Rosen (1982). *A Classical Introduction to Modern Number Theory*. New York: Springer-Verlag.

Isidori, A. (1989). *Nonlinear Control Systems* (2nd ed.). New York: Springer-Verlag.

Jacobson, N. (1985). *Basic Algebra*, Vols. I, II (2nd ed.). San Francisco: Freeman.

Janet, M. (1929). *Leçons sur les systèmes d'équations aux dérivées partielles*, Volume IV of *Cahiers Scientifiques*. Paris: Cahiers Scientifiques.

Jungnickel, D. (1993). *Finite Fields; Structure and Arithmetic.* Mannheim: Bibliographisches Institut.

Kahn, D. (1967). *The Codebreakers.* London: Weidenfeld & Nicholson.

Kalman, R. E., P. L. Falb & M. A. Arbib (1969). *Topics in Mathematical Systems Theory.* New York: McGraw-Hill.

Kalmbach, G. (1983). *Orthomodular Lattices.* London: Academic Press.

Kaltofen, E. (1992). Polynomial factorization 1987–1991. In I. Simon (ed.), *LATIN '92*, pp. 294–313. Berlin: Springer-Verlag.

Kanatani, K. (1990). *Group Theoretical Methods in Image Understanding.* Springer Series in Information Sciences. Berlin: Springer-Verlag.

Kautschitsch, H. (1991). Rings with feedback cyclization property. In *Contributions to General Algebra*, Volume 7, pp. 215–220. Stuttgart: Teubner.

Kerber, A. (1991). *Algebraic Combinatorics via Finite Group Actions.* Mannheim: Bibliographisches Institut.

Kieras, P. (1976). Finite automata and S-R-models. *J. Math. Psychology 13*, 127–147.

Kim, K. H. & F. W. Roush (1980). *Mathematics for Social Scientists.* New York: Elsevier.

Knobloch, H. W., A. Isidori & D. Flockerzi (1993). *Topics in Control Theory.* Basel: Birkhäuser.

Knobloch, H. W. & H. Kwakernaak (1985). *Lineare Kontrolltheorie.* Berlin: Springer-Verlag.

Knuth, D. E. (1970). Simple word problems in universal algebras. See Leech (1970), pp. 263–298.

Knuth, D. E. (1981). *The Art of Computer Programming, Vol. 2: Seminumerical Algorithms* (2nd ed.). Reading, Mass.: Addison-Wesley.

Koblitz, N. (1987). *A Course in Number Theory and Cryptography.* New York: Springer-Verlag.

Konheim, A. G. (1981). *Cryptography, A Primer.* New York: Wiley.

Krohn, K., R. Langer & J. Rhodes (1976). Algebraic principles for the analysis of a biochemical system. *J. Comput. Syst. Sci. 1*, 119–136.

Kronecker, L. (1887). Ein Fundamentalsatz der allgemeinen Arithmetik. *J. reine angew. Math. 100*, 490–510.

Kugi, A. & K. Schlacher (1996). Computer algebra algorithms for the differential geometric design of AI-systems. In *Workshop on Advanced Control Systems*, Vienna, pp. 179–188. Technical University of Vienna.

Kurosh, A. G. (1965). *Lectures on General Algebra*. New York: Chelsea.

Lallement, G. (1979). *Semigoups and Combinatorial Applications*. New York: Wiley.

Lam, C. W. H., L. Thiel & S. Swiercz (1989). The non-existence of finite projective planes of order 10. *Canad. J. Math. 41*, 1117–1123.

Lang, S. (1984). *Algebra* (2nd ed.). Reading, Mass.: Addison-Wesley.

Lausch, H. & W. Nöbauer (1973). *Algebra of Polynomials*. Amsterdam: North-Holland.

Lautrup, B. (1989). *The Theory of the Hopfield Model*. Singapore: World Scientific.

Le Chenadec, P. (1986). *Canonical Forms in Finitely Presented Algebras*. London: Pitman.

Leech, J. W. (1970). *Computational Problems in Abstract Algebra*. Oxford: Pergamon Press.

Leech, J. W. & D. J. Newman (1970). *How to Use Groups*. Birkenhead: Methuen.

Leiss, E. L. (1982). *Principles of Data Security*. New York: Plenum.

Lenz, R. (1990). *Group Theoretical Methods in Image Processing*. New York: Springer-Verlag.

Lidl, R. & H. Niederreiter (1983). *Finite Fields*. Addison-Wesley, now Cambridge University Press, Cambridge 1997.

Lidl, R. & H. Niederreiter (1994). *Introduction to Finite Fields and their Applications* (revised ed.). Cambridge: Cambridge University Press.

Lidl, R. & H. Niederreiter (1996). Finite fields and their applications. See Hazewinkel (1996), pp. 323–363.

Lidl, R. & G. Pilz (1984). *Applied Abstract Algebra* (1st ed.). New York: Springer-Verlag. Russian edition: Ekaterinenburg, 1997.

Lin, S. (1970). *An Introduction to Error-Correcting Codes*. Englewood Cliffs, N.J.: Prentice Hall.

Lindenmayer, A. (1968). Mathematical models for cellular interactions in development, I, II. *J. Theor. Biol. 18*, 280–315.

Lipschutz, S. (1976). *Discrete Mathematics*. Schaum's Outline Series. New York: McGraw-Hill.

Lipson, J. D. (1981). *Elements of Algebra and Algebraic Computing*. Reading, Mass.: Addison-Wesley.

Lüneburg, H. (1979). *Galoisfelder, Kreisteilungskörper und Schieberegisterfolgen*. Mannheim: Bibliographisches Institut.

MacLane, S. & G. Birkhoff (1967). *Algebra*. New York: Macmillan.

MacWilliams, F. J. & N. J. A. Sloane (1977). *The Theory of Error-Correcting Codes*, Vols. I, II. Amsterdam: North-Holland.

Mathiak, K. & P. Stingl (1968). *Gruppentheorie*. Braunschweig: Vieweg.

McDonald, B. R. (1974). *Finite Rings with Identity*. New York: Dekker.

McEliece, R. J. (1977). *Information Theory and Coding Theory*. Reading, Mass.: Addison-Wesley.

McEliece, R. J. (1987). *Finite Fields for Computer Scientists and Engineers*. Boston: Kluwer.

Mehrtens, H. (1979). *Die Entstehung der Verbandstheorie*. Hildcshcim: Gerstenberg Verlag.

Meldrum, J. D. P. (1995). *Wreath Products of Groups and Semigroups*. Marlow/Essex: Pitman.

Mendelson, E. (1970). *Boolean Algebras and Switching Circuits*. Schaum's Outline Series. New York: McGraw-Hill.

Menezes, A. J., I. F. Blake, X. Gao, R. C. Mullin, S. A. Vanstone & T. Yaghoobian (1993). *Applications of Finite Fields*. Boston: Kluwer.

Merkle, R. C. (1978). Secure communications over insecure channels. *Commun. ACM 21*, 294–299.

Merkle, R. C. (1982). *Secrecy, Authentication, and Public Key Systems*. Ann Arbor, Mich.: UMI Research Press.

Monk, J. D. & Bonnet, R. (Ed.) (1989). *Handbook of Boolean Algebras*, Vols. I, II. Amsterdam: North-Holland.

Mullen, G. L. & P. J. S. Shiue (Eds.) (1993). *Finite Fields, Coding Theory, and Advances in Communications in Computing*. New York: Dekker.

Myer, C. H. & S. M. Matyas (1982). *Cryptography: A New Dimension in Computer Data Security*. New York: Wiley.

Nagata, M. (1977). *Field Theory*. New York: Dekker.

Nagel, H. H. (1995). Arbeitsbericht der Forschungsgruppen. In *Festschrift 10 Jahre IAKS*, pp. 13–67. University of Karlsruhe (Germany).

Nahikian, H. M. (1964). *A Modern Algebra for Biologists*. Chicago: University of Chicago Press.

Niederreiter, H. (1993). A new efficient factorization algorithm for polynomials over small finite fields. *Appl. Alg. Engen. Comm. Comp. 4*, 81–87.

Noble, B. & J. W. Daniel (1977). *Applied Linear Algebra* (2nd ed.). Englewood Cliffs, N.J.: Prentice Hall.

Odlyzko, A. M. (1985). Discrete logarithms in finite fields and their cryptographic significance. In T. Beth, N. Cot, and I. Ingemarsson (Eds.), *Advances in Cryptology (Paris, 1984)*, Volume 209 of Lecture Notes in Computer Science, pp. 224–314. Berlin: Springer-Verlag.

Patterson, W. (1987). *Mathematical Cryptology for Computer Scientists and Mathematicians*. Totowa, N.J.: Rowman & Littlefield.

Perrin, J. P., M. Denquette & E. Dalcin (1972). *Switching Machines*, Vols. I,II. Dordrecht: Reidel.

Peschel, M. (1971). *Moderne Anwendungen algebraischer Methoden*. Berlin: Verlag Technik.

Peterson, W. W. & E. J. Weldon, Jr. (1972). *Error-Correcting Codes*. Cambridge, Mass.: MIT Press.

Petrich, M. (1973). *Introduction to Semigroups*. Columbus: Bell & Howell.

Pilz, G. (1983). *Near-Rings* (2nd ed.). Amsterdam: North-Holland.

Pilz, G. (1996). Near-rings and near-fields. See Hazewinkel (1996), pp. 463–498.

Pless, V. (1989). *Introduction to the Theory of Error-Correcting Codes* (2nd ed.). New York: Wiley.

Pohlmann, K. L. (1987). *Principles of Digital Audio*. New York: Macmillan.

Pomerance, C. (1996). A tale of two sieves. *Notices Amer. Math. Soc. 43*, 1473–1485.

Pommaret, J.-F. (1994). *Partial Differential Equations and Group Theory*. Dordrecht: Kluwer.

Prather, R. E. (1976). *Discrete Mathematical Structures for Computer Science*. Boston: Houghton Mifflin.

Preparata, F. P. & R. T. Yeh (1973). *Introduction to Discrete Structures*. Reading, Mass.: Addison-Wesley.

Rabin, M. O. (1980). Probabilistic algorithms in finite fields. *SIAM J. Comput. 9*, 273–280.

Rédei, L. (1967). *Algebra*. Oxford: Pergamon Press.

Reusch, B. (1975). Generation of prime implicants from subfunctions and a unifying approach to the covering problem. *IEEE Trans. Comput. 24*, 924–930.

Reusch, B. & L. Detering (1979). On the generation of prime implicants. *Fund. Inform. 2*, 167–186.

Rivest, R., A. Shamir & L. Adleman (1978). A method for obtaining digital signatures and public-key cryptosystems. *Commun. ACM 21*, 120–126.

Robinson, D. J. S. (1996). *A Course in the Theory of Groups* (2nd ed.). Graduate Texts in Mathematics. New York: Springer-Verlag.

Rorres, C. & H. Anton (1984). *Applications of Linear Algebra*. New York: Wiley.

Rosen, R. (1972, 1973). *Foundations of Mathematical Biology*, Vols. I–III. New York: Academic Press.

Rotman, J. J. (1995). *The Theory of Groups* (4th ed.). New York: Springer-Verlag.

Rowen, L. (1988). *Ring Theory*, Vols. I, II. London: Academic Press.

Rudeanu, S. (1974). *Boolean Functions and Equations*. Amsterdam: North-Holland.

Rueppel, R. A. (1986). *Analysis and Design of Stream Ciphers*. Berlin: Springer-Verlag.

Salomaa, A. (1969). *Theory of Automata*. Oxford: Pergamon Press.

Salomaa, A. (1981). *Jewels of Formal Language Theory*. London: Pitman.

Salomaa, A. (1990). *Public-Key Cryptography*. Berlin: Springer-Verlag.

Schlacher, K., A. Kugi & H. Irschik (1996). H_∞-control of nonlinear beam vibrations. In *Proceedings of the Third International Conference on Motion and Vibration Control*, Volume 3, Chiba, pp. 479–484. Japan Society of Mechanical Engineering.

Seberry, J. & J. Pieprzyk (1989). *Cryptography: An Introduction to Computer Security*. Englewood Cliffs, N.J.: Prentice Hall.

Selmer, E. S. (1966). *Linear Recurrence Relations over Finite Fields*. Bergen: University of Bergen.

Shamir, A. (1979). How to share a secret. *Comm. ACM 22*, 612–613.

Shannon, C. E. (1948). A mathematical theory of communication. *Bell System Tech. J. 27*, 379–423, 623–656.

Shparlinski, I. E. (1992). *Computational and Algorithmic Problems in Finite Fields*. Boston: Kluwer.

Simmons, G. J. (1979). Cryptology, the mathematics of secure communication. *Math. Intell. 1*, 233–246.

Simmons, G. J. (1992). *Contemporary Cryptology - The Science of Information Integrity*. Piscataway, N.J.: IEEE Press.

Sims, C. C. (1984). *Abstact Algebra, A Computational Approach*. New York: Wiley.

Sims, C. C. (1994). *Computation with Finitely Presented Groups*. Cambridge: Cambridge University Press.

Sinkov, A. (1978). *Elementary Cryptanalysis, a Mathematical Approach*. New York: Random House.

Sloane, N. J. A. & M. Harwit (1976). Masks for Hadamard transform, optics, and weighing designs. *Applied Optics 15*, 107–114.

Small, C. (1991). *Arithmetic of Finite Fields*. New York: Dekker.

Spindler, K. H. (1994). *Abstract Algebra with Applications*, Vols. I, II. New York: Dekker.

Steinitz, E. (1910). Algebraische Theorie der Körper. *J. reine angew. Math. 137*, 167–309.

Sternberg, S. (1994). *Group Theory and Physics*. Cambridge: Cambridge University Press.

Stone, H. S. (1973). *Discrete Mathematical Structures and Their Applications*. Chicago: Scientific Research Association.

Street, A. P. & D. J. Street (1987). *Combinatorics of Experimental Design*. Oxford: Oxford University Press.

Sturmfels, B. (1993). *Algorithms in Invariant Theory*. Wien: Springer-Verlag.

Sung, K. K. & G. Zimmermann (1986). Detektion und Verfolgung mehrerer Objekte in Bildfolgen. In G. Hartmann (Ed.), *Tagungsband zum 8. DAGM Symposium Paderborn (Germany)*, Informatik Fachberichte, pp. 181–184. Berlin: Springer-Verlag.

Suppes, D. (1969). Stimulus-response theory of finite automata. *J. Math. Psychology 6*, 327–355.

Szász, G. (1963). *Introduction to Lattice Theory*. New York: Academic Press.

Szidarovzky, F. & A. T. Bahill (1992). *Linear Systems Theory*. Boca Raton, FL: CRC Press.

Thomas, A. D. & G. V. Wood (1980). *Group Tables*. Orpington, Kent: Shiva.

Usmani, R. A. (1987). *Applied Linear Algebra*. Pure and Applied Mathematics. New York: Dekker.

van der Waerden, B. L. (1970). *Modern Algebra, I, II*. New York: Ungar.

van Lint, J. H. (1992). *Introduction to Coding Theory* (second ed.). New York: Springer-Verlag.

van Lint, J. H. (1994). Codes. See Graham et al. (1994), pp. 773–807.

van Lint, J. H. & R. M. Wilson (1992). *A Course in Combinatorics*. Cambridge: Cambridge University Press.

van Tilborg, H. C. A. (1988). *An Introduction to Cryptography*. Boston: Kluwer.

van Tilborg, H. C. A. (1993). *Error-Correction Codes, A First Course*. Browley: Chartwell-Bratt.

Vanasse, G. A. (1982). Infrared spectrometry. *Applied Optics 21*, 189–195.

Vanstone, S. A. & P. C. van Oorschot (1992). *An Introduction to Error Correcting Codes with Applications*. Boston: Kluwer.

Varadarajan, V. S. (1968). *Geometry of Quantum Theory*, Vols. I, II. Princeton, N.J.: Van Nostrand.

Vidyasagar, M. (1987). *Control System Synthesis; a Factorization Approach* (2nd ed.). Cambridge, Mass.: MIT Press.

Vidyasagar, M. (1989). *Nonlinear System Analysis*. Englewood Cliffs, N.J.: Prentice Hall.

von Neumann, J. (1955). *Mathematical Foundations of Quantum Mechanics*. Princeton, N.J.: Princeton University Press.

von zur Gathen, J. (1990). Functional decomposition: The tame case. *J. Symbolic Comput. 9*, 281–299.

Weber, H. (1893). Die allgemeinen Grundlagen der Galois'schen Gleichungstheorie. *Math. Ann. 43*, 521–549.

Weinstein, A. (1996). Groupoids: Unifying internal and external symmetry. *Notices Amer. Math. Soc. 434*, 744–752.

Wells, C. (1976). Some applications of the wreath product construction. *Amer. Math. Monthly 83*, 317–338.

Welsh, D. (1988). *Codes and Cryptography*. New York: Clarendon Press.

Weyl, H. (1952). *Symmetry*. Princeton, Mass.: Princeton University Press.

White, H. C. (1963). *An Anatomy of Kinship: Mathematical Models for Structures of Cumulated Roles*. Englewood Cliffs, N.J.: Prentice Hall.

White, H. C., S. A. Boorman & R. L. Breiger (1976). Social structure from multiple networks I: Blockmodels of roles and positions. *Amer. J. Sociol. 81*, 730–780.

Whitesitt, J. E. (1961). *Boolean Algebra and Its Applications*. Reading, Mass.: Addison-Wesley.

Wille, R. (1980). *Mathematische Sprache in der Musiktheorie*. Übungsbl. Math. Mannheim: Bibliographisches Institut.

Wille, R. (1982). Restructuring lattice theory: An approach based on hierarchies of concepts. In Rival (Ed.), *Ordered Sets*, pp. 445–470. Dordrecht: Reidel.

Winkler, F. (1996). *Polynomial Algorithms in Computer Algebra*. Wien: Springer-Verlag.

Winter, D. J. (1974). *The Structure of Fields*. Berlin: Springer-Verlag.

Wonham, W. M. (1974). *Linear Multivariable Control*. New York: Springer-Verlag.

Zelmer, V. & A. Stancu (1973). Mathematical approach on the behavior of biosystems. *Math. Cluj. 15*, 119–128.

Zierler, N. (1959). Linear recurring sequences. *J. SIAM 7*, 31–48.

Zippel, R. (1993). *Effective Polynomial Computation*. Boston: Kluwer.

Index

absorption law, 6
abstract data type, 453
acceptor, 353
acceptor language, 353
adder, 207
adjunction, 129
affine cipher, 243
affine geometry, 407
affine-input (AI) systems, 445
algebra, 205
  of events, 86
  σ-, 87
  universal, 448
algebraic, 130
algebraic equation, 427, 450
algebraic numbers, 132
algebraically closed, 135
alphabet, 351
alternating group, 291
A_n, 291
AND-gate, 59
antihomomorphism, 345
antisymmetric, 3
associative, 96
associative law, 6
asymptotically stable, 439
atom, 18
attack
  chosen-plaintext, 242
  ciphertext-only, 241
  known-plaintext, 242
automaton, 342
  finite, 342
  Mealy, 342
  Moore, 343
Axiom of Choice, 5
axis, 316

balanced block design, 399
base, 451
basis, 338
BCH code, 225
Berlekamp's Algorithm, 169

Bezout identity, 445
BIB-design, 400
binary code, 193
binary operation, 96
binary symmetric channel, 187
Birkhoff's Theorem, 450
block, 400
Boolean algebra, 19
Borel set, 88
bound
  lower, 5
  universal, 7
  upper, 5
bounded lattice, 7
Buchberger's algorithm, 430
Burnside's Lemma, 293

Caesar cipher, 243
cancellation rule, 17
canonical basic matrix, 194
canonical epimorphism, 336
cardinality, 7
Carmichael function, 262
Cayley's Theorem, 287
center, 107, 289
chain, 4
Chakrabarti's cell, 77
characteristic, 109
characteristic function, 21
characteristic polynomial, 369
check equations, 193
check polynomial, 211
check symbol, 187
Chinese Remainder Theorem, 167, 284
Church-Rosser property, 455
cipher, 240
circuit diagram, 56
circulating shift register, 377
class number, 107
code, 187
  BCH, 225
    narrow-sense, 225
    primitive, 225
  binary, 193
  binary Golay, 228
  binary simplex, 202
  constant weight, 410
  cyclic, 208
  dual, 195
  equal distance, 410
  equivalent, 195
  extended, 202
  first-order Reed-Muller, 202
  generalized Hamming, 201
  group, 193
  Hamming, 200
  irreducible, 213
  linear, 193
  maximal cyclic, 211
  MDS, 191
  minimal, 213
  orthogonal, 195
  perfect, 190
  quadratic residue, 233
  Reed-Solomon, 226
  repetition, 193
  self-orthogonal, 204
code polynomial, 209
code vector, 209
codeword, 187
coefficient assignable, 442
cofinite, 36
commutative law, 6
companion matrix, 151, 369
compatible, 102, 448
compensator, 441

complement, 17
  relative, 18
complementation, 80
complementation switch, 57
complete vocabulary, 351
complexity, 380
composition, 115
concatenation, 338
confluent, 455
congruence relation, 102, 336, 448
conjugate elements, 106
conjugates, 152, 153
conjunction, 80
constant, 115
constructible, 137
contact sketch, 56
context-free, 353
contradiction, 81
control symbols, 187
controllable, 438
controller, 441
convolution, 384
convolution property, 385
Cooley-Tuckey Algorithm, 380
coprime, 116
core, 44
coset leader, 198
cosets, 104
covering radius, 205
covers, 347
cryptogram, 240
cryptology, 239
cryptosystem, 240
  asymmetric, 241
  El Gamal, 268
  fair, 269
  no-key, 268
  public-key, 241
  single-key, 241
  symmetric, 241
crystallographic groups, 326
cycle, 288
cycle index, 296
cycle index polynomial, 296
cyclic, 100, 442
cyclic code, 208
cyclic shift, 205
cyclotomic coset, 156
cyclotomic decomposition, 146
cyclotomic field, 144
cyclotomic polynomial, 145

data type, 452
deciphering, 241
decomposable, 175
decomposition
  functional, 175
  into cycles, 288
Decomposition Theorem of Krohn-Rhodes, 348
defining relations, 339
degree, 115, 130, 131
  of extension, 130
delay, 207
De Morgan's Laws, 20
derivation, 352
derivative, 134
design, 400
designed distance, 225
DFT, 382
DFT matrix, 383
diamond, 16
difference equation, 366
differential operator, 134
Diffie-Hellman scheme, 267
dihedral group, 97

direct product, 13
direct sum, 100
Discrete Fourier Transform, 217
discrete logarithm, 142
disjunction, 80
distributive inequalities, 8
distributive laws, 16, 109
divides, 347
division ring, 109
domain of action, 288
Don't-care combinations, 66
Duality Principle, 7
Dyck's Theorem, 337

El Gamal digital signature, 268
El Gamal public key cryptosystem, 268
elementary cell, 326
elementary divisors, 106
elementary domain, 326
elementary event, 87
embedding, 100
empty word, 338
enciphering, 240
enciphering algorithm, 240
enciphering function, 240
encoding matrix, 194
encryption, 240
epimorphism
  group, 100
  lattice, 11
equation
  Boolean, 37
equationally definable, 450
equivalence
  class, 3
  of polynomials, 27
  relation, 3
equivalent codes, 195
error, 187
error correcting, 184
error-detecting, 184
error-location number, 216
error locator polynomial, 231
error vector, 187
error word, 187
Euclidean division, 116
Euler's officers problem, 389
Euler's phi-function, 144
event, 86
exponent, 159
extended code, 202
extension
  algebraic, 130
  finite, 130
  infinite, 130
  simple, 129
  transcendental, 130
extension field, 126

factor algebra, 448
  Boolean, 35
factor group, 102
factor ring, 112
factor semigroup, 336
faithful, 308
Fano geometry, 400
fast
  convolution, 384
  Fourier transfrom, 383
  Hadamard transform, 421
  multiplication, 384
feedback, 441, 442
Fermat's Little Theorem, 105, 151
FFT, 383
Fibonacci sequence, 378

field, 109
  cyclotomic, 144
  of formal power series, 126
  of quotients, 125
  of rational functions, 125
field-value, 251
filter, 36
final state, 353
finitely generated, 100
Fisher's inequality, 402
flip-flop, 343
floor function, 189
formal derivative, 134
formal power series, 114
Fourier coefficients, 382
Fourier transform, 382
  discrete, 382
  fast, 383
  general, 387
free, 451
free monoid, 338
free semigroup, 338
Frobenius automorphism, 144
full-adder, 74
fully reducible, 308
function ring, 110
Fundamental Theorem of Algebra, 115, 135

Galois field, 140
Galois group, 143
gate, 59
Gaussian integer, 136
gcd, 1, 116
general linear group, 97
generalized Hamming code, 201
generated
  equivalence relation, 338
  ideal, 114
  language, 352
  subgroup, 99
  variety, 452
generating function, 372
generation, 352
generator matrix, 194
generator polynomial, 209
geometrical crystal class, 326
Gilbert-Varshamov Bound, 196
Ginsburg's Theorem, 354
Golay codc
  binary, 228
  ternary, 233
grammar, 351
  phrase-structure, 351
  symbols, 351
greatest element, 4
Gröbner basis, 429
group, 96
  abelian, 96
  commutative, 96
  cyclic, 100
  dihedral, 97
  general linear, 97
  sub-, 99
  symmetric, 97
group algebra, 321, 387
group code, 193
group kernel, 335
groupoids, 448

Hadamard matrix, 413
Hadamard network, 423
Hadamard pattern, 423
Hadamard transform, 421
half-adder, 73
Hamming bound, 190

Hamming code, 200
Hamming distance, 188
Hamming weight, 188
Hasse diagram, 3, 9, 10
Hilbert's Basis Theorem, 428
Hilbert's Nullstellensatz, 428
Hill cipher, 248
homogeneous, 366, 453
homomorphism, 12, 448
  anti-, 345
  automata, 346
  Boolean, 21
  join, 11
  lattice, 11
  meet, 11
  of groups, 100
  order, 11
homomorphism theorem, 104, 113, 336, 449

ideal, 340
  dual, 36
  in Boolean algebra, 34
  maximal, 35, 114
  of a ring, 112
  principal, 35, 114
  proper, 35
idempotent, 334
idempotent law, 6
identity, 109
identity element, 334
Identity-Reset-Automaton, 344
image
  homomorphic, 11
  of group homomorphism, 101
image parameters, 303
image understanding, 303
implicant, 41
impulse response sequence, 376
incidence matrix, 401
incomplete, 400
indecomposable, 175
index, 142
  of subgroup, 104
infimum, 5, 7
information rate, 189
initial, 342
initial monomial, 429
initial state vector, 366
initial symbol, 351
initial values, 366
inner automorphisms, 292
input alphabet, 342
input variables, 57
integral, 109
integral domain, 109
interval, 11
invariants, 308
inverse, 96, 334
  relation, 4
inverse semigroups, 361
inverter, 59
invertible, 334
involutory, 248
IR flip-flop, 343
irreducible, 115, 308
irreducible characters, 321
ISB-number, 191
isomorphism
  group, 100
  of lattices, 11
  of posets, 15

Jacobi symbol, 266
join, 5

join-irreducible, 18
join-reducible, 18

Kalman's Theorem, 438
Karnaugh diagrams, 48
Kautschitsch's Theorem, 442
kernel
  of Boolean homomorphism, 34
  of group homomorphism, 101
key, 240, 243
key stream, 375
kinship system, 360
Kleene's Theorem, 354
knapsack problem, 273
Kronecker product, 393
Kronecker's algorithm, 176
Kronecker's Theorem, 128

Lagrange's interpolation formula, 120
Lagrange's Theorem, 104
language
  context-free, 353
  right linear, 353
Latin square, 388
lattice, 7, 326
  algebraic, 5
  Boolean, 19
  complemented, 17
  complete, 87
  distributive, 16
  orthocomplemented, 90
  orthomodular, 90
  sub-, 9
law, 83
lcm, 2
least period, 368
Legendre symbol, 415
length
  of a code, 187
lexicographic order, 15
linear (block) code, 193
linear complexity, 375
linear continuous system, 436
linear order, 4
linear recurrence relation, 366
linear recurring sequence, 366
linear recursive relation, 366
linear shift register, 207
loop, 389
lower bound, 5
Lüroth's Theorem, 135

Maschke's Theorem, 387
matrix representation, 308
matrix rings, 110
Mattson-Solomon polynomial, 217
maximal cyclic codes, 211
maximal element, 4
maximal ideal, 114
maximal period, 371
maximum distance separable (MDS) code, 191
MDS code, 191
Mealy automaton, 342
measurable space, 87
measure, 87
measure space, 88
meet, 5
message symbols, 187
metalinguistic symbols, 351
minimal element, 4
minimal polynomial, 131, 374

minimization
  of Boolean polynomials, 40
minimum distance, 188
Möbius function, 147
Möbius Inversion Formula, 147
model, 452
modular, 53
modular enciphering, 243
monic, 115
monoalphabetic, 242
monoid, 334
  free, 338
  of an automaton, 345
  syntactic, 345
monomorphic, 453
monomorphism
  group, 100
  lattice, 11
Moore automaton, 343
multiplicity, 120
multiplier, 207
multivariate polynomials, 426

NAND-gate, 59
near-ring, 402
nearest neighbor decoding, 188
negation, 80
net, 326
neural network, 423
neutral element, 96
next-state function, 342
Noetherian rings, 428
NOR-gate, 59
normal basis, 152
normal form, 456
  conjunctive, 33
  disjunctive, 29
  of Boolean polynomials, 28, 29
NOT-gate, 59
NP-complete problem, 274

object, 314
object parameters, 303
one-time pad, 247
one-way function, 258
operation, 448
operation table, 8
optimization
  of Boolean polynomials, 40
OR-gate, 59
orbit, 292
order, 7, 144, 159, 317
  lattice, 5
  lexicographic, 15
  linear, 4
  of a group, 96
  of group element, 105
  partial, 3
ordered field, 457
orthocomplemented lattice, 90
orthogonal, 91, 388
  code, 195
  group, 315
  idempotent, 214
  vectors, 195
orthogonal idempotents, 321
orthomodular identity, 90
output alphabet, 342
output function, 342
output map, 437

parallel, 346
parallel connection, 57
parity-check, 185
parity-check matrix, 193

Parseval identity, 385
partition, 289
Payley construction, 415
pentagon, 16
perfect, 190
period, 159, 368
periodic, 354, 368
periodic substitution cipher, 245
permutation
  even, 291
permutation polynomial, 252
phase property, 385
phase space, 89
phi-function, 144
phrase-structure grammar, 351
PID, 114
Pierce-operation, 59
plaintext, 240
planar near-ring, 403
Plancherel identity, 385
Plotkin Bound, 197
point, 400
pole assignment property, 442
polyalphabetic, 242
Pólya's Theorem, 298
Polybius square, 253
polymorphic, 453
polynomial
  Boolean, 26
  decomposable, 175
  indecomposable, 175
  irreducible, 115
  over a ring, 114
  primitive, 160
polynomial function, 119
  Boolean, 26
polynomially complete, 33, 121
poset, 3
power set, 3
predicate, 83
preperiod, 368
presentation, 339
prime field, 126
primitive, 139, 144, 160
principal ideal, 114
principal ideal domain, 114
Principal Theorem on Finite Abelian Groups, 106
probability, 88
probability space, 88
product
  of groups, 100
  of lattices, 13
production, 351
projective geometry, 407
proposition, 79
propositional algebra, 80
propositional form, 80
propositional variable, 80
pseudoprime, 266
  strong, 266
pseudotetrads, 72
public-key cryptosystem, 241

quadratic residue (QR codes) codes, 233
quasilattice, 53
quaternion group, 111
Quine-McCluskey method, 41
quotient field, 125
quotient semigroup, 336

random experiment, 86
rational function, 125
reciprocal polynomial, 212

recovery equations, 303
Redfield-Pólya Theorem, 299
reducible, 309
Reed-Muller code, 202
Reed-Solomon code, 226
Reed-Solomon (RS) code, 226
Rees congruence, 340
reflexive, 3
regular, 353
regulator, 441
relation, 3
  antisymmetric, 3
  compatible, 102
  congruence, 102
  equivalence, 3
  inverse, 4
  reflexive, 3
  symmetric, 3
  transitive, 3
relation product, 337
relation semigroup, 337
relational algebras, 452
relative complement, 18
relatively prime, 116
repetition code, 193
Representation Theorem
  Boolean algebras, 22
residue class, 102
rewriting rule, 351
right linear, 353
ring, 109
role structure, 363
root, 119, 127
  primitive, 139
root of unity, 144

S-matrix, 418
sample space, 86
Schur's Lemma, 313
sectionally complemented, 18
self-orthogonal code, 204
semigroup, 333
  commutative, 333
  finite, 334
  free, 338
  inverse, 361
  relation, 337
  table, 334
  word, 338
semigroup property, 437
semilattice, 53
separable, 133
series, 346
series connection, 57
Shannon's Theorem, 191
Sheffer-operation, 59
shift operation, 385
signal vector, 383
signature, 290, 453
Silver-Pohlig-Hellman algorithm, 270
simple
  extension, 129
  group, 107
  representation, 309
  ring, 112
  root, 120
simplex code, 202
simulate, 347
simultaneously verifiable, 91
single-key, 241
singleton bound, 190
skew field, 109
skew lattice, 53
smallest element, 4
society, 362

solution, 37, 427
solvable, 427
sorts, 453
specification, 454
spectral vector, 383
splitting field, 129
square-free, 134
stabilizer, 292
stable, 439
stable functions, 444
standard epimorphism, 337
standard form
  of a linear code, 193
state vector, 366
states, 342, 436
Steiner system, 407
Stone's Representation Theorem, 24
stream cipher, 375
strong prime, 261
subalgebra, 448
subautomaton, 346
subfield, 126
subgroup, 99
  generated, 99
  normal, 103
subgroup criterion, 99
subjunction, 81
subjunction-gate, 59
subring, 111
substitution cipher, 242
successor function, 452
sum
  of representations, 308
supremum, 5, 7
switching circuit, 56
switching diagram, 56
switching function, 57
  essentially similar, 300
switching network, 71
symmetric, 3, 400
symmetric difference, 83
symmetric group, 97
symmetry group, 314, 423
symmetry operation, 314
syndrome, 199
syntactic monoid, 345
system, 436
systematic linear code, 193
syzygy, 433

tautology, 81
term, 449
term algebra, 450
terminating, 455
tertium non datur, 79
tetrad, 72
theory, 452
three-pass algorithm, 268
time invariant, 437
totient function, 144
tournament, 393
transcendental, 130
transfer function, 443
transitive, 3
transitive hull, 337
translation, 385
transposition, 288
transposition cipher, 242
trapdoor, 258
truth function, 80
type, 448

ultimately periodic, 367
ultrafilters, 37
Unique Factorization Theorem, 118

unit, 7, 109
upper bound, 5

variety, 450
Vigenère cipher, 245

Wedderburn's Theorem, 111
Wilson's Theorem, 151
word problem, 454
word semigroup, 338
wreath product, 347

yield vector, 406

z-transformation, 382
Zech's Logarithm, 152
zero, 127
  of a code, 216
zero element, 7, 334
zero-ring, 110
Zorn's Lemma, 5

Undergraduate Texts in Mathematics

(continued from page ii)

James: Topological and Uniform Spaces.
Jänich: Linear Algebra.
Jänich: Topology.
Kemeny/Snell: Finite Markov Chains.
Kinsey: Topology of Surfaces.
Klambauer: Aspects of Calculus.
Lang: A First Course in Calculus. Fifth edition.
Lang: Calculus of Several Variables. Third edition.
Lang: Introduction to Linear Algebra. Second edition.
Lang: Linear Algebra. Third edition.
Lang: Undergraduate Algebra. Second edition.
Lang: Undergraduate Analysis.
Lax/Burstein/Lax: Calculus with Applications and Computing. Volume 1.
LeCuyer: College Mathematics with APL.
Lidl/Pilz: Applied Abstract Algebra. Second edition.
Macki-Strauss: Introduction to Optimal Control Theory.
Malitz: Introduction to Mathematical Logic.
Marsden/Weinstein: Calculus I, II, III. Second edition.
Martin: The Foundations of Geometry and the Non-Euclidean Plane.
Martin: Geometric Constructions.
Martin: Transformation Geometry: An Introduction to Symmetry.
Millman/Parker: Geometry: A Metric Approach with Models. Second edition.
Moschovakis: Notes on Set Theory.
Owen: A First Course in the Mathematical Foundations of Thermodynamics.
Palka: An Introduction to Complex Function Theory.
Pedrick: A First Course in Analysis.
Peressini/Sullivan/Uhl: The Mathematics of Nonlinear Programming.
Prenowitz/Jantosciak: Join Geometries.
Priestley: Calculus: An Historical Approach.
Protter/Morrey: A First Course in Real Analysis. Second edition.
Protter/Morrey: Intermediate Calculus. Second edition.
Roman: An Introduction to Coding and Information Theory.
Ross: Elementary Analysis: The Theory of Calculus.
Samuel: Projective Geometry. *Readings in Mathematics.*
Scharlau/Opolka: From Fermat to Minkowski.
Sethuraman: Rings, Fields, and Vector Spaces: An Approach to Geometric Constructability.
Sigler: Algebra.
Silverman/Tate: Rational Points on Elliptic Curves.
Simmonds: A Brief on Tensor Analysis. Second edition.
Singer: Geometry: Plane and Fancy.
Singer/Thorpe: Lecture Notes on Elementary Topology and Geometry.
Smith: Linear Algebra. Second edition.
Smith: Primer of Modern Analysis. Second edition.
Stanton/White: Constructive Combinatorics.
Stillwell: Elements of Algebra: Geometry, Numbers, Equations.
Stillwell: Mathematics and Its History.
Stillwell: Numbers and Geometry. *Readings in Mathematics.*
Strayer: Linear Programming and Its Applications.
Thorpe: Elementary Topics in Differential Geometry.
Toth: Glimpses of Algebra and Geometry.

Troutman: Variational Calculus and Optimal Control. Second edition.
Valenza: Linear Algebra: An Introduction to Abstract Mathematics.
Whyburn/Duda: Dynamic Topology.
Wilson: Much Ado About Calculus.